Studies in Classification, Data Analysis, and Knowledge Organization

For further volumes:
http://www.springer.com/series/1564

Berthold Lausen • Dirk Van den Poel
Alfred Ultsch
Editors

Algorithms from and for Nature and Life

Classification and Data Analysis

Editors
Berthold Lausen
Department of Mathematical Sciences
University of Essex
Colchester, United Kingdom

Dirk Van den Poel
Department of Marketing
Ghent University
Ghent, Belgium

Alfred Ultsch
Databionics, FB 12
University of Marburg
Marburg, Germany

ISSN 1431-8814
ISBN 978-3-319-00034-3 ISBN 978-3-319-00035-0 (eBook)
DOI 10.1007/978-3-319-00035-0
Springer Cham Heidelberg New York Dordrecht London

Library of Congress Control Number: 2013945874

Printed on acid-free paper

Springer is part of Springer Science+Business Media (www.springer.com)

Preface

Revised versions of selected papers presented at the Joint Conference of the German Classification Society (GfKl) – 35th Annual Conference – GfKl 2011 – , the German Association for Pattern Recognition (DAGM) – 33rd annual symposium – DAGM 2011 – and the Symposium of the International Federation of Classification Societies (IFCS) – IFCS 2011 – held at the University of Frankfurt (Frankfurt am Main, Germany) August 30 – September 2, 2011, are contained in this volume of "Studies in Classification, Data Analysis, and Knowledge Organization".

One aim of the conference was to provide a platform for discussions on results concerning the interface that data analysis has in common with other areas such as, e.g., computer science, operations research, and statistics from a scientific perspective, as well as with various application areas when "best" interpretations of data that describe underlying problem situations need knowledge from different research directions.

Practitioners and researchers – interested in data analysis in the broad sense – had the opportunity to discuss recent developments and to establish cross-disciplinary cooperation in their fields of interest. More than 420 persons attended the conference, more than 180 papers (including plenary and semiplenary lectures) were presented. The audience of the conference was very international.

Fifty-five of the papers presented at the conference are contained in this. As an unambiguous assignment of topics addressed in single papers is sometimes difficult the contributions are grouped in a way that the editors found appropriate. Within (sub)chapters the presentations are listed in alphabetical order with respect to the authors' names. At the end of this volume an index is included that, additionally, should help the interested reader.

The editors like to thank the members of the scientific program committee: D. Baier, H.-H. Bock, R. Decker, A. Ferligoj, W. Gaul, Ch. Hennig, I. Herzog, E. Hüllermeier, K. Jajuga, H. Kestler, A. Koch, S. Krolak-Schwerdt, H. Locarek-Junge, G. McLachlan, F.R. McMorris, G. Menexes, B. Mirkin, M. Mizuta, A. Montanari, R. Nugent, A. Okada, G. Ritter, M. de Rooij, I. van Mechelen, G. Venturini, J. Vermunt, M. Vichi and C. Weihs and the additional reviewers of the proceedings: W. Adler, M. Behnisch, C. Bernau, P. Bertrand, A.-L. Boulesteix,

A. Cerioli, M. Costa, N. Dean, P. Eilers, S.L. France, J. Gertheiss, A. Geyer-Schulz, W.J. Heiser, Ch. Hohensinn, H. Holzmann, Th. Horvath, H. Kiers, B. Lorenz, H. Lukashevich, V. Makarenkov, F. Meyer, I. Morlini, H.-J. Mucha, U. Müller-Funk, J.W. Owsinski, P. Rokita, A. Rutkowski-Ziarko, R. Samworth, I. Schmädecke and A. Sokolowski.

Last but not least, we would like to thank all participants of the conference for their interest and various activities which, again, made the 35th annual GfKl conference and this volume an interdisciplinary possibility for scientific discussion, in particular all authors and all colleagues who reviewed papers, chaired sessions or were otherwise involved. Additionally, we gratefully take the opportunity to acknowledge support by Deutsche Forschungsgemeinschaft (DFG) of the Symposium of the International Federation of Classification Societies (IFCS) – IFCS 2011.

As always we thank Springer Verlag, Heidelberg, especially Dr. Martina Bihn, for excellent cooperation in publishing this volume.

Colchester, UK Berthold Lausen
Ghent, Belgium Dirk Van den Poel
Marburg, Germany Alfred Ultsch

Contents

Part VI Text Mining, Social Networks and Clustering

Part VII Banking and Finance

Part VIII Marketing and Management

Contributors

Ulas Akkucuk Department of Management, Bogazici University, Istanbul, Turkey, ulas.akkucuk@boun.edu.tr

Alexander Albert Clinic of Cardiovascular Surgery, Heinrich-Heine University, 40225 Düsseldorf, Germany

Theodore Alexandrov Center for Industrial Mathematics, University of Bremen, 28359 Bremen, Germany

Grigory Alexandrovich Department of Mathematics and Computer Science, Marburg University, Marburg, Germany

Daniel Baier Institute of Business Administration and Economics, Brandenburg University of Technology Cottbus, Postbox 101344, 03013 Cottbus, Germany, daniel.baier@tu-cottbus.de

Hans-Georg Bartel Department of Chemistry, Humboldt University, Brook-Taylor-Straße 2, 12489 Berlin, Germany, hg.bartel@yahoo.de

Nadja Bauer Faculty of Statistics Chair of Computational Statistics, TU Dortmund, Dortmund, Germany, bauer@statistik.tu-dortmund.de

Christoph Bernau Institut für Medizinische Informationsverarbeitung, Biometrie und Epidemiologie, Universität München (LMU), Münich, Germany

Wolfgang Bessler Center for Finance and Banking, University of Giessen, Licher Strasse 74, 35394 Giessen, Germany, wolfgang.bessler@wirtschaft.uni-giessen.de

Holger Blume Institute of Microelectronic Systems, Appelstr. 4, 30167 Hannover, Germany, blume@ims.uni-hannover.de

Alix Boc Université de Montréal, C.P. 6128, succursale Centre-ville, Montréal, QC H3C 3J7 Canada, alix.boc@umontreal.ca

Anne-Laure Boulesteix Institut für Medizinische Informationsverarbeitung, Biometrie und Epidemiologie, Universität München (LMU), Münich, Germany, boulesteix@ibe.med.uni-muenchen.de

Nina Büchel European Research Center for Information Systems (ERCIS), University of Münster, Münster, Germany, buechel@ercis.de

Carlos Cuevas-Covarrubias Anahuac University, Naucalpan, State of Mexico, Mexico, ccuevas@anahuac.mx

J. Douglas Carroll Rutgers Business School, Newark and New Brunswick, Newark, NJ, USA

Andrea Cerioli Dipartimento di Economia, Universitàdi Parma, Parma, Italy, andrea.cerioli@unipr.it

Magdalena Chudy Centre for Digital Music, Queen Mary University of London, Mile End Road, London, E1 4NS UK, magdalena.chudy@eecs.qmul.ac.uk

Antonio D'Ambrosio Department of Mathematics and Statistics, University of Naples Federico II, Via Cinthia, M.te S. Angelo, Naples, Italy, antdambr@unina.it

Ines Daniel Institute of Business Administration and Economics, Brandenburg University of Technology Cottbus, Postbox 101344, 03013 Cottbus, Germany, ines.daniel@tu-cottbus.de

José G. Dias UNIDE, ISCTE – University Institute of Lisbon, Lisbon, Portugal

Edifício ISCTE, Av. das Forças Armadas, 1649-026 Lisboa, Portugal, jose.dias@iscte.pt

Simon Dixon Centre for Digital Music, Queen Mary University of London, Mile End Road, London, E1 4NS UK, simon.dixon@eecs.qmul.ac.uk

Jens Dolata Head Office for Cultural Heritage Rhineland-Palatinate (GDKE), Große Langgasse 29, 55116 Mainz, Germany, dolata@ziegelforschung.de

Florent Domenach Department of Computer Science, University of Nicosia, 46 Makedonitissas Avenue, PO Box 24005, 1700 Nicosia, Cyprus, domenach.f@unic.ac.cy

Plamen Dragiev Département d'Informatique, Université du Québec à Montréal, c.p. 8888, succ. Centre-Ville, Montreal, QC H3C 3P8 Canada

Department of Human Genetics, McGill University, 1205 Dr. Penfield Ave., Montreal, QC H3A-1B1 Canada

Kai Eckert KR & KM Research Group, University of Mannheim, Mannheim, Germany, Kai@informatik.uni-mannheim.de

Jorge Eduardo Ortiz Facultad de Estadística Universidad Santo Tomás, Bogotá, Colombia, jorgeortiz@usantotomas.edu.co

Thomas Fober Department of Mathematics and Computer Science, Philipps-Universität, 35032 Marburg, Germany, thomas@mathematik.uni-marburg.de

Stephen L. France Lubar School of Business, University of Wisconsin-Milwaukee, P. O. Box 742, Milwaukee, WI, 53201-0742 USA, france@uwm.edu

Sarah Frost Institute of Business Administration and Economics, Brandenburg University of Technology Cottbus, Postbox 101344, 03013 Cottbus, Germany, sarah.frost@tu-cottbus.de

Wolfgang Gaul Institute of Decision Theory and Management Science, Karlsruhe Institute of Technology (KIT), Kaiserstr. 12, 76128 Karlsruhe, Germany, wolfgang.gaul@kit.edu

Jan Gertheiss Department of Statistics, LMU Munich, Akademiestr. 1, 80799 Munich, Germany, jan.gertheiss@stat.uni-muenchen.de

Andreas Geyer-Schulz Information Services and Electronic Markets, IISM, Karlsruhe Institute of Technology, Kaiserstrasse 12, D-76128 Karlsruhe, Germany, andreas.geyer-schulz@kit.edu

Erhard Godehardt Clinic of Cardiovascular Surgery, Heinrich-Heine University, 40225 Düsseldorf, Germany, godehard@uni-duesseldorf.de

Isobel Claire Gormley University College Dublin, Dublin, Ireland, claire.gormley@ucd.ie

Bettina Grün Department of Applied Statistics, Johannes Kepler University Linz, Altenbergerstraße 69, 4040 Linz, Austria, bettina.gruen@jku.at

Dereje W. Gudicha Tilburg University, PO Box 50193, 5000 LE Tilburg, The Netherlands, d.w.gudicha@uvt.nl

Reinhold Hatzinger Institute for Statistics and Mathematics, WU Vienna University of Economics and Business, Augasse 2-6, 1090 Vienna, Austria, reinhold.hatzinger@wu.ac.at

Willem J. Heiser Institute of Psychology, Leiden University, P.O. Box 9555, 2300 RB Leiden, The Netherlands, Heiser@Fsw.Leidenuniv.nl

Irmela Herzog LVR-Amt für Bodendenkmalpflege im Rheinland, Bonn

Kay F. Hildebrand European Research Center for Information Systems (ERCIS), University of Münster, Münster, Germany, hildebrand@ercis.de

Stefanie Hillebrand Faculty of Statistics TU Dortmund, 44221 Dortmund, Germany

Paul Hofmarcher Institute for Statistics and Mathematics, WU (Vienna University of Economics and Business), Augasse 2-6, 1090 Wien, Austria, paul.hofmarcher@wu.ac.at

Christine Hohensinn Faculty of Psychology Department of Psychological Assessment and Applied Psychometrics, University of Vienna, Vienna, Austria, christine.hohensinn@univie.ac.at

Hajo Holzmann Department of Mathematics and Computer Science, Marburg University, Marburg, Germany

Fachbereich Math-ematik und Informatik, Philipps-Universit Marburg, Hans-Meerweinstr., D-35032 Marburg, Germany, holzmann@mathematik.uni-marburg.de

Kurt Hornik Institute for Statistics and Mathematics, WU (Vienna University of Economics and Business), Augasse 2-6, 1090 Wien, Austria, kurt.hornik@wu.ac.at

Eyke Hüllermeier Department of Mathematics and Computer Science, Philipps-Universität, 35032 Marburg, Germany, eyke@mathematik.uni-marburg.de

Eugene Kaciak Brock University, St. Catharines, ON, Canada, ekaciak@brocku.ca

Rebecca Klages Institute of Decision Theory and Management Science, Karlsruhe Institute of Technology (KIT), Kaiserstr. 12, 76128 Karlsruhe, Germany, rebecca.klages@kit.edu

Gerhard Klebe Department of Mathematics and Computer Science, Philipps-Universität, 35032 Marburg, Germany

Jan Hendrik Kobarg Center for Industrial Mathematics, University of Bremen, 28359 Bremen, Germany, jhkobarg@math.uni-bremen.de

Daniel Krausche Institute of Business Administration and Economics, Brandenburg University of Technology Cottbus, Postbox 101344, D-03013 Cottbus, Germany, daniel.krausche@TU-Cottbus.de

Klaus D. Kubinger Faculty of Psychology Department of Psychological Assessment and Applied Psychometrics, University of Vienna, Vienna, Austria, klaus.kubinger@univie.ac.at

Katarzyna Kuziak Department of Financial Investments and Risk Management, Wroclaw University of Economics, ul. Komandorska 118/120, Wroclaw, Poland, katarzyna.kuziak@ue.wroc.pl

Pierre Legendre Université de Montréal, C.P. 6128, succursale Centre-ville, Montréal, QC H3C 3J7 Canada, pierre.legendre@umontreal.ca

Caterina Liberati Economics Department, University of Milano-Bicocca, P.zza Ateneo Nuovo n.1, 20126 Milan, Italy, caterina.liberati@unimib.it

Artur Lichtenberg Clinic of Cardiovascular Surgery, Heinrich-Heine University, 40225 Düsseldorf, Germany

Loureiro Sandra Maria Correia Marketing, Operations and General Management Department, ISCTE-IUL Business School, Av., Forças Armadas, 1649-026 Lisbon, Portugal, sandra.loureiro@iscte.pt

Marco Maier Institute for Statistics and Mathematics, WU Vienna University of Economics and Business, Augasse 2-6, 1090 Vienna, Austria, marco.maier@wu.ac.at

Patrick Mair Institute for Statistics and Mathematics, WU (Vienna University of Economics and Business), Augasse 2-6, 1090 Wien, Austria, patrick.mair@wu.ac.at

Vladimir Makarenkov Département d'Informatique, Université du Québec à Montréal, C.P.8888, succursale Centre Ville, Montreal, QC H3C 3P8 Canada, makarenkov.vladimir@uqam.ca

Paolo Mariani Statistics Department, University of Milano-Bicocca, via Bicocca degli Arcimboldi, n.8, 20126 Milan, Italy, paolo.mariani@unimib.it

Verena Mattern Chair of Algorithm Engineering, TU Dortmund, Dortmund, Germany, verena.mattern@udo.edu

Damien McParland University College Dublin, Dublin, Ireland, damien.mcparland@ucd.ie

Miguel Angel Mendez-Mendez Universidad Anahuac, Mexico City, Mexico

Hans-Joachim Mucha Weierstrass Institute for Applied Analysis and Stochastics (WIAS), 10117 Berlin, Germany, mucha@wias-berlin.de

Ulrich Müller-Funk European Research Center for Information Systems (ERCIS), University of Münster, Münster, Germany, funk@ercis.de

Thomas Brendan Murphy School of Mathematical Sciences and Complex and Adaptive Systems Laboratory, University College Dublin, Dublin 4, Ireland, brendan.murphy@ucd.ie

Robert Nadon Department of Human Genetics, McGill University, 1205 Dr. Penfield Ave., Montreal, QC H3A-1B1 Canada

Akinori Okada Graduate School of Management and Information Sciences, Tama University, Tokyo, Japan, okada@rikkyo.ac.jp

Michael Ovelgönne Information Services and Electronic Markets, IISM, Karlsruhe Institute of Technology, Kaiserstrasse 12, D-76128 Karlsruhe, Germany, michael.ovelgoenne@kit.edu

Francesco Palumbo Università degli Studi di Napoli Federico II, Naples, Italy, francesco.palumbo@unina.it

Campo Elías Pardo Departamento de Estadística, Universidad Nacional de Colombia, Bogotá, Colombia, cepardot@unal.edu.co

Krzysztof Piontek Department of Financial Investments and Risk Management, Wroclaw University of Economics, ul. Komandorska 118/120, 53-345 Wroclaw, Poland, krzysztof.piontek@ue.wroc.pl

Surajit Ray Department of Mathematics and Statistics, Boston University, Boston, USA

Manuel Reif Faculty of Psychology Department of Psychological Assessment and Applied Psychometrics, University of Vienna, Vienna, Austria, manuel.reif@univie.ac.at

Marco Riani Dipartimento di Economia, Universitàdi Parma, Parma, Italy, mriani@unipr.it

Adrian Richter Institut für Medizinische Informationsverarbeitung, Biometrie und Epidemiologie, Universität München (LMU), Münich, Germany

Günther Rötter Institute for Music and Music Science, TU Dortmund, Dortmund, Germany, guenther.roetter@tu-dortmund.de

Günter Rudolph Chair of Algorithm Engineering, TU Dortmund, Dortmund, Germany, guenter.rudolph@udo.edu

Thomas Rusch84 Institute for Statistics and Mathematics, WU Vienna University of Economics and Business, Augasse 2-6, 1090 Vienna, Austria, thomas.rusch@wu.ac.at

Anna Rutkowska-Ziarko Faculty of Economic Sciences University of Warmia and Mazury, Oczapowskiego 4, 10-719 Olsztyn, Poland, aniarek@uwm.edu.pl

Adam Sagan Cracow University of Economics, Krakw, Poland, sagana@uek.krakow.pl

Michael Salter-Townshend School of Mathematical Sciences and Complex and Adaptive Systems Laboratory, University College Dublin, Dublin 4, Ireland, michael.salter-townshend@ucd.ie

Julia Schiffner Faculty of Statistics Chair of Computational Statistics, TU Dortmund, 44221 Dortmund, Germany, schiffner@statistik.tu-dortmund.de

Diana Schindler Department of Business Administration and Economics, Bielefeld University, Postbox 100131, 33501 Bielefeld, Germany, dschindler@wiwi.uni-bielefeld.de

Ingo Schmädecke Institute of Microelectronic Systems, Appelstr. 4, 30167 Hannover, Germany, schmaedecke@ims.uni-hannover.de

Ingo Schmitt Institute of Computer Science, Information and Media Technology, BTU Cottbus, Postbox 101344, D-03013 Cottbus, Germany, schmitt@tu-cottbus.de

Alexandra Schwarz German Institute for International Educational Research, Schloßstraße 29, D-60486 Frankfurt am Main, Germany, a.schwarz@dipf.de

Frank Siegmund Heinrich-Heine-Universität Düsseldorf, Düsseldorf

Martin Stein Information Services and Electronic Markets, IISM, Karlsruhe Institute of Technology, Kaiserstrasse 12, D-76128 Karlsruhe, Germany, martin.stein@kit.edu

Veronika Stelz Department of Statistics, LMU Munich, Akademiestr. 1, 80799 Munich, Germany

Dominik Stork KR & KM Research Group, University of Mannheim, Mannheim, Germany, dominik.stork@gmx.de

Heiner Stuckenschmidt KR & KM Research Group, University of Mannheim, Mannheim, Germany, Heiner@informatik.uni-mannheim.de

Mireille Gettler Summa CEREMADE, CNRS, Université Paris Dauphine, Paris, France, summa@ceremade.dauphine.fr

Ali Tayari Department of Computer Science, University of Nicosia, Flat 204, Democratias 16, 2370 Nicosia, Cyprus, a.tayari@hotmail.com

Myriam Thömmes Humboldt University of Berlin, Spandauer Str. 1, 10099 Berlin, Germany, thoemmem@hu-berlin.de

Francesca Torti Dipartimento di Economia, Universitàdi Parma, Parma, Italy

Dipartimento di Statistica, Università di Milano Bicocca, Milan, Italy, francesca.torti@nemo.unipr.it

Cristina Tortora Università degli Studi di Napoli Federico II, Naples, Italy

CEREMADE, CNRS, Université Paris Dauphine, Paris, France, cristina.tortora@unina.it

Matthias Trendtel Chair for Methods in Empirical Educational Research, TUM School of Education, Technische Universität München, München, Germany, matthias.trendtel@tum.de

Gerhard Tutz Department of Statistics, LMU Munich, Akademiestr. 1, 80799 Munich, Germany

Ali Ünlü Chair for Methods in Empirical Educational Research, TUM School of Education, Technische Universität München, München, Germany, ali.uenlue@tum.de

Igor Vatolkin Chair of Algorithm Engineering, TU Dortmund, Dortmund, Germany, igor.vatolkin@udo.edu; igor.vatolkin@tu-dortmund.de

Jeroen K. Vermunt Tilburg University, PO Box 50193, 5000 LE Tilburg, The Netherlands, j.k.vermunt@uvt.nl

Carmen Villar-Patiño Universidad Anahuac, Mexico City, Mexico, maria.villar@anahuac.mx

Sergio B. Villas-Boas Federal University of Rio de Janeiro, Rio de Janeiro, Brazil, sbvb@cos.ufrj.br

Dominique Vincent Institute of Decision Theory and Management Science, Karlsruhe Institute of Technology (KIT), Kaiserstr. 12, 76128 Karlsruhe, Germany, dominique.vincent@kit.edu

Adilson Elias Xavier Federal University of Rio de Janeiro, Rio de Janeiro, Brazil, adilson@cos.ufrj.br

Vinicius Layter Xavier Federal University of Rio de Janeiro, Rio de Janeiro, Brazil, vinicius@cos.ufrj.br

Sascha Voekler Institute of Business Administration and Economics, Brandenburg University of Technology Cottbus, Postbox 101344, D-03013 Cottbus, Germany, sascha.voekler@TU-Cottbus.de

Claus Weihs Faculty of Statistics Chair of Computational Statistics, TU Dortmund, 44221 Dortmund, Germany, weihs@statistik.tu-dortmund.de; claus.weihs@tu-dortmund.de

Peter Winker Department of Statistics and Econometrics, Justus-Liebig-University Giessen, Licher Str. 74, 35394 Giessen, Germany, peter.winker@wirtschaft.uni-giessen.de

Christoph Winkler Karlsruhe Institute of Technology (KIT), 76128 Karlsruhe, Germany, christoph.winkler@kit.edu

Dominik Wolff Center for Finance and Banking, University of Giessen, Licher Strasse 74, 35394 Giessen, Germany, dominik.wolff@wirtschaft.uni-giessen.de

Satoru Yokoyama Faculty of Economics Department of Business Administration, Teikyo University, Utsunomiya, Japan, satoru@main.teikyo-u.ac.jp

Part I
Invited

Size and Power of Multivariate Outlier Detection Rules

Andrea Cerioli, Marco Riani, and Francesca Torti

Abstract Multivariate outliers are usually identified by means of robust distances. A statistically principled method for accurate outlier detection requires both availability of a good approximation to the finite-sample distribution of the robust distances and correction for the multiplicity implied by repeated testing of all the observations for outlyingness. These principles are not always met by the currently available methods. The goal of this paper is thus to provide data analysts with useful information about the practical behaviour of some popular competing techniques. Our conclusion is that the additional information provided by a data-driven level of trimming is an important bonus which ensures an often considerable gain in power.

1 Introduction

Obtaining reliable information on the quality of the available data is often the first of the challenges facing the statistician. It is thus not surprising that the systematic study of methods for detecting outliers and immunizing against their effect has a long history in the statistical literature. See, e.g., Cerioli et al. (2011a), Hadi et al. (2009), Hubert et al. (2008) and Morgenthaler (2006) for recent reviews on this topic. We quote from Morgenthaler (2006, p. 271) that "Robustness of statistical methods in the sense of insensitivity to grossly wrong measurements is probably as old as the experimental approach to science". Perhaps less known is the fact that

A. Cerioli (✉) · M. Riani
Dipartimento di Economia, Università di Parma, Parma, Italy
e-mail: andrea.cerioli@unipr.it; mriani@unipr.it

F. Torti
Dipartimento di Economia, Università di Parma, Parma, Italy

Joint Research Centre, European Commission, Ispra (VA), Italy

B. Lausen et al. (eds.), *Algorithms from and for Nature and Life*, Studies in Classification, Data Analysis, and Knowledge Organization, DOI 10.1007/978-3-319-00035-0_1,

similar concerns were also present in the Ancient Greece more than 2,400 years ago, as reported by Thucydides in his History of The Peloponnesian War (III 20): "The Plataeans, who were still besieged by the Peloponnesians and Boeotians, ... made ladders equal in length to the height of the enemy's wall, which they calculated by the help of the layers of bricks on the side facing the town ... A great many counted at once, and, although some might make mistakes, the calculation would be oftener right than wrong; for they repeated the process again and again ... In this manner they ascertained the proper length of the ladders".[1]

With multivariate data outliers are usually identified by means of robust distances. A statistically principled rule for accurate multivariate outlier detection requires:

(a) An accurate approximation to the finite-sample distribution of the robust distances under the postulated model for the "good" part of the data;
(b) Correction for the multiplicity implied by repeated testing of all the observations for outlyingness.

These principles are not always met by the currently available methods. The goal of this paper is to provide data analysts with useful information about the practical behaviour of popular competing techniques. We focus on methods based on alternative high-breakdown estimators of multivariate location and scatter, and compare them to the results from a rule adopting a more flexible level of trimming, for different data dimensions. The present thus extends that of (Cerioli et al. 2011b), where only low dimensional data are considered. Our conclusion is that the additional information provided by a data-driven approach to trimming is an important bonus often ensuring a considerable gain in power. This gain may be even larger when the number of variables increases.

2 Distances for Multivariate Outlier Detection

2.1 *Mahalanobis Distances and the Wilks' Rule*

Let $y_1, \ldots, y_n$ be a sample of v-dimensional observations from a population with mean vector μ and covariance matrix Σ. The basic population model for which most of the results described in this paper were obtained is that

$$y_i \sim N(\mu, \Sigma) \qquad i = 1, \ldots, n. \tag{1}$$

[1]The Authors are grateful to Dr. Spyros Arsenis and Dr. Domenico Perrotta for pointing out this historical reference.

The sample mean is denoted by $\hat{\mu}$ and $\hat{\Sigma}$ is the unbiased sample estimate of Σ. The Mahalanobis distance of observation y_i is

$$d_i^2 = (y_i - \hat{\mu})'\hat{\Sigma}^{-1}(y_i - \hat{\mu}). \tag{2}$$

For simplicity, we omit the fact that d_i^2 is squared and we call it a distance.

Wilks (1963) showed in a seminal paper that, under the multivariate normal model (1), the Mahalanobis distances follow a scaled Beta distribution:

$$d_i^2 \sim \frac{(n-1)^2}{n} \text{Beta}\left(\frac{v}{2}, \frac{n-v-1}{2}\right) \qquad i = 1, \ldots, n. \tag{3}$$

Wilks also conjectured that a Bonferroni bound could be used to test outlyingness of the most remote observation without losing too much power. Therefore, for a nominal test size α, Wilk's rule for multivariate outlier identification takes the largest Mahalanobis distance among $d_1^2, \ldots, d_n^2$, and compares it to the $1 - \alpha/n$ quantile of the scaled Beta distribution (3). This gives an outlier test of nominal test size $\leq \alpha$.

Wilks' rule, adhering to the basic statistical principles (a) and (b) of Sect. 1, provides an accurate and powerful test for detecting a single outlier even in small and moderate samples, as many simulation studies later confirmed. However, it can break down very easily in presence of more than one outlier, due to the effect of masking. Masking occurs when a group of extreme outliers modifies $\hat{\mu}$ and $\hat{\Sigma}$ in such a way that the corresponding distances become negligible.

2.2 Robust Distances

One effective way to avoid masking is to replace $\hat{\mu}$ and $\hat{\Sigma}$ in (2) with high-breakdown estimators. A robust distance is then defined as

$$\tilde{d}_i^2 = (y_i - \tilde{\mu})'\tilde{\Sigma}^{-1}(y_i - \tilde{\mu}), \tag{4}$$

where $\tilde{\mu}$ and $\tilde{\Sigma}$ denote the chosen robust estimators of location and scatter. We can expect multivariate outliers to be highlighted by large values of $\tilde{d}_i^2$, even if masked in the corresponding Mahalanobis distances (2), because now $\tilde{\mu}$ and $\tilde{\Sigma}$ are not affected by the outliers.

One popular choice of $\tilde{\mu}$ and $\tilde{\Sigma}$ is related to the Minimum Covariance Determinant (MCD) criterion (Rousseeuw and Van Driessen 1999). In the first stage, we fix a coverage $\lfloor n/2 \rfloor \leq h < n$ and we define the MCD subset to be the sub-sample of h observations whose covariance matrix has the smallest determinant. The MCD estimator of μ, say $\tilde{\mu}_{(\text{MCD})}$, is the average of the MCD subset, whereas the MCD estimator of Σ, say $\tilde{\Sigma}_{(\text{MCD})}$, is proportional to the dispersion matrix of this subset (Pison et al. 2002). A second stage is then added with the aim of increasing efficiency, while preserving the high-breakdown properties of $\tilde{\mu}_{(\mathrm{MCD})}$ and $\tilde{\Sigma}_{(\mathrm{MCD})}$. Therefore, a one-step reweighting scheme is applied by giving weight $w_i = 0$ to observations whose first-stage robust distance exceeds a threshold value. Otherwise the weight is $w_i = 1$. We consider the Reweighted MCD (RMCD) estimator of μ and Σ, which is defined as

$$\tilde{\mu}_{\mathrm{RMCD}} = \frac{\sum_{i=1}^{n} w_i y_i}{w}, \quad \tilde{\Sigma}_{\mathrm{RMCD}} = \frac{\kappa \sum_{i=1}^{n} w_i (y_i - \tilde{\mu}_{(\mathrm{RMCD})})(y_i - \tilde{\mu}_{(\mathrm{RMCD})})'}{w-1},$$

where $w = \sum_{i=1}^{n} w_i$ and the scaling κ, depending on the values of m, n and v, serves the purpose of ensuring consistency at the normal model. The resulting robust distances for multivariate outlier detection are then

$$\tilde{d}^2_{i(\mathrm{RMCD})} = (y_i - \tilde{\mu}_{\mathrm{RMCD}})' \tilde{\Sigma}^{-1}_{\mathrm{RMCD}} (y_i - \tilde{\mu}_{\mathrm{RMCD}}) \qquad i = 1, \dots, n. \tag{5}$$

Multivariate S estimators are another common option for $\tilde{\mu}$ and $\tilde{\Sigma}$. For $\tilde{\mu} \in \Re^v$ and $\tilde{\Sigma}$ a positive definite symmetric $v \times v$ matrix, they are defined to be the solution of the minimization problem $|\tilde{\Sigma}| = \min$ under the constraint

$$\frac{1}{n} \sum_{i=1}^{n} \rho(\tilde{d}_i^2) = \zeta, \tag{6}$$

where $\tilde{d}_i^2$ is given in (4), $\rho(x)$ is a smooth function satisfying suitable regularity and robustness properties, and $\zeta = E\{\rho(z'z)\}$ for a v-dimensional vector $z \sim N(0, I)$. The ρ function in (6) rules the weight given to each observation to achieve robustness. Different specifications of $\rho(x)$ lead to numerically and statistically different S estimators. In this paper we deal with two such specifications. The first one is the popular Tukey's Biweight function

$$\rho(x) = \begin{cases} \frac{x^2}{2} - \frac{x^4}{2c^2} + \frac{x^6}{6c^4} & \text{if } |x| \le c \\ \frac{c^2}{6} & \text{if } |x| > c, \end{cases} \tag{7}$$

where $c > 0$ is a tuning constant which controls the breakdown point of S estimators; see Rousseeuw and Leroy (1987, pp.135–143) and Riani et al. (2012) for details. The second alternative that we consider is the slightly more complex Rocke's Biflat function, described, e.g., by Maronna et al. (2006, p. 190). This function assigns weights similar to (7) to distance values close to the median, but null weights outside a user-defined interval. Specifically, let

$$\eta = \min\left(\frac{\chi^2_{v,(1-\gamma)}}{v} - 1, 1\right), \tag{8}$$

where $\chi^2_{v,(1-\gamma)}$ is the $1-\gamma$ quantile of χ^2_v. Then, the weight under Rocke's Biflat function is 0 whenever a normalized version of the robust distance $\tilde{d}_i^2$ is outside the interval $[1-\eta, 1+\eta]$. This definition ensures better performance of S estimators when v is large. Indeed, it can be proved (Maronna et al. 2006, p. 221) that the weights assigned by Tukey's Biweight function (7) become almost constant as $v \to \infty$. Therefore, robustness of multivariate S estimators is lost in many practical situations where v is large. Examples of this behaviour will be seen in Sect. 3.2 even for v as small as 10.

Given the robust, but potentially inefficient, S estimators of μ and Σ, an improvement in efficiency is sometimes advocated by computing refined location and shape estimators which satisfy a more efficient version of (6) (Salibian-Barrera et al. 2006). These estimators, called MM estimators, are defined as the minimizers of

$$\frac{1}{n}\sum_{i=1}^{n} \rho_*(\tilde{\tilde{d}}_i^2), \tag{9}$$

where

$$\tilde{\tilde{d}}_i^2 = (y_i - \tilde{\tilde{\mu}})' \tilde{\tilde{\Sigma}}^{-1} (y_i - \tilde{\tilde{\mu}}) \tag{10}$$

and the function $\rho_*(x)$ provides higher efficiency than $\rho(x)$ at the null model (1). Minimization of (9) is performed over all $\tilde{\tilde{\mu}} \in \Re^v$ and all $\tilde{\tilde{\Sigma}}$ belonging to the set of positive definite symmetric $v \times v$ matrices with $|\tilde{\tilde{\Sigma}}| = 1$. The MM estimator of μ is then $\tilde{\tilde{\mu}}$, while the estimator of Σ is a rescaled version of $\tilde{\tilde{\Sigma}}$. Practical implementation of MM estimators is available using Tukey's Biweight function only (Todorov and Filzmoser 2009). Therefore, we follow the same convention in the performance comparison to be described in Sect. 3.

2.3 The Forward Search

The idea behind the Forward Search (FS) is to apply a flexible and data-driven trimming strategy to combine protection against outliers and high efficiency of estimators. For this purpose, the FS divides the data into a good portion that agrees with the postulated model and a set of outliers, if any (Atkinson et al. 2004). The method starts from a small, robustly chosen, subset of the data and then fits subsets of increasing size, in such a way that outliers and other observations not following the general structure are revealed by diagnostic monitoring. Let m_0 be the size of the starting subset. Usually $m_0 = v + 1$ or slightly larger. Let $S^{(m)}$ be the subset of data fitted by the FS at step m ($m = m_0, \ldots, n$), yielding estimates $\hat{\mu}(m)$, $\hat{\Sigma}(m)$ and distances

$$\hat{d}_i^2(m) = \{y_i - \hat{\mu}(m)\}' \hat{\Sigma}(m)^{-1} \{y_i - \hat{\mu}(m)\} \qquad i = 1, \ldots, n.$$

These distances are ordered to obtain the fitting subset at step $m+1$. Whilst $S^{(m)}$ remains outlier free, they will not suffer from masking.

The main diagnostic quantity computed by the FS at step m is

$$\hat{d}^2_{i_{\min}}(m): \qquad i_{\min} = \arg\min \hat{d}^2_i(m) \ \text{ for } i \notin S^{(m)}, \tag{11}$$

i.e. the distance of the closest observation to $S^{(m)}$, among those not belonging to this subset. The rationale is that the robust distance of the observation entering the fitting subset at step $m+1$ will be large if this observation is an outlier. Its peculiarity will then be revealed by a peak in the forward plot of $d^2_{i_{\min}}(m)$.

All the FS routines, as well as the algorithms for computing most of the commonly adopted estimators for regression and multivariate analysis, are contained in the FSDA toolbox for MATLAB and are freely downloadable from http://www.riani.it/MATLAB or from the web site of the Joint Research Centre of the European Commission. This toolbox also contains a series of dynamic tools which enable the user to link the information present in the different plots produced by the FS, such as the index or forward plot of robust Mahalanobis distances $\hat{d}^2_i(m)$ and the scatter plot matrix; see Perrotta et al. (2009) for details.

3 Comparison of Alternative Outlier Detection Rules

Precise outlier identification requires cut-off values for the robust distances when model (1) is true. If $\tilde{\mu} = \tilde{\mu}_{\text{RMCD}}$ and $\tilde{\Sigma} = \tilde{\Sigma}_{\text{RMCD}}$, Cerioli et al. (2009) show that the usually trusted asymptotic approximation based on the χ^2_v distribution can be largely unsatisfactory. Instead, Cerioli (2010) proposes a much more accurate approximation based on the distributional rules

$$\tilde{d}^2_{i(\text{RMCD})} \sim \frac{(w-1)^2}{w} \text{Beta}\left(\frac{v}{2}, \frac{w-v-1}{2}\right) \qquad \text{if} \qquad w_i = 1 \tag{12}$$

$$\sim \frac{w+1}{w}\frac{(w-1)v}{w-v} F_{v,w-v} \qquad \text{if} \qquad w_i = 0, \tag{13}$$

where w_i and w are defined as in Sect. 2.2. Cerioli and Farcomeni (2011) show how the same distributional results can be applied to deal with multiplicity of tests to increase power and to provide control of alternative error rates in the outlier detection process.

In the context of the Forward Search, Riani et al. (2009) propose a formal outlier test based on the sequence $\hat{d}^2_{i_{\min}}(m)$, $m = m_0, \ldots, n-1$, obtained from (11). In this test, the values of $\hat{d}^2_{i_{\min}}(m)$ are compared to the FS envelope

$$V^2_{m,\alpha}/\sigma_T(m)^2,$$

where $V_{m,\alpha}^2$ is the $100\alpha\,\%$ cut-off point of the $(m+1)$th order statistic from the scaled F distribution

$$\frac{(m^2-1)v}{m(m-v)}F_{v,m-v}, \tag{14}$$

and the factor

$$\sigma_T(m)^2 = \frac{P(X_{v+2}^2 < \chi_{v,m/n}^2)}{m/n} \tag{15}$$

allows for trimming of the $n-m$ largest distances. In (15), $X_{v+2}^2 \sim \chi_{v+2}^2$ and $\chi_{v,m/n}^2$ is the m/n quantile of χ_v^2.

The flexible trimming strategy enjoyed by the FS ensures a balance between the two enemy brothers of robust statistics: robustness against contamination and efficiency under the postulated multivariate normal model. This makes the Forward Search a valuable benchmark against which alternative competitors should be compared. On the other hand, very little is known about the finite sample behaviour of the outlier detection rules which are obtained from the multivariate S and MM estimators summarized in Sect. 2.2. In the rest of this section, we thus explore the performance of the alternative rules with both "good" and contaminated data, under different settings of the required user-defined tuning constants. We also provide comparison with power results obtained with the robust RMCD distances (5) and with the flexible trimming approach given by the FS.

3.1 Size

Size estimation is performed by Monte Carlo simulation of data sets generated from the v-variate normal distribution $N(0, I)$, due to affine invariance of the robust distances (4). The estimated size of each outlier detection rule is defined to be the proportion of simulated data sets for which the null hypothesis of no outliers, i.e. the hypothesis that all n observations follow model (1), is wrongly rejected. For S and MM estimation, the finite sample null distribution of the robust distances $\tilde{d}_i^2$ is unknown, even to a good approximation. Therefore, these distances are compared to the $1-\alpha/n$ quantile of their asymptotic distribution, which is χ_v^2. As in the Wilks' rule of Sect. 2.1, the Bonferroni correction ensures that the actual size of the test of no outliers will be bounded by the specified value of α if the χ_v^2 approximation is adequate.

In our investigation we also evaluate the effect on empirical test sizes of each of some user-defined tuning constants required for practical computation of multivariate S and MM estimators. See, e.g., Todorov and Filzmoser (2009) for details. Specifically, we consider:

- `bdp`: breakdown point of the S estimators, which is inherited by the MM estimators as well (the default value is 0.5);
- `eff`: efficiency of the MM estimators (the default value is 0.95);

- `effshape`[2]: dummy variable setting whether efficiency of the MM estimators is defined with respect to shape (`effshape` = 1) or to location (`effshape` = 0, the default value);
- `nsamp`: number of sub-samples of dimension $(p+1)$ in the resampling algorithm for fast computation of S estimators (our default value is 100);
- `refsteps`: maximum number of iterations in the Iterative Reweighted Least Squares algorithm for computing MM estimators (our default value is 20);
- `gamma`: tail probability in (8) for Rocke's Biflat function (the default value is 0.1).

Tables 1 and 2 report the results for $n = 200$, $v = 5$ and $v = 10$, when $\alpha = 0.01$ is the nominal size for testing the null hypothesis of no outliers and 5,000 independent data sets are generated for each of the selected combinations of parameter values. The outlier detection rule based on S estimators with Tukey's Biweight function (7) is denoted by ST. Similarly, SR is the S rule under Rocke's Biflat function. It is seen that the outlier detection rules based on the robust S and MM distances with Tukey's Biweight function can be moderately liberal, but with estimated sizes often not too far from the nominal target. As expected, liberality is an increasing function of dimension and of the breakdown point, both for S and MM estimators. Efficiency of the MM estimators (`eff`) is the only tuning constant which seems to have a major impact on the null behaviour of these detection rules. On the other hand, SR has the worst behaviour under model (1) and its size can become unacceptably high, especially when v grows. As a possible explanation, we note that a number of observations having positive weight under ST receive null weight with SR (Maronna et al. 2006, p. 192). This fact introduces a form of trimming in the corresponding estimator of scatter, which is not adequately taken into account. The same result also suggests that better finite-sample approximations to the null distribution of the robust distances $\tilde{d}_i^2$ with Rocke's Biflat function are certainly worth considering.

3.2 Power

We now evaluate the power of ST, SR and MM multivariate outlier detection rules. We also include in our comparison the FS test of Riani et al. (2009), using (14), and the finite-sample RMCD technique of Cerioli (2010), relying on (12) and (13). These additional rules have very good control of the size of the test of no outliers even for sample sizes considerably smaller than $n = 200$, thanks to their accurate cut-off values. Therefore, we can expect a positive bias in the estimated power of all the procedures considered in Sect. 3.1, and especially so in that of SR.

[2]In the RRCOV packege of the R software this option is called `eff.shape`

Table 1 Estimated size of the test of the hypothesis of no outliers for $n = 200$ and nominal test size $\alpha = 0.01$. ST is the outlier detection rule based on S estimators with Tukey's Biweight function (7); MM is the rule based on MM estimators, using again Tukey's Biweight function (7). Five thousand independent data sets are generated for each of the selected combinations of parameter values

		all parameters at default value	bdp		eff		effshape	nsamp		refsteps	
			0.15	0.25	0.8	0.98	0	10	500	10	500
$v = 5$	ST	0.023	0.010	0.014	0.023	0.023	0.023	0.026	0.024	0.023	0.023
	MM	0.021	0.019	0.020	0.023	0.015	0.023	0.021	0.020	0.022	0.023
$v = 10$	ST	0.033	0.005	0.007	0.033	0.033	0.033	0.031	0.036	0.033	0.033
	MM	0.038	0.035	0.028	0.068	0.019	0.038	0.029	0.030	0.034	0.036

Table 2 Estimated size of the test of the hypothesis of no outliers for $n = 200$ and nominal test size $\alpha = 0.01$, using S estimators with Rocke's Biflat function (SR), for different values of γ in (8). Five thousand independent data sets are generated for each of the selected combinations of parameter values

	gamma					
	0.15	0.10	0.05	0.025	0.01	0.001
$v = 5$	0.066	0.057	0.055	0.056	0.056	0.061
$v = 10$	0.089	0.080	0.079	0.078	0.077	0.081

Average power of an outlier detection rule is defined to be the proportion of contaminated observations rightly named to be outliers. We estimate it by simulation, in the case $n = 200$ and for $v = 5$ and $v = 10$. For this purpose, we generate v-variate observations from the location-shift contamination model

$$y_i \sim (1 - \delta)N(0, I) + \delta N(0 + \lambda e, I), \qquad i = 1, \ldots, n, \tag{16}$$

where $0 < \delta < 0.5$ is the contamination rate, λ is a positive scalar and e is a column vector of ones. The $0.01/n$ quantile of the reference distribution is our cut-off value for outlier detection. We only consider the default choices for the tuning constants in Tables 1 and 2, given that their effect under the null has been seen to be minor. We base our estimate of average power on 5,000 independent data sets for each of the selected combinations of parameter values.

It is worth noting that standard clustering algorithms, like g-means, are likely to fail to separate the two populations in (16), even in the ideal situation where there is a priori knowledge that $g = 2$. For instance, we have run a small benchmark study with $n = 200$, $v = 5$ and two overlapping populations by setting $\lambda = 2$ and $\delta = 0.05$ in model (16). We have found that the misclassification rate of g-means can be as high as 25 % even in this idyllic scenario where the true value of g is known and the covariance matrices are spherical. The situation obviously becomes much worse when g is unknown and must be inferred from the data. Furthermore, clustering algorithms based on Euclidean distances, like g-means, are not affine invariant and would thus provide different results on unstandardized data.

Tables 3–5 show the performance of the outlier detection rules under study for different values of δ and λ in model (16). If the contamination rate is small, it is seen that the four methods behave somewhat similarly, with FS often ranking first and MM always ranking last as λ varies. However, when the contamination rate increases, the advantage of the FS detection rule becomes paramount. In that situation both ST and MM estimators are ineffective for the purpose of identifying multivariate outliers. As expected, SR improves considerably over ST when $v = 10$ and $\delta = 0.15$, but remains ineffective when $\delta = 0.3$. Furthermore, it must be recalled that the actual size of SR is considerably larger, and thus power is somewhat biased.

Table 3 Estimated average power for different shifts λ in the contamination model (16), in the case $n = 200$, $v = 5$ and $v = 10$, when the contamination rate $\delta = 0.05$. Five thousand independent data sets are generated for each of the selected combinations of parameter values

		Mean shift λ					
		2	2.2	2.4	2.6	2.8	3
$v = 5$	ST	0.344	0.525	0.696	0.827	0.912	0.963
	SR	0.387	0.549	0.698	0.820	0.908	0.957
	MM	0.148	0.280	0.466	0.672	0.836	0.935
	RMCD	0.227	0.390	0.574	0.732	0.856	0.936
	FS	0.359	0.567	0.730	0.840	0.909	0.953
$v = 10$	ST	0.758	0.919	0.978	0.995	0.999	1
	SR	0.856	0.946	0.986	0.997	0.999	1
	MM	0.479	0.782	0.942	0.990	0.998	1
	RMCD	0.684	0.839	0.956	0.987	0.997	1
	FS	0.808	0.911	0.968	0.991	0.998	1

Table 4 Quantities as in Table 3, but now for $\delta = 0.15$

		Mean shift λ					
		2	2.4	2.6	2.8	3	3.4
$v = 5$	ST	0.073	0.532	0.772	0.901	0.960	0.996
	SR	0.275	0.433	0.594	0.742	0.854	0.925
	MM	0.006	0.010	0.012	0.016	0.026	0.397
	RMCD	0.096	0.428	0.652	0.815	0.913	0.988
	FS	0.580	0.803	0.878	0.935	0.965	0.993
$v = 10$	ST	0.006	0.007	0.008	0.01	0.013	0.041
	SR	0.696	0.825	0.895	0.923	0.931	0.946
	MM	0.001	0.001	0.001	0.001	0.003	0.030
	RMCD	0.530	0.938	0.959	0.993	1	1
	FS	0.887	0.938	0.974	0.991	0.998	1

A qualitative explanation for the failure of multivariate MM estimators is shown in Fig. 1 in the simple case $v = 2$. The four plots display bivariate ellipses corresponding to 0.95 probability contours at different iterations of the algorithm for computing MM estimators, for a data set simulated from the contamination model (16) with $n = 200$, $\delta = 0.15$ and $\lambda = 3$. The data can be reproduced using function `randn(200,2)` of MATLAB and putting the random number seed to 2. The contaminated units are shown with symbol $\circ$ and the two lines which intersect the estimate of the robust centroid are plotted using a dash-dot symbol. The upper left-hand panel corresponds to the first iteration (i1), where the location estimate is $\tilde{\mu} = (0.19, 0.18)'$ and the value of the robust correlation r derived from $\tilde{\Sigma}$ is 0.26. In this case the robust estimates are not too far from the true parameter values $\mu = (0, 0)'$ and $\Sigma = I$, and the corresponding outlier detection rule (i.e., the ST

Table 5 Quantities as in Table 3, but now for $\delta = 0.30$

		Mean shift λ						
		2	2.4	2.6	2.8	3	4	6
$\nu = 5$	ST	0.003	0.005	0.006	0.007	0.009	0.016	0.092
	SR	0.006	0.033	0.286	0.372	0.458	0.557	1
	MM	0.002	0.003	0.004	0.005	0.006	0.012	0.085
	RMCD	0.010	0.159	0.381	0.637	0.839	1	1
	FS	0.627	0.915	0.920	0.941	0.967	1	1
$\nu = 10$	ST	0.002	0.002	0.003	0.003	0.003	0.004	0.011
	SR	0.002	0.005	0.004	0.005	0.009	0.011	0.039
	MM	0.001	0.001	0.001	0.001	0.001	0.001	0.001
	RMCD	0.207	0.842	0.969	0.994	0.999	1	1
	FS	0.904	0.929	0.961	0.980	0.989	0.995	1

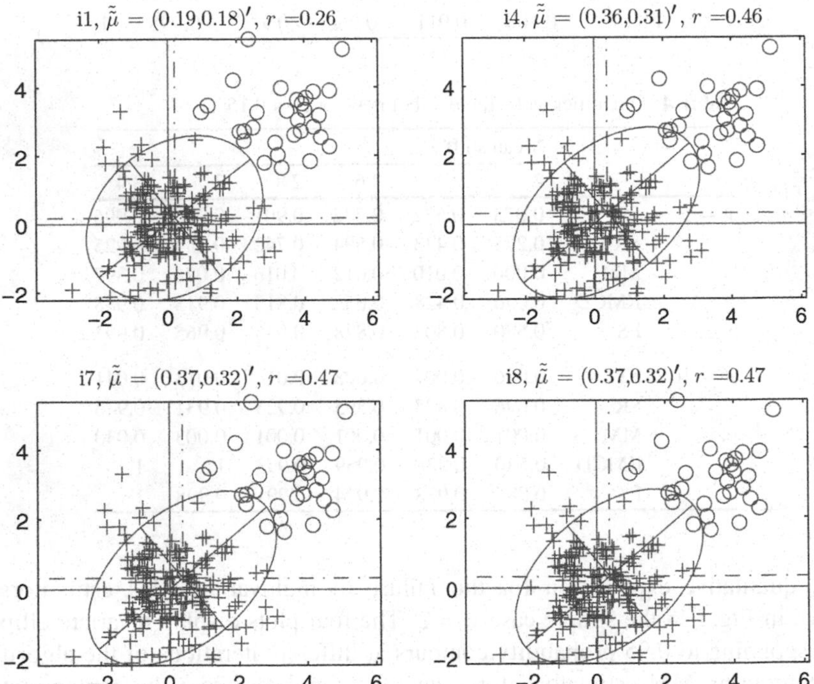

Fig. 1 Ellipses corresponding to 0.95 probability contours at different iterations of the algorithm for computing multivariate MM estimators, for a data set simulated from the contamination model (16) with $n = 200$, $\nu = 2$, $\delta = 0.15$ and $\lambda = 3$

rule in Tables 3–5) can be expected to perform reasonably well. On the contrary, as the algorithm proceeds, the ellipse moves its center far from the origin and the variables artificially become more correlated. The value of r in the final iteration (i8) is 0.47 and the final centroid $\tilde{\tilde{\mu}}$ is $(0.37; 0.32)'$. These features increase the bias

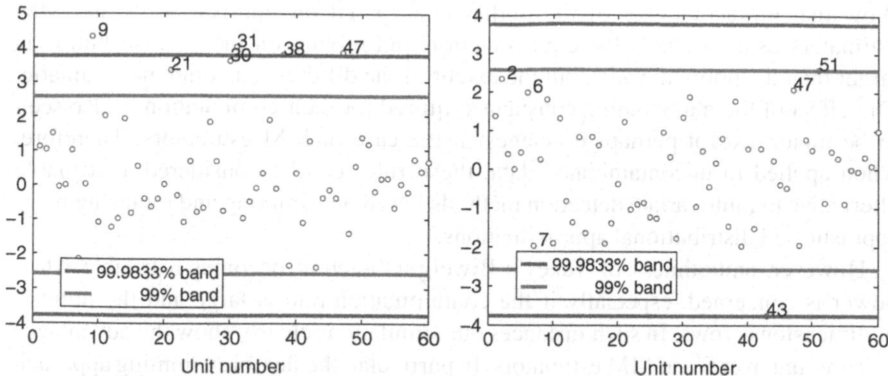

Fig. 2 Index plots of robust scale residuals obtained using MM estimation with a preliminary S-estimate of scale based on a 50 % breakdown point. *Left-hand panel*: 90 % nominal efficiency; *right-hand panel*: 95 % nominal efficiency. The *horizontal lines* correspond to the 99 % individual and simultaneous bands using the standard normal

of the parameter estimates and can contribute to masking in the supposedly robust distances (10).

A similar effect can also be observed with univariate ($v = 1$) data. For instance, Atkinson and Riani (2000, pp. 5–9) and Riani et al. (2011) give an example of a regression dataset with 60 observations on three explanatory variables where there are six masked outliers (labelled 9, 21 30, 31, 38 47) that cannot be detected using ordinary diagnostic techniques. The scatter plot of the response against the three explanatory variables and the traditional plot of residuals against fitted values, as well as the *qq* plot of OLS residuals, do not reveal observations far from the bulk of the data. Figure 2 shows the index plots of the scaled MM residuals. In the left-hand panel we use a preliminary S estimate of scale with Tukey's Biweight function (7) and 50 % breakdown point, and 90 % efficiency in the MM step under the same ρ function. In the right-hand panel we use the same preliminary scale estimate as before, but the efficiency is 95 %. As the reader can see, these two figures produce a very different output. While the plot on the right (which is similar to the masked index plot of OLS residuals) highlights the presence of a unit (number 43) which is on the boundary of the simultaneous confidence band, only the plot on the left (based on a smaller efficiency) suggests that there may be six atypical units (9, 21 30, 31, 38 47), which are indeed the masked outliers.

4 Conclusions

In this paper we have provided a critical review of some popular rules for identifying multivariate outliers and we have studied their behaviour both under the null hypothesis of no outliers and under different contamination schemes. Our results show that the actual size of the outlier tests based on multivariate S and MM estimators using Tukey's Biweight function and relying on the χ^2_v distribution is larger than the nominal value, but the extent of the difference is often not dramatic. The effect of the many tuning constants required for their computation is also seen to be minor, except perhaps efficiency in the case of MM estimators. Therefore, when applied to uncontaminated data, these rules can be considered as a viable alternative to multivariate detection methods based on trimming and requiring more sophisticated distributional approximations.

However, smoothness of Tukey's Biweight function becomes a trouble when power is concerned, especially if the contamination rate is large and the number of dimensions grows. In such instances our simulations clearly show the advantages of trimming over S and MM estimators. In particular, the flexible trimming approach ensured by the Forward Search is seen to greatly outperform the competitors, even the most liberal ones, in almost all our simulation scenarios and is thus to be recommended.

Acknowledgements The authors thank the financial support of the project MIUR PRIN MISURA - Multivariate models for risk assessment.

References

Atkinson, A. C., & Riani, M. (2000). *Robust diagnostic regression analysis*. New-York: Springer.

Atkinson, A. C., Riani, M., & Cerioli, A. (2004). *Exploring multivariate data with the forward search*. New York: Springer.

Cerioli, A. (2010). Multivariate outlier detection with high-breakdown estimators. *Journal of the American Statistical Association, 105*, 147–156.

Cerioli, A., & Farcomeni, A. (2011). Error rates for multivariate outlier detection. *Computational Statistics and Data Analysis, 55*, 544–553.

Cerioli, A., Riani, M., & Atkinson, A. C. (2009). Controlling the size of multivariate outlier tests with the MCD estimator of scatter. *Statistics and Computing, 19*, 341–353

Cerioli, A., Atkinson, A. C., & Riani, M. (2011a). Some perspectives on multivariate outlier detection. In S. Ingrassia, R. Rocci, & M. Vichi (Eds.), *New perspectives in statistical modeling and data analysis* (pp. 231–238). Berlin/Heidelberg: Springer.

Cerioli, A., Riani, M., & Torti, F. (2011b). Accurate and powerful multivariate outlier detection. *58th congress of ISI*, Dublin.

Hadi, A. S., Rahmatullah Imon, A. H. M., & Werner, M. (2009). Detection of outliers. *WIREs Computational Statistics, 1*, 57–70.

Hubert, M., Rousseeuw, P. J., & Van aelst, S. (2008). High-breakdown robust multivariate methods. *Statistical Science, 23*, 92–119.

Maronna, R. A., Martin, D. G., & Yohai, V. J. (2006). *Robust statistics*. New York: Wiley.

Morgenthaler, S. (2006). A survey of robust statistics. *Statistical Methods and Applications, 15*, 271–293 (Erratum 16, 171–172).

Perrotta, D., Riani, M., & Torti, F. (2009). New robust dynamic plots for regression mixture detection. *Advances in Data Analysis and Classification, 3*, 263–279.

Pison, G., Van aelst, S., & Willems, G. (2002). Small sample corrections for LTS and MCD. *Metrika, 55*, 111–123.

Riani, M., Atkinson, A. C., & Cerioli, A. (2009). Finding an unknown number of multivariate outliers. *Journal of the Royal Statistical Society B, 71*, 447–466.

Riani, M., Torti, F., & Zani, S. (2011). Outliers and robustness for ordinal data. In R. S. Kennet & S. Salini (Eds.), *Modern analysis of customer satisfaction surveys: with applications using R*. Chichester: Wiley.

Riani, M., Cerioli, A., & Torti, F. (2012). A new look at consistency factors and efficiency of robust scale estimators. Submitted.

Rousseeuw, P. J., & Leroy, A. M. (1987). *Robust regression and outlier detection*. New York: Wiley.

Rousseeuw, P. J. & Van Driessen, K. (1999). A fast algorithm for the minimum covariance determinant estimator. *Technometrics, 41*, 212–223.

Salibian-barrera, M., Van Aelst, S., & Willems, G. (2006). Principal components analysis based on multivariate mm estimators with fast and robust bootstrap. *Journal of the American Statistical Association, 101*, 1198–1211.

Todorov, V., & Filzmoser, P. (2009). An object-oriented framework for robust multivariate analysis. *Journal of Statistical Software, 32*, 1–47.

Wilks, S. S. (1963). Multivariate statistical outliers. *Sankhya A, 25*, 407–426.

Clustering and Prediction of Rankings Within a Kemeny Distance Framework

Willem J. Heiser and Antonio D'Ambrosio

Abstract Rankings and partial rankings are ubiquitous in data analysis, yet there is relatively little work in the classification community that uses the typical properties of rankings. We review the broader literature that we are aware of, and identify a common building block for both prediction of rankings and clustering of rankings, which is also valid for partial rankings. This building block is the Kemeny distance, defined as the minimum number of interchanges of two adjacent elements required to transform one (partial) ranking into another. The Kemeny distance is equivalent to Kendall's τ for complete rankings, but for partial rankings it is equivalent to Emond and Mason's extension of τ. For clustering, we use the flexible class of methods proposed by Ben-Israel and Iyigun (Journal of Classification 25: 5–26, 2008), and define the disparity between a ranking and the center of cluster as the Kemeny distance. For prediction, we build a prediction tree by recursive partitioning, and define the impurity measure of the subgroups formed as the sum of all within-node Kemeny distances. The median ranking characterizes subgroups in both cases.

1 Introduction

Ranking and classification are basic cognitive skills that people use every day to create order in everything that they experience. Many data collection methods in the life and behavioral sciences often rely on ranking and classification. Grouping and ordering a set of elements is also a major communication and action device in social

---

W.J. Heiser (✉)
Institute of Psychology, Leiden University, 2300 RB Leiden, The Netherlands
e-mail: Heiser@Fsw.Leidenuniv.nl

A. D'Ambrosio
Department of Industrial Engineering, University of Naples Federico II, Piazzale Tecchio, 80125, Naples, Italy
e-mail: antdambr@unina.it

B. Lausen et al. (eds.), *Algorithms from and for Nature and Life*, Studies in Classification, Data Analysis, and Knowledge Organization, DOI 10.1007/978-3-319-00035-0_2,

life, as is clear when we consider rankings of sport-teams, universities, countries, web-pages, French wines, and so on. Not surprisingly, the literature on rankings is scattered across many fields of science.

Statistical methods for the analysis of rankings can be distinguished in (1) data analysis methods based on badness-of-fit functions that try to describe the structure of rank data, (2) probabilistic methods that model the ranking process, and assume substantial agreement (or *homogeneity*) among the rankers about the underlying order of the rankings, and (3) probabilistic methods that model the population of rankers, assuming substantial disagreement (or *heterogeneity*) between them. Let us look at each of these in turn.

Two examples of data analysis methods based on badness-of-fit functions that have been applied to rankings are principal components analysis (PCA, see Cohen and Mallows 1980; Diaconis 1989; Marden 1995, Chap. 2), and multidimensional scaling (MDS) or unfolding (Heiser and de Leeuw 1981; Heiser and Busing 2004). In psychometrics, PCA on rankings was justified by what is called the *vector model for rankings,* going back to the independent contributions of Guttman (1946); Slater (1960) and Tucker (1960) and popularized by Carroll (1972, pp. 114–129) through his MDPREF method. It is also possible to perform a principal components analysis while simultaneously fitting some optimal transformation of the data that preserves the rank order (in a program called CATPCA, cf. Meulman et al. 2004). By contrast, the unfolding technique is based on the *ideal point model for rankings*, which originated with Coombs (1950, 1964, Chaps. 5–7), but his analytical procedures were only provisional and had been soon superseded by MDS methods (Roskam 1968; Kruskal and Carroll 1969). Unfortunately, however, MDS procedures for ordinal unfolding tended to suffer from several degeneracy problems for a long time (see Van Deun 2005; Busing 2009 for a history of these difficulties and state-of-the-art proposals to resolve them). One of these proposals, due to Busing et al. (2005), is available under the name PREFSCAL in the IBM-SPSS Statistics package.

Probabilistic modeling for the ranking process assuming homogeneity of rankers started with Thurstone (1927, 1931), who proposed that judgments underlying rank orders follow a multivariate normal distribution with location parameters corresponding to each ranked object. Daniels (1950) looked at cases in which the random variables associated with the ranked objects are independent. Examples of more complex *Thurstonian models* include Böckenholt (1992), Chan and Bentler (1998), Maydeu-Olivares (1999) and Yao and Böckenholt (1999). A second class of models assuming homogeneity of rankers started with Mallows (1957), and was also based upon a process in which pairs of objects are compared, but now according to the Bradley-Terry-Luce (BTL) model (Bradley and Terry 1952; Luce 1959), thus excluding intransitivities. These probability models amount to a negative exponential function of some distance between rankings, for example the distance related to Kendall's τ (see Sect. 3); hence their name *distance-based ranking models* (Fligner and Verducci 1986). A third class of models assuming homogeneity of rankers decompose the ranking process into a series of independent stages. The stages form a nested sequence, in each of which a Bradley-Terry-Luce choice process is assumed for selecting 1 out of j options, with $j = m, m - 1, \ldots, 2$; hence their name *multistage models* (Fligner and Verducci 1988). We refer to Critchlow et al. (1991) for an in-depth discussion of all of these models. Critchlow and Fligner (1991) demonstrated how both the Thurstonean models and the multistage BTL models can be seen as generalized linear models and be fitted with standard software.

Probabilistic models for the population of rankers assuming substantial heterogeneity of their rankings are of at least three types. First, there are probabilistic versions of the ideal point model involving choice data (Zinnes and Griggs 1974; Kamakura and Srivastava 1986), or rankings (Brady 1989; Van Blokland-Vogelesang 1989; Hojo 1997, 1998). Second, instead of assuming one probabilistic model for the whole population, we may move to (unknown) mixtures of subpopulations, characterized by different parameters. For example, mixtures of models of the BTL type were proposed by Croon (1989), and mixtures of distance-based models by Murphy and Martin (2003). Gormley and Murphy (2008a) provided a very thorough implementation of two multistage models with mixture components. Third, heterogeneity of rankings can also be accounted for by the introduction of *covariates*, from which we can estimate mixtures of *known* subpopulations. Examples are Chapman and Staelin (1982), Dittrich et al. (2000), Böckenholt (2001), Francis et al. (2002), Skrondal and Rabe-Hesketh (2003), and Gormley and Murphy (2008b). All of these authors use the generalized linear modeling framework.

Most methods that are mainstream in the classification community follow the first approach, that is, they use an algorithm model (e.g., hierarchical clustering, construction of phylogenetic trees), or try to optimize some badness-of-fit function (e.g., K-means, fuzzy clustering, PCA, MDS). Some of them analyze a rank ordering of dissimilarities, which makes the results order-invariant, meaning that order-preserving transformations of the data have no effect. However, there are very few proposals in the classification community directly addressing clustering of multiple rankings, or prediction of rankings based on explanatory variables characterizing the source of them (covariates). Our objective is to fill this gap, and to catch up with the statisticians.[1]

Common to all approaches is that they have to deal with the sample space of rankings, which has a number of very specific properties. Also, most methods either implicitly or explicitly use some measure of correlation or distance among rankings. Therefore, we start our discussion with a brief introduction in the geometry of rankings in Sect. 2, and how it naturally leads to measures of correlation and distance in Sect. 3. We then move to the median ranking in Sect. 4, give a brief sketch in Sect. 5 of how we propose to formulate a clustering procedure and to build a prediction tree for rankings, and conclude in Sect. 6.

---

[1]During the Frankfurt DAGM-GfKl-2011-conference, Eyke Hüllermeier kindly pointed out that there is related work in the computer science community under the name "preference learning" (in particular, Cheng et al. (2009), and more generally, Fürnkranz and Hüllermeier 2010).

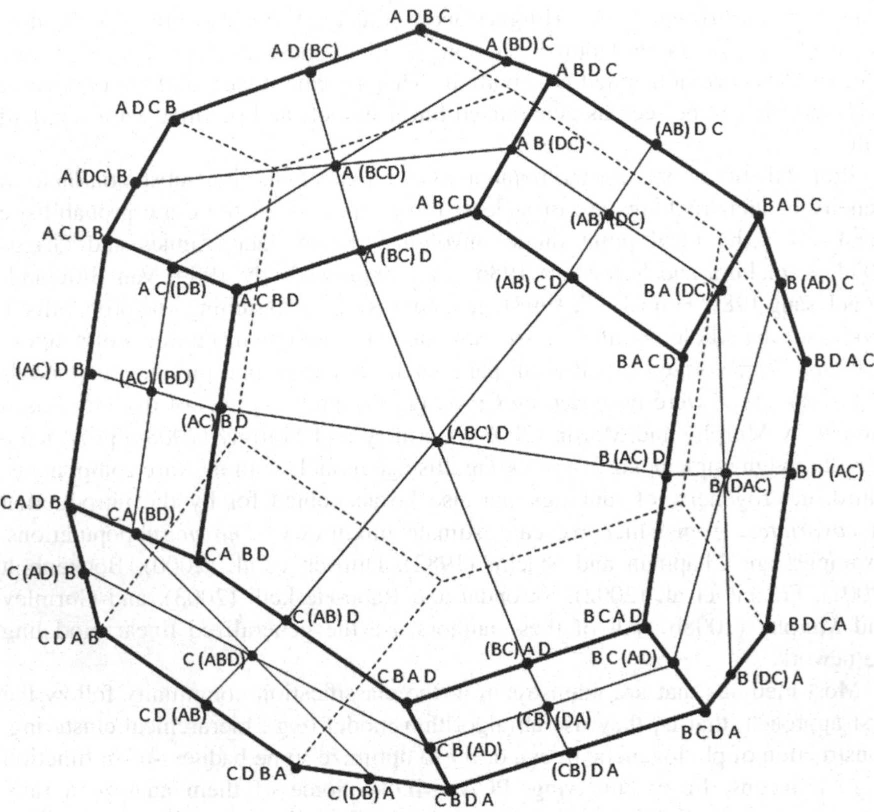

Fig. 1 Permutation polytope for all 24 full rankings of four objects, supplemented by all partial rankings with one tie-block of two or three objects, or two tie-blocks of two objects. Full rankings have equal distance towards the center; partial rankings lie strictly within this sphere. For clarity, mirror images at the back of the polytope are not labeled

2 Geometry of Rankings

The 24 full rankings that can be formed from four objects form a permutation polytope that has the shape of a truncated octahedron (cf. Thompson 1993; Heiser 2004). Thompson offered an thorough study of the permutation structure of partial rankings, showing that the 12 partial rankings with a tie in last position form a truncated tetrahedron, as do the 12 partial rankings with a tie in first position. The 12 partial rankings with a tie in middle position, however, are the intersection of a cube and an octahedron, forming a cuboctahedron. Then there are six partial rankings with two tie-blocks forming an octahedron, and finally four partial rankings with tie-blocks of three in last position or in first position, each forming a tetrahedron.

It should be noted that these generalized permutation polytopes can be connected with each other in a single graph if we introduce nodes in the original truncated octahedron that are half-way the nodes of the full rankings. This integrated graph of all full and partial rankings is given in Fig. 1. All lines in this graph now indicate a reversal or switch from one inequality to an equality, or vice versa, except for the lines in the hexagons that connect to partial rankings with tie-blocks of three, which represent two switches. The natural graphical distance in the integrated permutation polytope is the sum of the line segments that need to be traversed along the shortest path in going from one node to another, and this distance is equivalent to the count of the minimum number of interchanges of two adjacent elements required to transform one (partial) ranking into another.

More generally, it will be clear that the sample space of rankings has the following characteristic properties: it is finite and discrete, it has many symmetries (for every ranking there is a reverse ranking), it is endowed with a graphical metric, and it intersects with a hypersphere: all full rankings are equidistant towards the zero ranking in which all objects are tied. All partial rankings lie strictly within the hypersphere. For a discussion of the consequences of this geometry for various ranking and choice models, we refer to Zhang (2004). Rankings can also arise indirectly as a consequence of doing pairwise discriminant analyses among m populations (Kamiya and Takemura 1997, 2005). Under the unfolding model, only a limited amount of rankings can occur (Coombs 1964; Kamiya et al. 2006, 2011). The probabilistic models mentioned in the Introduction describe specific distributions across the polytope.

3 Kendall's τ and the Kemeny Distance

Although there was earlier relevant work (see Kruskal 1958, Sect. 17), Kendall (1938) marks the beginnings of the first wave of contributions to the study of rankings as a separate topic in statistics. Kendall defined τ as a coefficient that "measures the closeness of correspondence between two given rankings in the sense that it measures how accurate either ranking would be if the other were objective" (Kendall 1938, p. 85). He then derived its exact sampling distribution and standard error, assuming one given order and a universe in which all the possible rankings occur an equal number of times, and he showed that this distribution is already close to normal for relatively small sample size. In Kendall (1948), he also gave a second definition of τ as a "coefficient of disarray". Calling the minimum number of switches which transform any ranking into any other ranking of the same number of objects s, he showed that

$$\tau = 1 - \frac{2s}{\frac{1}{2}n(n-1)} .$$

This equivalence between τ and s establishes their connection with the permutation polytope, and thus their fundamental relevance for the study of rankings, because s is just the graphical distance defined in the previous section. The *minimum move metric* s is called the Kendall distance (cf. Marden 1995, p. 25).

Emond and Mason (2002) noted that there is a problem with the Kendall distance in the case of partial rankings. In that case, it is easy to show that it violates the triangle inequality (e.g., consider A(BC), ABC, and (AB)C), so it is not a proper metric. This anomaly is due to the way in which Kendall (1948, Chap. 3) defined τ when there are tied ranks.

Fortunately, there is a well-founded distance without these problems, called the Kemeny distance, conceived independently in the context of social choice theory (Kemeny 1959; Kemeny and Snell 1962). Kemeny had set up a set of reasonable axioms of which perhaps the most characteristic one is that the distance be *invariant under addition of equally ranked first and/or last objects*. The unique distance satisfying all axioms turns out to be:

$$d_{Kem}(R_s, R_t) = \tfrac{1}{2} \sum_{i=1}^{m} \sum_{j=1}^{m} \left| x_{(s)ij} - x_{(t)ij} \right|,$$

where R_s and R_t are any two rankings, m is the number of objects, and $x_{(s)ij}$ is defined as equal to 1 if object i is preferred to object j in ranking s, equal to -1 if the reverse is true, and equal to 0 if the two objects are tied. Clearly, the Kemeny distance is of the city-block type in the space of pair comparisons.

When there are no ties, the Kemeny distance is equal to the Kendall distance. From its definition, it is not hard to see that it counts the number of interchanges of pairs of elements required to transform one (partial) ranking into another, so it is equal to the graphical distance among any two elements in the integrated permutation polytope in Fig. 1.

4 Finding a Central Ranking: The Median Ranking

There is an extensive literature on finding a central ranking for a given set of individual rankings, also called the *social choice* problem, or the consensus problem. But when the Kemeny distance is the metric of choice, it will lead us to one specific central ranking. Consider a set of individual rankings R_s, with $s = 1, \ldots, n$, and let us indicate the center to be found by $\hat{S}$. Then we have

$$\hat{S} = \arg\min_{S} \sum_{s=1}^{n} w_s d_{Kem}(R_s, S).$$

Here we have used a weighted version, with weights w_s for ranking R_s (one obvious choice of weights is the relative frequency with which each unique ranking occurs). Center $\hat{S}$ so defined is usually called the *consensus ranking* in the social choice literature, as well as in discrete mathematics, and the *median ranking* in statistics. For a review of ranking models for the consensus problem, see Cook (2006).

Emond and Mason (2002) proposed a new rank correlation coefficient for the case of partial rankings, called τ_X (τ-extended), to resolve the difficulty with the Kendall distance mentioned in the previous section. It is equal to Kendall's τ for complete rankings, while for partial rankings $1 - \tau_X$ is equivalent to Kemeny distance. Maximizing the weighted sum of τ_X leads to the same median ranking. Now, it is well known that finding $\hat{S}$ is an NP-hard problem (Barthélemy et al. 1989). Emond and Mason's reformulation has the advantage that it allows a branch-and-bound algorithm that is practical up to about 20 objects and an unlimited number of rankers, and deals correctly with partial rankings.

5 Application to Clustering and Recursive Partitioning

We will now give a brief sketch of how we are using the Kemeny distance and the median ranking for classification of multiple rankings. First, we outline a non-hierarchical clustering algorithm and next we show how to use explanatory variables (covariates) to build a prediction tree. For clustering, we follow a generalized K-means method, and for building the prediction tree, we use standard CART methodology (Breiman et al. 1984) involving a binary segmentation procedure that recursively partitions the set of rankings, with a specific impurity measure in the splitting rule. But of course, other choices are possible.

Ben-Israel and Iyigun's (2008) *probabilistic distance clustering* framework allows for probabilistic allocation of cases to classes. So it is a form of fuzzy clustering, rather than hard clustering. It is based on the principle that probability and distance are inversely related. Shepard (1987) accumulated lots of evidence for a similar principle governing contingencies of behavior. Under this principle, we define a loss function for *K-Median Cluster Component Analysis* (*CCA*) as follows:

$$CCA\left(\mathbf{P}, S_1, \cdots, S_K\right) = \sum_{s=1}^{n} \sum_{k=1}^{K} p_k^2\left(R_s\right) d_{Kem}\left(R_s, S_k\right),$$

where $p_k(R_s)$ is the probability of allocating ranking s to cluster component k, S_k is the center of component k for $k = 1, \ldots, K$, and $\mathbf{P}$ is the $n \times K$ matrix of allocation probabilities. If we differentiate the *CCA* function with respect to $p_k(R_s)$, subject to the constraint that allocation probabilities for a given ranking sum to one, we obtain the stationary equation $p_k(R_s)\ d_{Kem}(R_s, S_k) = \textit{constant depending on } R_s$. So the stationary equations of the *CCA* optimization problem are consistent with the principle of probability being inversely related to distance. Since the CCA function splits into K parts, finding S_k given some given values of the allocation probabilities $\mathbf{P}$ reduces to finding a median ranking using the *kth* column of $\mathbf{P}$. For finding $\mathbf{P}$ given K median rankings an explicit formula is available. A more detailed description and evaluation of K-median cluster component analysis is in preparation (Heiser and D'Ambrosio 2011).

Now consider the case in which we have a set of explanatory variables (or covariates) giving one point z_s in predictor space for each ranking R_s. The aim is to predict the differences between the rankings. Tree-based methods partition the predictor space into a set of rectangular regions parallel to the coordinate axes (i.e., the explanatory variables), and fit a simple model in each of them (Hastie et al. 2001). During the recursive partitioning process in which we form a nested sequence of subsamples, we have to determine, for each possible split along the coordinate axis of any variable, the impurity of the subsamples formed. The impurity measure $Q_l(T)$ that we choose for a subsample in subtree T at node l representing a region G_l containing the profiles of n_l rankings is

$$Q_l(T) = \frac{1}{\frac{1}{2}n_l(n_l-1)} \sum_{z_s \in G_l}^{n_l} \sum_{z_t \in G_l}^{n_l} d_{Kem}(R_s, R_t), \text{ with } s > t.$$

Alternatively, we could have chosen the weighted sum of Kemeny distances towards the median ranking, but that would force us to solve a hard combinatorial problem many times when growing the tree. Our pruning strategy is cost-complexity pruning (Hastie et al. 2001, p. 270; also see: Mingers (1989); Cappelli et al. 2002). For the pruned tree, we calculate in each terminal node the consensus ranking as described in Sect. 4 and its corresponding τ_X, and determine for the internal nodes of the tree the weighted average τ_X. For a more detailed description and evaluation of our distance-based prediction tree, we refer to D'Ambrosio and Heiser (2011), which is based on earlier work of D'Ambrosio (2007).

In one of our test applications, on a real dataset with 500 rankings of 15 objects and 128 explanatory variables, we first obtained a maximum tree with 24 terminal nodes. In Fig. 2, the top panel shows how the impurity in the training sample (bottom line) goes down monotonically, while in the test sample (upper line) the impurity goes up when tree size passes 11, which is the size of the pruned tree. The bottom panel of Fig. 2 shows the average τ_X weighted by node size, which gives a better interpretable scale. At the root node, overall $\tau_X = 0.387$, a moderate correlation, which reaches $\tau_X = 0.489$ on average for the maximum tree. Some of the terminal nodes in the pruned tree even reach $\tau_X = 0.510$, but others are lower.

6 Concluding Remarks

Kemeny distance is the natural graphical distance on the permutation polytope, which is the sample space of rankings. The polytope can be extended to accommodate partial rankings. It provides a standard for other approaches that use more assumptions or proceed by first embedding the polytope in Euclidean space. Minimizing the sum of Kemeny distances leads to the median ranking as a center. For full rankings, one minus Kendall's τ is equivalent to the Kemeny distance. Often the median ranking has ties, or the data are partial rankings to start with. In that

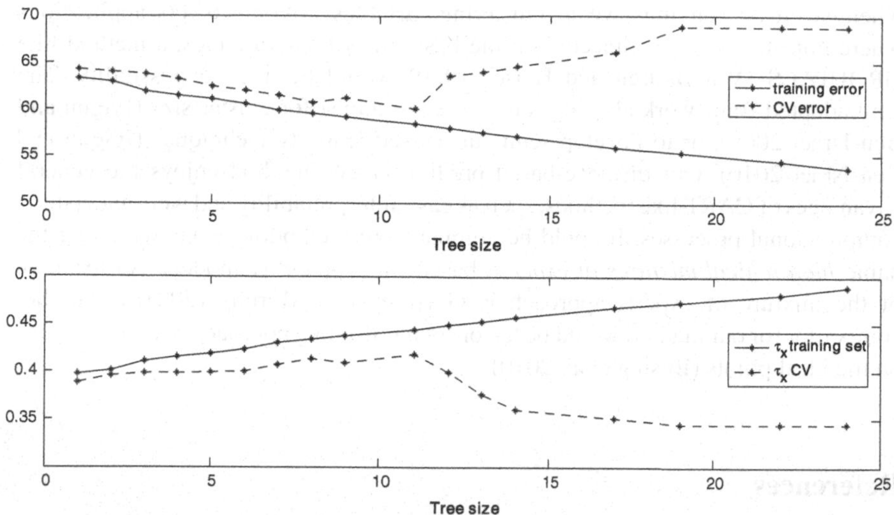

Fig. 2 Pruning sequence to decide on the depth of the tree. Training error rate is based on 350 rankings, cross-validated error rate is based on 150 rankings (using tenfold cross-validation). *Top panel* gives error rates (impurity), *bottom panel* gives the average τ_X

case, one minus Kendall's τ is faulted as a distance, because it no longer satisfies the metric axioms. Emond and Mason (2002) provided a different definition of τ for partial rankings, called τ_X, for which $1 - \tau_X$ is equal to the Kemeny distance. The new definition is welcome, because the scale of τ_X is easier to interpret than a distance scale: it is comparable across different numbers of objects.

We believe that loss-function based methods enjoy general advantages compared to methods based on probability models. They do not depend on assumptions that may be unrealistic for certain data. For rankings, in particular, the probability rationale often refers to replicated judgment processes, which is not so relevant for ranking the States of the United States (O'Leary Morgan and Morgan 2010), where the raw data are rates or percentages in the population. Note that in our use of probabilistic distance clustering, the term "probabilistic" merely expresses the uncertainty in the allocation of rankings to clusters, and does not imply an assumption about the data generating process, as in probability models.

Loss-function based methods generally tend to lead to better understood computational processes. Inclusion of weights in loss functions allows greater flexibility and generality, and in our case we profit from it in the median ranking and in the clustering algorithm. But weights can also be useful to emulate maximum likelihood estimation or to down-weight unreliable parts of the data. Some people hold, for example, that the beginning and the end of a ranking is more reliable than the middle.

Our clustering method could be compared with probabilistic models like Croon (1989), Murphy and Martin (2003), and Gormley and Murphy (2008a). Note that when we cluster rankings, we are clustering variables, not objects. For applications where objects are to be clustered on the basis of ordinal variables, a method like GROUPALS (Van Buuren and Heiser 1989) would be a good possibility. The here adopted framework also gives us a way to adjust for cluster size (Iyigun and Ben-Israel 2008), or to develop semi-supervised learning techniques (Iyigun and Ben-Israel 2010). Our distance-based prediction tree method enjoys the general advantages of CART-like methods, such as easy interpretability and well-understood computational processes. It could be compared to methodology known under the name *hierarchical mixtures of experts*, based on probability models. An example of the mixture of experts approach is Gormley and Murphy (2008b). Another competitor for our method would be the ordinal unfolding approach with restrictions on the ideal points (Busing et al. 2010).

References

Barthelémy, J. P., Guénoche, A., & Hudry, O. (1989). Median linear orders: Heuristics and a branch and bound algorithm. *European Journal of Operational Research, 42*, 313–325.

Ben-Israel, A., & Iyigun, C. (2008). Probabilistic distance clustering. *Journal of Classification, 25*, 5–26.

Böckenholt, U. (1992). Thurstonian representation for partial ranking data. *British Journal of Mathematical and Statistical Psychology, 45*, 31–49.

Böckenholt, U. (2001). Mixed-effects analysis of rank-ordered data. *Psychometrika, 77*, 45–62.

Bradley, R. A., & Terry, M. A. (1952). Rank analysis of incomplete block designs, I. *Biometrika, 39*, 324–345.

Brady, H. E. (1989). Factor and ideal point analysis for interpersonally incomparable data. *Psychometrika, 54*, 181–202.

Breiman, L., Froedman, J.H., Olshen, R.A., & Stone, C.J. (1984). *Classification and regression trees.* Wadsworth Publishing Co., Inc, Belmont, CA.

Busing, F. M. T. A. (2009). *Some advances in multidimensional unfolding*. Doctoral Dissertation, Leiden, The Netherlands: Leiden University.

Busing, F. M. T. A., Groenen, P., & Heiser, W. J. (2005). Avoiding degeneracy in multidimensional unfolding by penalizing on the coefficient of variation. *Psychometrika, 70*, 71–98.

Busing, F. M. T. A., Heiser, W. J., & Cleaver, G. (2010). Restricted unfolding: Preference analysis with optimal transformations of preferences and attributes. *Food Quality and Preference, 21*, 82–92.

Cappelli, C., Mola, F., & Siciliano, R. (2002). A statistical approach to growing a reliable honest tree. *Computational Statistics and Data Analysis, 38*, 285–299.

Carroll, J. D. (1972). Individual differences and multidimensional scaling. In R. N. Shepard et al. (Eds.), *Multidimensional scaling, Vol. I theory* (pp. 105–155). New York: Seminar Press.

Chan, W., & Bentler, P. M. (1998). Covariance structure analysis of ordinal ipsative data. *Psychometrika, 63*, 369–399.

Chapman, R. G., & Staelin, R. (1982). Exploiting rank ordered choice set data within the stochastic utility model. *Journal of Marketing Research, 19*, 288–301.

Cheng, W., Hühn, J., & Hüllermeier, E. (2009). *Decision tree and instance-based learning for label ranking.* In: Proceedings of the 26th international conference on machine learning (pp. 161–168). Montreal. Canada.

Cohen, A., & Mellows, C. L. (1980). *Analysis of ranking data* (*Tech. Rep.*). Murray Hill: Bell Telephone Laboratories.

Cook, W. D. (2006). Distance-based and ad hoc consensus models in ordinal preference ranking. *European Journal of Operational Research, 172*, 369–385.

Coombs, C. H. (1950). Psychological scaling without a unit of measurement. *Psychological Review, 57*, 145–158.

Coombs, C. H. (1964). *A theory of data*. New York: Wiley.

Critchlow, D. E., & Fligner, M. A. (1991). Paired comparison, triple comparison, and ranking experiments as generalized linear models, and their implementation on GLIM. *Psychometrika, 56*, 517–533.

Critchlow, D. E., Fligner, M. A., & Verducci, J. S. (1991). Probability models on rankings. *Journal of Mathematical Psychology, 35*, 294–318.

Croon, M. A. (1989). Latent class models for the analysis of rankings. In G. De Soete et al. (Eds.) *New developments in psychological choice modeling* (pp. 99–121). North-Holland, Elsevier.

D'ambrosio, A. (2007). *Tree-based methods for data editing and preference rankings*. Doctoral dissertation. Naples, Italy: Department of Mathematics and Statistics.

D'ambrosio, A., & Heiser, W. J. (2011). Distance-based multivariate trees for rankings. Technical report.

Daniels, H. E. (1950). Rank correlation and population models. *Journal of the Royal Statistical Society, Series B, 12*, 171–191.

Diaconis, P. (1989). A generalization of spectral analysis with application to ranked data. *The Annals of Statistics, 17*, 949–979.

Dittrich, R., Katzenbeisser, W., & Reisinger, H. (2000). The analysis of rank ordered preference data based on Bradley-Terry type models. *OR-Spektrum, 22*, 117–134.

Emond, E. J., & Mason, D. W. (2002). A new rank correlation coefficient with application to the consensus ranking problem. *Journal of Multi-Criteria Decision Analysis, 11*, 17–28.

Fligner, M. A., & Verducci, J. S. (1986). Distance based ranking models. *Journal of the Royal Statistical Society, Series B, 48*, 359–369.

Fligner, M. A., & Verducci, J. S. (1988). Multistage ranking models. *Journal of the American Statistical Association, 83*, 892–901.

Francis, B., Dittrich, R., Hatzinger, R., & Penn, R. (2002). Analysing partial ranks by using smoothed paired comparison methods: An investigation of value orientation in Europe. *Applied Statistics, 51*, 319–336.

Fürnkranz, J., & Hüllermeier, E. (Eds.). (2010). *Preference learning*. Heidelberg: Springer.

Gormley, I. C., & Murphy, T. B. (2008a). Exploring voting blocs within the Irish electorate: A mixture modeling approach. *Journal of the American Statistical Association, 103*, 1014–1027.

Gormley, I. C., & Murphy, T. B. (2008b). A mixture of experts model for rank data with applications in election studies. *The Annals of Applied Statistics, 2*, 1452–1477.

Guttman, L. (1946). An approach for quantifying paired comparisons and rank order. *Annals of Mathematical Statistics, 17*, 144–163.

Hastie, T., Tibshirani, R., & Friedman, J. (2001). *The elements of statistical learning: Data mining, inference, and prediction*. New York: Springer.

Heiser, W. J. (2004). Geometric representation of association between categories. *Psychometrika, 69*, 513–546.

Heiser, W. J., & Busing, F. M. T. A. (2004). Multidimensional scaling and unfolding of symmetric and asymmetric proximity relations. In D. Kaplan (Ed.), *The SAGE handbook of quantitative methodology for the social sciences* (pp. 25–48). Thousand Oaks: Sage.

Heiser, W. J., & D'ambrosio, A. (2011). *K*-Median cluster component analysis. Technical report.

Heiser, W. J., & De Leeuw, J. (1981). Multidimensional mapping of preference data. *Mathématiques et Sciences Humaines, 19*, 39–96.

Hojo, H. (1997). A marginalization model for the multidimensional unfolding analysis of ranking data. *Japanese Psychological Research, 39*, 33–42.

Hojo, H. (1998). Multidimensional unfolding analysis of ranking data for groups. *Japanese Psychological Research, 40*, 166–171.

Iyigun, C., & Ben-Israel, A. (2008). Probabilistic distance clustering adjusted for cluster size. *Probability in the Engineering and Informational Sciences, 22*, 603–621.

Iyigun, C., & Ben-Israel, A. (2010). Semi-supervised probabilistic distance clustering and the uncertainty of classification. In A. Fink et al. (Eds.), *Advances in data analysis, data handling and business intelligence* (pp. 3–20). Heidelberg: Springer.

Kamakura, W. A., & Srivastava, R. K. (1986). An ideal-point probabilistic choice model for heterogeneous preferences. *Marketing Science, 5*, 199–218.

Kamiya, H., & Takemura, A. (1997). On rankings generated by pairwise linear discriminant analysis of *m* populations. *Journal of Multivariate Analysis, 61*, 1–28.

Kamiya, H., & Takemura, A. (2005). Characterization of rankings generated by linear discriminant analysis. *Journal of Multivariate Analysis, 92*, 343–358.

Kamiya, H., Orlik, P., Takemura, A., & Terao, H. (2006). Arrangements and ranking patterns. *Annals of Combinatorics, 10*, 219–235.

Kamiya, H., Takemura, A., & Terao, H. (2011). Ranking patterns of unfolding models of codimension one. *Advances in Applied Mathematics, 47*, 379–400.

Kemeny, J. G. (1959). Mathematics without numbers. *Daedalus, 88*, 577–591.

Kemeny, J. G., & Snell, J. L. (1962). Preference rankings: An axiomatic approach. In J. G. Kemeny & J. L. Snell (Eds.), *Mathematical models in the social sciences* (pp. 9–23). New York: Blaisdell.

Kendall, M. G. (1938). A new measure of rank correlation. *Biometrika, 30*, 81–93.

Kendall, M. G. (1948). *Rank correlation methods*. London: Charles Griffin.

Kruskal, W. (1958). Ordinal measures of association. *Journal of the American Statistical Association, 53*, 814–861.

Kruskal, J. B., & Carroll, J. D. (1969). Geometrical models and badness-of-fit functions. In P. R. Krishnaiah (Ed.), *Multivariate analysis* (Vol. 2, pp. 639–671). New York: Academic.

Luce, R. D. (1959). *Individual choice behavior*. New York: Wiley.

Mallows, C. L. (1957). Non-null ranking models, I. *Biometrika, 44*, 114–130.

Marden, J. I. (1995). *Analyzing and modeling rank data*. New York: Chapman & Hall.

Maydeu-Olivares, A. (1999). Thurstonian modeling of ranking data via mean and covariance structure analysis. *Psychometrika, 64*, 325–340.

Meulman, J. J., Van Der Kooij, A. J., & Heiser, W. J. (2004). Principal components analysis with nonlinear optimal scaling transformations for ordinal and nominal data. In D. Kaplan (Ed.), *The SAGE handbook of quantitative methodology for the social sciences* (pp. 49–70). Thousand Oaks: Sage.

Mingers, J. (1989). An empirical comparison of pruning methods for decision tree induction. *Machine Learning, 4*, 227–243.

Morgan, K. O., & Morgan, S. (2010). *State rankings 2010: A statistical view of America*. Washington, DC: CQ Press.

Murphy, T. B., & Martin, D. (2003). Mixtures of distance-based models for ranking data. *Computational Statistics and Data Analysis, 41*, 645–655.

Roskam, Ed. E. C. I. (1968). *Metric analysis of ordinal data in psychology: Models and numerical methods for metric analysis of conjoint ordinal data in psychology*. Doctoral dissertation, Voorschoten, The Netherlands: VAM.

Shepard, R. N. (1987). Toward a universal law of generalization for psychological science. *Science, 237*, 1317–1323.

Skrondal, A., & Rabe-Hesketh, S. (2003). Multilevel logistic regression for polytomous data and rankings. *Psychometrika, 68*, 267–287.

Slater, P. (1960). The analysis of personal preferences. *British Journal of Statistical Psychology, 13*, 119–135.

Thompson, G. L. (1993). Generalized permutation polytopes and exploratory graphical methods for ranked data. *The Annals of Statistics, 21*, 1401–1430.

Thurstone, L. L. (1927). A law of comparative judgment. *Psychological Review, 34*, 273–286.

Thurstone, L. L. (1931). Rank order as a psychophysical method. *Journal of Experimental Psychology, 14*, 187–201.

Tucker, L. R. (1960). Intra-individual and inter-individual multidimensionality. In H. Gulliksen & S. Messick (Eds.), *Psychological scaling: Theory and applications* (pp. 155–167). New York: Wiley.

Van Blokland-Vogelesang, A. W. (1989). Unfolding and consensus ranking: A prestige ladder for technical occupations. In G. De Soete et al. (Eds.), *New developments in psychological choice modeling* (pp. 237–258). The Netherlands\North-Holland: Amsterdam.

van Buuren, S., & Heiser, W. J. (1989). Clustering *n* objects into *k* groups under optimal scaling of variables. *Psychometrika, 54*, 699–706.

Van Deun, K. (2005). *Degeneracies in multidimensional unfolding*. Doctoral dissertation, Leuven, Belgium: Catholic University of Leuven.

Yao, G., & Böckenholt, U. (1999). Bayesian estimation of Thurstonian ranking models based on the Gibbs sampler. *British Journal of Mathematical and Statistical Psychology, 52*, 79–92.

Zhang, J. (2004). Binary choice, subset choice, random utility, and ranking: A unified perspective using the permutahedron. *Journal of Mathematical Psychology, 48*, 107–134.

Zinnes, J. L., & Griggs, R. A. (1974). Probabilistic, multidimensional unfolding analysis. *Psychometrika, 39*, 327–350.

Solving the Minimum Sum of L1 Distances Clustering Problem by Hyperbolic Smoothing and Partition into Boundary and Gravitational Regions

Adilson Elias Xavier, Vinicius Layter Xavier, and Sergio B. Villas-Boas

Abstract The article considers the minimum sum of distances clustering problem, where the distances are measured through the L1 or Manhattan metric (MSDC-L1). The mathematical modelling of this problem leads to a *min-sum-min* formulation which, in addition to its intrinsic bi-level nature, has the significant characteristic of being strongly non differentiable.

We propose the AHSC-L1 method to solve this problem, by combining two techniques. The first technique is Hyperbolic Smoothing Clustering (HSC), that adopts a smoothing strategy using a special C^∞ completely differentiable class function. The second technique is the partition of the set of observations into two non overlapping groups: "data in frontier" and "data in gravitational regions". We propose a classification of the gravitational observations by each component, which simplifies of the calculation of the objective function and its gradient. The combination of these two techniques for MSDC-L1 problem drastically simplify the computational tasks.

1 Introduction

Cluster analysis deals with the problems of classification of a set of patterns or observations. In general the observations are represented as points in a multidimensional space. The purpose of cluster analysis is to define the clusters to that each observation belongs, following two basic and simultaneous objectives: patterns in the same clusters must be similar to each other (homogeneity objective) and different from patterns in other clusters (separation objective) Hartigan (1975) and Späth (1980).

A.E. Xavier (✉) · V.L. Xavier · S.B. Villas-Boas
Federal University of Rio de Janeiro, Rio de Janeiro, Brazil
e-mail: adilson@cos.ufrj.br; vinicius@cos.ufrj.br; sbvb@cos.ufrj.br

B. Lausen et al. (eds.), *Algorithms from and for Nature and Life*, Studies in Classification, Data Analysis, and Knowledge Organization, DOI 10.1007/978-3-319-00035-0_3,

In this paper, a particular clustering problem formulation is considered. Among many criteria used in cluster analysis, a frequently adopted criterion is the minimum sum of L1 distances clustering (MSDC-L1); see for example Bradley and Mangasarian (1996). This criterion corresponds to the minimization of the sum of distances of observations to their centroids, where the distances are measured through the L1 or Manhattan metric. As broadly recorded by the literature, the Manhattan distance is more robust against outliers.

For the sake of completeness, we present first the Hyperbolic Smoothing Clustering Method (HSC), Xavier (2010). Basically the method performs the smoothing of the non differentiable *min-sum-min* problem engendered by the modelling of a broad class of clustering problems, including the minimum sum of L1 distances clustering (MSDC-L1) formulation. This technique was developed through an adaptation of the hyperbolic penalty method originally introduced by Xavier (1982). By smoothing, we fundamentally mean the substitution of an intrinsically non differentiable two-level problem by a C^∞ unconstrained differentiable single-level alternative.

Additionally, the paper presents an accelerated methodology applied to the specific considered problem. The basic idea is to partition the set of observations into two non overlapping parts. By using a conceptual presentation, the first set corresponds to the observation points relatively close to two or more centroids. The second set corresponds to observation points significantly closer to a single centroid in comparison with others. The same partition scheme was presented first by Xavier and Xavier (2011) in order to solve the specific minimum sum of squares clustering (MSSC) formulation. In this paper, specific features of the minimum sum of L1 distances clustering (MSDC-L1) formulation are explored in order to take additional advantages of the partition scheme.

2 The Minimum Sum of L1 Distances Clustering Problem

Let $S = \{s_1, \ldots, s_m\}$ denote a set of m patterns or observations from an Euclidean n-space, to be clustered into a given number q of disjoint clusters. To formulate the original clustering problem as a $min - sum - min$ problem, we proceed as follows. Let $x_i, i = 1, \ldots, q$ be the centroids of the clusters, where each $x_i \in \mathbb{R}^n$. The set of these centroid coordinates will be represented by $X \in \mathbb{R}^{nq}$.

Given a point s_j of S, we initially calculate the L1 distance from s_j to the nearest center. This is given by $z_j = \min_{i=1,\ldots,q} \|s_j - x_i\|_1$. A frequent measurement of the quality of a clustering associated to a specific position of q centroids is provided by the sum of the L1 distances, which determines the MSDC-L1 problem:

$$minimize \sum_{j=1}^{m} z_j \tag{1}$$

$$\text{subject to} \quad z_j = \min_{i=1,\ldots,q} \|s_j - x_i\|_1, \qquad j = 1, \ldots, m$$

3 The Hyperbolic Smoothing Clustering Method

Considering its definition, each z_j must necessarily satisfy the following set of inequalities: $z_j - \|s_j - x_i\|_1 \leq 0, i = 1, \ldots, q$. Substituting these inequalities for the equality constraints, Problem (1) produces the relaxed problem:

$$\begin{aligned} & \textit{minimize} \sum_{j=1}^{m} z_j \\ & \text{subject to} \quad z_j - \|s_j - x_i\|_1 \leq 0, \quad j = 1, \ldots, m, \quad i = 1, \ldots, q. \end{aligned} \tag{2}$$

Since the variables z_j are not bounded from below, the optimization procedure will determine $z_j \to \infty,\ j = 1, \ldots, m$. In order to obtain the desired equivalence, we must, therefore, modify Problem (2). We do so by first letting $\varphi(y)$ denote $\max\{0, y\}$ and then observing that, from the set of inequalities in (2), it follows that $\sum_{i=1}^{q} \varphi(z_j - \|s_j - x_i\|_1) = 0, j = 1, \ldots, m$. In order to bound the variables $z_j,\ j = 1, \ldots, m$ we include an $\varepsilon > 0$ perturbation.

$$\begin{aligned} & \textit{minimize} \sum_{j=1}^{m} z_j \\ & \text{subject to} \sum_{i=1}^{q} \varphi(z_j - \|s_j - x_i\|_1) \geq \varepsilon\,, \quad j = 1, \ldots, m \end{aligned} \tag{3}$$

Since the feasible set of Problem (1) is the limit of that of (3) when $\varepsilon \to 0_+$, we can then consider solving (1) by solving a sequence of problems like (3) for a sequence of decreasing values for ε that approaches 0.

Analysing the Problem (3), the definition of function φ and the definition of L1 distance endows it with an extremely rigid non differentiable structure, which makes its computational solution very hard. In view of this, the numerical method we adopt for solving Problem (1), takes a smoothing approach. From this perspective, let us define the approximation functions below:

$$\phi(y, \tau) = \left(y + \sqrt{y^2 + \tau^2} \right) / 2 \tag{4}$$

$$\theta_1(s_j\,, x_i\,, \gamma) = \sum_{l=1}^{n} \sqrt{(s_j^l - x_i^l)^2 + \gamma^2} \tag{5}$$

By using the asymptotic approximation properties of the functions θ_1 and ϕ, the following completely differentiable problem is now obtained:

$$\text{minimize} \sum_{j=1}^{m} z_j \tag{6}$$

$$\text{subject to} \quad \sum_{i=1}^{q} \phi(z_j - \theta_1(s_j, x_i, \gamma), \tau) \geq \varepsilon, \qquad j = 1, \ldots, m.$$

So, the properties of functions ϕ and θ_1 allow us to seek a solution to Problem (3) by solving a sequence of subproblems like Problem (6), produced by the decreasing of the parameters $\gamma \to 0$, $\tau \to 0$ and $\varepsilon \to 0$.

On the other side, the constraints will certainly be active and Problem (6) will at last be equivalent to problem:

$$\text{minimize} \sum_{j=1}^{m} z_j \tag{7}$$

$$\text{subject to} \quad h_j(z_j, x) = \sum_{i=1}^{q} \phi(z_j - \theta_1(s_j, x_i, \gamma), \tau) - \varepsilon = 0, \quad j = 1, \ldots, m.$$

Problem (7) has a separable structure, because each variable z_j appears only in one equality constraint. Therefore, as the partial derivative of $h(z_j, x)$ with respect to z_j, $j = 1, \ldots, m$ is not equal to zero, it is possible to use the Implicit Function Theorem to calculate each component z_j, $j = 1, \ldots, m$ as a function of the centroid variables x_i, $i = 1, \ldots, q$. In this way, the unconstrained problem

$$\text{minimize} \ f(x) = \sum_{j=1}^{m} z_j(x) \tag{8}$$

is obtained, where each $z_j(x)$ results from the calculation of a zero of each equation

$$h_j(z_j, x) = \sum_{i=1}^{q} \phi(z_j - \theta_1(s_j, x_i, \gamma), \tau) - \varepsilon = 0, \quad j = 1, \ldots, m. \tag{9}$$

Again, due to the Implicit Function Theorem, the functions $z_j(x)$ have all derivatives with respect to the variables x_i, $i = 1, \ldots, q$, and therefore it is possible to calculate the gradient of the objective function of Problem (8),

$$\nabla f(x) = \sum_{j=1}^{m} \nabla z_j(x) \tag{10}$$

where

$$\nabla z_j(x) = -\nabla h_j(z_j, x) \ / \ \frac{\partial h_j(z_j, x)}{\partial z_j}, \tag{11}$$

while $\nabla h_j(z_j, x)$ and $\partial h_j(z_j, x)/\partial z_j$ are obtained from Eqs. (4), (5) and (9).

In this way, it is easy to solve Problem (8) by making use of any method based on first or second order derivative information. At last, it must be emphasized that Problem (8) is defined on an (nq)–dimensional space, so it is a small problem, since the number of clusters, q, is, in general, very small for real applications.

The solution of the original clustering problem can be obtained by using the Hyperbolic Smoothing Clustering Algorithm, described below in a simplified form.

4 The Simplified HSC-L1 Algorithm

Initialization Step:

Choose initial values: x^0, γ^1, τ^1, ε^1.
Choose values $0 < \rho_1 < 1$, $0 < \rho_2 < 1$, $0 < \rho_3 < 1$; let $k = 1$.
Main Step: Repeat until a stopping rule is attained
Solve Problem (8) with $\gamma = \gamma^k$, $\tau = \tau^k$ and $\varepsilon = \varepsilon^k$, starting at the initial point x^{k-1} and let x^k be the solution obtained.
Let $\gamma^{k+1} = \rho_1 \gamma^k$, $\tau^{k+1} = \rho_2 \tau^k$, $\varepsilon^{k+1} = \rho_3 \varepsilon^k$, $k := k + 1$. ■

Just as in other smoothing methods, the solution to the clustering problem is obtained, in theory, by solving an infinite sequence of optimization problems. In the HSC-L1 algorithm, each problem to be minimized is unconstrained and of low dimension.

Notice that the algorithm causes τ and γ to approach 0, so the constraints of the subproblems as given in (6) tend to those of (3). In addition, the algorithm causes ε to approach 0, so, in a simultaneous movement, the solved Problem (3) gradually approaches the original MSDC-L1 Problem (1).

5 The Accelerated Hyperbolic Smoothing Clustering Method

The calculation of the objective function of the Problem (8) demands the determination of the zeros of m Eq. (9), one equation for each observation point. This is a relevant computational task associated to HSC-L1 Algorithm.

In this section, it is presented a faster procedure. The basic idea is the partition of the set of observations into two non overlapping regions. By using a conceptual presentation, the first region corresponds to the observation points that are relatively close to two or more centroids. The second region corresponds to the observation points that are significantly close to a unique centroid in comparison with the other ones.

So, the first part J_B is the set of boundary observations and the second is the set J_G of gravitational observations. Considering this partition, Eq. (8) can be expressed in the following way:

$$\textit{minimize } f(x) = \sum_{j=1}^{m} z_j(x) = \sum_{j \in J_B} z_j(x) + \sum_{j \in J_G} z_j(x), \tag{12}$$

so that the objective function can be presented in the form:

$$\textit{minimize } f(x) = f_B(x) + f_G(x), \tag{13}$$

where the two components are completely independent.

The first part of expression (13), associated with the boundary observations, can be calculated by using the previously presented smoothing approach, see (8) and (9). The second part of expression (13) can be calculated by using a faster procedure, as we will show right away.

Let us define the two parts in a more rigorous form. Let be $\overline{x}_i$, $i = 1, \ldots, q$ be a referential position of centroids of the clusters taken in the iterative process.

The boundary concept in relation to the referential point $\overline{x}$ can be easily specified by defining a δ band zone between neighbouring centroids. For a generic point $s \in \mathbb{R}^n$, we define the first and second nearest distances from s to the centroids:

$$d_1(s, \overline{x}) = \| s - \overline{x}_{i_1} \| = \min_i \| s - \overline{x}_i \| \tag{14}$$

$$d_2(s, \overline{x}) = \| s - \overline{x}_{i_2} \| = \min_{i \neq i_1} \| s - \overline{x}_i \| , \tag{15}$$

where i_1 and i_2 are the labelling indexes of these two nearest centroids.

By using the above definitions, let us define precisely the δ boundary band zone:

$$Z_\delta(\overline{x}) = \{ s \in \mathbb{R}^n \mid d_2(s, \overline{x}) - d_1(s, \overline{x}) < 2\delta \} \tag{16}$$

and the gravity region, this is the complementary space:

$$G_\delta(\overline{x}) = \{ s \in \mathbb{R}^n - Z_\delta(\overline{x}) \} . \tag{17}$$

Figure 1 illustrates in $\mathbb{R}^2$ the $Z_\delta(\overline{x})$ and $G_\delta(\overline{x})$ partitions. The central lines form the Voronoi polygon associated with the referential centroids $\overline{x}_i$, $i = 1, \ldots, q$. The region between two parallel lines to Voronoi lines constitutes the boundary band zone $Z_\delta(\overline{x})$.

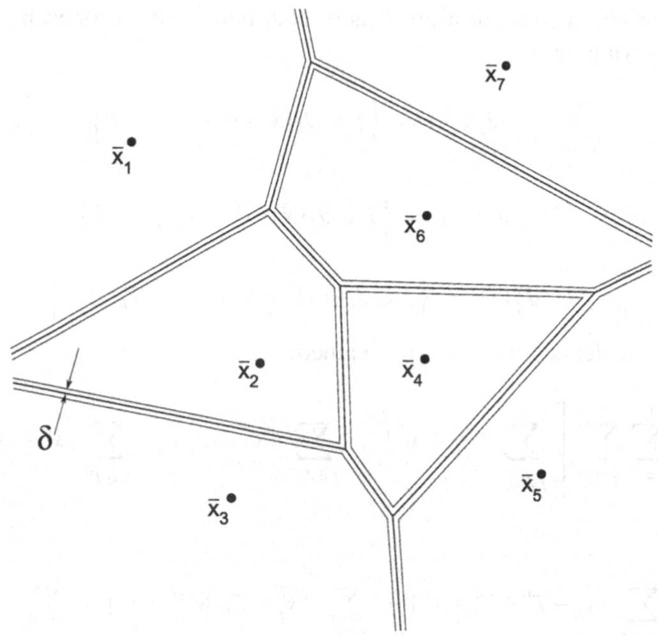

Fig. 1 The $Z_\delta(\overline{x})$ and $G_\delta(\overline{x})$ partitions

Now, the sets J_B and J_G can be defined in a precise form:

$$J_B(\overline{x}) = \{j = 1, \ldots, m \mid s_j \in Z_\delta(\overline{x})\}, \tag{18}$$

$$J_G(\overline{x}) = \{j = 1, \ldots, m \mid s_j \in G_\delta(\overline{x})\}. \tag{19}$$

In Xavier and Xavier (2011) it is shown the proof of proposition below.

Proposition 1. *Let s be a generic point belonging to the gravity region $G_\delta(\overline{x})$, with nearest centroid i_1. Let x be the current position of the centroids. Let $\Delta x = \max_i \|x_i - \overline{x}_i\|$ be the maximum displacement of the centroids. If $\Delta x < \delta$ then s will continue to be nearer to centroid x_{i_1} than to any other one.* ■

Since $\delta \geq \Delta x$, Proposition 1 makes it possible to calculate exactly expression (12) in a very fast way. First, let us define the subsets of gravity observations associated with each referential centroid:

$$J_i(\overline{x}) = \left\{ j \in J_G \mid \min_{p=1,\ldots,q} \| s_j - \overline{x}_p \| = \| s_j - \overline{x}_i \| \right\} \tag{20}$$

Let us consider the second sum in expression (12).

$$f_G(x) = \sum_{j \in J_G} z_j(x) = \sum_{i=1}^{q} \sum_{j \in J_i} \|s_j - x_i\|_1 = \sum_{i=1}^{q} \sum_{j \in J_i} \sum_{l=1}^{n} |s_j^l - x_i^l| = \sum_{i=1}^{q} \sum_{l=1}^{n} \sum_{j \in J_i} |s_j^l - x_i^l|.$$

Let us now perform the partition of each set J_i into 3 subsets for each component l in the following form:

$$J_{il}^{+}(\overline{x}) = \left\{ j \in J_i(\overline{x}) \,|\, s_j^l - \overline{x}_i^l \geq \delta \right\} \tag{21}$$

$$J_{il}^{-}(\overline{x}) = \left\{ j \in J_i(\overline{x}) \,|\, s_j^l - \overline{x}_i^l \leq -\delta \right\} \tag{22}$$

$$J_{il}^{0}(\overline{x}) = \left\{ j \in J_i(\overline{x}) \,|\, -\delta < s_j^l - \overline{x}_i^l < \delta \right\} \tag{23}$$

By using the defined subsets, it is obtained:

$$f_G(x) = \sum_{i=1}^{q} \sum_{l=1}^{n} \left[\sum_{j \in J_{il}^{+}} |s_j^l - x_i^l| + \sum_{j \in J_{il}^{-}} |s_j^l - x_i^l| + \sum_{j \in J_{il}^{0}} |s_j^l - x_i^l| \right] =$$

$$\sum_{i=1}^{q} \sum_{l=1}^{n} \left[\sum_{j \in J_{il}^{+}} |s_j^l - \overline{x}_i^l + \overline{x}_i^l - x_i^l| + \sum_{j \in J_{il}^{-}} |s_j^l - \overline{x}_i^l + \overline{x}_i^l - x_i^l| + \sum_{j \in J_{il}^{0}} |s_j^l - x_i^l| \right]$$

Let us define the component displacement of centroid $\Delta x_i^l = x_i^l - \overline{x}_i^l$. Since $|\Delta x_i^l| < \delta$, from the above definitions of the subsets, it follows that:

$$|s_j^l - x_i^l| = |s_j^l - \overline{x}_i^l| - \Delta x_i^l \quad \text{for } j \in J_{il}^{+} \tag{24}$$

$$|s_j^l - x_i^l| = |s_j^l - \overline{x}_i^l| + \Delta x_i^l \quad \text{for } j \in J_{il}^{-}$$

So, it follows:

$$f_G(x) = \sum_{i=1}^{q} \sum_{l=1}^{n} \left[\sum_{j \in J_{il}^{+}} \left(|s_j^l - \overline{x}_i^l| - \Delta x_i^l \right) + \sum_{j \in J_{il}^{-}} \left(|s_j^l - \overline{x}_i^l| + \Delta x_i^l \right) + \sum_{j \in J_{il}^{0}} |s_j^l - x_i^l| \right] =$$

$$\sum_{i=1}^{q} \sum_{l=1}^{n} \left[\sum_{j \in J_{il}^{+}} |s_j^l - \overline{x}_i^l| - |J_{il}^{+}| \Delta x_i^l + \sum_{j \in J_{il}^{-}} |s_j^l - \overline{x}_i^l| + |J_{il}^{-}| \Delta x_i^l + \sum_{j \in J_{il}^{0}} |s_j^l - x_i^l| \right] \tag{25}$$

where $|J_{il}^{+}|$ and $|J_{il}^{-}|$ are the cardinalities of two first subsets.

When the position of centroids $x_i, i = 1, \ldots, q$ moves within the iterative process, the value of the first two sums of (25) assumes a constant value, since the values s_j^l and $\overline{x}_i^l$ are fixed. So, to evaluate $f_G(x)$ it is only necessary to calculate the displacements Δx_i^l , $i = 1,\ldots,q$, $l = 1,\ldots,n$, and evaluate the last sum, that normally has only a few number of terms because δ assumes in general a relatively small value.

The function $f_G(x)$ above specified is non differentiable due the last sum, so in order to use gradient information, it is necessary to use a smooth approximation:

$$f_G(x) = \sum_{i=1}^{q} \sum_{l=1}^{n} \Bigg[\sum_{j \in J_{il}^+} | s_j^l - \overline{x}_i^l | - | J_{il}^+ | \Delta x_i^l + \sum_{j \in J_{il}^-} | s_j^l - \overline{x}_i^l | + | J_{il}^- | \Delta x_i^l + \sum_{j \in J_{il}^0} \sigma(s_j^l, x_i^l, \gamma) \Bigg] \tag{26}$$

where σ is the smoothing function for each unidimensional distance: $\sigma(s_j^l, x_i^l, \gamma) = ((s_j^l - x_i^l)^2 + \gamma^2)^{1/2}$.

So, the gradient of the smoothed second part of objective function is easily calculated by:

$$\nabla f_G(x) = \sum_{i=1}^{q} \sum_{l=1}^{n} \Bigg[- | J_{il}^+ | + | J_{il}^- | + \sum_{j \in J_{il}^0} -(s_j^l - x_i^l) / \sigma(s_j^l, x_i^l, \gamma) \Bigg] e_{il} \tag{27}$$

where e_{il} stands for a unitary vector with the component l of centroid i equal to 1.

Therefore, if $\delta \geq \Delta x$ was observed within the iterative process, the calculation of the expression $\sum_{j \in J_G} z_j(x)$ and its gradient can be exactly performed by very fast procedures, Eqs. (26) and (27).

By using the above results, it is possible to construct a specific method, the Accelerated Hyperbolic Smoothing Method Applied to the Minimum of Sum of L1 Distances Clustering Problem, which has conceptual properties to offer a faster computational performance for solving this specific clustering problem given by formulation (13), since the calculation of the second sum ($f_G(x)$) is very simple.

A fundamental question is the proper choice of the boundary parameter δ. Moreover, there are two main options for updating the boundary parameter δ, inside the internal minimization procedure or after it. For simplicity sake, the AHSC-L1 method connected with the partition scheme presented below adopts the second option, which offers a better computational performance, in spite of an eventual violation of the $\delta \geq \Delta x$ condition, which gets corrected in the next partition update.

6 The Simplified AHSC-L1 Algorithm

Initialization Step:

Choose initial start point: x^0;
Choose parameter values: $\gamma^1, \tau^1, \varepsilon^1$;
Choose reduction factors:
$0 < \rho_1 < 1,\ 0 < \rho_2 < 1,\ 0 < \rho_3 < 1$;
Specify the boundary band width: δ^1;
Let $k = 1$.

Main Step: Repeat until an arbitrary stopping rule is attained

For determining the $Z_\delta(\overline{x})$ and $G_\delta(\overline{x})$ partitions, given by (16) and (17), use $\overline{x} = x^{k-1}$ and $\delta = \delta^k$.

Determine the subsets J_{il}^+, J_{il}^- and J_{il}^0 and calculate the cardinalities of two first sets: $|J_{il}^+|$ and $|J_{il}^-|$.

Solve Problem (13) starting at the initial point x^{k-1} and let x^k be the solution obtained:

For solving the equations associated to the first part given by (9), take the smoothing parameters:

$$\gamma = \gamma^k, \tau = \tau^k \text{ and } \varepsilon = \varepsilon^k .$$

For solving the second part, given by (26), use the above determined subsets and their cardinalities.

Updating procedure:
Let $\gamma^{k+1} = \rho_1 \gamma^k$, $\tau^{k+1} = \rho_2 \tau^k$, $\varepsilon^{k+1} = \rho_3 \varepsilon^k$.
If necessary redefine the boundary value: δ^{k+1}.
Let $k := k + 1$. ■

The efficiency of the AHSC-L1 algorithm depends strongly on the parameter δ. A choice of a small value for it will imply an improper definition of the set $G_\delta(\overline{x})$, and frequent violation of the basic condition $\Delta x < \delta$, for the validity of Proposition 1. Otherwise, a choice of a large value will imply a decrease in the number of gravitational observation points and, therefore, the computational advantages given by formulation (26) will be reduced.

As a general strategy, within first iterations, larger δ values must be used, because the centroid displacements are more expressive. The δ values must be gradually decreased in the same proportion of the decrease of these displacements.

7 Computational Results

The numerical experiments have been carried out on a PC Intel Celeron with 2.7 GHz CPU and 512 MB RAM. The programs are coded with Compac Visual FORTRAN, Version 6.1. The unconstrained minimization tasks were carried out by means of a Quasi-Newton algorithm employing the BFGS updating formula from the Harwell Library, obtained in the site: (http://www.cse.scitech.ac.uk/nag/hsl/).

Table 1 Results of AHSC-L1 applied to TSPLIB-1060 and TSPLIB-3038 Instance

	TSPLIB 1060				TSPLIB 3038			
q	$f_{AHSC-L1_{Best}}$	Occur	E_{Mean}	T_{cpu}	$f_{AHSC-L1_{Best}}$	Occur	E_{Mean}	T_{cpu}
2	0.386500E7	2	0.00	0.10	0.373171E7	2	0.88	0.76
3	0.313377E7	1	0.22	0.14	0.300708E7	1	1.33	0.76
4	0.258205E7	5	0.43	0.21	0.254499E7	1	0.67	0.79
5	0.231098E7	1	0.76	0.29	0.225571E7	1	1.28	0.95
6	0.213567E7	1	0.79	0.35	0.206006E7	1	1.12	0.95
7	0.196685E7	1	1.41	0.48	0.189650E7	1	1.34	1.06
8	0.183280E7	1	2.22	0.53	0.176810E7	1	1.14	1.13
9	0.168634E7	1	3.42	0.60	0.164559E7	1	2.13	1.21
10	0.155220E7	1	2.74	0.68	0.154550E7	1	1.97	1.28

Table 2 Results of AHSC-L1 applied to D15112 and Pla85900 Instance

	D15112				Pla85900			
q	$f_{AHSC-L1_{Best}}$	Occur	E_{Mean}	T_{cpu}	$f_{AHSC-L1_{Best}}$	Occur	E_{Mean}	T_{cpu}
2	0.822872E8	6	0.97	10.35	0.883378E10	2	0.00	242.74
3	0.655831E8	1	1.43	7.56	0.667961E10	1	0.21	166.48
4	0.567702E8	1	1.60	6.21	0.551287E10	2	0.06	129.30
5	0.511639E8	1	1.17	5.73	0.482328E10	1	1.43	112.95
6	0.462612E8	1	1.67	5.33	0.432972E10	1	2.47	103.02
7	0.425722E8	1	2.30	4.96	0.401388E10	1	1.87	98.25
8	0.398389E8	1	1.83	5.02	0.373878E10	1	3.47	92.14
9	0.376863E8	1	1.60	5.03	0.355741E10	1	2.40	82.33
10	0.354762E8	1	2.41	5.01	0.341472E10	1	1.77	87.41

In order to exhibit the distinct performance of the AHSC-L1 algorithm, Tables 1 and 2 present the computational results of AHSC-L1 applied to four benchmark problems, all from TSPLIB (Reinelt 1991; http://www.iwr.uni-heidelberg.de/groups/comopt/software). Table 1 represent two instances frequently used as benchmark clustering problems. Table 2 left contains data of 15,112 German cities. Table 2 right is the largest symmetric problem of TSPLIB.

The AHSC-L1 is a general framework that bears a broad number of implementations. In the initialization steps the following choices were made for the reduction factors: $\rho_1 = 1/4, \rho_2 = 1/4$ and $\rho_3 = 1/4$. The specification of initial smoothing and perturbation parameters was automatically tuned to the problem data. So, the initial max function smoothing parameter (4) was specified by $\tau^1 = \sigma/10$ where σ^2 is the variance of set of observation points: $S = \{s_1, \ldots, s_m\}$. The initial perturbation parameter (3) was specified by $\epsilon^1 = 4\tau^1$ and the Euclidian distance smoothing parameter by $\gamma^1 = \tau^1/100$.

All experiments where done using ten initial points. The adopted stopping criterion was the execution of the main step of the AHSC-L1 algorithm in a fixed number of six iterations.

Table 3 Speed-up of AHSC-L1 compared to HSC-L1 (the larger the better)

q	TSP1060	TSP3038	D15112	Pla85900
2	2.60	1.33	0.94	0.82
3	3.79	1.84	1.43	0.95
4	3.24	3.08	1.95	1.30
5	4.41	3.16	3.17	1.76
6	5.23	4.69	4.60	2.58
7	4.96	5.84	5.94	3.80
8	6.94	7.09	8.44	4.44
9	6.30	7.97	10.34	6.78
10	7.66	12.07	13.26	8.75

The meaning of the columns Tables 1 and 2 is as follows. q = the number of clusters. $f_{AHSC-L1-Best}$ = the best results of cost function using points obtained from AHSC-L1 method out of all the ten random initial points. Occur. = number of times the same best result was obtained from all the tenl random initial points. E_{Mean} = the average error of the ten solutions in relation to the best solution obtained ($f_{AHSC-L1-Best}$). Finally, T_{cpu} = the average execution time per trial, in seconds.

The "A" of AHSC-L1 means "accelerated", that is, the technique that partitions the set of observations into two non overlapping groups: "data in frontier" and "data in gravitational regions". The sample problems were solved using HSC-L1 and AHSC-L1 methods. Both algorithms obtain the same results with three decimal digits of precision. The Table 3 shows the speed-up produced by the acceleration technique. The meaning of the columns of Table 3 is as follows. q = the number of clusters. Speed-up for TSPLIB-1060 Instance. Speed-up for TSPLIB-3038 Instance. Speed-up for D15112 Instance. Speed-up for Pla85900 Instance.

The speed-up was calculated as the ratio between execution times T_{HSC-L1} and $T_{AHSC-L1}$, as shown in Eq. (28).

$$Speed - up = \frac{T_{HSC-L1}}{T_{AHSC-L1}} \tag{28}$$

The results in the Table 3 show that in most cases the "accelerated" technique produces speed-up of the computation effort. In some cases, the speed-up is > 10. In a few cases (e.g. q = 2, 85,900), the gains produced by the acceleration do not compensate the fixed costs introduced by the calculus of partition. In these cases the speed-up is less than one, that is, AHSC-L1 takes longer to run when compared to HSC-L1.

8 Conclusions

In this paper, a new method for the solution of the minimum sum of L1 Manhattan distances clustering problem is proposed, called AHSC-L1 (Accelerated Hyperbolic Smoothing Clustering – L1). It is a natural development of the original HSC method and its descendant AHSC-L2 method, linked to the minimum sum-of-squares clustering (MSSC) formulation, presented respectively by Xavier (2010) an by Xavier and Xavier (2011).

The special characteristics of L1 distance were taken into account to adapt inside the AHSC-L1 method from the AHSC-L2. The main idea proposed in this paper is the acceleration of the AHSC-L1 method by the partition of the set of observations into two non overlapping parts – gravitational and boundary. The classification of gravitational observations by each component, implemented by Eqs. (21)–(23), simplifies of the calculation of the objective function (26) and its gradient (27). This classification produces a drastic simplification of computational tasks. The computational experiments confirm the speed-up, as shown in Table 3.

The computational experiments presented in this paper were obtained by using a particular and simple set of criteria for all specifications. The AHSC-L1 algorithm is a general framework that can support different implementations.

We could not find in the literature any reference mentioning the solution of cluster L1 problem with instances of sizes similar to those presented in this paper. So, our results represent a challenge for future works.

The most relevant computational task associated with the AHSC-L1 algorithm remains the determination of the zeros of the Eq. (9), for each observation in the boundary region, with the purpose of calculating the first part of the objective function. However, since these calculations are completely independent, they can be easily implemented using parallel computing techniques.

It must be observed that the AHSC-L1 algorithm, as presented here, is firmly linked to the MSDC-L1 problem formulation. Thus, each different problem formulation requires a specific methodology to be developed, in order to apply the partition into boundary and gravitational regions.

Finally, it must be remembered that the MSDC-L1 problem is a global optimization problem with several local minima, so both HSC-L1 and AHSC-L1 algorithms can only produce local minima. The obtained computational results exhibit a deep local minima property, which is well suited to the requirements of practical applications.

Acknowledgements The author would like to thank Cláudio Joaquim Martagão Gesteira of Federal University of Rio de Janeiro for the helpful review of the work and constructive comments.

References

Bradley, P. S., & Mangasarian, O. L. (1997) Clustering via concave minimization (Mathematical Programming Technical Report 96-03), May 1996. *Advances in neural information processing systems* (Vol. 9, pp. 368–374). Cambridge: MIT.

Hartigan, J. A. (1975) *Clustering algorithms*. New York: Wiley.

Reinelt, G. (1991). TSP-LIB – a traveling salesman library. *ORSA Journal on Computing, 3*, 376–384.

Späth, H. (1980). *Cluster analysis algorithms for data reduction and classification.* Upper Saddle River: Ellis Horwood.

Xavier, A. E. (1982). *Penalização hiperbólica: um novo método para resolução de problemas de otimização*. M.Sc. Thesis, COPPE – UFRJ, Rio de Janeiro.

Xavier, A. E. (2010). The hyperbolic smothing clustering method. *Pattern Recognition, 43*, 731–737.

Xavier A. E., & Xavier, V.L. (2011). Solving the minimum sum-of-squares clustering problem by hyperbolic smoothing and partition into boundary and gravitational regions. *Pattern Recognition, 44*, 70–77.

Part II
Clustering and Unsupervised Learning

On the Number of Modes of Finite Mixtures of Elliptical Distributions

Grigory Alexandrovich, Hajo Holzmann, and Surajit Ray

Abstract We extend the concept of the ridgeline from Ray and Lindsay (Ann Stat 33:2042–2065, 2005) to finite mixtures of general elliptical densities with possibly distinct density generators in each component. This can be used to obtain bounds for the number of modes of two-component mixtures of t distributions in any dimension. In case of proportional dispersion matrices, these have at most three modes, while for equal degrees of freedom and equal dispersion matrices, the number of modes is at most two. We also give numerical illustrations and indicate applications to clustering and hypothesis testing.

1 Introduction

Finite mixtures are a popular tool for modeling heterogenous populations. In particular, multivariate finite mixtures are often used in cluster analysis, see e.g. McLachlan and Peel (2000). Here, analysis is mainly based on mixtures with multivariate normal components. However, mixtures of multivariate t-distributions offer an attractive, more flexible and more robust alternative, see McLachlan and Peel (2000).

G. Alexandrovich
Department of Mathematics and Computer Science, Marburg University, Marburg, Germany

H. Holzmann (✉)
Department of Mathematics and Computer Science, Marburg University, Marburg, Germany

Fachbereich Mathematik und Informatik, Philipps-Universität Marburg, Hans-Meerweinstr., D-35032 Marburg, Germany
e-mail: holzmann@mathematik.uni-marburg.de

S. Ray
Department of Mathematics and Statistics, Boston University, Boston, MA, USA

B. Lausen et al. (eds.), *Algorithms from and for Nature and Life*, Studies in Classification, Data Analysis, and Knowledge Organization, DOI 10.1007/978-3-319-00035-0_4,

An important feature of these mixtures are their analytic properties, in particular their modality structure. Modes are essential for a proper interpretability of the resulting density. For example, in cluster analysis, when there are less modes than components in a mixture, it is reasonable to merge several components into a single cluster based on their modality structure, see Hennig (2010). On the other hand, having more modes than components in a mixtures as can happen in dimensions >1 is an undesirable feature.

The most important tools for assessing the number of modes of finite mixtures of multivariate normal distributions are the concepts of the ridgeline and the Π-function as introduced in Ray and Lindsay (2005). Recently, Ray and Ren (2012) showed that for two-component mixtures of normals in dimension D, the number of modes is at most $D + 1$, and further constructed examples which achieved these bounds.

Here, we extend their concept of the ridgeline to finite mixtures of general elliptical densities with possibly distinct density generators in each component. This can be used to obtain bounds for the number of modes of two-component mixtures of t distributions with possibly distinct degrees of freedom in any dimension. In case of proportional dispersion matrices, we show that these have at most three modes, while for equal degrees of freedom and equal dispersion matrices, the number of modes is at most two.

The paper is structured as follows. In Sect. 2 we introduce the concept of the ridgeline and the Π-function for mixtures of general elliptical distributions, and state some basic properties. These are used in Sect. 3 to assess the model structure of two-component t-mixtures. In Sect. 4 we give numerical illustrations and indicate some statistical applications to clustering and hypothesis testing.

2 Ridgeline Theory for General Elliptical Distributions

As indicated in Ray and Lindsay (2005), several of their results extend from finite mixtures of multivariate normal distributions to finite mixtures of general elliptical densities. In this section we formulate the relevant statements, for the proofs see Alexandrovich (2011).

First, we introduce some notation. A nonnegative measurable function $\varphi : [0, \infty) \to [0, \infty)$ for which $c_\varphi := \int_{\mathbb{R}^D} \varphi(x^T x)\, dx < \infty$ is finite is called a density generator of a D-dimensional spherical distribution. Evidently, $f(x) = c_\varphi^{-1} \varphi(x^T x)$ is then a D-dimensional density w.r.t. Lebesgue measure. If $\mu \in \mathbb{R}^D$ and $\Sigma > 0$ is a positive definite $D \times D$ matrix, then

$$f(x; \mu, \Sigma) = k\, \varphi\big((x-\mu)^{\mathsf{T}} \Sigma^{-1} (x-\mu)\big), \qquad k = \big(c_\varphi \det(\Sigma)^{1/2}\big)^{-1}$$

is a density from the associated family of elliptical distributions. For further details on elliptical distributions and their density generators see Fang et al. (1989). We consider general finite mixtures of elliptical densities with possibly distinct density generators in each component, i.e. densities of the form

$$g(x; \mu_i, \Sigma_i, \pi_i, \varphi_i, i = 1, \ldots, K) = \sum_{i=1}^{K} \pi_i \, k_i \, \varphi_i((x - \mu_i)^{\mathsf{T}} \Sigma_i^{-1} (x - \mu_i)), \quad (1)$$

where $\mu_i \in \mathbb{R}^D$, $\Sigma_i > 0$ are positive definite $D \times D$ matrices, φ_i are density generators with $k_i = \left(c_{\varphi_i} \det(\Sigma_i)^{1/2}\right)^{-1}$ the appropriate normalizing constant, and $\pi_i \in [0, 1]$ with $\sum_{i=1}^{K} \pi_i = 1$. Typically, the density generators φ_i will all be equal as in case of normal mixtures, or at least belong to a parametric family of density generators such as t-distributions with distinct degrees of freedom. Set

$$\mathcal{S}_K := \left\{\alpha = \left(\alpha_1, \ldots, \alpha_K\right)^T \in \mathbb{R}^K : \alpha_i \in [0, 1], \sum_{i=1}^{K} \alpha_i = 1\right\}.$$

Ray and Lindsay (2005) introduced the map $x^* : \mathcal{S}_K \to \mathbb{R}^D$,

$$x^*(\alpha) = \left[\alpha_1 \Sigma_1^{-1} + \ldots + \alpha_K \Sigma_K^{-1}\right]^{-1} \left[\alpha_1 \Sigma_1^{-1} \mu_1 + \ldots + \alpha_K \Sigma_K^{-1} \mu_K\right],$$

the so-called *ridgeline function*. The next theorem summarizes the connection between the modes of the finite mixture g in (1) and the ridgeline. For the proof in this general setting see Alexandrovich (2011).

Theorem 1. *Suppose that the density generators φ_i in the finite mixture g (see (1)) are continuously differentiable and strictly decreasing. Then*

1. *All critical points of g as defined in* (1) *are contained in $x^*(\mathcal{S}_K)$, the image of $\mathcal{S}_K$ under the mapping x^*.*
2. *Set $h(\alpha) = g(x^*(\alpha))$, $\alpha \in \mathcal{S}_K$. Then α_{crit} is a critical point (resp. local maximum) of h if and only if $x^*(\alpha_{crit})$ is a critical point (resp. local maximum) of g.*
3. *If $D > K - 1$, then g has no local minima, only local maxima and saddle points.*

Thus, looking for modes of g it is sufficient to look for modes of h.

For a two component mixture, setting

$$\delta(x, i) = (x - \mu_i)^{\mathsf{T}} \Sigma_i^{-1} (x - \mu_i), \qquad i = 1, 2, \quad (2)$$

we can write

$$g(x; \pi, \mu_1, \mu_2, \Sigma_1, \Sigma_2, \varphi_1, \varphi_2) = \pi \, k_1 \varphi_1\big(\delta(x, 1)\big) + (1 - \pi) \, k_2 \varphi_2\big(\delta(x, 2)\big).$$

For the ridgeline, we write in slightly different notation than in the above section

$$x^*(\alpha) = S_\alpha^{-1} \left((1 - \alpha)\Sigma_1^{-1} \mu_1 + \alpha \Sigma_2^{-1} \mu_2\right), \qquad S_\alpha = (1 - \alpha)\Sigma_1^{-1} + \alpha \Sigma_2^{-1}. \quad (3)$$

As above, set $h(\alpha) = g\big(x^*(\alpha)\big)$. Then solving

$$\partial_\alpha h(\alpha) = \pi k_1\, \partial_\alpha \varphi_1\big(\delta\big(x^*(\alpha), 1\big)\big) + (1-\pi)k_2\, \partial_\alpha \varphi_2\big(\delta\big(x^*(\alpha), 2\big)\big) = 0$$

for π, where ∂_α is the derivative w.r.t. the real parameter α, we get

$$\pi = \frac{k_2\, \partial_\alpha \varphi_2\big(\delta\big(x^*(\alpha), 2\big)\big)}{k_2\, \partial_\alpha \varphi_2\big(\delta\big(x^*(\alpha), 2\big)\big) - k_1\, \partial_\alpha \varphi_1\big(\delta\big(x^*(\alpha), 1\big)\big)} =: \Pi(\alpha),$$

the so-called Π-*function*. Note that the Π-function depends on parameters $\mu_i, \Sigma_i, \varphi_i,\ i = 1, 2$, but not on the weight π. For given π, it can be used to find the critical points of g. Further, it provides general bounds on the number of modes as follows.

Theorem 2. *(a)* $\Pi(0) = 1$, $\Pi(1) = 0$ *and* $\Pi(\alpha) \in [0, 1]$.
Let N be the number of zeros of the derivative $\partial_\alpha \Pi(\alpha)$ of $\Pi(\alpha)$ w.r.t. α within the interval $[0, 1]$. *Then*
(b) N is even, and for any $\pi \in [0, 1]$ the equation $\Pi(\alpha) = \pi$ has at most $N + 1$ solutions, the smallest of which, α_1, gives a mode $x^*(\alpha_1)$ of g.
(c) For any π, g has at most $1 + N/2$ modes.

We can compute general expressions for the Π-function and its derivative as follows. This will be refined for the t distribution in the next section.

Proposition 1. *Let $\varphi_i'(t) = d\varphi_i/dt(t)$, $t \in \mathbb{R}$, $i = 1, 2$ be the derivatives of the density generators. Then for* $0 < \alpha < 1$

$$\Pi(\alpha) = \frac{(1-\alpha)\, k_2\, \varphi_2'}{(1-\alpha)\, k_2\, \varphi_2' + \alpha\, k_1\, \varphi_1'}$$

$$\partial_\alpha \Pi(\alpha) = -k_1 k_2 \frac{\varphi_1'\, \varphi_2' + 2\alpha(1-\alpha)\, p(\alpha)\big((1-\alpha)\varphi_1'\varphi_2'' + \alpha\varphi_2'\varphi_1''\big)}{\big((1-\alpha)\, k_2\, \varphi_2' + \alpha\, k_1\, \varphi_1'\big)^2} \tag{4}$$

where φ_2' and φ_2'' are evaluated at $\delta\big(x^*(\alpha), 2\big)$ (see (2)), while φ_1' and φ_1'' are evaluated at $\delta\big(x^*(\alpha), 1\big)$, and

$$p(\alpha) = (\mu_2 - \mu_1)^\top \Sigma_1^{-1} S_\alpha^{-1} \Sigma_2^{-1} S_\alpha^{-1} \Sigma_2^{-1} S_\alpha^{-1} \Sigma_1^{-1} (\mu_2 - \mu_1). \tag{5}$$

3 Modes of Two Components Mixtures of *t* Distributions

In this section, based on the results of the previous section we give bounds on the number of modes of two-component t-mixtures. Observe that from Theorem 2(c), for given parameters $\mu_i, \Sigma_i, i = 1, 2$ (and degrees of freedom n_i in case of the t distribution), the number of modes of the resulting mixture g for any weight π can be bounded by the number of zeros of $\partial_\alpha \Pi$ in $[0, 1]$. Thus, if we can bound this number of zeros in $[0, 1]$ for any parameter combination μ_i, Σ_i (and n_i), we obtain bounds in the number of modes of the mixture g.

For mixtures of t-distributions, the density generators are given by

$$\varphi(t; n_i) = k_i \Big(1 + \frac{x}{n_i}\Big)^{-(n_i+D)/2}, \qquad k_i = \frac{\Gamma\big(\frac{n_i+D}{2}\big)}{|\Sigma_i|^{\frac{1}{2}}\Gamma\big(n_i/2\big)(n_i\pi)^{D/2}}, \qquad i = 1, 2,$$

where n_i denotes the degrees of freedom in the ith component. The general two-component t-mixture is given by

$$g(x; \pi, \mu_1, \mu_2, \Sigma_1, \Sigma_2, n_1, n_2) = \pi k_1 \varphi\big(\delta(x, 1); n_1\big) + (1 - \pi) k_2 \varphi\big(\delta(x, 2); n_2\big), \tag{6}$$

Lemma 1. *Consider a general t-mixture as in* (6). *Set*

$$\Sigma^* = \Sigma_2^{-1/2}\Sigma_1\Sigma_2^{-1/2}, \qquad \mu^* = \Sigma_2^{-1/2}(\mu_1 - \mu_2) \tag{7}$$

and let $\Sigma^* = QD^*Q^T$, *where* $D = \operatorname{diag}(\lambda_1^*, \ldots, \lambda_D^*)$ *and* Q *is an orthogonal matrix, denote the spectral decomposition of* Σ^*. *Then the number of modes of* $g(x; \pi, \mu_1, \mu_2, \Sigma_1, \Sigma_2, n_1, n_2)$ *is the same as that of* $g(x, \pi, Q^T\mu^*, 0, D^*, I_D, n_1, n_2)$.

This follows along similar lines as Theorem 4 in Ray and Ren (2012). Using this simplification, by bounding the number of zeros of $\partial_\alpha \Pi$-function one can obtain

Theorem 3. *1. Let* $g(x; \pi, \mu_1, \mu_2, \Sigma_1, \Sigma_2, n_1, n_2) = \pi k_1 \varphi\big(\delta(x, 1); n_1\big) + (1 - \pi) k_2 \varphi\big(\delta(x, 2); n_2\big)$ *be a two-component mixture of t distributions in dimension D, and let d be the number of distinct eigenvalues of the matrix* $\Sigma_2^{-1/2}\Sigma_1\Sigma_2^{-1/2}$. *Then the number of modes of g is at most* $1 + 2d$.

2. Let $g(x; \pi, \mu_1, \mu_2, \Sigma, \Sigma/\lambda, n_1, n_2)$, $\lambda > 0$, *be a two-component mixture of t distributions in dimension D with proportional covariance matrices. Then the number of modes of g in any dimension is at most three.*

3. A two-component t-mixture with equal degrees of freedom and dispersion matrices, $g(x; \pi, \mu_1, \mu_2, \Sigma, n)$ *has at most two modes in any dimension D.*

4 Illustrations and Applications

4.1 Numerical Illustrations

We start by giving some numerical illustrations of some of the results in the paper.

1. First, we investigate the effect of varying the degrees of freedom in a mixture of two t-distributions while keeping the covariances of components fixed. We also

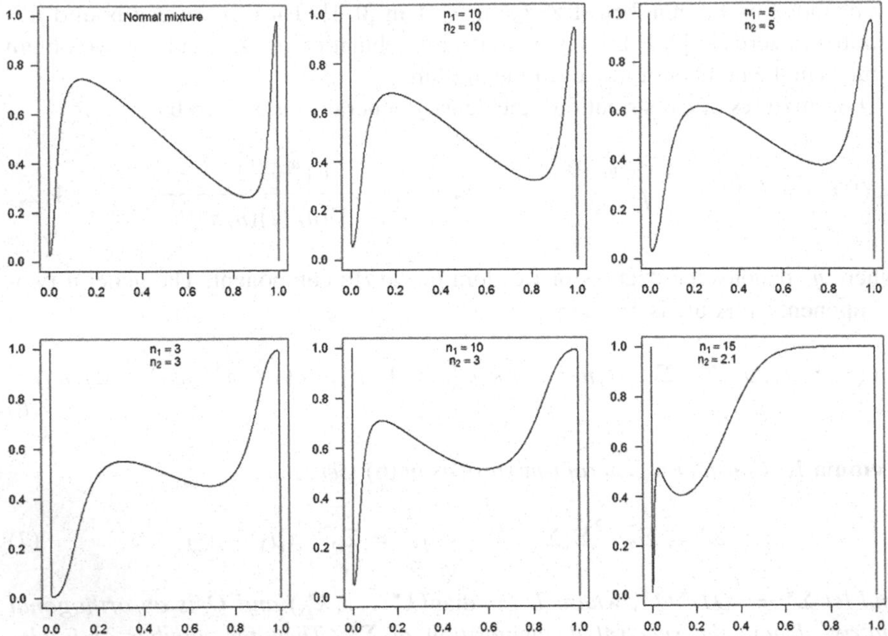

Fig. 1 Π-functions for Gauss- and t-mixtures with various degrees of freedom

consider a corresponding Gauss mixtures which can be considered as a limit case in which the degrees of freedom tend to ∞. Specifically, the parameters of the mixtures are

$$\mu_1 = \begin{pmatrix} 0 \\ 0 \end{pmatrix}, \mu_2 = \begin{pmatrix} 1 \\ 1 \end{pmatrix}, \Sigma_1 = \begin{pmatrix} 1 & 0 \\ 0 & 0.05 \end{pmatrix}, \Sigma_2 = \begin{pmatrix} 0.05 & 0 \\ 0 & 1 \end{pmatrix}. \tag{8}$$

In the case of t-mixtures we scale the matrices Σ_i, $i = 1, 2$ with the factors $\frac{n_i - 2}{n_i}$ in order to retain equal covariances in each constellation of degrees of freedom.

Figure 1 contains plots of the Π-functions for various combinations of degrees of freedom, while Fig. 2 has the corresponding for the weight $\pi = 0.65$. From Fig. 1 we see that with decreasing degrees of freedoms, the range of mixture weights for which the mixture has three modes decreases as well. For the choice $\pi = 0.65$, the first (normal), second ($n_1 = n_2 = 10$) and forth ($n_1 = 10, n_2 = 3$) have three modes, otherwise there are only two.

2. Second, we consider the transformation in Lemma 1 to diagonal dispersion matrices for a two-component t-mixture with 15 degrees of freedom and $\pi = 0.5$ for a special parameter combination. Specifically, consider

$$\mu_1 = \begin{pmatrix} 0.5 \\ 0.5 \end{pmatrix}, \mu_2 = \begin{pmatrix} 1.5 \\ 1.5 \end{pmatrix}, \Sigma_1 = \begin{pmatrix} 1 & 0.14 \\ 0.14 & 0.06 \end{pmatrix}, \Sigma_2 = \begin{pmatrix} 0.06 & 0.14 \\ 0.14 & 1 \end{pmatrix}.$$

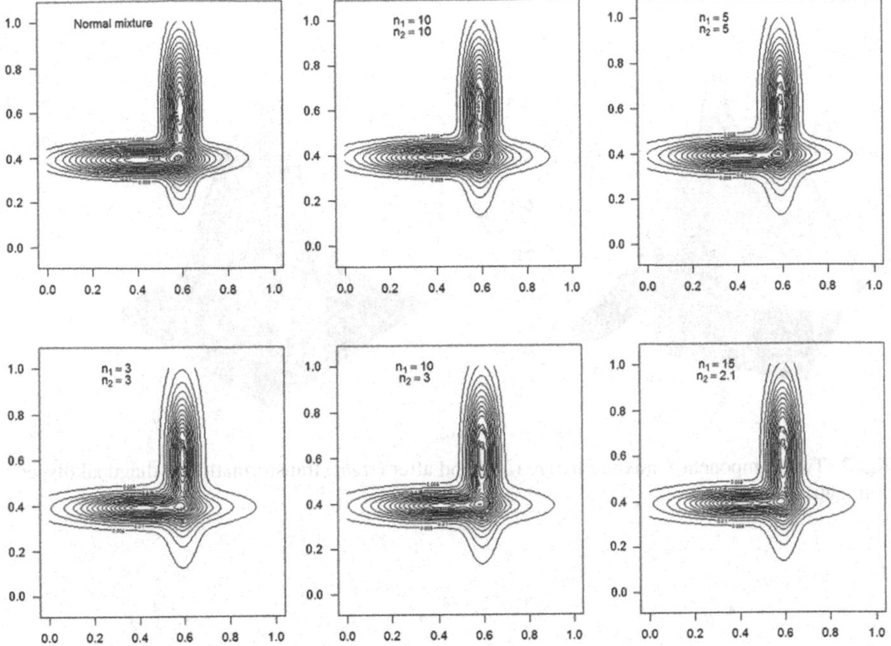

Fig. 2 The corresponding contours for the mixtures for $\pi = 0.65$

Then the transformed parameters are given by

$$\mu_1 = \begin{pmatrix} 0 \\ 0 \end{pmatrix}, \mu_2 = \begin{pmatrix} -4.39 \\ -1.11 \end{pmatrix}, \Sigma_1 = \begin{pmatrix} 1 & 0 \\ 0 & 1 \end{pmatrix}, \Sigma_2 = \begin{pmatrix} 24.19 & 0 \\ 0 & 0.041 \end{pmatrix}.$$

Figure 3 contains plots of the corresponding densities, which look quite distinct. Thus, it is not apparent that the transformation keeps the number of modes

3. Third, we investigate the effect when rotating one component while keeping everything else fixed. We consider a two-component t-mixture with 15 degrees of freedom in each component, and parameters as in (8). We rotate the second component clockwise, with angles ranging from 45 % up to 135 % in equidistant steps. The corresponding densities are plotted in Fig. 4. In the process a third mode appears at an angle around 90 % and vanishes again for higher angles.

4.2 *Statistical Applications*

Finally, we indicate two potential statistical application of the above theory.

1. Merging components in mixtures of t-distributions. McLachlan and Peel (2000) recommend the use of finite mixtures of t-distributions as a more robust

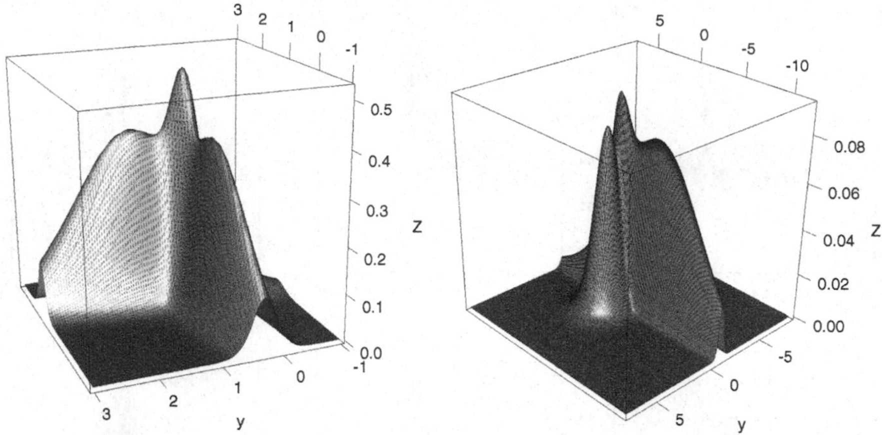

Fig. 3 Two-component t-mixture before (*left*) and after (*right*) transformation to diagonal dispersion matrices

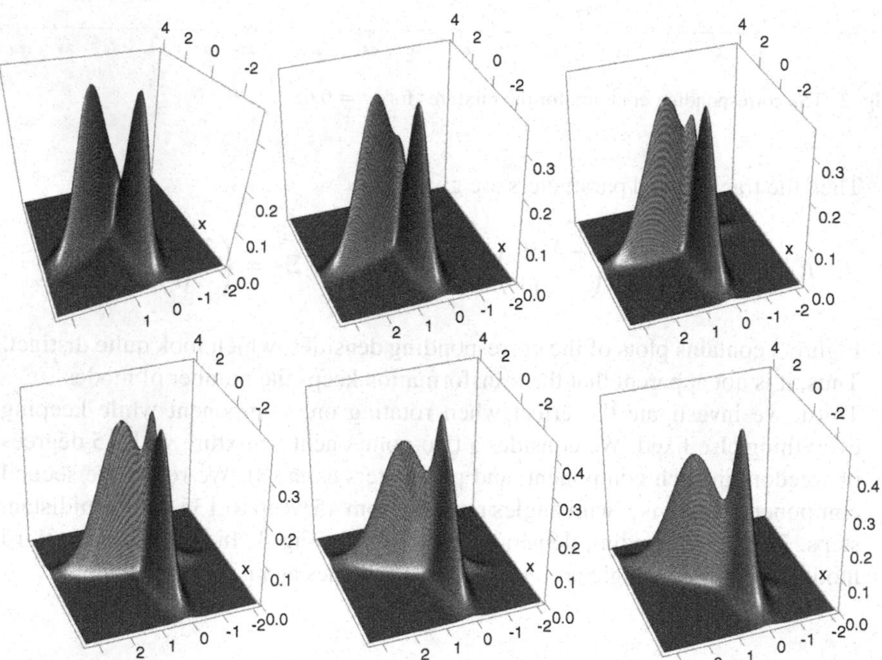

Fig. 4 The clockwise rotation of one mixture component

alternative to normal mixtures. While t-mixtures allow for heavier tails of the components, asymmetry can still not be dealt with, and thus, the number of components may exceed the actual number of clusters in the data. Thus, modality-based merging algorithms like in Hennig (2010) for normal mixtures, based on the ridgeline as in Theorem 1, can be employed.
2. Testing for the number of modes. If two-component mixtures under suitable parameter restrictions allow at most two modes, such as two-component normals with proportional covariances, or t-mixtures with equal degrees of freedom and covariances, one can use parametric methods to test for one against two modes in such a model by likelihood-ratio based methods, see Holzmann and Vollmer (2008) for univariate normal and von Mises mixtures. This requires explicit characterizations of the parameter constellations which yield unimodal or bimodal mixtures. For two-component normals with proportional covariances, these are given in Ray and Lindsay (2005), Corollary 4, while corresponding characterizations based on Theorem 3 (2) and (3) still need to be derived.

References

Alexandrovich, G. (2011). *Analytische Eigenschaften von Mischungen elliptischer Verteilungen und deren Anwendung in der Clusteranalyse*. Diploma thesis, Philipps Universität Marburg.

Fang, K. T., Kotz, S., & Ng, K. W. (1989). *Symmetric multivariate and related distributions*. London: Chapman & Hall.

Hennig, C. (2010). Methods for merging Gaussian mixture components. In *Advances in data analysis and classification*. doi: 10.1007/s11634-010-0058-3. http://dx.doi.org/10.1007/s11634-010-0058-3.

Holzmann, H., & Vollmer, S. (2008). A likelihood ratio test for bimodality in two-component mixtures – With application to regional income distribution in the EU. *AStA – Advances in Statistical Analysis, 92*, 57–69.

McLachlan, G. J., & Peel, D. (2000). *Finite mixture models*. New York: Wiley.

Ray, S., & Lindsay, B. G. (2005). The topography of multivariate normal mixtures, *The Annals of Statistics, 33*, 2042–2065.

Ray, S., & Ren, D. (2012). On the upper bound of the number of modes of a multivariate normal mixture, *Journal of Multivariate Analysis, 108*, 41–52.

Implications of Axiomatic Consensus Properties

Florent Domenach and Ali Tayari

Abstract Since Arrow's celebrated impossibility theorem, axiomatic consensus theory has been extensively studied. Here we are interested in implications between axiomatic properties and consensus functions on a profile of hierarchies. Such implications are systematically investigated using Formal Concept Analysis. All possible consensus functions are automatically generated on a set of hierarchies derived from a fixed set of taxa. The list of implications is presented and discussed.

1 Introduction

The problem of combining rival structures into a representative one is the central focus of the consensus problem. Arrow's celebrated work (Arrow 1951), followed by May's (1952), opened the door to impossibility results for linear orders, where consensus satisfying desirable properties were characterized as dictatorial (see Powers and White (2008) for some impossibility results on hierarchies). Although several consensus methods were developed for supertrees (Semple and Steel 2000), i.e. when the phylogenetic trees have distinct (but overlapping) sets of taxa, in this paper we will focus on the classical case where the consensus tree has the same taxa set as every input tree. An exhaustive survey on consensus theories can be found in Day and McMorris (2003) and Hudry and Monjardet (2010).

F. Domenach (✉) · A. Tayari
Department of Computer Science, University of Nicosia, 46 Makedonitissas Avenue, 24005, 1700 Nicosia, Cyprus
e-mail: domenach.f@unic.ac.cy; a.tayari@hotmail.com

B. Lausen et al. (eds.), *Algorithms from and for Nature and Life*, Studies in Classification, Data Analysis, and Knowledge Organization, DOI 10.1007/978-3-319-00035-0_5,

Despite the fact that several studies had investigated the statistical behavior of consensus methods (Degnan et al. 2009) or their relationships (see Bryant (2003) for a classification based on refinement), none were interested in their inherent structure in conjunction with fundamental axioms. Our study doesn't aim to exhaustively enumerate all existing consensus functions: we focused on a selection of 13 functions, either for their commonality or their distinctiveness, and 9 axioms.

The rest of the paper is organized as follows: in Sect. 2 we recall fundamental definitions of consensus functions and axiomatic properties on such functions. We introduce Formal Concept Analysis together with implications in Sect. 3. Finally, in Sect. 4, we discuss the results obtained by systematically generating any possible input trees and any possible consensus tree and testing axioms and consensus functions on it.

2 Consensus Methods

Let S be a finite set with n elements, usually called *taxa* or operational taxonomic units. A *hierarchy* H on S, also called n-tree, is a family of nonempty subsets of S (called the *clusters* of H) such that $S \in H$, $\{s\} \in H$ for all $s \in S$, and $A \cap B \in \{\emptyset, A, B\}$ for all $A, B \in H$. We will denote the set of all hierarchies on S by $\mathscr{H}$. All the hierarchies considered here are defined on the same set S.

A series of properties can be defined on hierarchies. Two sets A and B are *compatible* if $A \cap B \in \{\emptyset, A, B\}$, and a set A is compatible with a hierarchy H if it is compatible with every cluster of H (or, equivalently, if $A \cup H \in \mathscr{H}$). Hierarchies can also be defined (Colonius and Schulze 1981) through triplets $ab|c$, $a, b, c \in S$, denoting the grouping of a and b relative to c. We say that $ab|c$ in H if there exists a cluster $X \in H$ such that $a, b \in X$ but $c \notin X$. Adams (1986) extended that idea to *nestings*, where X nests in Y in H, denoted as $X <_H Y$ iff $X \subset Y$ and there is $Z \in H$ such that $X \subseteq Z$ and $Y \nsubseteq Z$. The *canonical height* $\eta_0(X)$ of a cluster $X \subseteq S$ is defined as $\eta_0(S) = 0$ and $\eta_0(X) = h$ iff there is a maximal sequence $S \supset X_1 \supset \ldots \supset X_{h-1} \supset X_h = X$. $\pi(H)$ is the maximal cluster partition for H with blocks equal to the maximal clusters of H.

Let $H^* = (H_1, H_2, \ldots, H_k)$ be a profile of hierarchies on S, and K will denote the set of indices of the hierarchies of H^*, $K = \{1, \ldots, k\}$. A *consensus function* on $\mathscr{H}$ is a map $c : \mathscr{H}^k \to \mathscr{H}$ with $k \geq 2$ and $\mathscr{H}^k$ the k cartesian product, which associate to any profile H^* a unique hierarchy consensus, $c(H^*)$. We will denote the set of hierarchies of the profile H^* containing the cluster X by $K_X(H^*)$, i.e. $K_X(H^*) = \{i \in K : X \in H_i\}$. Set $K_{\overline{X}}(H^*) = \{i \in K : X \cup H_i \notin \mathscr{H}\}$.

Given a profile H^* of hierarchies, many different consensus functions can be defined. The most famous one is the strict consensus, where the consensus tree is only the common clusters. The majority consensus (Margush and McMorris 1981) considers clusters appearing in at least half of the trees, while the loose consensus (Barthélemy 1992) (originally called combinable component (Bremer 1990)) will consider subsets as long as they are compatible with all trees. They were recently (Dong et al. 2011) extended to the majority-rule (+) by adding some compatible clusters. As noted in the literature, those rules may miss some structural features of the hierarchies, particularly the fact that two elements could be closer than a third one. It might be desirable for these elements to be separated in the consensus hierarchy – which is what Adams' function (1972) achieves. Other functions can be based on clusters' frequency (Nelson-Page (Nelson 1979; Page 1990), frequency difference), on height assignment (Durchschnitt (Neumann 1983)) or distance between trees (median, asymmetric median (Phillips and Warnow 1996)). One can refer to Bryant (2003) for a discussion about their respective advantages and drawbacks. Following is the list of consensus functions we have implemented:

(Str) Strict: $Str(H^*) = \bigcap_{i \in K} H_i$

(Prj) Projection: $\exists j \in K : Prj(H^*) = H_j$

(Ol) Oligarchy: $\exists J \subseteq K : Ol(H^*) = \bigcap_{j \in J} H_j$

(Maj) Majority: $Maj(H^*) = \{X \subseteq S : |K_X(H^*)| > \frac{k}{2}\}$

(Lo) Loose: $L(H^*) = \bigcup\{X \subseteq S : \exists j \in K, X \in H_j$ and $\forall i \in K, X \cup H_i \in \mathscr{H}\}$

(LM) Loose and Majority Function Property: $LM(H^*) = Maj(H^*) \cup L(H^*)$

(Maj+) Majority-rule (+) : $Maj^+(H^*) = \{X \subseteq S : |K_X(H^*)| > |K_{\overline{X}}(H^*)|\}$

(NlP) Nelson-Page: The consensus tree is made of maximum weight ($w(X) = |K_X(H^*)| - 1$) compatible clusters. If there is a tie, take the intersection.

(FD) Frequency Difference:
$FD(H^*) = \{X : |K_X(H^*)| > max\{|K_Y(H^*)| : Y$ not compatible with $X\}\}$

(Dur) Durchschnitt: $Dur(H^*) = \bigcup_{j=1}^{\omega}\{\bigcap_{i \in K} X_i : X_i \in H_i$ and $\eta_0(X_i) = j\}$, with $\omega = min_{i \in K} max_{X \in H_i} \eta_0(X)$

(Ad) Adams (from Bryant (2003)): Procedure AdamsTree($H_1, \ldots, H_k$)
Construct $\pi(H)$, the product of $\pi(H_1), \ldots, \pi(H_k)$.
For each block B of $\pi(H)$ do AdamsTree($H_1|_B, \ldots, H_k|_B$)

(Med) Median: $Med(H^*) = \{H \in \mathscr{H} : \sum_{i=1}^{k} |H \Delta H_i|$ is minimum$\}$

(AM) Asymmetric Median: $AMed(H^*) = \{H \in \mathscr{H} : \sum_{i=1}^{k} |H_i - H|$ is minimum$\}$

Arrow's result (for linear orders) characterize consensus functions satisfying some desirable properties. We have considered the following, taken from Day and McMorris (2003):

(PO) Pareto Optimality: $(\forall X \subseteq S)(X \in \bigcap_{i=1}^{k} H_i \Rightarrow X \in c(H^*))$

(Dct) Dictatorship: $(\exists j \in K)(\forall X \subseteq S)(X \in H_j \Rightarrow X \in c(H^*))$

(cPO) co-Pareto Optimality: $(c(H^*) \subseteq \bigcup_{i=1}^{k} H_i)$

(TPO) Ternary Pareto Optimality:
$(\forall x, y, z \in S)((\forall i \in K)(xy|z \in H_i) \Rightarrow xy|z \in c(H^*))$

(NP) Nesting Preservation:
$(\forall \emptyset \neq X, Y \subseteq S)((\forall i \in K)(X <_{H_i} Y) \Rightarrow (X <_{c(H^*)} Y))$

(SP) Strong Presence: $(\forall \emptyset \neq X, Y \subseteq S)(X <_{c(H^*)} Y \Rightarrow (\forall i \in K)(X <_{H_i} Y))$

(QSP) Qualified Strong Presence:
$(\forall X, Y \in c(H^*))(X <_{c(H^*)} Y \Rightarrow (\forall i \in K)(X <_{H_i} Y))$

(USP) Upper Strong Presence:
$(\forall X \in c(H^*))(X <_{c(H^*)} S \Rightarrow (\forall i \in K)(X <_{H_i} S))$

(Btw) Betweenness: (for any family $(X_i)_{i \in K}$ with $X_i \in H_i)(\exists Y \in c(H^*))(\bigcap_{i=1}^{k} X_i \subseteq Y \subseteq \bigcup_{i=1}^{k} X_i)$

3 Formal Concept Analysis

Formal Concept Analysis (FCA) (Ganter and Wille 1996) was developed in Darmstadt as a mathematical theory for modeling the notion of "concept". It starts from a *formal context* (G, M, I), with a set G of objects, a set M of attributes, and a binary relation $I \subseteq G \times M$. $(g, m) \in I$ is read as "object g has attribute m". To this formal context, one can associate to a set of objects $A \subseteq G$ its intension $A' = \{m \in M : \forall g \in A, (g, m) \in I\}$ of all properties shared by A. Dually, we can define $B' = \{g \in G : \forall m \in B, (g, m) \in I\}$, the extension of a set of properties $B \subseteq M$. A pair (A, B), $A \subseteq G$, $B \subseteq M$, is a *formal concept* if $A' = B$ and $B' = A$.

The set of all formal concepts, ordered by inclusion of their intent, forms a lattice (Barbut and Monjardet 1970), called *concept lattice*. It generates and visualizes hierarchies of concepts. For more terms and definitions on lattice theory, one can refer to Birkhoff (1967) and Davey and Priestley (2002). FCA is intensively used in data mining, together with the (equivalent) *implicational system*. An implication $X \rightarrow Y$ represents the fact that every object satisfying the set of attributes X will also satisfy the set of attributes Y, or, equivalently in FCA terminology, $X \subseteq Y''$. This set of implications can be reduced to the Duquenne-Guigues canonical basis (Guigues and Duquenne 1986), a minimal set of implications from which any implication can be generated.

4 Results and Discussion

Our simulation has been implemented using C++, as it takes advantage of low level optimization. Initially, it generates all possible hierarchies based on a given set of n taxa. Then it exhaustively traverses through all possible profiles of k hierarchies, together with all possible consensus trees, creating what we called a configuration. A *configuration* (H^*, H) is a pair of input trees (a profile) together with a consensus tree. Each configuration was compared against axiomatic properties and consensus functions in order to create the formal context (G, M, I), with G the set of configurations, M the set of axiomatic properties and consensus functions, and $(g, m) \in I$ if a configuration g has the (axiomatic) property m. For example, $((H^*, H), (PO)) \in I$ if H contains all common clusters of H^*; likewise, $((H^*, H), (Str)) \in I$ if $H = Str(H^*)$.

A first issue arose from the number of possible configurations. Given the number of n-trees (Felsenstein 1978), and the NP-hard nature of some consensus functions (Phillips and Warnow 1996), we were able to run our application only for $n \leq 5$. For $n = 4$ (resp. $n = 5, n = 6$) and $k = 3$, we have 73,125 configurations (resp. 514,807,450, 9.57×10^{12}), and, for each, 22 properties (9 axioms and 13 functions) were tested on a laptop intel-core i5, 2.3 Gh. In order to significantly improve the running time, the consensus trees set was reduced to a more compact set of structure-based trees. All the trees were divided into equivalent classes such that all trees in a class are isomorphic up to a permutation of their labels. Consider two consensus trees H and H' in the same class and σ the permutation between their labels, the configuration (H^*, H) has the same properties satisfied as the configuration $(\sigma(H^*), H')$.

Since the running time of the simulation increases exponentially with slight addition to n or k, in order to have partial results from otherwise computationally impossible simulations, randomly selected profiles were chosen for every unique representative consensus tree in order to have a more accurate context and so a more precise set of implications.

Figure 1 shows the overall concept lattice, having 2,821 concepts. Although such a huge lattice is hard to read, it is strongly well-structured. There are only 82 implications on the canonical basis (Table 1). The lower (in the lattice) a property is, the less specific it is: the atoms define four big (overlapping) families of functions: (USP), (NIP), (cPO) and (PO), setting Nelson-Page function apart. Under (PO) and (cPO), we can find the family of consensus functions satisfying both: (LM), (FD), (Med), (Maj+), (Ol).

A few (well known) implications arise from the lattice. The meet of (NP) and (QSP) is the Adams' consensus rule, thus uniquely defining it (Adams 1986). (USP) is a weakening of (QSP), which is a weakening of (SP). Relationships amongst axioms (Fig. 2, left) are becoming clearer too: (PO) is satisfied if we have (Btw) (Neumann 1983), which is satisfied if we have (Dct). While considering the lattice of consensus functions (Fig. 2, right), it is similarly well-structured. Apart from obvious special cases ((Str) and (Prj) implying (Ol), (Lo) implying (LM)) and previously known implications ((Maj) implying (Med)), all consensus functions are clearly independent and well-defined.

Our main result is a negative one: there are few unknown implications, and the consensus functions studied are independent. Unfortunately, a drawback of our approach is that we cannot implement fundamental (and desirable) axioms like *Independence* or *Neutrality* by construction as these properties are on two different profiles. We are planning to code more consensus functions (such as MRP, local, ...) in order to reach some exhaustive,or as close as it can be, study

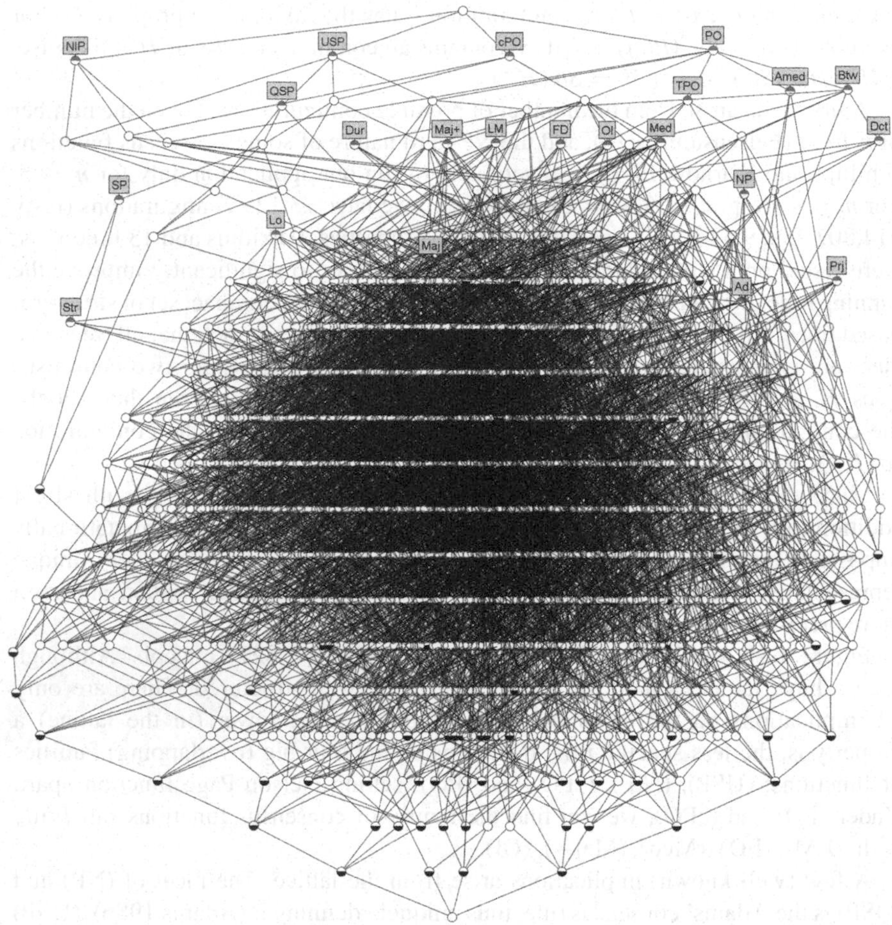

Fig. 1 Concept lattice on 9 axioms and 13 consensus functions (Drawn with ConExp Yevtushenko (2000))

of consensus functions on hierarchies. Similar work is scheduled to extend the simulation software to more general structures, such as weak hierarchies (Bandelt and Dress 1989), 2–3 hierarchies (Bertrand 2000), pyramids (Bertrand and Diday 1985), and different classes of lattices.

Table 1 Implications associated with the lattice

1. Btw → PO	42. PO cPO SP → Str
2. TPO → PO	43. Dct TPO USP Med → Prj
3. Dct → Btw	44. Dct USP NlP Maj+ → Ol
4. NP → TPO Btw	45. TPO Dur Maj+ → QSP
5. QSP → USP	46. Dct USP NlP FD → Ol
6. SP → cPO QSP	47. Maj+ Amed → Btw FD
7. Med → PO cPO	48. TPO USP NlP Maj+ → QSP
8. Ol → PO cPO	49. Btw Med LM Amed → Maj
9. Maj+ → PO cPO	50. Str Med → LM
10. FD → PO cPO	51. TPO USP NlP FD → QSP
11. Prj → Dct cPO Ol	52. TPO USP Med FD → QSP
12. Lo → LM	53. Dct TPO USP NlP → cPO Prj Ad Dur
13. Maj → Med LM;	54. TPO QSP Ol Dur Maj+ → Ad
14. Dur → USP Btw	55. Maj Ol Lo NlP Maj+ → FD
15. Amed → PO	56. Dct TPO Lo Med → NP
16. Ad → NP QSP Btw	57. TPO Dur FD → QSP
17. LM → PO cPO	58. Btw NlP LM Amed → Dct
18. Str → SP Ol	59. Str TPO Dur → Ad
19. Lo FD → Maj+	60. Str Maj+ → Lo Med
20. NP USP Btw → QSP Ad	61. Dct TPO Lo NlP → NP
21. Dct USP Dur → cPO Prj	62. TPO QSP Ol Dur FD → Ad
22. Dct QSP → Prj Dur	63. USP Amed → Btw
23. Dct TPO USP Ol → Prj	64. Btw NlP Maj+ FD Amed → Dct
24. Dct NlP Med → Ol	65. Str FD → Lo Med Maj+
25. FD Amed → Btw	66. TPO QSP Maj FD → Ol
26. Maj NlP FD → Ol	67. Btw NlP Med Amed → Prj
27. TPO Dur Med → Ad	68. PO cPO Btw Lo LM Amed → Dct Maj+ FD
28. Dct USP NlP LM → Ol	69. Btw Med Maj+ FD Amed → Maj
29. TPO Ol Dur LM → Ad	70. Str NlP → Med LM
30. USP Maj NlP → Ol	71. TPO Btw Maj+ FD Amed → Dct
31. LM Amed → Btw	72. Str Lo NlP Med → Maj+ FD
32. TPO USP Ol NlP → QSP	73. USP Btw LM Amed → Dct
33. TPO USP NlP LM → QSP	74. USP Btw FD Amed → Dct
34. Dct TPO Ol NlP LM → Prj	75. Str Btw NlP Med LM → TPO
35. Ol Amed → Prj	76. USP Btw NlP Amed → Prj FD
36. TPO NlP Dur → Ad	77. USP Btw Med Amed → Maj Prj
37. TPO USP NlP Med → QSP	78. Dct TPO Lo Maj+ FD Amed → NP
38. Dct TPO Ol NlP Med → Prj	79. Str SP Prj Lo Dur Med Maj+ → Maj
39. TPO USP Lo Med → QSP	80. Str TPO SP Prj NlP Ad Dur Med LM → Maj
40. Dct TPO Ol Lo → Prj	81. QSP Btw Amed → Str Maj Prj Lo Dur Maj+ FD
41. NlP Amed → Btw	82. USP Btw Dur Amed → Str QSP SP Maj Prj Lo Maj+ FD

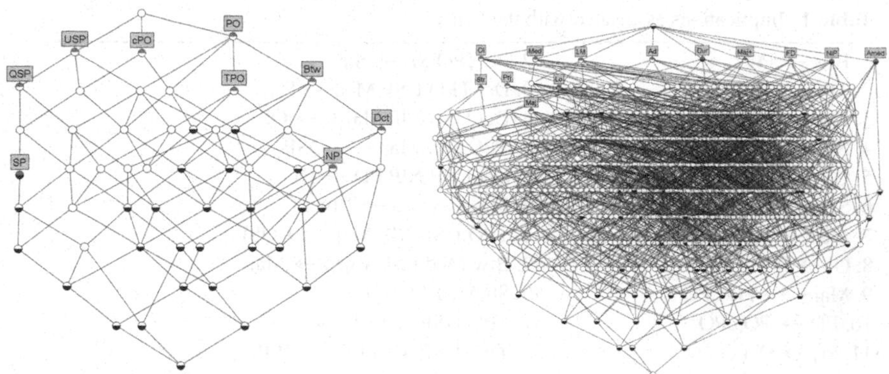

Fig. 2 Concept lattice associated with axioms (*left*) and consensus functions (*right*) (Drawn with ConExp Yevtushenko (2000))

Acknowledgements The authors would like to thank the referees for their useful comments and references.

References

Adams, E. N., III. (1972). Consensus techniques and the comparison of taxonomic trees. *Systematic Zoology, 21*, 390–397.

Adams, E. N., III. (1986). N-trees as nestings: complexity, similarity, and consensus. *Journal of Classification, 3*, 299–317.

Arrow, K. J. (1951). *Social choice and individual values*. New York: Wiley.

Bandelt, H.-J., & Dress, A. (1989). Weak hierarchies associated with similarity measures: an additive clustering technique. *Bulletin of Mathematical Biology, 51*, 133–166.

Barbut, M., & Monjardet, B. (1970). *Ordres et classification: algèbre et combinatoire (tome II)*. Paris: Hachette.

Barthélemy, J.-P., McMorris, F. R., & Powers, R. C. (1992). Dictatorial consensus functions on n-trees. *Mathematical Social Sciences, 25*, 59–64.

Bertrand, P. (2000). Set systems and dissimilarities. *European Journal of Combinatorics, 21*, 727–743.

Bertrand, P., & Diday, E. (1985). A visual representation of compatibility between an order and a dissimilarity index: the pyramids. *Computer Statistics Quarterly, 2*, 31–44.

Birkhoff, G. (1967). *Lattice theory* (3rd ed.). Providence: American Mathematical Society.

Bremer, K. (1990). Combinable component consensus. *Cladistics, 6*, 369–372.

Bryant, D. (2003). A classification of consensus methods for phylogenetics. In M. Janowitz, F. J. Lapointe, F. McMorris, B. Mirkin, & F. Roberts (Eds.), *Bioconsensus, DIMACS* (pp. 163–184). Providence: DIMACS-AMS.

Colonius, H., & Schulze, H.-H. (1981). Tree structure for proximity Data. *British Journal of Mathematical and Statistical Psychology, 34*, 167–180.

Day, W. H. E., & McMorris, F. R. (2003). *Axiomatic consensus theory in group choice and biomathematics*. Philadelphia: Siam.

Davey, B. A., & Priestley, H. A. (2002). *Introduction to lattices and order* (2nd ed.). Cambridge: Cambridge University Press.

Degnan, J. H., DeGiorgio, M., Bryant, D., & Rosenberg, N. A. (2009). Properties of consensus methods for inferring species trees from gene trees. *Systems Biology, 58*, 35–54.

Dong, J., Fernández-Baca, D., McMorris, F. R., & Powers, R. C. (2011). An axiomatic study of majority-rule (+) and associated consensus functions on hierarchies. *Discrete Applied Mathematics, 159*, 2038–2044.

Felsenstein, J. (1978). The number of evolutionary trees. *Systematic Zoology, 27*, 27–33.

Ganter, B., & Wille, R. (1996). *Formal concept analysis: mathematical foundations*. Heidelberg: Springer.

Guigues, J.-L., & Duquenne, V. (1986). Familles minimales d'implications informatives résultant d'un tableau de données binaires. *Mathématiques et Sciences Humaines, 95*, 5–18.

Hudry, O., & Monjardet, B. (2010). Consensus theories. An oriented survey. *Mathématiques et Sciences Humaines, 190*, 139–167.

Margush, T., & McMorris, F. R. (1981). Consensus n-trees. *Bulletin of Mathematical Biology, 43*, 239–244.

May, K. O. (1952). A set of independent necessary and sufficient conditions for simple majority decision. *Econometrica, 20*, 680–684.

Nelson, G. (1979). Cladistic analysis and synthesis: principles and definitions, with a historical note on adanson's famille des plantes (1763–1764). *Systematic Zoology, 28*, 1–21.

Neumann, D. A. (1983). Faithful consensus methods for n-trees. *Mathematical Biosciences, 63*, 271–287.

Page, R. D. M. (1990). Tracks and trees in the antipodes: a reply to humphries and seberg. *Systematic Zoology, 39*, 288–299.

Phillips, C., & Warnow, T. J. (1996). The aymmetric median tree – a new model for building consensus trees. *Discrete Applied Mathematics, 71*, 311–335.

Powers, R. C., & White, J. M. (2008). Wilson's theorem for consensus functions on hierarchies. *Discrete Applied Mathematics, 156*, 1321–1329.

Semple, M., & Steel, C. (2000). A supertree method for rooted trees. *Discrete Applied Mathematics, 105*, 147–158.

Yevtushenko, S. A. (2000). System of data analysis "Concept Explorer". In *Proceedings of the 7th national conference on Artificial Intelligence KII-2000*, Russia, (pp. 127–134).

Comparing Earth Mover's Distance and its Approximations for Clustering Images

Sarah Frost and Daniel Baier

Abstract There are many different approaches to measure dissimilarities between images on the basis of color histograms. Some of them operate fast but generate results in contradiction to human perception. Others yield better results, especially the Earth Mover's Distance (EMD) (Rubner et al., Int J Comput Vis, 40(2): 99–121, 2000), but its computational complexity prevents its usage in large databases (Ling et al., IEEE Trans Pattern Anal Mach Intell, 29(5):840–853, 2007). This paper presents a new intuitive intelligible approximation of EMD. The empirical study tries to answer the question whether the good results of EMD justify its long computation time. We tested several distances with images that were changed by normally-distributed failures and evaluate their results by means of the adjusted Rand index (Hubert et al., J Classif, 2:193–218, 1985).

1 Introduction

The aim of our research is to cluster image databases. One purpose will be in marketing, e.g. it could be used for clustering consumers based on their favorite holiday pictures (Baier and Daniel 2011). There will also be a private use, e.g users would be able to organize their image databases automatically. Therefore we are searching for reliable distance measures to cluster images with the smallest number of misclassifications according to some prespecified criteria.

The Earth Mover's Distance is already known in the area of image retrieval but it has not been used for image clustering. There are already some empirical evaluations (e.g. Puzicha et al. (1999) or Ling and Okada (2007)) verifying that EMD is one of the most suitable distances for content based image retrieval (CBIR).

S. Frost (✉) · D. Baier
Institute of Business Administration and Economics, Brandenburg University of Technology Cottbus, 03013 Cottbus, Germany
e-mail: sarah.frost@tu-cottbus.de; daniel.baier@tu-cottbus.de

B. Lausen et al. (eds.), *Algorithms from and for Nature and Life*, Studies in Classification, Data Analysis, and Knowledge Organization, DOI 10.1007/978-3-319-00035-0_6,

In CBIR applications the EMD between the query image and every contemplable image of the database has to be computed. Therefore $n-1$ distances are required for a database with n images. For clustering images of the same database, usually $\frac{1}{2}\left(n^2-n\right)$ distances have to be calculated. The high number of required distances is one more reason to search for a faster metric for the area of image clustering.

The rest of this paper is organized as follows: Sect. 2 will briefly summarize the theoretical foundations for our study. In Sect. 3 the mathematical background of the original EMD and our own approximation algorithm will be presented. The results of our experiments are illustrated and declared in Sect. 4. Finally we conclude our findings and particularize our plans for future work.

2 Comparing Images Using Color Histograms

2.1 *Color Histograms for Images*

In the following we assume that an image A (or $B, C, \ldots$) can be summarized as a collection of T^A pixels with color measurements $C^A = \{\mathbf{c}_1^A, \ldots, \mathbf{c}_{T^A}^A\}$. The measurement $\mathbf{c}_t^A$ $(t = 1, \ldots, T^A)$ takes values from $X = \{\mathbf{x}_1, \ldots, \mathbf{x}_M\} \subset \mathbb{R}^K$ which reflects the possible intensities in a underlying prespecified K-dimensional color space. In digital image processing, $K = 3$ has proven to be useful e.g. when the 8-bit-coded RGB (Red-Green-Blue) color space $X = \{(0,0,0), (0,0,1), \ldots, (255,255,255)\}$ is used with $M = 256^3 = 16{,}777{,}216$ different colors (= RGB intensity triples).

For characterizing the distribution of these measurements across the image (and for comparing images), (color) histograms can be used. A histogram is a fixed-size discrete distributional function with an a priori declared number N of disjoint color ranges $X_i \subset X$ $(i = 1, \ldots, N)$, the so-called bins. So, e.g., if each possible color of the 8-bit-coded RGB space would be declared as a bin, we would have $N = M = 16{,}777{,}216$ bins. Alternatively, if 8 subsequent intensities in each of the $K = 3$ dimensions would be summarized, one would receive only $N = \left(\frac{256}{8}\right)^3 = 32^3 = 32{,}768$ bins. For each bin i of histogram $\mathbf{h}^A = \{(\mathbf{p}_1^A, h_1^A), \ldots, (\mathbf{p}_N^A, h_N^A)\}$ we can calculate its number of colors in its range N_i, its position (mean color) in the color space $\mathbf{p}_i$ and the corresponding number of pixels h_i^A in image A according to

$$N_i = \sum_{m=1}^{M} 1_{\{\mathbf{x}_m \in X_i\}}, \ \mathbf{p}_i = \frac{1}{N_i} \sum_{\mathbf{x}_m \in X_i} \mathbf{x}_m, \ h_i^A = \sum_{t=1}^{T^A} 1_{\{\mathbf{c}_t^A \in X_i\}}. \tag{1}$$

Usually, to make images comparable with different numbers of pixels $(T^A, T^B, \ldots)$, the histogram values are transformed into shares of pixels, i.e. the frequency h_i^A is replaced by the weight (share) $w_i^A = h_i^A / T^A$. Thus we get the normalized histogram $\mathbf{w}^A = \{(\mathbf{p}_1^A, w_1^A), \ldots, (\mathbf{p}_N^A, w_N^A)\}$. Also, instead of histograms w.r.t. to the same color ranges across all images, so-called signatures can be used,

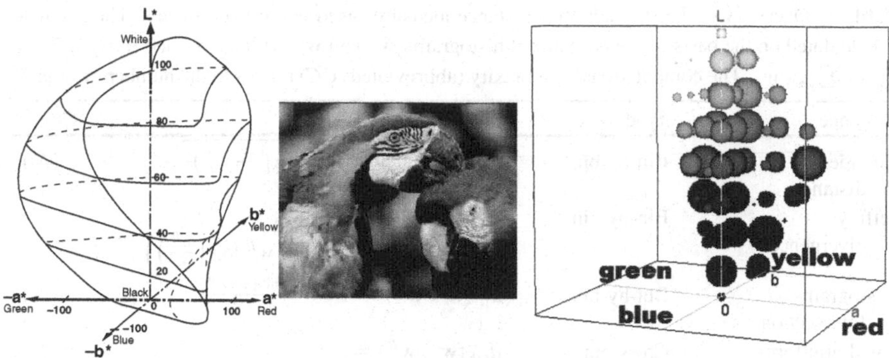

Fig. 1 The CIE L*a*b* color space (*left*) and an example of a color intensity histogram (*right*) of a macaw picture (*middle*)

which describe the color distribution of an image via image-specific non-empty bins. An adequate coding for a signature for image A with N^A bins could be $\mathbf{q}^A = \{(\mathbf{p}_1^A, w_1^A), \ldots, (\mathbf{p}_{N^A}^A, w_{N^A}^A)\}$ $(N^A \leq N)$. The positions and weights are calculated in the same way as above. This is especially useful when parts of the color space are expected to have no weights, e.g. when comparing underwater-images without red or yellow colored pixels. However, in the following, we restrict our discussions to histograms and weights.

2.2 The CIE L*a*b* Color Space

In our experiments we use distances between color histograms to compare images. For humans, the similarity of colors has proven to be very important (see Swain and Ballard 1991). So, e.g. humans use colors to interpret traffic signs and to decide whether food is healthy or not, animals use colors to send messages or warnings.

There are many different color spaces. Probably the best known is the above mentioned RGB space which uses an additive mixing of primary color intensities. Initially it was used for photographic experiments by James Clerk Maxwell in 1861 using three color-filtered separate takes. Later it became popular for TV, computer displays and other electronic devices. It bases on the human processing of color stimuli via three types of retinal cones in the eye. But it was not designed to produce color codings whose distances reflect perceived color differences. Here, the CIE (Commission Internationale de l'Eclairage) L*a*b* color space – also a three-dimensional color space – has proven to be more adequate (see, e.g. the experiments conducted by Schwarz et al. (1987)). The first dimension represents the lightness (L*), the second and third are opponent red-green (a*) and yellow-blue (b*) color-axes (see Fig. 1). Every visible color is represented in this space, thus no output-device can display all colors of the L*a*b* color space. The color

Table 1 Overview of the six traditional distance measures used in our experiments. The distance is calculated on the basis of the normalized histograms $\mathbf{w}^A$ and $\mathbf{w}^B$ with $m_i = (w_i^A + w_i^B)/2$ and $\hat{w}_i = \sum_{j=1}^{i} w_j$. The computational complexity (abbreviated: *CC*) relates to the number of bins N

Distance	Type	Formula	CC
Euclidean distance	Bin-by-bin	$d_{L_2}(\mathbf{w}^A, \mathbf{w}^B) = \sqrt{\sum_i (w_i^A - w_i^B)^2}$	$\mathcal{O}(N)$
Jeffrey divergence	Bin-by-bin	$d_J(\mathbf{w}^A, \mathbf{w}^B) = \sum_i \left(w_i^A \log\left(\frac{w_i^A}{m_i}\right) + w_i^B \log\left(\frac{w_i^B}{m_i}\right)\right)$	$\mathcal{O}(N)$
Histogram-Intersection	Bin-by-bin	$d_{\cap}(\mathbf{w}^A, \mathbf{w}^B) = 1 - \frac{\sum_i \min(w_i^A, w_i^B)}{\sum_i w_i^B}$	$\mathcal{O}(N)$
Quadratic-form	Cross-bin	$d_{QF}(\mathbf{w}^A, \mathbf{w}^B) = \sqrt{(\mathbf{w}^A - \mathbf{w}^B)^T \mathbf{S} (\mathbf{w}^A - \mathbf{w}^B)}$	$\mathcal{O}(N^2)$
Match distance	Cross-bin	$d_M(\mathbf{w}^A, \mathbf{w}^B) = \sum_i \mid \hat{w}_i^A - \hat{w}_i^B \mid$	$\mathcal{O}(N)$
Kolmogorov-Smirnov-dist.	Cross-bin	$d_{KS}(\mathbf{w}^A, \mathbf{w}^B) = (\max_i \mid \hat{w}_i^A - \hat{w}_i^B \mid)$	$\mathcal{O}(N)$

similarity functions developed by the CIE are based on so called 'supplementary standard colorimetric observers' to get a relation between the perceived color and the physical reason for a color-stimulus (Wyszecki and Stiles 1982). It is one of the common color feature spaces in computer vision (see: Rubner et al. (1997) or Puzicha et al. (1999)) and even professional image processing software like Adobe Photoshop uses the L*a*b* color space for internal color processing. The L*a*b* coordinates can be obtained by transforming RGB coordinates into the XYZ system, which is able to normalize the colors with respect to the source of illumination. The second step is a nonlinear transformation from the normalized XYZ-system into the L*a*b*-system. In this special color space the calculated Euclidean distance between two colors corresponds to the human perception of color dissimilarity. As a result the *ground distance* (Sect. 3.1) for the EMD will be the Euclidean distance, while using this color space in our experiments.

2.3 Traditional Distance Measures

Puzicha et al. (1999) compared the EMD with seven other distance measures on histograms w.r.t. image retrieval. They compared the number of correct retrieved images relative to total number of retrieved images. Unfortunately, their results don't solve our problem. For example, Kullback-Leibler divergence yielded a small percentage of classification errors but because of its asymmetry it is not usable for hierarchical clustering approaches. Table 1 lists the traditional distance measures we used in our evaluation. All distances between image A and B make use of the corresponding normalized histograms $\mathbf{w}^A, \mathbf{w}^B$ w.r.t. the L*a*b* color space and the same color ranges (bins). The quadratic-form distance additionally uses the symmetric similarity matrix $\mathbf{S} \in \mathbb{R}^{N \times N}$ with $s_{ij} = 1 - \frac{g_{ij}}{\max g_{ij}}$ (Rubner et al. 2000), where g_{ij} is the distance (we used Euclidean dist.) between the positions $\mathbf{p}_i^A$ and $\mathbf{p}_j^B$ of the bins i and j.

3 EMD and its Approximations

3.1 Theoretical Background of Earth Mover's Distance

The basis for comparing images via minimum histogram transformation costs was firstly presented by Peleg et al. (1989). Rubner et al. (1997) than introduced signatures to shrink the problem and firstly named this measure *Earth Mover's Distance* (EMD) according to a common transportation problem, which searches for the 'cheapest' way to put earth from a set of regionally spread piles (suppliers) to a set of regionally spread holes (demanders). That also makes clear, why empty bins can be removed; since suppliers without earth or demanders without demands are unimportant for this problem. Assuming that the color signature of one image is the available earth distributed across a set of regions (= bins) and the signature of the second image is the distributed demand across the same set of regions, the EMD computes the minimum cost that has to be paid to distribute the available earth according to the demands.

The units of earth transported from a supplier (pile) i to demander (hole) j are called *flow* f_{ij} and the distance the flow has to take is called *ground distance*. The ground distance g_{ij} represents the costs to transport one unit of earth from supplier i to demander j. Usually the Euclidean distance between the corresponding bin positions is used. Rubner et al. (1997) formally defined the problem as follows: For two given signatures $\mathbf{q}^A = \{(\mathbf{p}_1^A, w_1^A), \ldots, (\mathbf{p}_{N^A}^A, w_{N^A}^A)\}$ and $\mathbf{q}^B = \{(\mathbf{p}_1^B, w_1^B), \ldots, (\mathbf{p}_{N^B}^B, w_{N^B}^B)\}$, where $\mathbf{p}_i^A$ and $\mathbf{p}_j^B$ are the positions and w_i^A and w_j^B the weights of the bins, with $g_{ij} = \| \mathbf{p}_i^A - \mathbf{p}_j^B \|_2$ (if the Euclidean distance is used as ground distance) we search for the flow, that minimizes transportation costs

$$EMD(\mathbf{q}^A, \mathbf{q}^B) = \min \sum_{i=1}^{N^A} \sum_{j=1}^{N^B} f_{ij}\, g_{ij} \quad \text{s.t.} \tag{2}$$

$$f_{ij} \geq 0; \quad \sum_{j=1}^{N^B} f_{ij} \leq w_i^A \quad \forall 1 \leq i \leq N^A; \quad \sum_{i=1}^{N^A} f_{ij} \leq w_j^B \quad \forall 1 \leq i \leq N^B;$$

$$\sum_{i=1}^{N^A} \sum_{j=1}^{N^B} f_{ij} = min \left(\sum_{i=1}^{N^A} w_i^A, \sum_{j=1}^{N^B} w_j^B \right).$$

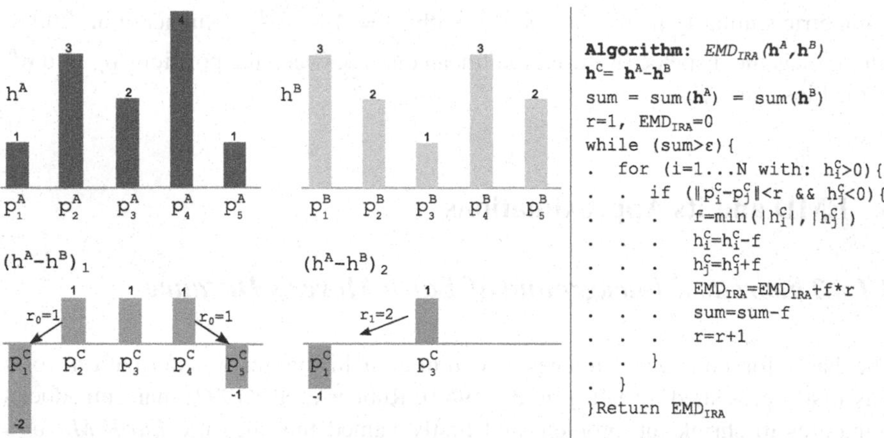

Fig. 2 A visualization of the IRA algorithm using a sample calculation (*left*) and the pseudo code (*right*). $\mathbf{h}^A$ and $\mathbf{h}^B$ are two one-dimensional histograms. $(\mathbf{h}^A - \mathbf{h}^B)_1$ is the difference-histogram. In the first step, where the radius equals one, there is a hole in a circuit of the pile at position $\mathbf{p}_2^C$ and bin $\mathbf{p}_4^C$. After filling this holes there is one unit left. In $(\mathbf{h}^A - \mathbf{h}^B)_2$ the radius is increased. The last hole can be filled because it is in a circuit of the pile at $\mathbf{p}_3^C$

The theoretical computation complexity of the traditional EMD is $\mathcal{O}(z^3 \log z^3)$ where z is the number of non-empty bins if signatures are used. Empirical running times are given by Rubner et al. (2000). The main advantage of EMD for image clustering is that it is a cross-bin distance with adjustable ground distances. That means, the ground distance can be adapted to the corresponding feature space (e.g. L*a*b*-color space, texture-distributions, etc.). So, in contrast to bin-by-bin distances, that only compare corresponding bins, the EMD notices if there is a small color shift between two images and calculates only a small difference between these images.

3.2 *Increasing Radius Algorithm (IRA)*

Now we want to introduce our own approximation, the Increasing Radius Algorithm (IRA), which is much more intelligible than other approximations and fast to implement. The first step is to calculate the difference-histogram (Fig. 2). Positive values now will be the piles and negative ones the holes of earth.

The first radius $r_0 = 1$. The algorithm runs through the difference histogram and adds as much mass as possible to all negative bins that are located within a circuit (with radius $r_0 = 1$) around a positive bin. After every iteration the radius will be increased ($r_1 = 2$). In the example in Fig. 2 there are two units of earth that are transported over a ground distance of $r_0 = 1$ and one unit that is transported over a ground distance of $r_1 = 2$. The result is a distance between the histograms of

$2 * 1 + 1 * 2 = 4$, which still has to be divided by the total weight 11. After at most N^2 iterations the algorithm finishes. The results are presented in Sect. 4. At the moment the Increasing Radius Algorithm (IRA) only works with histograms up to three dimensions. But we are going to extend of the algorithm for higher-dimensional feature spaces, where the radius will determine a hyper sphere around positive bins.

4 Experimental Comparison of Distance Measures

We compared the traditional EMD with six other common distance measures, and the referred Increasing Radius Algorithm. In image retrieval mostly the number of correct retrieved images is counted. To make the clustering result comparable we used the adjusted Rand index (Hubert and Arabie 1985).

For our tests we used 15 holiday pictures from three categories: sunset, mountains, and cities at night. We did the test with low-level, medium-level, and high-level failures (see: Table 2). In each test set-up every of the 15 images was copied 10 times. The copies were changed by normally-distributed failures. In each case we did one of the following distortions: decreasing resolution, decreasing contrast, decreasing the red color-band, increasing brightness, and adding Gaussian noise. We picked out these distortions, because they seem to be common in digital photography. To make an empirical evaluation we repeated every test set-up five times with new random values to make a Monte Carlo analysis. We decreased the resolution e.g. in high-level distortion up to 95 %, but nearly all metrics were robust to this change. We also decreased the contrast from the original (0:255) range down to a minimum of (108:147). The brightness was increased up to 192, where the maximum 255 would be a totally white image. In the case of color distortion we decreased the percentage of the red color band, what means that the image became more cyan. The last distortion was to add Gaussian noise with a standard deviation up to 800 (in high-level case) (Bordese and Alini 2010, pp. 37–38).

In average the EMD yielded the highest adj. Rand indices but also the IRA, quadratic-form, histogram-intersection, and Jeffrey-div. achieved high values. With an Intel core i7 2,6 GHz PC (Windows 7) with 4 GB of RAM the average computation times for one distance are shown in Fig. 3 (software was implemented by our own with C++ using Qt application framework).

5 Conclusion

We found out, that distances that produce good CBIR results are not necessarily usable for image clustering. But the EMD proved to be an appropriate measure to cluster images because it is robust to changes in color, resolution, and noise, it yielded the best average adjusted Rand index (0.91), and for a limited number

Table 2 Empirical comparison of eight distance measures using a Monte Carlo Analysis with five image distortion factors each with three levels. For each factor-level 150 modified sunset, mountain, and city night images are clustered and the adjusted Rand indices were calculated, comparing the expected and the derived clustering. The values are the mean adjusted Rand indices after five repetitions with new random values. 1.0 means perfect clustering

	Resolution			Contrast			Brightness			Color			Noise			
	Low	Med	High	Low	Med	High	Low	Med	High	Low	Med	High	Low	Med	High	
Distance	100–50	100–25	100–5	0–36	0–72	0–108	0–64	0–128	0–192	100–83	100–66	100–50	0–100	0–500	0–800	Mean
EMD	1.0	1.0	1.0	1.0	0.96	0.68	1.0	0.71	0.36	1.0	1.0	1.0	1.0	1.0	1.0	0.91
IRA	1.0	0.86	0.93	0.93	0.89	0.52	0.90	0.76	0.56	1.0	0.98	0.97	1.0	1.0	1.0	0.89
QF-distance	1.0	1.0	1.0	0.82	0.86	0.69	1.0	0.72	0.39	0.82	0.87	0.87	1.0	1.0	1.0	0.87
Hist.-Inters.	1.0	1.0	1.0	0.93	0.81	0.61	0.83	0.70	0.46	1.0	0.83	0.72	1.0	1.0	1.0	0.86
Jeffrey-div.	1.0	1.0	1.0	1.0	0.85	0.52	0.83	0.72	0.39	0.96	0.84	0.67	1.0	1.0	1.0	0.85
Euclidean	0.69	0.69	0.69	0.49	0.49	0.40	0.35	0.44	0.38	0.73	0.68	0.60	0.69	0.67	0.68	0.58
Match	0.58	0.58	0.58	0.58	0.41	0.30	0.34	0.19	0.08	0.58	0.55	0.56	0.58	0.58	0.58	0.47
KS-distance	0.52	0.52	0.52	0.52	0.40	0.29	0.31	0.21	0.09	0.55	0.54	0.54	0.52	0.50	0.50	0.44

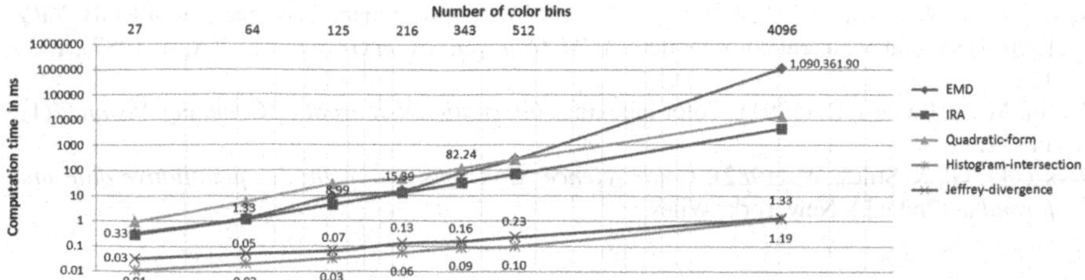

Fig. 3 A log-log plot of average computation times of the five best distance measures of our experiment in relation to the number of histogram bins

of colors it is even faster than quadratic-form. There are other distances, that are almost as good as EMD and only take a fraction of the time EMD needed. For example histogram-intersection and Jeffrey-divergence are bin-by-bin distances and yielded an average adjusted Rand index of about 0.85. But they only took 0.10 ms to calculate one distance. Our own Increasing Radius Algorithm achieved the second best results but it admits of improvement.

In the next time we will check more approximations for their usability in image clustering and extend our own algorithm for multi-dimensional features and different ground distances.

Acknowledgements This research is funded by Federal Ministry for Education and Research under grants 03FO3072. The author is responsible for the content of this paper.

References

Baier, D., & Daniel, I. (2011). Image clustering for marketing purposes. In *Proceedings 34th annual conference of the gfkl*. Heidelberg/Berlin: Springer.

Bordese, M., & Alini, W. (2010). Package "biOps". http://cran.r-project.org/web/packages/biOps/biOps.pdf.

Hubert, L., & Arabie, P. (1985). Comparing partitions. *Journal of Classification, 2*, 193–218.

Ling, H., & Okada, K. (2007). An efficient earth mover's distance algorithm for robust histogram comparison. *IEEE Transactions on Pattern Analysis and Machine Intelligence, 29*(5), 840–853.

Peleg, S., Werman, M., & Rom, H. (1989). A unified approach to the change of resolution: space and gray-level. *IEEE Transactions on Pattern Analysis and Machine Intelligence, 11*(7), 739–742.

Puzicha, J., Buhmann, J. M., Rubner, Y., & Tomasi, C. (1999). Empirical evaluation of dissimilarity measures for color and texture. In *Proceedings ot the Seventh IEEE International Conference on Computer Vision, 2*, 1165–1172.

Rubner, Y., Guibas, L., & Tomasi, C. (1997). The earth mover's distance, multi-dimensional scaling, and color-based image retrieval. In *Proceedings of the ARPA image understanding workshop*, New Orleans, LA, pp. 661–668.

Rubner, Y., Tomasi, C., & Guibas, L. (2000). The earth mover's distance as a metric for image retrieval. *International Journal of Computer Vision, 40*(2), 99–121.

Schwarz, M. W., Cowan, W. B., & Beatty, J. C. (1987). An experimental comparison of RGB, YIQ, LAB, HSV, and opponent color models. *ACM Transactions on Graphics 6,* 2(April 1987), 123–158.

Swain, M., & Ballard, D. (1991). Color indexing. *International Journal of Computer Vision,* 7(1), 11–32.

Wyszecki, G., & Stiles, W. (1982). *Color science. Concepts and methods, quantitative data and formulae* (2nd ed.). New York: Wiley.

A Hierarchical Clustering Approach to Modularity Maximization

Wolfgang Gaul and Rebecca Klages

Abstract The problem of uncovering clusters of objects described by relationships that can be represented with the help of graphs is an application, which arises in fields as diverse as biology, computer science, and sociology, to name a few. To rate the quality of clusterings of undirected, unweighted graphs, modularity is a widely used goodness-of-fit index. As finding partitions of a graph's vertex set, which maximize modularity, is NP-complete, various cluster heuristics have been proposed. However, none of these methods uses classical cluster analysis, where clusters based on (dis-)similarity data are sought. We consider the lengths of shortest paths between all vertex pairs as dissimilarities between the pairs of objects in order to apply standard cluster analysis methods. To test the performance of our approach we use popular real-world as well as computer generated benchmark graphs with known optimized cluster structure. Our approach is simple and compares favourably w.r.t. results known from the literature.

1 Introduction

Graph clustering, sometimes also referred to as community structure detection in graphs, combines the research areas of graph theory (where binary, symmetric relations between objects are illustrated by undirected, unweighted edges between the vertices of a graph which represent the objects) and cluster analysis (where groups of vertices revealing special graph structures have to be found).

While in standard cluster analysis of dissimilarity data homogeneous clusters are sought that are heterogeneous among each other, in graphs we try to find tightly knit groups of vertices with few edges between these groups.

W. Gaul (✉) · R. Klages
Institute of Decision Theory and Management Science, Karlsruhe Institute of Technology (KIT), Kaiserstr. 12, 76128 Karlsruhe, Germany
e-mail: wolfgang.gaul@kit.edu; rebecca.klages@kit.edu

B. Lausen et al. (eds.), *Algorithms from and for Nature and Life*, Studies in Classification, Data Analysis, and Knowledge Organization, DOI 10.1007/978-3-319-00035-0_7,

A popular goodness-of-fit index to estimate and compare the quality of this kind of clusterings in graphs is called modularity, which was suggested in 2004 (see, e.g., Newman and Girvan 2004; Newman 2004a,b; Clauset et al. 2004). Other suggestions to measure graph clustering solutions are known (see, e.g., Brandes and Erlebach (Eds.) 2005), but will not be addressed here. A definition of modularity as well as a discussion concerning approaches using modularity are presented in Sect. 2. The application of shortest path dissimilarities in order to cluster graphs is motivated and explained in Sect. 3, where we also propose our approach, which consists of the following main steps: (1) computation of all shortest path lengths, (2) application of standard hierarchical clustering, (3) search for possible local improvements with the help of a vertex exchange algorithm. In Sect. 4 we show how our approach performs on benchmark graphs from the literature with known cluster structure. Finally, we give a summary as well as a brief outlook in Sect. 5.

2 Modularity as a Goodness-of-Fit Index

By $G = (V, E)$ we denote an undirected, unweighted graph with a set V of n vertices and a set E of m edges that link pairs of vertices, i.e., $e = (i, j) \in E$ with $i, j, \in V$, where no parallel edges (for each pair of vertices at most one edge exists) and no loops ($e = (i, i), i \in V$) are considered. $A = (A_{ij})$ with $A_{ij} = 1$, if $e = (i, j) \in E$, and $A_{ij} = 0$, otherwise, describes the adjacency information of the graph.

Formally, modularity is defined using the entries A_{ij} of the adjacency matrix A, the degrees k_i of vertices i (number of edges incident to i), and the underlying graph clustering, where c_i denotes the cluster that contains vertex i. In modularity calculations only relationships between vertices in the same cluster are considered which is achieved by using the Kronecker-Delta $\delta(c_i, c_j)$ (equal to 1 if $c_i = c_j$, and equal to 0, otherwise). Then, as formula for the modularity Q one can use

$$Q = \frac{1}{2m} \sum_{i,j} \left(A_{ij} - \frac{k_i \cdot k_j}{2m} \right) \cdot \delta(c_i, c_j). \quad (1)$$

Note that every edge is incident to exactly two vertices, so $2m$ is equal to the sum of all vertex degrees. $[-0.5; 1]$ is a theoretical interval for values of Q (see, e.g., Brandes et al. 2007), which depend on the graph's structure, i.e., two clusterings in different graphs cannot be compared using modularity. In real-world graphs (Newman and Girvan 2004) state that optimized values of Q are often elements of the interval $[0.3; 0.7]$.

Modularity has been applied in quite a number of contributions, which shows the importance of this measure in scientific context. Originally introduced by Newman and Girvan (2004) as a new goodness-of-fit index along with a divisive hierarchical graph clustering procedure, Newman (2004a) suggested an agglomerative clustering method, which merges those two clusters in each agglomerative step whose fusion causes the largest increase or smallest decrease of modularity. Brandes et al. (2008) showed that modularity maximization over all partitions of the vertex set V is NP-complete. Therefore, various heuristics to tackle this problem have been proposed. A modification of Newman's agglomerative algorithm was given by Schuetz and Caflisch (2008), which enables the fusion of more than two communities in each iteration step. Other hierarchical approaches were given by Radicchi et al. (2004), Xiang et al. (2008) as well as Mann et al. (2008). Algorithms similar to hierarchical clustering have been proposed by Arenas et al. (2007), Djidjev (2008), Zhu et al. (2008), and Blondel et al. (2008), who developed different procedures to coarsen the graph in question. Then, the coarsened copies of the original graph are either clustered or the coarsest version of the graph is defined as a clustering, whose cluster solutions are conveyed and refined to fit the original graph using iterative uncoarsening. Also, approaches using heuristics known from other fields of research have been applied, for example mathematical optimization (Duch and Arenas (2005) employ an extremal optimization procedure, Agarwal and Kempe (2008) express the problem with the help of linear and vector programming). The application of probabilitic flows on random walks in graphs (Rosvall and Bergstrom 2008) and matrix factorization (Ma et al. 2010) have also been suggested. A significant number of authors (see e.g.,Newman 2006) use spectral clustering algorithms (see Nascimento and de Carvalho (2010) for a recent overview).

Besides the development of various procedures that aim to find a partition of the vertex set with highest possible modularity, the concept has also been criticized. Fortunato and Barthélemy (2007) showed that there is a lower bound to the sizes of clusters that can be detected using heuristics which strive to maximize modularity. This lower bound depends on the number of vertices n and the interconnectedness of the clusters. Variations of modularity were proposed to avoid this weakness (see e.g., Li et al. (2008), who suggested a local measure that takes the density of subgraphs into account). However, the original definition of modularity is still widely used and extensions to weighted graphs (Newman 2004b) as well as to directed graphs (e.g., Arenas et al. 2007; Leicht and Newman 2008; Kim et al. 2010) have been presented. Good et al. (2010) review the performance of modularity maximization in practical contexts.

3 A New Heuristic to Find Clusters in Graphs

Given the many contributions in which modularity was used for community structure detection we considered the following idea to cluster a graph into subgraphs with closely connected vertices and comparatively few edges between different subgraphs: In a graph the important information is stored in the adjacency matrix $A = (A_{ij})$. While $A_{ij} = 1$ might be a reason to put vertex i and j into the same cluster, for a pair of vertices i and j with $A_{ij} = 0$ there is no information how similar i and j might be. They could have a common neighbour but they could also be in completely different areas of the graph. Therefore, we define the dissimilarity between two vertices as length of a shortest path that connects them in the graph. If no path exists a sufficiently large constant is used to indicate this situation. Now, the solution of the underlying problem in terms of standard cluster analysis is straightforward and, as writing restrictions do not allow to describe the single steps of the new heuristic in mathematical terms, we give the following verbal explanations: First, we determine the lengths of the shortest paths between all pairs of vertices in the graph (see, e.g., Floyd 1962; Warshall 1962). Second, we apply a hierarchical clustering method to the shortest path length matrix computed in the first step and calculate the modularity for the clusterings given by the hierarchy. (Average and Weighted Average Linkage were well-suited for our data.) Third, an exchange algorithm called vertex mover (Schuetz and Caflisch 2008) is applied to the clustering in the hierarchy with highest modularity, which also provides the number of clusters needed to check whether improvements by exchange operations are still possible.

4 Performance on Benchmark Graphs

To test the performance of our approach we used several well-known undirected, unweighted real-world as well as computer generated benchmark graphs.

Example 1. The first example is a well-known real-world graph by Zachary (1977), who examined the relations between 34 members of a karate club. Two vertices of the graph are adjacent, if the corresponding people spent a significant amount of time together during the examination (see Fig. 1).

By chance there was a dispute between the principal karate teacher (vertex 33) and the administrator (vertex 1) of the club while Zachary studied the relations in this group, which caused the club to spilt into two subgroups. This real-life partition is depicted in Fig. 1 by a black line. The modularity of this spilt is 0.3715. Interestingly, a better Q for a solution with two clusters is 0.3718. Not only is this value only slightly larger, the clusters are also almost the same, just for vertex 10 the group membership has to be changed. This shows that modularity can successfully be used to predict the clusters of the split of this social network which separated into two groups. Our approach finds the two-cluster-solution with the better modularity mentioned above, which was also detected by Newman and Girvan (2004).

Additionally, our method finds the largest known modularity value for a clustering of this graph which is $Q = 0.4198$ as also reported by Duch and Arenas (2005). This value is obtained for the division into four clusters, which is also shown in Fig. 1, where the four different colors of the vertices indicate the four groups. These clusters happen to be subgroups of the real-life decomposition that Zachary (1977) observed.

Example 2. As real-world graphs known from the literature can be very specific and sometimes need lengthy explanations of the relationships that underly the

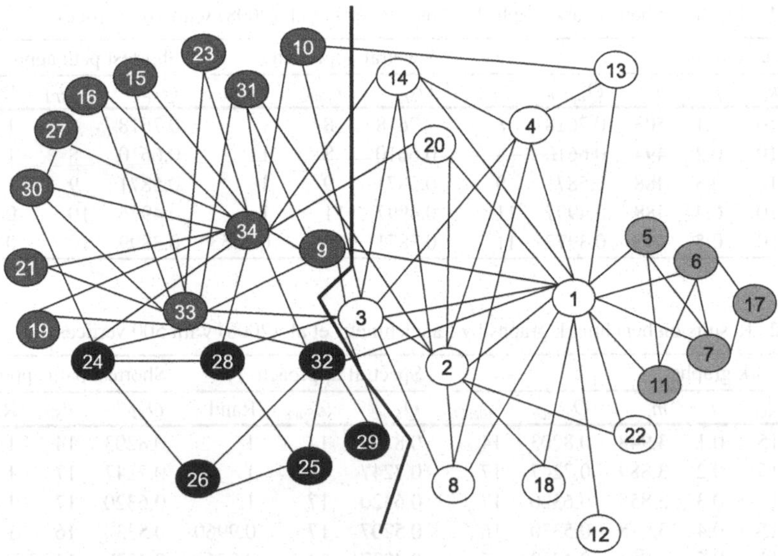

Fig. 1 The Zachary network of friendship ties between members of a karate club

described situation we tested our new method on a class of computer generated benchmark graphs with built-in community structures introduced by Lancichinetti et al. (2008). Although the authors argue that there is no guarantee that the bulit-in community structures constitute solutions with highest modularity these graphs provide the up-to-now best known benchmarks. They can be constructed for any choice of the following three parameters: n (the number of vertices), k_{av} (the average vertex degree), and μ (a mixing parameter which denotes the fraction of a vertex v's neighbours not in c_v). Heterogeneous vertex degrees are modeled by a power law distribution with parameter τ_1, heterogeneous cluster sizes are modeled by a power law distribution with parameter τ_2. To take into account restrictions for real-life graphs the authors propose $\tau_1 \in [2,3]$, $\tau_2 \in [1,2]$, and report on results of testgraphs with all four combinations of the extreme cases of τ_1 and τ_2 for which a very similar behaviour of the modularity calculation was found in all cases. Therefore we chose $\tau_1 = 2$ and $\tau_2 = 1$. So far we used different numbers of vertices $n \in \{100; 500; 1{,}000\}$ with adequate average degrees $k_{av} \in \{10; 15; 20\}$ and mixing parameters $\mu \in \{0.1; 0.2; 0.3; 0.4; 0.5\}$. Note that $\mu = 0.1$ indicates a strong cluster structure as 90 % of the neighbours of each vertex v are in the same cluster c_v, while in graphs with $\mu = 0.5$ half of the neighbours of each vertex are in other clusters than v. A minimal and a maximal value for the cluster sizes can also be selected. The software to construct these benchmarks is explained in a read-me file provided by the authors, in which the cluster sizes are chosen to be in the interval [20, 50], so we also used these values. In order to better analyze the results found by our method, we did not only compare our findings with the built-in community structures given by Lancichinetti et al. (2008), but also implemented the spectral approach proposed by

Table 1 Results on benchmark graphs by Lancichinetti et al. (2008) with 100 vertices

Benchmark graphs						Spectral approach			Shortest path approach		
n	k_{av}	μ	m	Q_{Bench}	$\mathscr{C}_{Bench}$	Q_{Spec}	$\mathscr{C}_{Spec}$	Rand	Q_{SP}	$\|\mathscr{C}_{SP}\|$	Rand
100	10	0.1	503	0.7618	8	0.7618	8	1	0.7618	8	1
100	10	0.2	494	0.6610	8	0.6610	8	1	0.6610	8	1
100	10	0.3	488	0.5871	9	0.5871	9	1	0.5871	9	1
100	10	0.4	488	0.4997	11	0.4997	11	1	0.4976	10	0.9836
100	10	0.5	495	0.3977	11	0.3871	8	0.9453	0.3899	10	0.9675

Table 2 Results on benchmark graphs by Lancichinetti et al. (2008) with 500 vertices

Benchmark graphs						Spectral approach			Shortest path approach		
n	k_{av}	μ	m	Q_{Bench}	$\mathscr{C}_{Bench}$	Q_{Spec}	$\mathscr{C}_{Spec}$	Rand	Q_{SP}	$\mathscr{C}_{SP}$	Rand
500	15	0.1	3,948	0.8203	14	0.8203	14	1	0.8203	14	1
500	15	0.2	3,889	0.7247	17	0.7247	17	1	0.7247	17	1
500	15	0.3	3,855	0.6320	17	0.6320	17	1	0.6320	17	1
500	15	0.4	3,918	0.5330	16	0.5207	17	0.9960	0.5330	16	1
500	15	0.5	3,853	0.4262	15	0.4037	12	0.9563	0.4229	14	0.9921
500	20	0.1	4,663	0.8187	15	0.8187	15	1	0.8187	15	1
500	20	0.2	5,041	0.7262	16	0.7262	16	1	0.7262	16	1
500	20	0.3	4,801	0.6167	13	0.6167	13	1	0.6167	13	1
500	20	0.4	5,065	0.5238	14	0.5238	14	1	0.5238	14	1
500	20	0.5	4,906	0.4202	13	0.4202	13	1	0.4170	12	0.9913

Newman (2006) to see which results a known algorithm obtains on these graphs. Of course, comparisons to other modularity optimizing techniques could be performed. We selected spectral clustering because of the recent overview of Nascimento and de Carvalho (2010).

In the Tables 1–3 we present a comparison between the modularity value Q_{Bench} of the built-in community structures $\mathscr{C}_{Bench}$ to the modularity values Q_{Spec} of solutions $\mathscr{C}_{Spec}$ found by our implementation of Newmans spectral method and to Q_{SP} of clusterings $\mathscr{C}_{SP}$ constructed by our own SP (Shortest Path) approach. With $|\mathscr{C}|$ as cardinality of a clustering $\mathscr{C}$, the numbers $|\mathscr{C}_{Bench}|$, $|\mathscr{C}_{Spec}|$, and $|\mathscr{C}_{SP}|$ of the clusters of the three solutions are given along with the Rand indices (see, e.g., Hubert and Arabie 1985) comparing the clusterings of the spectral procedure and of our method with the benchmark solution.

From the 25 (n, k_{av}, μ, m) benchmark graphs in the Tables 1–3 in three cases ((100, 10, 0.4, 488), (500, 20, 0.5, 4,906), and (1,000, 20, 0.4, 9,731)) the spectral approach performed slightly better while in seven cases ((100, 10, 0.5, 495), (500, 15, 0.4, 3,918), (500, 15, 0.5, 3,853), (1,000, 15, 0.3, 7,609), (1,000, 15, 0.4, 7,631), (1,000, 15, 0.5, 7,571), and (1,000, 20, 0.5, 9,581)) our shortest path approach was in front. In the other 15 cases both approaches showed identical outcomes. These are convincing results that the enrichment of the adjacency information by shortest path

Table 3 Results on benchmark graphs by Lancichinetti et al. (2008) with 1,000 vertices

Benchmark graphs						Spectral approach			Shortest path approach		
n	k_{av}	μ	m	Q_{Bench}	$\mathscr{C}_{Bench}$	Q_{Spec}	$\mathscr{C}_{Spec}$	Rand	Q_{SP}	$\mathscr{C}_{SP}$	Rand
1,000	15	0.1	7,930	0.8594	29	0.8594	29	1	0.8594	29	1
1,000	15	0.2	7,858	0.7624	31	0.7624	31	1	0.7624	31	1
1,000	15	0.3	7,609	0.6617	30	0.6516	30	0.9969	0.6617	30	1
1,000	15	0.4	7,631	0.5615	29	0.5522	28	0.9965	0.5583	25	0.9901
1,000	15	0.5	7,571	0.4639	30	0.4293	16	0.9396	0.4555	21	0.9755
1,000	20	0.1	9,834	0.8585	30	0.8585	30	1	0.8585	30	1
1,000	20	0.2	10,017	0.7582	29	0.7582	29	1	0.7582	29	1
1,000	20	0.3	9,765	0.6622	30	0.6622	30	1	0.6622	30	1
1,000	20	0.4	9,731	0.5659	31	0.5659	31	1	0.5596	25	0.9868
1,000	20	0.5	9,581	0.4633	30	0.4581	24	0.9875	0.4609	23	0.9880

lengths and the application of standard cluster analysis leads to a useful alternative to known community structure detection techniques.

5 Conclusion

Against the background that finding a partition of a graph's vertex set with maximal modularity is a NP-complete problem, we proposed the application of standard cluster analysis methods developed for dissimilarity data to the problem of graph clustering. As dissimilarities between pairs of objects, in our case vertices, are needed, we transformed the adjacency matrix into a matrix of shortest path lengths between all vertex pairs of the graph. From all clusterings of the hierarchy that was computed by a standard agglomerative cluster procedure we chose the one with highest modularity as starting solution for an exchange algorithm. On several benchmark graphs we obtained promising results showing that our approach compares favorably with findings from the literature. A next challenge is to transfer the ideas presented in this paper to directed graphs.

References

Agarwal, G., & Kempe, D. (2008). Modularity-maximizing graph communities via mathematical programming. *European Physical Journal B, 66*, 409–418.

Arenas, A., Duch, J., Fernández, A., & Gómez, S. (2007). Size reduction of complex networks preserving modularity. *New Journal of Physics, 9*, 176.

Blondel, V. D., Guillaume, J.-L., Lambiotte, R., & Lefebvre, E. (2008). Fast unfolding of community hierarchies in large networks. *Journal of Statistical Mechanics, 10*, P10008.

Brandes, U., & Erlebach, T. (Eds.). (2005). Network analysis: methodological foundations. In *Lecture notes in computer science* (Vol. 3418). Berlin/Heidelberg: Springer.

Brandes, U., Delling, D., Gaertler, M., Goerke, R., Hoefer, M., Nikoloski, Z., & Wagner, D. (2007). On finding graph clusterings with maximum modularity. In *Lecture notes in computer science* (Vol. 4769, pp. 121–132). Berlin/Heidelberg: Springer.

Brandes, U., Delling, D., Gaertler, M., Goerke, R., Hoefer, M., Nikoloski, Z., & Wagner, D. (2008). On modularity clustering. *IEEE Transactions on Knowledge and Data Engineering, 20*(2), 172–188.

Clauset, A., Newman, M. E., & Moore, C. (2004). Finding community structure in very large networks. *Physical Review E, 70*, 066111.

Djidjev, H. N. (2008). A scalable multilevel algorithm for graph clustering and community structure detection. In *Lecture notes in computer science* (Vol. 4936, pp. 117–128). Berlin/Heidelberg: Springer.

Duch, J., & Arenas, A. (2005). Community detection in complex networks using extremal optimization. *Physical Review E, 72*, 027104.

Floyd, R. W. (1962). Algorithm 97: shortest path. *Communications of the ACM, 5*(6), 345–345.

Fortunato, S., & Barthélemy, M. (2007). Resolution limit in community detection. *Proceedings of the National Academy of Sciences, 104*(1), 36–41.

Good, B. H., de Montjoye, Y.-A., & Clauset, A. (2010). The performance of modularity maximization in practical contexts. *Physical Review E, 81*, 046106.

Hubert, L., & Arabie, P. (1985). Comparing partitions. *Journal of Classification, 2*, 193–218.

Kim, Y., Son, S.-W., & Jeong, H. (2010). LinkRank: finding communities in directed networks. *Physical Review E, 81*, 016103.

Lancichinetti, A., Fortunato, S., & Radicchi, F. (2008). Benchmark graphs for testing community detection algorithms. *Physical Review E, 78*, 046110.

Leicht, E. A., & Newman, M. E. (2008). Community structure in directed networks. *Physical Review Letters, 100*, 118703.

Li, Z., Zhang, S., Wang, R.-S., Zhang, X.-S., & Chen, L. (2008). Quantitative function for community detection. *Physical Review E, 77*, 036109.

Ma, X., Gao, L., Yong, X., & Fu, L. (2010). Semi-supervised clustering algorithm for community structure detection in complex networks. *Physica A, 389*, 187–197.

Mann, C. F., Matula, D. W., & Olinick, E. V. (2008). The use of sparsest cuts to reveal the hierarchical community structure of social networks. *Social Networks, 30*, 223–234.

Nascimento, M. C., & de Carvalho, A. C. (2010). Spectral methods for graph clustering – a survey. *European Journal of Operational Research, 211*(2), 221–231.

Newman, M. E. (2004a). Fast algorithm for detecting community structure in networks. *Physical Review E, 69*, 066133.

Newman, M. E. (2004b). Analysis of weighted networks. *Physical Review E, 70*, 056131.

Newman, M. E. (2006). Finding community structure in networks using the eigenvectors of matrices. *Physical Review E, 74*, 036104.

Newman, M. E., & Girvan, M. (2004). Finding and evaluating community structure in networks. *Physical Review E, 69*, 026113.

Radicchi, F., Castellano, C., Cecconi, F., Loreto, V., & Parisi, D. (2004). Defining and identifying communities in networks. *Proceedings of the National Academy of Sciences, 101*(9), 2658–2663.

Rosvall, M., & Bergstrom, C. T. (2008). Maps of random walks on complex networks reveal community structure. *Proceedings of the National Academy of Sciences, 105*, 1118–1123.

Schuetz, P., & Caflisch, A. (2008). Efficient modularity optimization by multistep greedy algorithm and vertex mover refinement. *Physical Review E, 77*, 046112.

Warshall, S. (1962). A theorem on Boolean matrices. *Journal of the ACM, 9*(1), 11–12.

Xiang, J., Hu, K., & Tang, Y. (2008). A class of improved algorithms for detecting communities in complex networks. *Physica A, 387*, 3327–3334.

Zhu, Z., Wang, C., Ma, L., Pan, Y., & Ding, Z. (2008). Scalable community discovery of large networks. In *Proceedings of the 2008 ninth international conference on web-age information management*, Zhangjiajie, China, pp. 381–388.

Zachary, W. W. (1977). An information flow model for conflict and fission in small groups. *Journal of Anthropological Research, 33*(4), 452–473.

Mixture Model Clustering with Covariates Using Adjusted Three-Step Approaches

Dereje W. Gudicha and Jeroen K. Vermunt

Abstract When using mixture models, researchers may investigate the associations between cluster membership and covariates by introducing these variables in a (logistic) regression model for the prior class membership probabilities. However, a very popular alternative among applied researchers is a three-step approach in which after estimating the mixture model (step 1) and assigning subjects to clusters (step 2), the cluster assignments are regressed on covariates (step 3). For mixture models for categorical responses, (Bolck et al., Political Anal 12:3–27, 2004) and (Vermunt, Political Anal 18:450–469, 2010) showed this approach may severely downward bias covariate effects, and moreover showed how to adjust for this bias. This paper generalizes their corrections methods to be applicable also with mixture models for continuous responses, where the main complicating factor is that a complex multidimensional integral needs to be solved to obtain the classification errors needed for the corrections. We propose approximating this integral by a summation over the empirical distribution of the response variables. The simulation study showed that the approaches work well, except for the combination of very badly separated components and a small sample size.

1 Introduction

Most applied researchers using mixture models not only aim at finding a meaningful set of clusters, but also wish to investigate which factors are associated with the cluster membership of subjects. This profiling of clusters (or latent classes) as a function of external variables (covariates) can either be achieved using a one-step approach or a three-step approach (Bolck et al. 2004). In the one-step approach,

---

D.W. Gudicha (✉) · J.K. Vermunt
Tilburg University, 5000 LE Tilburg, The Netherlands
e-mail: d.w.gudicha@uvt.nl; j.k.vermunt@uvt.nl

B. Lausen et al. (eds.), *Algorithms from and for Nature and Life*, Studies in Classification, Data Analysis, and Knowledge Organization, DOI 10.1007/978-3-319-00035-0_8,

the mixture model is expanded by including the relevant covariates in a regression model for the prior class membership probabilities (Bandeen-Roche et al. 1997; Dayton and Macready 1988). The parameters defining the mixture components – the cluster specific means and (co)variances – and the covariate effects on the cluster membership are estimated simultaneously. Alternatively, in the much more popular three-step approach, the analysis is done in a stepwise manner. First, a standard mixture model clustering is performed without covariates; second, the class membership is predicted, typically using the Bayes modal rule; third, the association between external variables and the predicted class membership is assessed, for example, via a logistic regression analysis. Bolck et al. (2004) and Vermunt (2010) showed for latent class models with categorical responses that this three-step approach may yield severely downward biased estimates for the covariate effects. These authors also showed how to adjust for this bias in step 3 by using information on the classification errors introduced in step 2. Bolck et al. proposed a weighted analysis with the inverse of the classification errors as weights whereas Vermunt proposed a maximum likelihood method that takes the classification errors into account.

While the use of mixture models with continuous response variables is very common, it is not immediately clear how the adjusted three-step methods should be implemented when the response variables used in the mixture model are not categorical but continuous. The aim of the current paper is to come up with such a generalization. The main complicating factor is that the computation of the classification error matrix needed in step 3 requires solving a complex multidimensional integral. We propose using Monte Carlo integration for this purpose, which if the model holds can be replaced by a summation over the observed data points. The performance of this approach is investigated in a simulation study.

The remainder of this paper is organized as follows. First, the mixture model of normal distributions is introduced and the estimation of the class memberships and the quantification of the classification errors is discussed. Subsequently, the relevant one- and three-step approaches for investigating the association between external variables and class membership are presented. These approaches are evaluated in a simulation study. The paper ends with conclusions and practical recommendations.

2 Mixture Modeling and Classification

2.1 *Mixture Models*

The first step of the three-step approach involves estimating the parameters of a mixture model without covariates (i.e., the class proportions and the cluster specific means and (co)variances). Suppose that we have information on p response variables and that the interest lies in clustering of n observations into k exhaustive and mutually exclusive homogeneous subgroups (latent classes). Let T be an unobservable random variable containing the labels of the k subpopulations with realizations

$t = 1, 2, 3, \ldots, k$ and let $y_{i1}, y_{i2}, \ldots, y_{ip}$ be the p-dimensional continuous random variable of interest with joint probability density function $f(y_i, \theta)$ on $\Re^P$ for $i = 1, 2, 3, \ldots n$, where θ represents the vector of unknown parameters. The joint density of y_i can be defined as:

$$f(y_i, \theta) = \sum_{t=1}^{k} \pi_t f(y_i, \theta_t), \tag{1}$$

where $\pi_t = P(T = t)$, with $\sum_{t=1}^{k} \pi_t = 1$ and $\pi_t > 0 \, \forall \, t$, and where θ_t denote the vector of unknown parameters for cluster t. Each of the component density functions are assumed to come from a multivariate normal distribution parameterized by mean vector μ_t and variance covariance matrix Σ_t; that is, $\theta_t = (\mu_t, \Sigma_t)$. The unknown parameters are typically estimated using maximum likelihood, using an algorithm for finding the maximum of the likelihood such as the Expectation-Maximization algorithm (McLachlan and Peel 2000). Various software packages implementing mixture of normals are currently available (e.g., Latent GOLD; Vermunt and Magidson 2005).

2.2 *Classification Rules and Classification Errors*

Once the cluster-specific parameters of the mixture distribution are estimated, the second step in the three-step approach involve allocating each subject to one of the k classes. We will denote the predicted class membership by W, with realization $s = 1, 2, 3, \ldots, k$. The prediction for observation i is based on the cluster membership probabilities which can be obtained using Bayes' theorem:

$$P(T = t | y_i, \theta) = \frac{\pi_t f(y_i, \theta_t)}{f(y_i, \theta)}. \tag{2}$$

Let $w_{is} = P(W = s | y_i, \theta)$ be the likelihood of being assigned to class s given the assignment rule that is used. The most common rule is modal assignment, in which case w_{is} is a hard indicator; that is,

$$w_{is} = \begin{cases} 1 \text{ if } P(T = s | y_i, \theta) > P(T = t | y_i, \theta) \forall s \neq t \\ 0 \text{ otherwise} \end{cases} \tag{3}$$

An alternative rule is proportional assignment, in which case $w_{is} = P(T = s | y_i, \theta)$ (Vermunt 2010).

Except for the situation in which $P(T = t | y_i, \theta)$ is either 0 or 1 for all i, there will be misclassifications. As discussed in more detail below, the total amount of classification errors can be quantified as the probability that a respondent belonging to cluster t is assigned to cluster s, which can be expressed as follows:

$$P(W = s|T = t) = \frac{P(W = s, T = t)}{P(T = t)} \tag{4}$$

The numerator of Eq. (4) is the joint marginal probability of W and T, which can be expressed in terms of the mixture model density. This yields:

$$\begin{aligned} P(W = s|T = t) &= \int \frac{P(W = s, T = t|y, \theta) f(y, \theta) dy}{P(T = t)} \\ &= \int \frac{P(T = t|y, \theta) P(W = s|y, \theta) f(y, \theta) dy}{P(T = t)}. \end{aligned} \tag{5}$$

The last step follows from the fact that W is independent of T conditional on y.

A complication factor in the computation of $P(W = s|T = t)$ is that the expression in Eq. (5) contain an intractable higher-dimensional integral. We propose solving this integral using Monte Carlo integration, which implies sampling say m units from $f(y, \theta)$ and computing the average of this sample. It should, however, be noted that if the mixture model holds, the sample used to solve integral can also be the n data points in the sample used to estimate the mixture model. This implies that $P(W = s|T = t)$ is approximated as follows:

$$P(W = s|T = t) \approx \frac{1}{n} \sum_{i=1}^{n} \frac{P(T = t|y_i, \theta) P(W = s|y_i, \theta)}{P(T = t)}. \tag{6}$$

3 Relationship Between Class Membership and Covariates

3.1 One-Step Full Information ML Approach

Let z_i denote the vector with covariate values for subject i. In the one-step approach, inclusion of covariates involves expanding the standard mixture model defined in Eq. (1) as follows (Vermunt and Magidson 2005; Dayton and Macready 1988):

$$f(y_i, \theta|z_i) = \sum_{t=1}^{k} P(T = t|z_i) f(y_i, \theta_t). \tag{7}$$

As can be seen, the prior class membership probabilities are now a function of covariates. These probabilities are typically modelled using a logistic regression equation; that is,

$$P(T = t|z_i) = \frac{\exp(\gamma_{0t} + \sum_{q=1}^{Q} \gamma_{qt} z_{iq})}{\sum_{m=1}^{k} \exp(\gamma_{0m} + \sum_{q=1}^{Q} \gamma_{qm} z_{iq})}.$$

The parameters of the mixture distribution and the covariate effects on the latent cluster membership can be estimated simultaneously using maximum likelihood estimation.

3.2 Standard Three-Step Approach

An alternative is to use a three-step procedure. After estimating a standard mixture model and assigning individuals to classes, the relationship between the predicted class (W) and external variables is investigated using a standard multinomial logistic regression model:

$$P(W = s|z_i) = \frac{\exp(\gamma_{0s} + \sum_{q=1}^{Q} \gamma_{qs} z_{iq})}{\sum_{m=1}^{k} \exp(\gamma_{0m} + \sum_{q=1}^{Q} \gamma_{qm} z_{iq})}. \tag{8}$$

The γ parameters are estimated by maximizing the log-likelihood function:

$$\log L_{step3} = \sum_{i=1}^{n} \sum_{s=1}^{k} w_{is} \log P(W = s|z_i), \tag{9}$$

where in the case of modal assignment w_{is} is the hard indicator defined in Eq. (3).

3.3 Two Adjusted Three-Step Approaches

The standard three-step approach defines a model for the relationship between external variables and the predicted cluster membership W instead of the true cluster membership T, which results in downward biased estimates for the covariate effects. However, Bolck et al. (2004) and Vermunt (2010) showed how to adjust for this bias by making use of the known relationship between $P(W = s|z_i)$ and $P(T = t|z_i)$.

More precisely, the adjustment methods described below are based on the following simple relationship:

$$P(W = s|z_i) = \sum_{t=1}^{k} P(T = t|z_i) P(W = s|T = t), \tag{10}$$

where $P(W = s|T = t)$ was defined in Eqs. (4)–(6). It can be seen that $P(W = s|z_i)$ is a weighted sum of $P(T = t|z_i)$ where the $P(W = s|T = t)$ serve as weights.

The logic of the correction method proposed by Bolck et al. (2004) – which we refer to as *the BCH approach* – is that if (10) holds, $P(T = t|z_i)$ can also be expressed as a weighted sum of $P(W = s|z_i)$; that is,

$$P(T = t|z_i) = \sum_{s=1}^{k} P(W = s|z_i)d_{st}, \tag{11}$$

where d_{st} is an element of the inverse of the k-by-k matrix with elements $P(W = s|T = t)$. Bolck et al. (2004) proposed re-weighting the data on W (the class assignment weights w_{is}) by d_{st} to obtain approximate data on T. As shown by Vermunt (2010), the BCH approach can be implemented by creating an expanded data matrix containing k records per individual. The weight associated with the tth record equals $w^*_{it} = \sum_{s=1}^{k} w_{is}d_{st}$. A logistic regression model for T can now be estimated by maximizing the following weighted log-likelihood function:

$$\log L_{BCH} = \sum_{i=1}^{n}\sum_{t=1}^{k} w^*_{it} \log P(T = t|z_i). \tag{12}$$

Vermunt (2010) proposed using a sandwich variance estimator to take into account the weighting and the multiple observations per individual.

Vermunt (2010) proposed another simpler adjusted three-step method. It is based on the observation that Eq. (10) is in fact the equation of a latent class model with a single response variable W and with covariates. Since $P(W = s|T = t)$ is estimated step 2, it can be treated as known in step 3. Because this three-step approach involves maximizing a standard log-likelihood function in step 3, we refer to it as *the ML approach*. More specifically, the parameters for the effects of the covariates on cluster membership can be estimated by maximizing the following log-likelihood function:

$$\log L_{ML} = \sum_{i=1}^{n}\sum_{s=1}^{k} w_{is} \log \sum_{t=1}^{k} P(T = t|z_i)P(W = s|T = t) \tag{13}$$

4 Simulation Study

4.1 Simulation Design

A simulation study was conducted to evaluate the performance of the various approaches for dealing with covariates in mixture models for continuous responses; i.e., the one-step ML, standard three-step, three-step BCH, and three-step ML method. Data sets were generated from a three-class mixture model for six

Table 1 Results averaged over the nine combinations of sample size and separation between components for a γ parameter with a true value of 2

Model	Estimate	SE	SD	MSE
Standard	1.013	0.079	0.096	0.852
ML correction	1.962	0.210	0.223	0.231
One step	2.043	0.187	0.194	0.200
BCH	1.982	0.551	0.390	0.393

continuous response variables and three covariates. The six responses were assumed to come from univariate normal distributions within classes. The residual variance was assumed to be equal across clusters and was used to manipulate the level of separation between clusters. The two independent factors that were manipulated were the separation between components, quantified using an entropy based R^2 measure, with low ($R^2 = 0.43$), medium ($R^2 = 0.66$) and high ($R^2 = 0.86$) separation, and sample size ($n = 500$; $n = 1{,}000$; $n = 10{,}000$). We looked at the averages of the covariate effects across replications, their standard deviations across replications, the averages of the standard errors of the estimates, and the mean square errors of the estimates. The Latent GOLD program Vermunt and Magidson 2005 was used in all stages of the simulation study such as generating data, estimating parameters in the various modeling approaches, getting classifications, and preparing expanded data sets. For this purpose, the program was called in a loop from a batch file.

4.2 Results

Table 1 presents the results averaged over all nine conditions for one of the covariate effect having a true value of 2. It can be seen that the standard three-step approach severely underestimates the parameter of interest, whereas both the ML and BCH correction method reduce the bias considerably. The correction methods drop the percentage of bias from 50 % in the standard three-step approach to less than 2 %, which is similar to the bias of the one-step approach. The mean square error (MSE) of the estimates indicates that the ML correction method is almost as accurate as the one step method, whereas the BCH method is much less stable.

The performances of the various methods across the different simulation conditions were also investigated. The results reveal that except for the small sample size (n = 500) and low separation (entropy = 0.43) combination, the BCH correction is found to have less bias than the ML correction and the one-step method. Consistent with the results of Table 1, the ML correction method is almost as efficient as the one-step method especially for the better separation and larger sample size conditions. In sum, the BCH method substantially reduces bias but is less efficient than the one-step and the ML correction method, while the ML correction provides estimates of a quality similar to the one-step approach for the more favorable conditions (larger sample sizes and higher separation levels).

5 Conclusions

This paper showed how to generalize the correction method for three-step latent class analysis with categorical response variables proposed by Bolck et al. (2004) and Vermunt (2010) to be applicable also in mixture models with continuous response variables. In agreement with theory on Monte Carlo integration, it was proposed to approximate the integral over the population density for the response variables by a summation over the observations in the data set at hand. What is clearly understood from the simulation results is that both the BCH and ML correction method performs quite well, except when the separation between components is extremely low and the sample size is small. This is in agreement with simulation results by Vermunt (2010) for mixture models with categorical responses. The practical advise to applied research is that one should not to use an uncorrected three-step method, but instead use one of the adjusted method. Only with extremely low separation levels combined with small samples, the one-step approach is clearly the best choice.

One issue requires further research, that is, finding an explanation for the instability of the BCH method and its overestimation of the SEs when used under the least favorable conditions. Further extensions of the correction methods would include situations where other categorical or continuous latent variables are used to explain the class membership, or more in general to any situation in which results from a mixture model clustering are used in subsequent analyses. These kinds of extension seems to be more straightforward with the ML method than with the BCH method.

References

Bandeen-Roche, K., Miglioretti, D. L., Zeger, S. L., & Rathouz, P. J. (1997). Latent variable regression for multiple discrete outcomes. *Journal of the American Statistical Association, 92*, 1375–1386.

Bolck, A., Croon, M. A., & Hagenaars, J. A. (2004). Estimating latent structural models with categorical variables: one – step versus three-step estimators. *Political Analysis, 12*, 3–27.

Dayton, C. M., & Macready, G. B. (1988). Concomitant-variable latent class models. *Journal of the American Statistical Association, 83*, 173–178.

McLachlan, G., & Peel, D. (2000). *Finite mixture models*. New York: John Wiley.

Vermunt, J. K.(2010). Latent class modeling with covariates: two improved three-step approaches. *Political Analysis, 18*, 450–469.

Vermunt, J. K., & Magidson, J. (2005). *Latent GOLD 4.0 user's guide*. Belmont: Statistical Innovations Inc.

Efficient Spatial Segmentation of Hyper-spectral 3D Volume Data

Jan Hendrik Kobarg and Theodore Alexandrov

Abstract Segmentation of noisy hyper-spectral imaging data using clustering requires special algorithms. Such algorithms should consider spatial relations between the pixels, since neighbor pixels should usually be clustered into one group. However, in case of large spectral dimension (p), cluster algorithms suffer from the curse of dimensionality and have high memory requirements as well as long run-times.

We propose to embed pixels from a window of $w \times w$ pixels to a feature space of dimension pw^2. The effect of implicit denoising due to the window is controlled by weights depending on the spatial distance. We propose either using Gaussian weights or data-adaptive weights based on the similarity of pixels. Finally, any vectorial clustering algorithm, like k-means, can be applied in this feature space. Then, we use the FastMap algorithm for dimensionality reduction.

The proposed algorithm is evaluated on a large simulated imaging mass spectrometry dataset.

J.H. Kobarg (✉)
Center for Industrial Mathematics, University of Bremen, 28359 Bremen, Germany
e-mail: jhkobarg@math.uni-bremen.de

T. Alexandrov
Center for Industrial Mathematics, University of Bremen, 28359 Bremen, Germany

Steinbeis Innovation Center SCiLS Research, 28211 Bremen, Germany
e-mail: theodore@math.uni-bremen.de

B. Lausen et al. (eds.), *Algorithms from and for Nature and Life*, Studies in Classification, Data Analysis, and Knowledge Organization, DOI 10.1007/978-3-319-00035-0_9,

1 Introduction

Clustering is an excellent tool to divide a dataset into distinct groups. Especially if no prior knowledge of the underlying structure is given, a representative can be found for each group. Clustering of image pixels is a method of image segmentation, where spatial smoothness of the segmentation map is often desired.

Image segmentation is used for mining of *matrix assisted laser desorption/ionisation* (MALDI) *imaging mass spectrometry* (IMS) datasets, to highlight spatial regions of similar chemical composition (e.g. a tumor region). Given a thin flat sample, like a biological tissue slice, IMS measures high-dimensional mass spectra at their spatial points, providing a hyper-spectral image. Each channel in this hyper-spectral image represents numbers of particles of the corresponding mass-to-charge ratio (m/z). With a mass spectrum measured at each pixel, IMS produces large datasets which are considerable difficult to process with most clustering algorithms: the number of spectra (pixels) is around 10^4 to 10^5 (e. g. 200×100 pixels), the length of each spectrum is around 10,000. Figure 1 shows the intensity plot of a typical mass spectrum. For processing such a large dataset, one should use suitable dimensionality reduction and incorporate denoising to suppress the noise. Several strategies have been proposed for the spatial segmentation of an IMS dataset. Popular choices are feature extraction with principal components analysis (PCA) and then either hierarchical clustering (Deininger et al. 2008) or clustering with k-means (McCombie et al. 2005) of features obtained by PCA. However, clustering can be inefficient due to the number of spectra or returns implausible results, because spatial relations are neglected. Recently, Alexandrov et al. (2010) proposed denoising by spatial smoothing of each channel prior to clustering.

We propose a noise-suppressing efficient segmentation based on a spatially aware embedding approach. This merges denoising and dimensionality reduction into one step. We define our embedding function Φ based on a window of $w \times w$ pixels and weights $\{\alpha_{ij}\}$ used in Φ. The embedding function and the weights are used to project n spectra of length p into a high dimensional feature space of dimension pw^2. The points in the feature space can be processed with standard clustering algorithms. Furthermore, we show how to apply the efficient dimensionality reduction algorithm *FastMap* (Faloutsos and Lin 1995) to speed up the procedure and to reduce the memory requirements. We will extend the basic principle of our mapping strategy to make the procedure edge-preserving. The concept of embedding the spatial information in the data was reported earlier (Alexandrov and Kobarg 2011), where we applied it for segmentation of 2D MALDI IMS data. Here, we extend the proposed procedure to 3D and apply it to a simulated 3D MALDI IMS dataset.

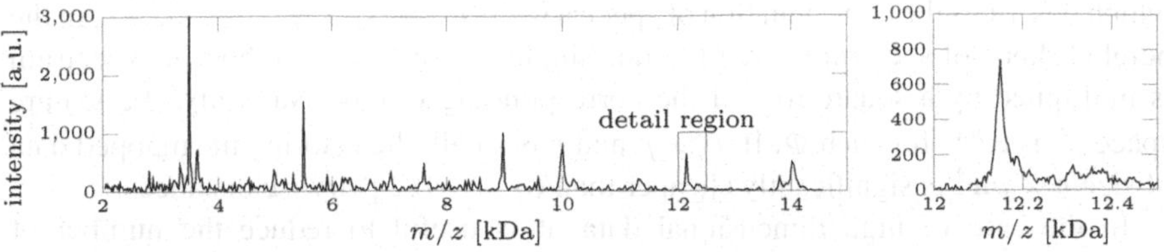

Fig. 1 Measured mass spectrum after compensating baseline effect

2 Methods

Throughout this paper we will denote $s_i = s(x_i, y_i) \in \mathbb{R}^p$ as an intensity vector of a mass spectrum measured at spatial points $(x_i, y_i) \in \mathbb{Z}^2$ for $i = 1, \ldots, n$ spectra. Later we will extend the spatial point to $(x_i, y_i, z_i) \in \mathbb{Z}^3$.

Probably, the most apparent way to embed the spatial relations between pixels into a clustering algorithm is to use a distance-based clustering. With distance-based clustering it is easy to replace the default distance function, such as the Euclidean distance $d(s_1, s_2) = \|s_1 - s_2\|_2$ between the intensity vectors, by one working with a filter window of width $w = 2r + 1$. A distance function, that uses information from neighboring spectra in small window of radius r then looks like

$$d_{r,\{\alpha_{ij}\}}(s_1, s_2)^2 = \sum_{-r \le i,j \le r} \alpha_{ij} \left\| s\left(x_1 + i, y_1 + j\right) - s\left(x_2 + i, y_2 + j\right) \right\|_2^2, \quad (1)$$

where $\{\alpha_{ij}\}$ are factors weighting the influence of pixels from the neighborhood. It is natural to choose weights $\{\alpha_{ij}\}$ which decrease with increasing $i^2 + j^2$. For pixels distant from the neighborhood center the weights will be small. In a neighborhood of radius r, we define the *Gaussian weights* as $\alpha_{ij} = \exp\left((-i^2 - j^2)/(2\sigma^2)\right)$, with $\sigma = (2r + 1)/4$ selected according to the two-sigma rule.

Unfortunately, this approach is both time and memory-consuming for datasets with many spectra, since it requires calculating a distance matrix of size $(n^2 - n)/2$ and storage space for each of those values. Therefore, we propose to map the spectra of length p into a Euclidean feature space $\mathcal{F}$ using a mapping $\Phi\colon \mathbb{R}^p \to \mathcal{F}$. The feature space is selected such that within $\mathcal{F}$ the standard Euclidean distance

$$\|\Phi(s_1) - \Phi(s_2)\|_2 = d_{r,\{\alpha_{ij}\}}(s_1, s_2) \quad (2)$$

equals the desired distance (1). This can be achieved by using the mapping

$$\Phi(s) = \Phi(s(x, y)) = \left[\sqrt{\alpha_{-r,-r}}\, s^{\mathrm{T}}(x - r, y - r), \ldots, \sqrt{\alpha_{0,0}}\, s^{\mathrm{T}}(x, y), \ldots, \sqrt{\alpha_{r,r}}\, s^{\mathrm{T}}(x + r, y + r)\right]^{\mathrm{T}}, \quad (3)$$

which describes the concatenation of spectra $s(x+i, y+j)$, $i, j = -r, \dots, r$, in the neighborhood of spectrum $s(x, y)$ to one single vector. Each neighboring spectrum is multiplied by a square root of the corresponding weight. Naturally, the feature space $\mathcal{F}$ is $\mathbb{R}^{pw^2}$ for such Φ. If $n \gg p$ and r is small, then storing the mapped data of size $n \times pw^2$ is significantly cheaper than $(n^2 - n)/2$ pairwise distances.

In the case of high dimensional data, it is useful to reduce the number of dimensions and preserve some of its properties i.e. the distance between each spectrum. Distance preserving projection can be achieved with multidimensional scaling (MDS, Hastie et al. 2009). However, the original distance matrix D of size $n \times n$ is needed and an eigenvalue decomposition has to be computed. The FastMap algorithm (Faloutsos and Lin 1995) is a related method better suited in this context as the computational cost is much smaller. Compared to MDS, FastMap does not need the full distance matrix D, but only a small subset, so that implementation wise it can work on the dataset itself. This allows FastMap to be much more memory preserving than MDS. Instead of $O(n^2)$ operations for computing the entire distance matrix in advance only $O(3n)$ computations per iteration are needed. The iterative projection of the high-dimensional data into a lower dimensional hyperspace returns pseudo-Euclidean vectors. The distances between these new vectors are similar to the inter-distances originally present in the dataset.

The basic idea of FastMap is to use the two p dimensional spectra s_a and s_b with greatest inter-distance $d_{a,b} = d(s_a, s_b)$ as pivot elements to form a new axis. Computing the scale

$$z_i = \frac{d_{a,i}^2 - d_{b,i}^2 + d_{a,b}^2}{2d_{a,b}}, \quad i = 1, \dots, n, \tag{4}$$

on this new axis exploits only the distances from the two rows $\{d(s_a, s_i)\}$ and $\{d(s_b, s_i)\}$. These rows and the inter-distance $d_{a,b}$ are the only parts of the distance matrix needed in Eq. (4), explaining why the dataset itself is sufficient to preserve the distances. By design of the algorithm, these two rows are even the same ones needed to find the pivot elements s_a and s_b and will be computed on-the-fly.

Before proceeding to a new iteration, the spectra's projections $\tilde{s}_i$ into a $p-1$ dimensional hyperspace $\mathcal{H}$ orthogonal to the s_a-s_b-axis are calculated. In this new iteration pairwise distances $\tilde{d}_{ij} = d(\tilde{s}_i, \tilde{s}_j)$ between the projected spectra $\tilde{s}_i$, $\tilde{s}_j$ in $\mathcal{H}$ will be needed. However, as s_a-s_b-axis is orthogonal to $\mathcal{H}$ Pythagoras' theorem is true and $\tilde{d}_{i,j}^2 = d_{i,j}^2 - (z_j - z_i)^2$ holds. Being dependent only on the scales for each spectrum, this again makes full computation unnecessary. After finishing q iterations of FastMap, the scales z_i^ν, $\nu = 1, \dots, q$, correspond to the new coordinates for all mapped spectra $\tilde{s}_i = (z_i^1, \dots, z_i^q)$, $i = 1, \dots, n$.

Using FastMap's ability to create a vectorial representation of data while preserving specified distances between the objects can be exploited in another setting. Tomasi and Manduchi (1998) proposed *bilateral filtering* with data adaptive weights for edge-preserving greyscale image denoising. Being applied in our context, the weights $\tilde{\alpha}_{ij}(x, y) = \alpha_{ij} \cdot \exp\left(-\frac{1}{2}\lambda^{-2}\|s(x+i, y+j) - s(x, y)\|^2\right)$not only respect the spatial relation, but adjust for differences between spectra, too. The parameter λ controls the strength of smoothing. Due to the usage of the norm, the effect of λ depends on the number of image channels p. Usually p varies from experiment to experiment, which is not the case for grey or RGB images considered in bilateral filtering. Therefore, we propose to choose λ in each neighborhood such that $\min_{i,j} \tilde{\alpha}_{ij} = e^2$, instead of finding a global parameter for each experiment. If the weights $\{\tilde{\alpha}_{ij}\}$ are chosen to become structure adaptive, the weights are no longer constant for two different neighborhoods. However, without constant weights, the mapping function (3) for embedding spatial information directly into the data does not work. Consequently, there is no set of vectors that can be clustered directly. However, FastMap can be used to find vectors which possess pairwise distances (1) with adaptive weights and still allows to employ the concept of clustering in a higher, structure aware dimensional space.

All of the employed algorithms scale linearly to the number of present image pixels. For datasets with three spatial dimensions this is extremely important, as with each layer the total number of pixels increases. Our proposed algorithm (Alexandrov and Kobarg 2011) was designed to embed the spatial information directly into the data. Furthermore, the calculation of the weights and the mapping can be extended to allow the mapping of three dimensional hyper-spectral data. Hence, the segmentation problem is separated from the number of spatial dimensions.

3 Results

The proposed method was already applied to 2D real-life imaging mass spectrometry (IMS) data. Here, we will demonstrate the performance of our method with a simulated 3D IMS dataset. For a detailed description of biological application and IMS measurement, see Alexandrov et al. (2010). Next step will be to solve several technical aspects of real-life 3D IMS data, which we will discuss at the end of this paper.

We simulated a dataset with $n = 45 \times 45 \times 15$ spectra located at three spatial dimensions. The spectra belong to $k = 5$ classes which form simple geometric shapes, see Fig. 2. These objects are background (1), sticks (2), upper rectangle (3), ball (4), pyramid (5a), and lower rectangle (5b), with the last two belonging to the same class. Based on the true class assigned to the voxel, each spectrum was simulated independently of other voxels. For each class a template spectrum with $d = 6{,}972$ channels was selected from a real-life dataset. The dataset used is a rat kidney, which is well known for its simplicity. The selection of the template spectra was based on an initial segmentation of the unsmoothed real-life dataset into five classes. We selected those five spectra as templates that were closest to the computed class means. The abundances at $p = 145$ peak positions found in the template spectra were used within this class. As in the real world, the peak positions vary by small differences in atom weight which add up to mass offset errors. This effect is simulated by using the physical model of particles traversing a flight tube

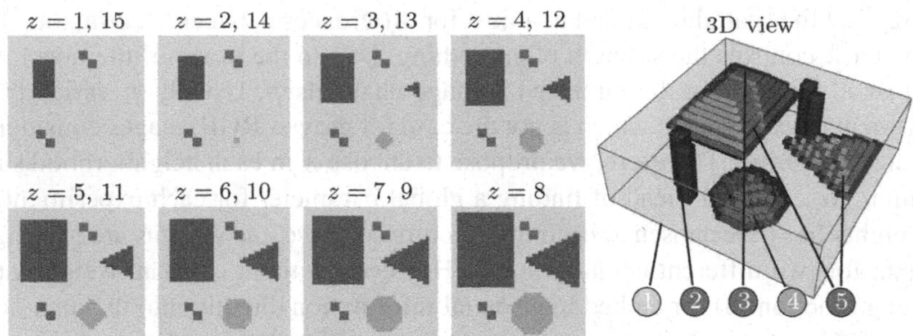

Fig. 2 Simulated dataset, $45 \times 45 \times 15 = 30{,}375$ voxels, 5 classes, $d = 6{,}639$ channels

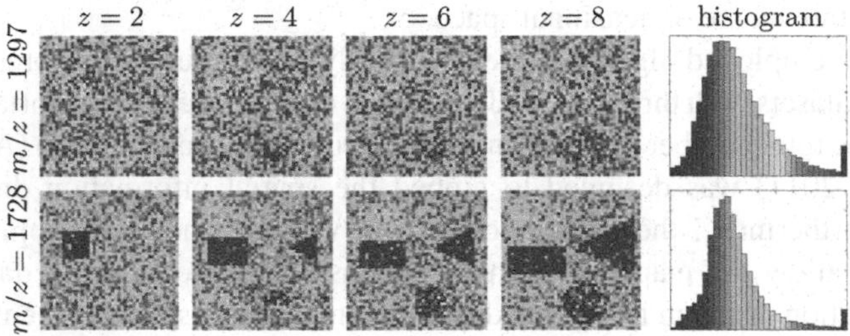

Fig. 3 Two selected channel images, along with histogram of intensities

(Coombes et al. 2005). MALDI spectra have the peaks superimposed to a noise level that forms a baseline. Such a baseline can be approximated by the sum of two exponential functions which in turn are characterized by four parameters in total. For the baselines found in the five template spectra estimation of their parameters is done according to the method described by House et al. (2011). These parameters are taken as estimates of the true parameters for baselines in the dataset. In the last step the function is used as the estimate of the general noise level in the m/z channels of the spectrum. The effect of noise can be seen in Fig. 3, where two channels are displayed.

The simulated data was treated like any raw IMS data in the way that standard preprocessing routines were applied before segmentation was computed. Total ion count normalization was applied to the data—such that each intensity vector has the same area under its curve—followed by baseline estimation and their subtraction (Alexandrov and Kobarg 2011). Furthermore, the number of image channels is reduced to those that contain peaks in the mass spectra (Alexandrov et al. 2010). Standard k-means was then applied directly, with constant weights and adaptive weights, each with a neighborhood of width $w = 5$.

As can be seen in Fig. 4, the objects cannot be discriminated if the dataset is clustered directly and is rather split within groups. This prevents the upper rectangle class to be detected even after increasing the number of clusters. While in the case of $k = 6$ three structures are well separated from background, the background itself is

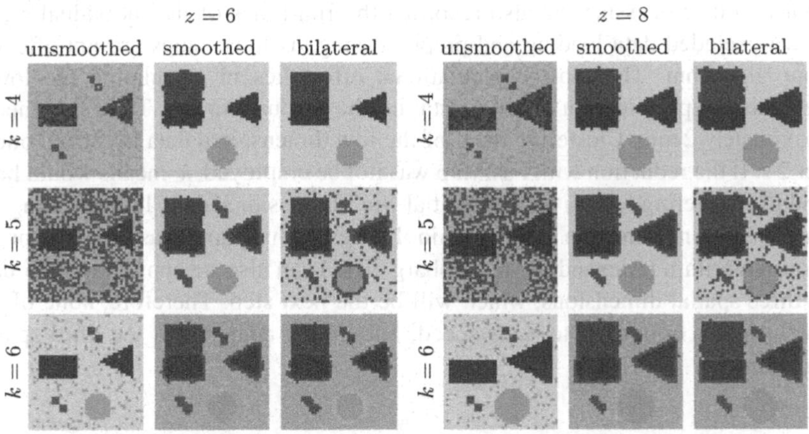

Fig. 4 k-means clustering, without smoothing vs. smoothing ($w = 5$)

Table 1 Balanced accuracy for clustering results against ground truth (in percent)

Classes	Unsmoothed	Smoothed	Bilateral
$k = 4$	72.32	98.72	99.38
$k = 5$	56.71	96.30	89.30
$k = 6$	67.40	75.52	76.24

further divided into three groups, none of them related to the rectangle. The class of stick pixels is lost within the noise for $k = 5$ even though it can be clearly identified with $k = 4$ or $k = 6$.

Once the approach with spatial information embedding is employed, the segmentation map is not affected by noise. In the case of $k = 4$, the isolated stick structures cannot be found by k-means, because the number of classes is too restrictive. As soon as $k \geq 5$, the pixels belonging to this class form their own segment. Also for $k \geq 5$ an artifact starts to appear, namely that the pixels which are located next to two different classes are all put into the same group. Bilateral filtering seems to work perfect in the case of $k = 4$ and $z = 6$, as there appear no misclassifications, however $z = 8$ shows small errors on the edge of both rectangle classes. Furthermore, for $k = 5$ classes the weights are too adaptive to the data and the segmentation map is again affected by noise and an even stronger boundary effect.

As the true classes are known, the comparison of the clustering results is possible with standard classification measures based on confusion matrices (Hastie et al. 2009). As each class has different number of members we use the balanced accuracy $ba = \frac{1}{2}(tp/(tp + fn) + tn/(tn + fp))$, which is the mean of sensitivity and specificity. As can be seen in Table 1, both types of smoothing outperform direct clustering. The visual deficits of bilateral with $k = 5$ are also visible in the score. Even with $k = 6$, i.e. more classes than truly exist, they outperform direct clustering.

During our processing we also recorded the run-times of the individual steps, here we excluded data loading and preprocessing, as those steps are not affected by our algorithm. The most computational effort lies in calculating the lower dimensional representation of the data in the feature space. This step needs approximately 2 min. Clustering itself of the low dimensional data is carried out in under 2 s. If the reduction with FastMap was not be employed, k-means would have to find the clustering of data with the initial pw^3 dimensional data. In this setting the algorithm does not finish in under 15 min. However, this paper was aimed to prove that our algorithm (Alexandrov and Kobarg 2011) can also be applied to IMS data with three spatial dimensions, which will be the next step. Therefore, none of the algorithms was optimized towards speed, but simply adapted for the third spatial dimension.

4 Conclusion

In this paper we showed how to produce smooth 3D segmentation maps for a noisy hyper-spectral image if spatial relations between pixels are exploited. In comparison with a plain spectra clustering, our segmentation maps reveal more spatial features and have higher accuracy with fewer misassignments as compared with the gold standard. We successfully showed that the embedding of spatial information into the dataset does not depend on the number of spatial dimensions. The proposed procedures based on the FastMap dimensionality reduction have linear computational complexity and linear memory requirements (both in the number of spectra) and does not depend on the length of spectra. The use of FastMap allows the integration of spatial and structural information, permitting data adaptive weights. Additionally it avoids the need to store large data matrices and it saves computation time during clustering, both by finding a low dimensional representation of the data in advance.

Even though it has been shown that the segmentation results have high accuracy compared against the underlying ground truth, several technical issues have to be solved before the method can be applied to real-life 3D data, where the gold standard is not given. In the current setting, the distance (1) assumes perfect alignment of all pixels in a grid. This is usually not the case for real life datasets where each slice of tissue has to be converted in a common world coordinate system of the unsliced object. Such an interpolation might cause further, unwanted smoothing to the data. Furthermore, selection of the smoothing level based on the window size becomes a problem. While most parts of our algorithm scale linear in the number of inputs, the window size grows cubic in a three dimensional setting. This implementation issue has to be solved, before application to real-life data is feasible.

Acknowledgements The authors gratefully acknowledge the financial support of the European Union Seventh Framework Programme (project "UNLocX", grant 255931).

References

Alexandrov, T., Becker, M., Deininger, S. O., Ernst, G., Wehder, L., Grasmair, M., von Eggeling, F., Thiele, H., & Maass, P. (2010). Spatial segmentation of imaging mass spectrometry data with edge-preserving image denoising and clustering. *Journal of Proteome Research, 9*(12), 6535–6546.

Alexandrov, T., & Kobarg, J. H. (2011). Efficient spatial segmentation of large imaging mass spectrometry datasets with spatially aware clustering. *Bioinformatics, 27*(13), i230–i238.

Coombes, K. R., Koomen, J. M., Baggerly, K. A., Morris, J. S., & Kobayashi, R. (2005). Understanding the characteristics of mass spectrometry data through the use of simulation. *Cancer Informatics, 1*, 41–52.

Deininger, S. O., Ebert, M. P., Fütterer, A., Gerhard, M., & Röcken, C. (2008). MALDI imaging combined with hierarchical clustering as a new tool for the interpretation of complex human cancers. *Journal of Proteome Research, 7*(12), 5230–5236.

Faloutsos, C., & Lin, K. I. (1995). FastMap: a fast algorithm for indexing, data-mining and visualization of traditional and multimedia datasets. In *Proceedings SIGMOD, international conference on management of data* (pp. 163–174). San Jose: ACM.

Hastie, T., Tibshirani, R., & Friedman, J. (2009). *The elements of statistical learning: data mining, inference, and prediction*. New York: Springer.

House, L. L., Clyde, M. A., & Wolpert, R. L. (2011). Bayesian nonparametric models for peak identification in MALDI-TOF mass spectroscopy. *Annals of Applied Statistics, 5*(2B), 1488–1511.

McCombie, G., Staab, D., Stoeckli, M., & Knochenmuss, R. (2005). Spatial and spectral correlations in MALDI mass spectrometry images by clustering and multivariate analysis. *Analytical Chemistry, 77*(19), 6118–6124.

Tomasi, C., & Manduchi, R. (1998). Bilateral filtering for gray and color images. In *Proceedings of sixth international conference on computer vision* (pp. 839–846). Bombay: IEEE Computer Society.

Cluster Analysis Based on Pre-specified Multiple Layer Structure

Akinori Okada and Satoru Yokoyama

Abstract Cluster analysis can be divided into two categories; hierarchical and non-hierarchical cluster analyses. In the present study, a method of cluster analysis which does not utilize hierarchical nor non-hierarchical procedures is introduced. The present cluster analysis pre-specifies a structure having multiple layers, e.g., the species, the genus, the family, and the order. The highest layer or layer 0 has one cluster which all objects belong to. The cluster at layer 0 has the pre-specified number of clusters at the next lower layer or layer 1. Each cluster at layer 1 has the pre-specified number of clusters at the next lower layer or layer 2, and so on. The cluster analysis classifies the object into one of the clusters at all layers simultaneously. While the cluster structure is hierarchical, the procedure is not hierarchical which is different from that of the agglomerative or divisive algorithms of the hierarchical cluster analysis. The algorithm tries to optimize the fitness measure at all layers simultaneously. The cluster analysis is applied to the data on whisky molts.

1 Introduction

There are two categories of cluster analysis; one is the hierarchical cluster analysis and the other is the non-hierarchical cluster analysis. In the hierarchical cluster analysis, two clusters are agglomerated into one cluster at each stage in the case

---

A. Okada (✉)
Graduate School of Management and Information Sciences, Tama University, Tokyo, Japan
e-mail: okada@rikkyo.ac.jp

S. Yokoyama
Faculty of Economics, Department of Business Administration, Teikyo University, Utsunomiya, Japan
e-mail: satoru@main.teikyo-u.ac.jp

B. Lausen et al. (eds.), *Algorithms from and for Nature and Life*, Studies in Classification, Data Analysis, and Knowledge Organization, DOI 10.1007/978-3-319-00035-0_10,

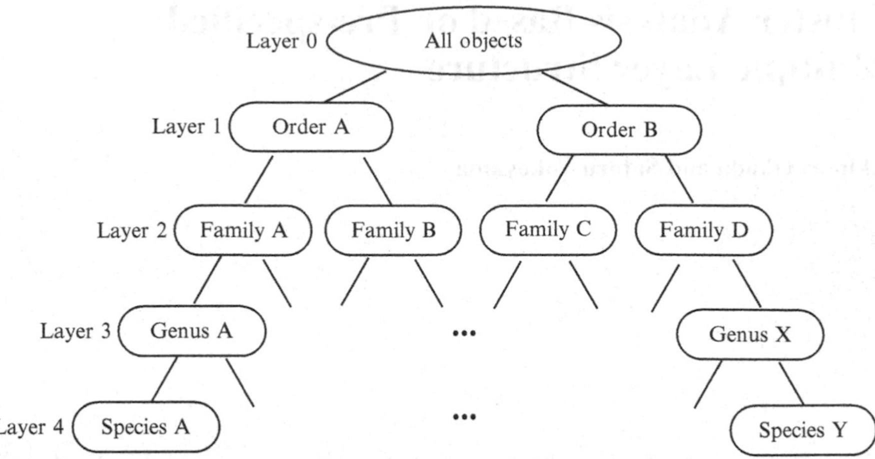

Fig. 1 Cluster structure having multiple layers

of the agglomerative procedure, or one cluster is divided into two clusters at each stage in the case of the divisive procedure. At each stage, the object is classified into one of the clusters. In the non-hierarchical cluster analysis or the partitioning, the object is classified into one of the clusters of pre-specified number. The algorithms of both the hierarchical and the non-hierarchical cluster analyses allocate the object to one of the clusters at each stage or at the pre-specified number of clusters. The algorithm does not deal with more than one stage nor more than one pre-specified number of clusters simultaneously.

The cluster analysis introduced in the present study pre-specifies a cluster structure having layers (Okada and Yokoyama 2010). The highest layer or layer 0 has one cluster which all objects belong to. The cluster at layer 0 consists of clusters at the next lower layer or layer 1. The cluster at layer 1 consists of clusters at the next lower layer or layer 2, $\cdots$, the cluster at layer n consists of clusters at layer $(n + 1)$, and so on. The cluster structure having multiple layers is similar to the structure comprises the species, the genus, the family, and the order (Gordon 1999; Mirkin 1996) as shown in Fig. 1.

While the cluster structure is hierarchical, the structure is not same as that derived by the hierarchical cluster analysis (cf. Arabie and Hubert 1994) and its algorithm is not hierarchical. The object is allocated to one of the clusters at each of all layers simultaneously. This means that the object is classified into the hierarchical structure consists of multiple layers, but the algorithm itself is not hierarchical as that of the hierarchical cluster analysis. The purpose of the present study is to introduce a cluster analysis method pre-specifying the cluster structure which has multiple layers, and to apply the cluster analysis to the data on whisky molts.

2 The Algorithm

The present cluster analysis deals with two-mode two-way data; such as object × variable or subject × attribute data. The algorithm to fit the cluster structure to two-mode two-way data is described below. The objective of the algorithm is to allocate the object to one of the clusters at each layer of the cluster structure specified in advance, where the sum of squared deviations within the cluster at all layers is minimized, i.e. the sum of the within cluster sum of squared deviations from the centroid of the cluster at all layers is minimized. Let the sum of the within cluster sum of squared deviations from the centroid of the cluster at all layers be *SSW*;

$$SSW = \sum_{n=1}^{N} SSWn, \tag{1}$$

where *SSWn* is the sum of the within cluster sum of squared deviations from the centroid of the cluster at layer n, and N is the number of layers. *SSWn* is defined by

$$SSWn = \sum_{m=1}^{Mn} SSWmn, \tag{2}$$

where *SSWmn* is the sum of the within cluster sum of squared deviations from the centroid of cluster m at layer n, and *Mn* is the number of clusters at layer n. *SSWmn* is defined by

$$SSWmn = \sum_{j=1}^{Jmn} \sum_{t=1}^{P} (x_{jt} - \bar{x}_{(mn)t})^2, \tag{3}$$

where *Jmn* is the number of objects in cluster m at layer n, x_{jt} is the value of object j along variable t, P is the number of variables, and $\bar{x}_{(mn)t}$ is the mean value of objects in cluster m at layer n along variable t.

The algorithm to minimize *SSW* of Eq. (1) is iterative, which comprises nine steps. The algorithm below is described when the cluster structure has three layers; layers 0, 1 and 2, which is the most simple case of the multiple layer cluster structure (e.g., Fig. 2).

- Step 1: Determine the initial center of each cluster at layer 1 by randomly selecting an object.
- Step 2: Classify the object into the cluster at layer 1 whose center is nearest to the object.
- Step 3: In each cluster at layer 1, determine the center of each cluster at layer 2 by randomly selecting an object in each cluster.
- Step 4: Classify the object into the cluster at layer 2 whose center (centroid) is nearest to the object. In this step, the object can be classified into the cluster at

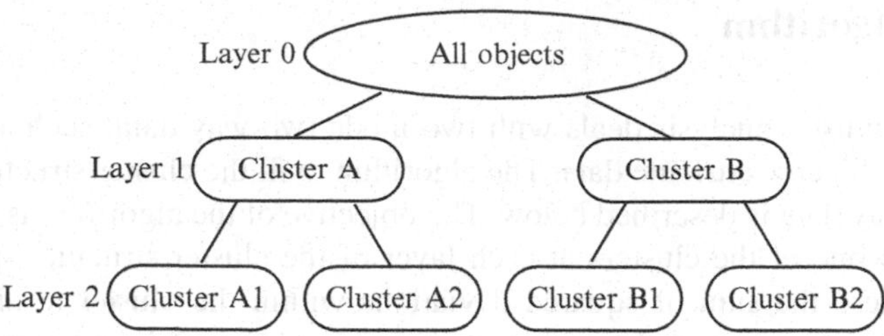

Fig. 2 Cluster structure of the present study

layer 2 which belongs to the different cluster at layer 1 from the cluster the object originally belonged to at layer 1.

- Step 5: Determine the centroid of objects in each cluster at layer 1.
- Step 6: Classify the object into the cluster at layer 1 whose centroid is nearest to the object.
- Step 7: In each cluster at layer 1, determine the centroid of objects in each cluster at layer 2.
- Step 8: Return to Step 4, and iterate Steps 4 through 7.
- Step 9: Stop the iteration of Steps 4 through 7 when the allocation of the object to the cluster at all layers stabilized.

3 The Application

In this section, the present cluster analysis is applied to the data on whisky molts shown in Wishart (2002).

3.1 The Data

The data consists of 35 whisky molts measured on 12 features describing the whisky molt (body, sweetness, $\cdots$, and, floral) on a rating scale. The 35 whisky molts consists of four types (types D, E, F, and G). They consist of 10 whisky molts of type D, 9 whisky molts of type E, 10 whisky molts of type F, and 6 whisky molts of type G. The data are two-mode two-way (whisky molt $\times$ feature), and are arrayed in a 35×12 table. The 35 whisky molts are represented in the first and the seventh columns, and their types are represented in the second and the eighth columns of Table 1.

Table 1 The results of the analysis

Whisky molt	Type	Layer 1	Layer 2	$k = 2$	$k = 4$	Whisky molt	Type	Layer 1	Layer 2	$k = 2$	$k = 4$
An Cnoc	D	A	A2	2	1	Tomintoul	E	A	A1	1	3
Auchentoshan	D	A	A2	2	1	Ardmore	F	B	B2	1	2
Aultmore	D	A	A2	2	1	Auchroisk	F	B	B2	1	2
Cardhu	D	A	A2	2	1	Danston	F	B	B2	1	2
Glengoyne	D	A	A2	2	1	Glen Deveron	F	B	B2	1	2
Glen Grant	D	A	A2	2	1	Glen Keith	F	B	B2	1	2
Mannochmore	D	A	A2	2	1	Glenrothes	F	B	B2	1	2
Speyside	D	A	A2	2	1	Old Fettercairn	F	B	B2	1	2
Tamdhu	D	A	A2	2	1	Tomatin	F	B	B2	1	2
Tobermory	D	A	A2	2	1	Tormore	F	B	B2	1	2
Bladnoch	E	A	A1	2	3	Tullibadine	F	B	B2	1	2
Bunnahabhain	E	A	A1	2	3	Isle of Arran	G	B	B1	1	4
Glenallachie	E	A	A1	2	3	Dufftown	G	B	B1	2	4
Glenkinchie	E	A	A1	2	3	Glenfiddich	G	A	B1	2	4
Glenlossie	E	A	A1	2	3	Glen Spey	G	B	B1	1	4
Glen Moray	E	A	A1	2	3	Miltonduff	G	B	B1	1	4
Inchgower	E	A	A1	1	3	Speyburn	G	B	B1	1	4
Loch Lomond	E	A	A1	2	3						

Table 2 The correspondence between clusters at layer 2 of the present cluster analysis and the type of the whisky molt

	Type			
Layer 2	D	E	F	G
A1	0	9	0	0
A2	10	0	0	0
B1	0	0	0	6
B2	0	0	10	0

3.2 The Analysis and the Result

The analysis was done by pre-specifying the model which have one cluster at layer 0, two clusters at layer 1 (clusters A and B), and four clusters at layer 2 where each cluster at layer 1 consists of two clusters at layer 2 (cluster A at layer 1 consists of clusters A1 and A2 at layer 2, and cluster B at layer 1 consists of clusters B1 and B2 at layer 2). The pre-specified cluster structure is shown in Fig. 2.

The data were analyzed without any normalization. The resulting clusters at layers 1 and 2 are shown in Table 1. The third and the ninth columns show the cluster at layer 1, and the fourth and the 10th columns show the cluster at layer 2. Table 1 also shows the result of the analysis by k-means cluster analysis (MacQueen 1967) at the 5th, the 6th, the 11th and the 12th columns which will be discussed later.

From Table 2 we can see that there is a perfect correspondence between four clusters at layer 2 and the type of the whisky molt. At layer 2, cluster A1 corresponds to type E, cluster A2 corresponds to type D, cluster B1 corresponds to type G, and cluster B2 corresponds to type F. At layer 1 the correspondence is almost perfect; cluster A corresponds to types D and E, and cluster B corresponds to types F and G. The hierarchical relationships between clusters (A and B) at layer 1 and clusters (A1, A2, B1, and B2) at layer 2 are coincide with the relationships among these types suggested in Wishart (2002).

4 Discussion

A model of cluster analysis pre-specifying a cluster structure which has multiple layers and an associated algorithm to fit the model were introduced. While the algorithm of the cluster analysis is not hierarchical, the resulting tree diagram of clusters is hierarchical. The application of the model to the data on whisky molts was done using the model which has two clusters at layer 1, where each of them consists of two clusters at layer 2.

The present cluster analysis can be regarded as one sort of the constrained classification (Everitt et al. 2011; Gordon 1996; Murtagh 1985), where the constraint is given not on the contiguity among objects but on the tree diagram. The constraint on the tree diagram seems different from those suggested in the past studies, but assumes a priori hierarchical structure (Arabie and Hubert 1994). The present cluster analysis also can be regarded as an equivalence of the external analysis of multidimensional scaling (Borg and Groenen 2005; Coombs 1964) in cluster analysis, where the externally given cluster structure is utilized in allocating the object to the cluster. In the case of multidimensional scaling, the external analysis is done to facilitate the interpretation of the resulting configuration, for example by imbedding the ideal point or the ideal vector, or to derive a configuration of sources based on an externally given configuration of objects (the external analysis of INDSCAL by Carroll and Chang 1970). In the present cluster analysis, the pre-specified cluster structure can make it easy to interpret the resulting tree diagram.

The advantageous aspect of the present cluster analysis is that the cluster structure having multiple layers seems easy to interpret, because layer 1 provides the classification of objects at the most coarse classification, and the next lower layer provides the next less coarse or finer classification of objects to the cluster at layer 2 which is formed by dividing the cluster at layer 1, and so on. One possible usage of it is to find the primary concern, e.g., brand primary or form primary, of a market (Arabie and Hubert 1994) by assuming specific cluster structures and by comparing the goodness of fit of the results.

The algorithm described as steps 1 through 9 in Sect. 2 is similar to k-means algorithm or the simple hill-climbing algorithm of Ball and Hall (1967), where objects are simultaneously relocated but not singly. But it does not comprise k-mean algorithms (one k-means algorithm at each layer except layer 0). The algorithm deals with multiple layers which causes some discrepancies from k-means algorithm. At lower layers (not layer 0), objects are relocated across all clusters including those belong to different clusters at the upper layer. The simulated annealing algorithm might be useful to improve the algorithm (cf. Everitt et al. 2011, p. 123; Gordon 1999, p. 43). While the algorithm determined the initial center randomly, in the case of examining brand primacy of a market, it is not difficult to select the initial center based on the supposed structure reflecting brand primacy of a market.

The data were analyzed by k-means cluster analysis (MacQueen 1967) for $k = 2$ and $k = 4$ (two and four clusters). The resulting classification when the number of clusters is two ($k = 2$) is shown in the 5th and the 11th columns of Table 1. That when the number of clusters is four ($k = 4$) is shown in the 6th and the 12th columns of Table 1. Table 1 tells that the classification at layer 2 given by the present cluster analysis and that given by k-means cluster analysis when $k = 4$ completely coincide.

Table 1 tells that cluster 1 of k-means cluster analysis when $k = 4$ corresponds with cluster A2 and with type D, cluster 2 corresponds with cluster B2 and with type F, cluster 3 corresponds with cluster A1 and with type E, and cluster 4 corresponds with cluster B1 and with type G. Cluster 1 derived by the k-means cluster analysis when $k = 2$ corresponds to types F and G (four anomalies). Cluster 2 corresponds to types D an E (four anomalies). While layers 1 (clusters A and B) and 2 (clusters A1, A2, B1, and B2) derived by the present cluster analysis are inherently hierarchical, clusters 1 and 2 derived by the k-means cluster analysis when $k = 2$ and clusters 1,

Table 3 The correspondence between clusters when $k = 2$ and $k = 4$ of the k-means cluster analysis

	$k = 4$			
$k = 2$	1	2	3	4
1	0	10	2	4
2	10	0	7	2

2, 3, and 4 derived by the k-means cluster analysis when $k = 4$ are not hierarchical as shown in Table 3.

The sum of the within cluster sum of squared deviations from the centroid of the cluster at layer 2 $SSW_2 = 104.3$ (22.9(cluster A1) + 25.0(A2) + 17.7(B1) + 38.7(B2)). The sum of the within cluster sum of squared deviations from the centroid of the cluster derived by the k-means cluster analysis when $k = 4$ was also 104.3, because the layer 2 of the present cluster analysis and the k-means cluster analysis when $k = 4$ resulted in exactly the same clusters. The sum of the within cluster sum of squared deviations from the centroid of the cluster at layer 1 $SSW_1 = 135.7$ (67.6(cluster A) + 68.1(B)). The sum of the within cluster sum of squared deviations from the centroid of the cluster derived by the k-means cluster analysis when $k = 2$ was 132.3 (71.6(cluster1) + 60.7(2)), which is smaller than SSW_1 given by the present cluster analysis. This is natural, because two clusters given by the k-means cluster analysis when $k = 2$ was obtained so that the sum of the within cluster sum of squared deviations from the centroid of the cluster for the two clusters was minimized, while the present cluster analysis derived the hierarchical structure where $SSW_1 + SSW_2$ was minimized. The hierarchical structure, which the latter has to subject, deteriorated the goodness of fit but makes the interpretation of the resulting tree diagram easier.

The algorithm was executed only on the simplest cluster structure having three layers. While in principle the algorithm can deal with the cluster structure having more than three layers, whether the algorithm is effective in dealing with the structure having more than three layers has to be examined. The algorithm is valid only for the cluster structure where all clusters at a layer consist of the same number of clusters at the next lower layer. This is the reason of using the model shown in Fig. 2 in analyzing the data. This limitation restricts the application of the present cluster analysis. To relax the limitation so that clusters at a layer consist of different number of clusters at the next lower layer is desirable, which will increase the versatility of the present cluster analysis.

References

Arabie, P., & Hubert, L. (1994). Cluster analysis in marketing research. In R. P. Bgozzi (Ed.), *Advanced methods of marketing research* (pp. 160–189). Cambridge: Blackwell.

Ball, G. H., & Hall, D. J. (1967). A clustering technique for summarizing multivariate data. *Behavioral Science, 12*, 153–155.

Borg, I., & Groenen, P. J. F. (2005). *Modern multidimensional scaling* (2nd ed.). New York: Springer.

Carroll, J. D., & Chang, J. J. (1970). Analysis of individual differences in multidimensional scaling via an N-way generalization of "Eckart-Young" decomposition. *Psychometrika, 35*, 283–319.

Coombs, C. H. (1964). *A theory of data*. New York: Wiley.

Everitt, B. S., Landau, S., Leese, M., & Stahl, D. (2011). *Cluster analysis* (5th ed.). West Sussex: Wiley.

Gordon, A. D. (1996). A survey of constrained classification. *Computational Statistics & Data Analysis, 21*, 17–29.

Gordon, A. D. (1999). *Classification* (2nd ed.). Boca Raton: Chapman & Hall/CRC.

MacQueen, J. B. (1967). Some methods for classification and analysis of multivariate observations. In *Proceeding of the 5th Berkeley symposium on mathematical statistics and probability*, Berkeley (Vol. 2, PP. 281–297).

Mirkin, B. (1996). *Mathematical classification and clustering*. Dordrecht: Kluwer.

Murtagh, F. (1985). A survey of algorithms for contiguity-constrained clustering and related problems. *The Computer Journal, 28*, 82–88.

Okada, A., & Yokoyama, S. (2010). Multi-layer cluster analysis. In *Proceeding of the 28th meeting of the Japanese classification society*, Tokyo, Japan (pp. 11–12).

Wishart, D. (2002). *Whisky classified: Choosing single molts by flavours*. London: Pavillion.

Factor PD-Clustering

Cristina Tortora, Mireille Gettler Summa, and Francesco Palumbo

Abstract Probabilistic Distance (PD) Clustering is a non parametric probabilistic method to find homogeneous groups in multivariate datasets with J variables and n units. PD Clustering runs on an iterative algorithm and looks for a set of K group centers, maximising the empirical probabilities of belonging to a cluster of the n statistical units. As J becomes large the solution tends to become unstable. This paper extends the PD-Clustering to the context of Factorial clustering methods and shows that Tucker3 decomposition is a consistent transformation to project original data in a subspace defined according to the same PD-Clustering criterion. The method consists of a two step iterative procedure: a linear transformation of the initial data and PD-clustering on the transformed data. The integration of the PD Clustering and the Tucker3 factorial step makes the clustering more stable and lets us consider datasets with large J and let us use it in case of clusters not having elliptical form.

1 Introduction

Organising data into homogeneous groups is one of the most fundamental tasks in many research domains. To cope with many different analysis conditions, several clustering approaches and thousands of clustering algorithms have been proposed

C. Tortora (✉)
Università degli Studi di Napoli Federico II, Naples, Italy

CEREMADE, CNRS, Université Paris Dauphine, Paris, France
e-mail: cristina.tortora@unina.it

F. Palumbo
Università degli Studi di Napoli Federico II, Naples, Italy
e-mail: francesco.palumbo@unina.it

M. Gettler Summa
CEREMADE, CNRS, Université Paris Dauphine, Paris, France
e-mail: summa@ceremade.dauphine.fr

B. Lausen et al. (eds.), *Algorithms from and for Nature and Life*, Studies in Classification, Data Analysis, and Knowledge Organization, DOI 10.1007/978-3-319-00035-0_11,

in literature. Cluster analysis is commonly defined as a multivariate problem where the aim is to identify homogeneous groups of units. In a geometrical view, given a set of n statistical units represented as points in the multidimensional space spanned by J variables, clustering aims at finding dense regions in the original feature space or embedded in a properly defined subspace. We do not consider hierarchical algorithms in this paper, but only probabilistic and geometric non hierarchical strategies. Probabilistic approaches assume that data are generated from a mixture distribution, where each cluster is generated by one (or more) mixture component. Two main types of methods have been developed in this context, parametric approaches and non parametric ones. On the other hand, in the geometric framework, dense regions are defined on the basis of proper distance measures; the most known algorithm is the k-means (Jain 2010).

In the last two decades, the increased data storage capacity has made even more large datasets available, where the number of units and the number of variables is very large. In some web mining applications, for example, there are hundreds of thousands of variables .

The clustering in high dimensional spaces issue is drawing many researchers' attention.

A larger number of variables does not necessarily ensure better results and can mask the existing clusters structure. Several different preprocessing approaches have been proposed to reduce the dimensionality problem before performing a cluster analysis. They can be divided into two main strategies: the first one consists in selecting the most relevant attributes by removing the redundant variables from the analysis, the other one linearly combines the variables into a reduced number of latent variables according to the task of the overall analysis. See Parsons et al. for more details (Parsons et al. 2004). However, strategies combining the dimensionality reduction and the clustering into one consistent iterative algorithm have recently been proposed in the literature, both in the probabilistic case (Montanari and Viroli 2011) and in the geometric framework (Vichi and Kiers 2001).

In the probabilistic and non parametric framework, this paper proposes an integrated clustering approach that combines the Probabilistic Distance (PD) clustering algorithm of Ben-Israel And Iyigun (2008) and the Tucker3 dimensionality reduction (Kroonenberg 2008) to consistently combine the clustering and the dimensionality reduction into one iterative procedure.

The paper consists of the following sections. Section 2 provides a description of Probabilistic Distance Clustering. In Sect. 3 Factor Probabilistic Distance Clustering is presented and developed. Section 4 contains an application on a real dataset.

2 Probabilistic Distance Clustering

Given a set of n statistical units described by J continuous variables, PD-clustering is a non hierarchical clustering algorithm that assigns the n units to K clusters according to the probability of them belonging to the cluster.

We introduce PD-clustering (Ben-Israel and Iyigun 2008) according to Ben-Israel and Iyigun notation. Given X a generic data matrix with n units and J variables, given K clusters that are assumed not empty, PD-Clustering is based on two quantities: the distance of each data point x_i from the K cluster centers c_k, $d(x_i, c_k)$, and the probabilities for each point to belong to a cluster, $p(x_i, c_k)$ with $k = 1, \ldots, K$ and $i = 1, \ldots, n$. The relation between them is the basic assumption of the method, the probability of any point to belong to each class is assumed to be inversely proportional to the distance from the centers of the clusters. Let us consider the general term x_{ij} of X and a center matrix C, of elements c_{kj} with $k = 1, \ldots, K$, $i = 1, \ldots, n$ and $j = 1, \ldots, J$, the distances between each point and all centers can be computed according to different criteria; the squared norm is one of the most commonly used. The quantity $d(x_i, c_k)$ represents the distance of the generic point i to the generic center k. The probability $p(x_i, c_k)$ of each point to belong to a cluster is computed according to the following assumption: for any k, given i the product between the distance $d(x_i, c_k)$ and the probability $p(x_i, c_k)$ is a constant $F(x_i)$ depending on x_i. For short we use $p_{ik} = p(x_i, c_k)$ and $d_k(x_i) = d(x_i, c_k)$. PD-clustering basic assumption is:

$$p_{ik} d_k(x_i) = F(x_i). \tag{1}$$

Looking at the (1) we notice that as the distance of the point from the cluster center decreases, the belonging probability of the point to the cluster increases. The constant depends only on the point and does not depend on the cluster k.

The quantity $F(x_i)$, also called *Joint Distance Function* (JDF), is a measure of the closeness of x_i from all cluster centers. It measures the classificability of the point x_i with respect to the centers c_k, with $k = 1, \ldots, K$. The point coincides with one of the cluster centers if it is equal to zero; in that case the point belongs to the class with probability 1. If all the distances between the point x_i and the centers of the classes are equal to d_i, $F(x_i) = d_i / k$ and all the belonging probabilities to each class are equal: $p_{ik} = 1/K$. The smaller the JDF value, the higher the probability for the point to belong to one cluster. The whole clustering problem consists in the identification of the centers that minimise the JDF. The function in (1) is nonsmooth, a smoothed version of it is: $p_{ik}^2 d_k(x_i) = F(x_i)$ (Iyigun 2007). Without loss of generality the PD-Clustering optimality criterion can be demonstrated according to $K = 2$.

PD-Clustering aims at finding cluster centers such that:

$$\min \left(d_1(x_i) p_{i1}^2 + d_2(x_i) p_{i2}^2 \right) \qquad \text{s.t.}: p_{i1} + p_{i2} = 1; \quad p_{i1}, p_{i2} \geq 0. \tag{2}$$

The Lagrangian of this problem is:

$$\mathscr{L}(p_{i1}, p_{i2}, \lambda) = d_1(x_i) p_{i1}^2 + d_2(x_i) p_{i2}^2 - \lambda(p_{i1} + p_{i2} - 1). \tag{3}$$

Setting to zero the partial derivates with respect to p_{i1} and p_{i2} and considering the principle $p_{i1}d_1(x_i) = p_{i2}d_2(x_i)$ Ben-Israel and Iyigun obtain the optimal value of the Lagrangian (See Ben-Israel and Iyigun 2008 for the proof.):

$$\mathcal{L}(p_{i1}, p_{i2}, \lambda) = \frac{d_1(x_i)d_2(x_i)}{d_1(x_i) + d_2(x_i)}. \tag{4}$$

This value coincides with the JDF. The matrix of centers that minimises this principle minimises the JDF too. Substituting the generic value $d_k(x_i)$ with $\|x_i - c_k\|$, we can find the equations of the centers that minimise the JDF (and maximize the probability of each point to belong to only one cluster):

$$c_k = \sum_{i=1,\ldots,N} \left(\frac{u_k(x_i)}{\sum_{j=1,\ldots,N} u_k(x_j)} \right) x_i, \tag{5}$$

where $u_k(x_i) = \frac{p_{ik}^2}{d_k(x_i)}$.

As shown before, the value of JDF at any center k is equal to zero and it is necessarily positive elsewhere. So centers are the global minimiser of the JDF. Other stationary points may exist because the function is not convex or even quasi-convex, but they are saddle points.

For the sake of brevity we don't go into distance choice details; in this paper we consider the squared form:

$$d_k(x_i) = \sum_{j=1}^{J} (x_{ij} - c_{kj})^2, \tag{6}$$

where $k = 1, \ldots, K$ and $i = 1, \ldots, N$. Starting from the (6) the distance matrix D of order $n \times K$ is defined, where the general element is $d_k(x_i)$. Indicating with c_k the generic center, the final solution $\widehat{JDF}$ is obtained minimising the quantity:

$$\text{JDF} = \sum_{i=1}^{n} \sum_{k=1}^{K} d_k(x_i) p_{ik}^2 = \sum_{i=1}^{n} \sum_{j=1}^{J} \sum_{k=1}^{K} (x_{ij} - c_{kj})^2 p_{ik}^2,$$

$$\widehat{\text{JDF}} = \arg\min_{C,P} \sum_{i=1}^{n} \sum_{j=1}^{J} \sum_{k=1}^{K} (x_{ij} - c_{kj})^2 p_{ik}^2. \tag{7}$$

An iterative algorithm allows one to compute the solution of PD-clustering problem. The algorithm properties are illustrated in Iyigun (2007), where the Author demonstrates the convergences, too. Each unit is then assigned to the kth cluster according to the highest probability that is computed a posteriori.

3 Factor PD-Clustering

PD-clustering is stable dealing with a large number of units but it becomes unstable as the number of variables increases. A linear transformation of original variables into a reduced number of orthogonal ones can significantly improve the algorithm performance. The linear transformation of variables and the PD-Clustering need to optimize a common criterion.

The FPDC consists in an integrated procedure based on the Tucker3 decomposition (Kroonenberg 2008) and PD-Clustering; an algorithm is then proposed to perform the method.

The minimization problem in (7) corresponds to the Tucker3 decomposition of the 3-way distance matrix G of general element $g_{ijk} = |x_{ij} - c_{kj}|$, where $i = 1, \ldots, n$ indicates the units, $j = 1, \ldots, J$ the variables and $k = 1, \ldots, K$ the occasions. For any c_k a $n \times J$ a G_k distance matrix is defined. In matrix notation:

$$G_k = X - hc_k, \tag{8}$$

where h is an $n \times 1$ column vector with all terms equal to 1; X and c_k $(k = 1, \ldots, K)$ has already been defined in Sect. 2.

The Tucker3 method decomposes the matrix G in three components, one for each mode, in a full core array Λ, and in an error term E:

$$g_{ijk} = \sum_{r=1}^{R} \sum_{q=1}^{Q} \sum_{s=1}^{S} \lambda_{rqs}(u_{ir} b_{jq} v_{ks}) + e_{ijk}, \tag{9}$$

where λ_{rqs} and e_{ijk} are respectively the general term of the three way matrix Λ of order $R \times S \times Q$ and E of order $n \times J \times K$; u_{ir}, b_{jq} and v_{ks} are respectively the general term of the matrix U of order $n \times R$, B of order $J \times Q$ and V of order $K \times S$, with $i = 1, \ldots, n$, $j = 1, \ldots, J$, $k = 1, \ldots, K$.

As in all factorial methods, factorial axes in the Tucker3 model are sorted according to their explained variability. The first factorial axes explain the greatest part of the variability; the latest factors represent the ground noise. According to Kiers and Kinderen (2003), the choice of the parameters R, Q and S is a ticklish problem as they define the overall explained variability. Interested readers are referred to Kroonenberg (2008) for the theoretical aspects concerning that choice. We propose an heuristic approach based on the eigenvalues scree plot to cope with this crucial issue.

The coordinates x^*_{iq} of the generic unit x_i into the space of variables obtained through a Tucker3 decomposition are obtained by the following expression:

$$x^*_{iq} = \sum_{j=1}^{J} x_{ij} b_{jq}. \tag{10}$$

Finally, on the x^*_{iq} coordinates a PD-Clustering is applied in order to solve the clustering problem (2). Let us start considering the expression (7); it is worth noting that minimising the quantity:

$$\text{JDF} = \sum_{i=1}^{n} \sum_{j=1}^{J} \sum_{k=1}^{K} (x_{ij} - c_{kj})^2 p_{ik}^2 \qquad \text{s.t.} \quad \sum_{i=1}^{n} \sum_{k=1}^{K} p_{ik}^2 \leq n, \quad (11)$$

is equivalent to computing the maximum of $-\sum_{i=1}^{n} \sum_{j=1}^{J} \sum_{k=1}^{K} (x_{ij} - c_{kj})^2 p_{ik}^2$, under the same constraints.

In Tortora et al. (2011) it is demonstrated that, given p_{ik} and c_{ik}, b_{jq} is obtained according to a Tucker3 transformation of the distance matrix G minimising the JDF.

An iterative algorithm alternatively calculates c_{kj} and p_{ik} on one hand, and b_{jq} on the other hand, until the convergence is reached. It can be summarized in the following steps: (i) random initialization of center matrix C; (ii) computation of distance matrix G; (iii) Tucker3 decomposition of $G = U\Lambda(V' \otimes B')$; (iv) projection of data point in the new space $X^* = XB$; (v) PD-clustering of X^* uploading C. Steps ii–v are iterated until the convergence is reached. The minimisation of the quantity in the formula (11) converges at least to local minima, it can be empirically demonstrated.

A simulation study (Tortora 2011) has demonstrated that difference of variance among clusters does not affect the algorithm efficiency. Classic PD-clustering becomes less efficient when the number of elements in each cluster is different, FPDC results are not affected by this issue. The method performs well in presence of outliers, it can detect the right clustering structure in presence of 20 % of outliers. Using FPDC, unit weights are inversely proportional to the distance from the cluster center, thanks to this characteristic, outliers have a low weight in the determination of the centers. The Tucker3 method looks for the decomposition that divides clusters better, according to the partition obtained in the PD-clustering step. Therefore FPDC, that is based on the two methods, is not affected by outliers.

4 Application on a Real Dataset

The method has been applied to Water Treatment Plant dataset.[1] This dataset comes from the daily measures of sensors in a urban waste water treatment plant. The objective is to classify the operational state of the plant in order to predict faults through the state variables of the plant at each of the stages of the treatment process. It is composed of 527 units and 38 variables. The number of clusters K has been chosen equal to 4. To appreciate the FPDC, a comparative study graphically compares the following methods: k-means, PD-clustering and Factorial k-means. Results are shown in Fig. 1. Each method has been iterated 100 times. Applying k-means at each iteration the value of the within variance has been measured. Results obtained show that there are two minima, the first is obtained in 39 % of the

[1] http://archive.ics.uci.edu/ml/index.html

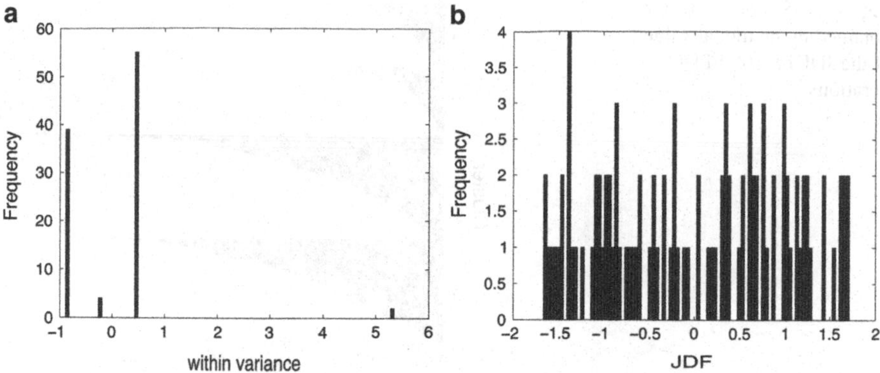

Fig. 1 (**a**) Value of standardized heterogeneity index at each iteration of the k-means algorithm on 100 iterations; (**b**) value of standardized JDF at each iteration of the PD-clustering algorithm on 100 iterations

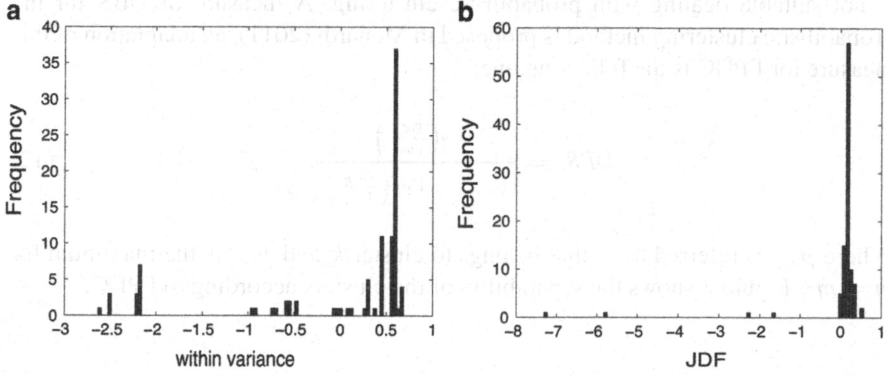

Fig. 2 (**a**) Value of standardized heterogeneity index at each iteration of the Factorial k-means algorithm; (**b**) value of standardized JDF at each iteration of the FPDC algorithm (on 100 iterations)

cases and the second in 54 %. When PD-clustering is applied, in order to measure the stability of the method, the JDF has been measured. In this case results are very unstable.

Applying factorial clustering methods the stability of the results improve, Fig. 2. In the Factorial k-means output (Vichi and Kiers 2001) there is only one minimum reached in 37 % of the cases. Factor PD-clustering presents a high improvement of stability, the modal value is obtained in 58 % and in other cases the value of the JDF is not far from the modal value.

A Density Based Silhouette plot (DBS) is helpful to evaluate the cluster partition. According to this method the DBS index is measured for all the observations x_i, all the clusters are sorted in a decreasing order with respect to DBS and

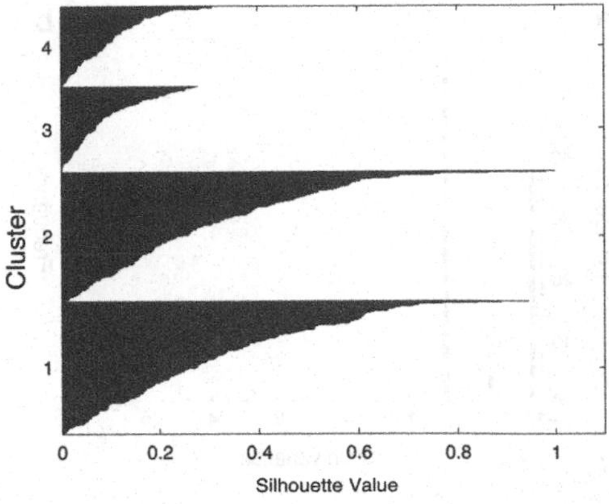

Fig. 3 DBS plot on clusters obtained in the modal value of the JDF on 100 FPDC iterations

plotted on a bar graph, Fig. 3. Usually Euclidean distance is used to measure the distance between cluster centers and each datapoint; however Euclidean distance is not suitable dealing with probabilistic clustering. A measure of DBS for the probabilistic clustering method is proposed in Menardi (2011), an adaptation of this measure for FPDC is the following one:

$$DBS_i = \frac{log\left(\frac{p_{im_k}}{p_{im_1}}\right)}{\max_{i=1,\ldots,n} |log\left(\frac{p_{im_k}}{p_{im_1}}\right)|}. \tag{12}$$

Where p_{im_k} is referred to x_i that belongs to cluster k and p_{im_1} is the maximum for $m \neq m_k$. Figure 3 shows the separability of the clusters according to FPDC.

5 Conclusion

Integrated strategies for dimensionality reduction and non hierarchical clustering have been receiving wide interest as a tool for performing clustering and dimension reduction simultaneously. In this framework, the present paper proposes a new strategy that combines the Tucker3 analysis with the Probabilistic Distance Clustering. Simulation studies, whose results are not reported here, have demonstrated that the algorithm ensures good, and in some cases excellent, improvement in the cluster solution. We have performed some comparative studies with the Vichi and Kiers' Factorial k-means algorithm and we have outlined the conditions in which the Factor PD-clustering outperforms the k-means. At the current developing state, the procedure has a semi-automatic workflow. Assuming that the parameter K is known or already defined, Factor PD-clustering requires the choice of the sub-dimensions to be performed by the analyst using an exploratory strategy.

References

Ben-Israel, A., & Iyigun, C. (2008). Probabilistic d-clustering. *Journal of Classification, 25*(1), 5–26.

Iyigun, C. (2007). *Probabilistic distance clustering*. Ph.D. thesis, New Brunswick Rutgers, The State University of New Jersey.

Jain, A. K. (2009). Data clustering: 50 years beyond K-means. *Pattern Recognition Letters, 31*, 651–666.

Kiers, H., & Kinderen, A. (2003). A fast method for choosing the numbers of components in tucker3 analysis. *British Journal of Mathematical and Statistical Psychology, 56*(1), 119–125.

Kroonenberg, P. (2008). *Applied multiway data analysis*. Ebooks Corporation, Baarn, Nederland.

Menardi, G. (2011). Density-based Silhouette diagnostics for clustering methods. *Statistics and Computing, 21*, 295–308.

Montanari, A., & Viroli, C. (2011). Maximum likelihood estimation of mixtures of factor analyzers. Computational Statistics and Data Analysis, 55, 2712–2723.

Parsons, L., Haque, E., & Liu, H. (2004). Subspace clustering for high dimensional data: A review *SIGKDD Explorations Newsletter, 6*, 90–105.

Tortora, C. (2011). Non-hierarchical clustering methods on factorial subspaces. Ph.D. thesis at Universitá di Napoli Federico II, Naples.

Tortora, C., Palumbo, F., & Gettler Summa, M. (2011). Factorial PD-clustering. Working paper. arXiv:1106.3830v1.

Vichi, M., & Kiers, H. (2001). Factorial k-means analysis for two way data. *Computational Statistics and Data Analysis, 37*, 29–64.

Part III
Statistical Data Analysis, Visualization and Scaling

Clustering Ordinal Data via Latent Variable Models

Damien McParland and Isobel Claire Gormley

Abstract Item response modelling is a well established method for analysing ordinal response data. Ordinal data are typically collected as responses to a number of questions or items. The observed data can be viewed as discrete versions of an underlying latent Gaussian variable. Item response models assume that this latent variable (and therefore the observed ordinal response) is a function of both respondent specific and item specific parameters. However, item response models assume a homogeneous population in that the item specific parameters are assumed to be the same for all respondents. Often a population is heterogeneous and clusters of respondents exist; members of different clusters may view the items differently. A mixture of item response models is developed to provide clustering capabilities in the context of ordinal response data. The model is estimated within the Bayesian paradigm and is illustrated through an application to an ordinal response data set resulting from a clinical trial involving self-assessment of arthritis.

1 Introduction

Ordinal data arise naturally in many different fields and are typically collected as responses to a number of questions or items. A common approach to analysing such data is to view the observed ordinal data as discrete versions of an underlying latent Gaussian 'generating' variable. Many models such as graded response models and ordinal regression models (Albert and Chib 1993) make use of this concept of latent generating variables.

Item response modelling (Fox 2010) is an established method for analysing ordinal response data. It is assumed that the observed ordinal response to an

D. McParland (✉) · I.C. Gormley
University College Dublin, Dublin, Ireland
e-mail: damien.mcparland@ucd.ie; claire.gormley@ucd.ie

B. Lausen et al. (eds.), *Algorithms from and for Nature and Life*, Studies in Classification, Data Analysis, and Knowledge Organization, DOI 10.1007/978-3-319-00035-0_12,

item will be level k, say, if the underlying latent variable lies within a specified interval. Item response models further assume that the latent generating variable (and therefore the observed ordinal response) is a function of both respondent specific and item specific parameters. The respondent specific parameters are often called *latent traits*. The probability of a certain response from a respondent is related to both the value of their latent trait and also some item specific parameters.

Item response models assume that the item specific parameters are the same for all respondents, i.e. a homogeneous population is assumed. Often a population is heterogeneous however and clusters of respondents exist; members of different clusters may view the items differently. Here an item response model is embedded in a mixture modelling framework to facilitate clustering of respondents in the context of ordinal response data. Under the mixture of item response models the probability that a respondent gives a certain response depends on their latent trait and on group specific item parameters. An alternative approach to this problem is given in Von Davier and Yamamoto (2004).

The mixture of item response models is developed and estimated within the Bayesian paradigm using Markov chain Monte Carlo methods. A key issue is choosing the optimal model or equivalently, the number of components in the optimal mixture model. The marginal likelihood is employed here to choose between models and a bridge sampling approach to estimating the marginal likelihood is used.

The model is illustrated through an application to an ordinal response data set resulting from a clinical trial involving self-assessment of arthritis pain levels.

The paper proceeds as follows. In Sect. 2 the arthritis pain levels data set used to demonstrate the model is introduced. Item response models and their extension to a mixture of item response models are considered in Sect. 3. Section 4 is concerned with Bayesian model estimation and also model selection. The results from fitting the model to the illustrative data set are presented in Sect. 5. Finally, discussion of the model takes place in Sect. 6.

2 Arthritis Pain Data

An ordinal data set is employed to illustrate the mixture of item response models. The data come from a clinical trial in which patients suffering from rheumatoid arthritis are randomly assigned to a treatment group or a placebo group. The patients self-assess their arthritis related pain as 1 (poor), 2 (fair) or 3 (good) at 1 and 5 month examinations. Some covariate information associated with each patient such as their age and sex are also recorded. Further details are given in Lipsitz and Zhao (1994) and Agresti (2010).

Here only the ordinal response data are analysed. Interest lies in determining if there is an underlying group structure among the group of 289 patients in the clinical trial. Members of the same group would be expected to have similar arthritis pain profiles. In particular, whether or not patients in the treatment group are differentiated from the patients in the placebo group is of interest.

3 Item Response Models and Mixtures of Item Response Models

The concepts behind item response models and the proposed extension to a mixture of item response models are explained in this section.

3.1 *Item Response Models for Ordinal Data*

Suppose the responses of N individuals to each of J items are observed. Since the data are ordinal, the set of possible responses to item j is $\{1, 2, \ldots, K_j\}$ where K_j denotes the number of possible responses to item j. Thus the data can be represented by an $N \times J$ matrix, Y, where y_{ij} is the response of individual i to item j.

Corresponding to each ordinal response, y_{ij}, is a latent Gaussian variable, z_{ij}. A Gaussian link function is used here but other link functions, such as the logit (Fox 2010), can be employed. For each item there exists a vector of threshold parameters $\underline{\gamma}_j = (\gamma_{j,0}, \gamma_{j,1}, \ldots, \gamma_{j,K_j})$. This vector is subject to the constraint:

$$-\infty = \gamma_{j,0} \leq \gamma_{j,1} \leq \ldots \leq \gamma_{j,K_j} = \infty$$

The observed ordinal response, y_{ij}, serves as an indicator to the latent variable z_{ij}:

$$y_{ij} = k \quad \Rightarrow \quad \gamma_{j,k-1} \leq z_{ij} \leq \gamma_{j,k}$$

In addition to the latent variable, z_{ij}, it is assumed that there exists a latent trait vector, $\underline{\theta}_i$, of dimension q corresponding to each individual. Here q is user specified. The mean of the conditional distribution of z_{ij} is related to this latent trait:

$$z_{ij}|\underline{\theta}_i \sim N(\underline{\lambda}_j^T \underline{\theta}_i - b_j, 1)$$

In the item response literature the item parameters $\underline{\lambda}_j$ and b_j are usually termed *item discrimination* parameters and *item difficulty* parameters respectively. The conditional probability that a response takes a certain ordinal value can then be expressed as the difference between two standard Gaussian cumulative density functions:

$$P(y_{ij} = k|\underline{\lambda}_j, b_j, \underline{\gamma}_j, \underline{\theta}_i) = \Phi[\gamma_{j,k} - (\underline{\lambda}_j^T \underline{\theta}_i - b_j)] - \Phi[\gamma_{j,k-1} - (\underline{\lambda}_j^T \underline{\theta}_i - b_j)]$$

3.2 *A Mixture of Item Response Models (MIRM) for Ordinal Data*

A mixture modelling framework can be imposed on the item response model for cases where there is an underlying group structure in the data. The aim of this mixture of item response models is to cluster individuals into their unobservable groups. Under the MIRM, the latent variable z_{ij} is a mixture of G Gaussian densities:

$$f(z_{ij}) = \sum_{g=1}^{G} \pi_g N(\underline{\lambda}_{gj}^T \underline{\theta}_i - b_{gj}, 1)$$

The probability of belonging to group g is denoted π_g while $\underline{\lambda}_{gj}$ and b_{gj} are *group specific* item discrimination and difficulty parameters respectively.

A latent indicator variable, $\underline{\ell}_i = (\ell_{i1}, \ldots, \ell_{iG})$ is introduced for each individual i. This binary vector indicates to which group individual i belongs i.e. $l_{ig} = 1$ if i belongs to group g; all other entries in the vector are 0. Thus, conditional on $\underline{\ell}_i$, the probability of observing a particular ordinal response is:

$$P(y_{ij} = k | \underline{\lambda}_{gj}, b_{gj}, \underline{\gamma}_j, \underline{\theta}_i, l_{ig} = 1) = \Phi\left[\gamma_{jk} - (\underline{\lambda}_{gj}^T \underline{\theta}_i - b_{gj})\right] \\ -\Phi\left[\gamma_{j,k-1} - (\underline{\lambda}_{gj}^T \underline{\theta}_i - b_{gj})\right]$$

The augmented likelihood, $\mathscr{L}(\Lambda, B, \Gamma, \Theta, L, Z | Y)$, is given by:

$$\prod_{i=1}^{N}\prod_{g=1}^{G}\prod_{j=1}^{J}\left\{\left[\sum_{k=1}^{K_j} \mathbf{1}\left(\gamma_{j,k-1} \le z_{ij} \le \gamma_{j,k}\right) \mathbf{1}\left(y_{ij} = k\right)\right] N\left(\underline{\lambda}_{gj}^T \underline{\theta}_i - b_{gj}, 1\right)\right\}^{\ell_{ig}}$$

An assumption of local independence is implicit here, i.e. conditional on the latent trait $\underline{\theta}_i$ the J responses by individual i are independent. The responses of different individuals are also regarded as independent.

4 Parameter Estimation and Model Selection

The Bayesian framework in which the model is estimated, the Markov chain Monte Carlo (MCMC) algorithm used to fit the model and the bridge sampling algorithm which facilitates model selection are all described in what follows.

4.1 Prior and Posterior Distributions

To implement the model described above in a Bayesian framework prior distributions must be specified for all unknown parameters. Priors are required for the threshold parameters $\underline{\gamma}_j$, the item parameters, $\underline{\lambda}_{gj}$ and $\underline{b}_g$, and for the mixing weights

$\underline{\pi}$ (for $j = 1, \ldots, J$ and $g = 1, \ldots, G$). Specifically, a uniform prior is specified for the threshold parameters and for the other parameters the prior distributions are:

$$p(\underline{\lambda}_{gj}) = MVN_q(\underline{\mu}_{\lambda}, \Sigma_{\lambda}) \qquad p(\underline{b}_g) = MVN_J(\underline{\mu}_b, s_b^2\mathbf{I}) \qquad p(\underline{\pi}) = Dir(\underline{\alpha})$$

The posterior distribution is:

$$p(\Lambda, B, \Gamma, \underline{\pi}, \Theta, L, Z|Y) \propto \mathcal{L}(\Lambda, B, \Gamma, \Theta, L, Z|Y)p(\Lambda)p(B)p(\Gamma)p(\Theta)p(L|\underline{\pi})p(\underline{\pi})$$

where $p(\Lambda)$, $p(B)$, $p(\Gamma)$ and $p(\underline{\pi})$ are the prior distributions detailed above. The latent trait variable $\underline{\theta}_i$ is assumed to have a standard multivariate Gaussian distribution; the latent indicator variables $\underline{l}_i$ follow a *Multinomial*$(1, \underline{\pi})$ distribution.

This model suffers from non-identifiability. To identify the model (as in Johnson and Albert (1999)) the second element of each of the threshold vectors, $\underline{\gamma}_j$ for $j = 1, \ldots, J$, is fixed at 0. The model is also rotationally invariant. Therefore, a specific form is imposed on each matrix of discrimination parameters Λ_g for $g = 1, \ldots, G$. As in Geweke and Zhou (1996), the first q rows of this matrix are constrained to have a lower triangular form. In what follows, the free and fixed elements of the jth row of Λ_g are denoted by $\underline{\lambda}_{gj}^{\circ}$ and $\underline{\lambda}_{gj}^{\bullet}$ respectively.

4.2 Estimation via a Markov Chain Monte Carlo Algorithm

The marginal distributions of the unknown parameters cannot be obtained analytically for this model so a MCMC algorithm is used to produce estimates of the model parameters. The algorithm used here is similar to the algorithm proposed in Cowles (1996). A Gibbs sampler is used to sample all latent variables and parameters, except the threshold parameters, $\underline{\gamma}_j$. These are sampled using a Metropolis-Hastings step.

Full conditional distributions for the model parameters and latent variables are:

- $\underline{\ell}_i | \ldots \sim$ Multinomial$(1, \underline{p} = (p_1, \ldots, p_G))$ where

$$p_g \propto \pi_g \prod_{j=1}^{J} \left[\sum_{k=1}^{K_j} \mathbf{1}\left(\gamma_{j,k-1} \leq z_{ij} \leq \gamma_{j,k}\right) \mathbf{1}(y_{ij} = k) \right] \mathrm{N}(\underline{\lambda}_{gj}^T \underline{\theta}_i - b_{gj}, 1)$$

- $\underline{\pi} | \ldots \sim$ Dirichlet$(n_1 + \alpha_1, \ldots, n_G + \alpha_G)$ where $n_g = \sum_{i=1}^{N} l_{ig}$.
- $z_{ij} | \ldots \sim \mathrm{N}^T\left(\underline{\lambda}_{gj}^T \underline{\theta}_i - b_{gj}, 1\right)$ where the distribution is truncated to $[\gamma_{j,(y_{ij}-1)}, \gamma_{j,(y_{ij})}]$.
- $\underline{\theta}_i | \ldots \sim \mathrm{MVN}_q\left[D_g^{-1} \Lambda_g^T \left(\underline{z}_i + \underline{b}_g\right), D_g^{-1}\right]$ where, $\underline{z}_i = (z_{i1}, \ldots, z_{iJ})^T$ and $D_g = \Lambda_g^T \Lambda_g + \mathbf{I}_q$.

- $\underline{\lambda}^{\circ}_{gj}|\ldots \sim \text{MVN}\left\{S^{-1}\left[\Theta^{\circ T}_{g}\left(\underline{z}_{gj} - \Theta^{\bullet}_{g}\underline{\lambda}^{\bullet}_{gj} + b_{gj}\underline{1}\right) + \Sigma^{-1}_{\lambda}\underline{\mu}_{\lambda}\right], S^{-1}\right\}$ where $S = \left[\Sigma^{-1}_{\lambda} + \Theta^{\circ T}_{g}\Theta^{\circ}_{g}\right]$ and $\underline{1} = (1, \ldots, 1)^T$. The ith row of Θ°_{g} consists of the elements of $\underline{\theta}_i$ which multiply $\underline{\lambda}^{\circ}_{gj}$ for all individuals i in group g. Similarly $\Theta^{\bullet}_{g}$ consists of the elements which multiply $\underline{\lambda}^{\bullet}_{gj}$. The elements of the jth column of the $N \times J$ matrix Z corresponding to individuals in group g are denoted by $\underline{z}_{gj}$.
- $b_{gj}|\ldots \sim N\left[(n_g + s_b^{-2})^{-1}(\underline{1}^T\Theta_g\underline{\lambda}_{gj} + s_b^{-2}\mu_{bj} - \underline{z}^T_{gj}\underline{1}), (n_g + s_b^{-2})^{-1}\right]$ where the rows of Θ_g are the latent trait vectors $\underline{\theta}_i$ for all individuals i in group g.

The posterior full conditional distribution of each of the threshold parameters, $\gamma_{j,k}$ can be shown to be uniform. When there are a large number of observations in adjacent categories this interval tends to be small which results in minimal movement of the Gibbs sampler. The algorithm therefore converges slowly. This difficulty is overcome by sampling from the posterior of the threshold parameters using a Metropolis-Hastings step, as in Cowles (1996) and Johnson and Albert (1999). Candidate values $v_{j,k}$ are proposed for $\gamma_{j,k}$ from the Gaussian distribution $\text{N}^T(\gamma^{(t-1)}_{j,k}, \sigma^2_{MH})$, truncated to the interval $(v_{j,k-1}, \gamma^{(t-1)}_{j,k+1})$ where $\gamma^{(t-1)}_{j,k+1}$ is the value of $\gamma_{j,k+1}$ at iteration $(t-1)$. The tuning parameter σ^2_{MH} is selected to achieve appropriate acceptance rates.

4.3 Model Selection via the Bridge Sampler

Since the proposed model is a finite mixture model, the number of components G in the mixture must be chosen. A bridge sampling algorithm (Meng and Wong 1996; Frühwirth-Schnatter 2004) is employed to approximate the marginal likelihood of a G component model. The marginal likelihood is evaluated for a range of models with different values of G and the model with the highest marginal likelihood is chosen as optimal. Here, the posterior mean of the latent Gaussian variable Z is treated as the 'observed data'. This approach removes the need to work with the intractable marginal distribution of the ordinal data, Y, and also the posterior distribution of the threshold parameters.

In order to use bridge sampling to approximate the marginal likelihood it is important that the MCMC algorithm mixes well over all posterior modes. The random permutation MCMC sampler (Frühwirth-Schnatter 2001) is used to achieve this. For more details on the bridge sampling estimator of the marginal likelihood of a mixture model see Frühwirth-Schnatter (2006).

5 Arthritis Pain Data: Results

The mixture of item response models (MIRM) was fitted to the ordinal arthritis pain data described in Sect. 2. A number of mixture of item response models were fitted to the data with the number of components G ranging from one to five, and

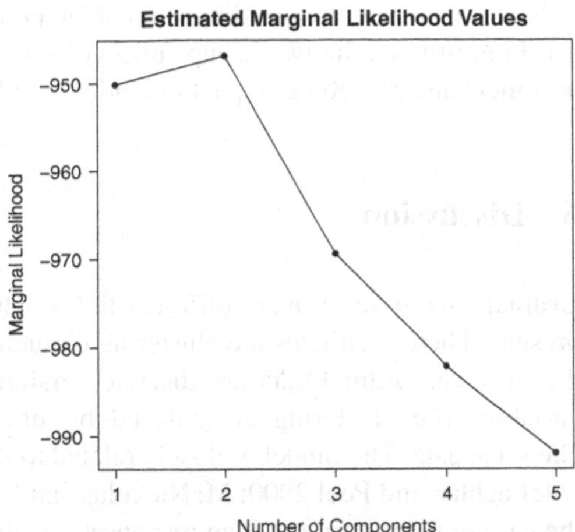

Fig. 1 Estimated marginal likelihood values for a range of mixture of item response models with a one dimensional latent trait

Table 1 Posterior mean estimates (and 95 % quantile-based confidence regions) for the optimal model

Parameter	Posterior mean	
b_{11}	−0.18	[−1.02, 0.47]
b_{12}	−0.20	[−1.11, 0.70]
b_{21}	−2.29	[−3.35, −1.50]
b_{22}	−2.27	[−3.35, −1.44]
λ_{11}	0.59	[−0.25, 1.49]
λ_{12}	0.81	[−0.32, 1.71]
λ_{21}	0.96	[−0.02, 1.79]
λ_{22}	0.75	[−0.10, 1.54]
$\gamma_{1,2}$	2.10	[1.59, 2.97]
$\gamma_{2,2}$	1.78	[1.30, 2.63]
π_1	0.40	[0.22, 0.60]
π_2	0.60	[0.40, 0.78]

with a user specified $q = 1$ dimensional latent trait. The marginal likelihood of each of the models was estimated using the bridge sampling technique described in Sect. 4.3. The values obtained are illustrated in Fig. 1. The highest marginal likelihood value is obtained when a two component MIRM is fitted. Posterior mean parameter estimates for the optimal model are detailed in Table 1.

Inspection of the responses of individuals in each cluster suggests that the patients have been partitioned into a group (group 1) who judge the state of their arthritis to be poor to fair and a group (group 2) who consider the state of their arthritis to be fair to good. Although the item difficulty parameters for both groups are negative, the parameters for group 1 [$\underline{b}_1 = (-0.18, -0.20)$] are smaller in magnitude than those for group 2 [$\underline{b}_2 = (-2.29, -2.27)$]. This difference means that the values of the latent Gaussian variable Z (with marginal mean $-\underline{b}_g$ for $g = 1, 2$) are lower in group 1, reflecting the generally lower observed ordinal responses found in group 1. The confidence regions for the discrimination parameters include 0 which indicates that even the one dimensional latent trait may be unnecessary for this data set. Interestingly, the two groups uncovered by the model do not correspond to the treatment and placebo group (Rand index = 0.51, Adjusted Rand index = 0.015).

6 Discussion

Ordinal data arise in many different fields. The mixture of item response models presented here facilitates the clustering of such data. This is achieved by assuming the observed ordinal data are discrete versions of an underlying latent Gaussian variable. The clustering is achieved by fitting a mixture model to the latent Gaussian data. The model is closely related to the mixture of factor analysers model (McLachlan and Peel 2000; McNicholas and Murphy 2008) for continuous data; in the case of the mixture of item response models however, only a discrete version of the data are observed.

Bridge sampling was employed for model selection. Simulation studies and the illustrative data example suggest that the bridge sampling approach works well in the context of the mixture of item response models. However, it should be noted that as the bridge sampler relies on the posterior mean of the latent Gaussian data Z, the same 'data' are not used when evaluating the marginal likelihood for different models. Again, simulation studies suggest that given a sufficiently large data set (both in terms of number of observations and cell counts for the ordinal variables) the results are not very sensitive to this approximation to Y.

There are a number of ways in which the model could be extended. The model selection technique employed here is used only to choose the number of components in the mixture. Extending the bridge sampling technique to determine the optimal number of dimensions (q) for the latent trait would be very beneficial (Lopes and West 2004). Additionally, in the illustrative data set used here covariate data were available. Incorporating these data in the model would be potentially informative and could be achieved within a mixture of experts framework (Jacobs et al. 1991; Gormley and Murphy 2008). Finally, as with most clustering models, the set of variables on which the clustering is based strongly influences the MIRM; incorporating a variable selection step while clustering would potentially improve clustering performance.

Acknowledgements This work has emanated from research conducted with the financial support of Science Foundation Ireland under Grant Number 09/RFP/MTH2367.

References

Agresti, A. (2010). *Analysis of ordinal categorical data*. Hoboken: Wiley.

Albert, J. H., & Chib, S. (1993). Bayesian analysis of binary and polychotomous response data. *Journal of the American Statistical Association, 88*, 669–679.

Cowles, M. K. (1996). Accelerating Monte Carlo Markov chain convergence for cumulative-link generalized linear models. *Journal of the American Statistical Association, 6*, 101–111.

Fox, J. P. (2010). *Bayesian item response modeling*. New York: Springer.

Frühwirth-Schnatter, S. (2001). Markov chain Monte Carlo estimation of classical and dynamic switching and mixture models. *Journal of the American Statistical Association, 96*, 194–209.

Frühwirth-Schnatter, S. (2004). Estimating marginal likelihoods for mixture and Markov switching models using bridge sampling techniques. *Statistica Sinica, 6*, 831–860.

Frühwirth-Schnatter, S. (2006). *Finite mixture and Markov switching models*. New York: Springer.

Geweke, J., & Zhou, G. (1996). Measuring the price of arbitrage theory. *The Review of Financial Studies, 9*, 557–587.

Gormley, I. C., & Murphy, T. B. (2008). A mixture of experts model for rank data with applications in election studies. *The Annals of Applied Statistics, 2*(4), 1452–1477.

Jacobs, R. A., Jordan, M. I., Nowlan, S. J., & Hinton, G. E. (1991). Adaptive mixture of local experts. *Neural Computation, 3*, 79–87.

Johnson, V. E., & Albert, J. H. (1999). *Ordinal data modeling*. New York: Springer.

Lipsitz, S. R., & Zhao, L. (1994). Analysis of repeated categorical data using generalized estimating equations. *Statistics in Medicine, 13*, 1149–1163.

Lopes, H. F., & West, M. (2004). Bayesian model assessment in factor analysis. *Statistica Sinica, 14*, 41–67.

McLachlan, G. J., & Peel, D. (2000). *Finite mixture models*. New York: Wiley.

McNicholas, P. D., & Murphy, T. B. (2008). Parsimonious Gaussian mixture models. *Statistics and Computing, 18*(3), 285–296.

Meng, X. L., & Wong, W. H. (1996). Simulating ratios of normalizing constants via a simple identity: a theoretical exploration. *The Econometrics Journal, 7*, 143–167.

Von Davier, M. & Yamamoto, K. (2004). Partially observed mixtures of IRT models: an extension of the generalized partial credit model. *Applied Psychological Measurement, 28*(6), 389–406.

Sentiment Analysis of Online Media

Michael Salter-Townshend and Thomas Brendan Murphy

Abstract A joint model for annotation bias and document classification is presented in the context of media sentiment analysis. We consider an Irish online media data set comprising online news articles with user annotations of negative, positive or irrelevant impact on the Irish economy. The joint model combines a statistical model for user annotation bias and a Naive Bayes model for the document terms. An EM algorithm is used to estimate the annotation bias model, the unobserved biases in the user annotations, the classifier parameters and the sentiment of the articles. The joint modeling of both the user biases and the classifier is demonstrated to be superior to estimation of the bias followed by the estimation of the classifier parameters.

1 Introduction

Sentiment analysis involves extracting contextual information from documents (Pang and Lee 2008). Media sentiment has been shown to be of importance in economic contexts (Tetlock 1139–1168). We examine a corpus of Irish news articles that have been annotated by a number of inexpert volunteers as having a sentiment which has positive, negative or irrelevant impact on the Irish economy. The aim of the analysis is to develop a classification method that can estimate the correct labelling of the articles in the corpus as well as the correct classification for other news articles. A core goal is to increase the accuracy of both the annotation based labelling and the classifier. Whilst the methods outlined herein are developed in the context of the media sentiment, they are readily applicable in any context where a classifier is trained on (potentially) biased and noisy annotations.

M. Salter-Townshend (✉) · T.B. Murphy
School of Mathematical Sciences and Complex and Adaptive Systems Laboratory, University College Dublin, Dublin 4, Ireland
e-mail: michael.salter-townshend@ucd.ie; brendan.murphy@ucd.ie

B. Lausen et al. (eds.), *Algorithms from and for Nature and Life*, Studies in Classification, Data Analysis, and Knowledge Organization, DOI 10.1007/978-3-319-00035-0_13,

The media sentiment analysis involves a classification task where the sample labels are noisy and biased user annotations. Many existing classifiers do not take into account user (annotator) bias in reporting and a simple majority vote is used to select the true article type from the observed annotations; this majority vote labelling is particularly problematic in the presence of user bias. Some previous work has been proposed to help address the annotator bias issue. Smyth et al. (1994) applied the method of Dawid and Skene (1979) to correct for annotator bias and estimate the true labeling before developing a classifier in an object recognition problem. Most recently Rogers et al. (2010) and Raykar et al. (2010) propose methods to address the problem of multiple imperfect annotations and classification. Rogers et al. (2010) deals with the labelling of clinical reports and uses Bayesian models with Gaussian processes for classification and ordinal regression. Raykar et al. (2010) address the problem of training a classifier with multiple imperfect annotations by extending the model of Dawid and Skene (1979) to learn a classifier at the same time as the annotator biases via maximum likelihood; this work is similar to the approach developed herein. Specifically, they train a logistic regression classifier and learn the sensitivity and specificity of the annotators in the context of binary labelling. The model that we present differs from that paper in that we explore a trinary labelling system (an arbitrary finite number of categories is possible) and train a Naive Bayes classifier. The contribution of our work is to demonstrate the method with another classifier, a greater number of potential labels and to report upon the comparative effectiveness of our approach in the context of the Irish online media sentiment analysis.

We validate our approach on a simulated dataset and calculate performance scores for both the decoupled estimator (learn the biases and then train the classifier) and the joint estimator model. We demonstrate the superiority of the joint estimator for various levels of bias and then apply it to the media dataset.

1.1 Sentiment Data

The Irish media dataset that we analyze is a subset of the data described in detail in Brew et al. (2010a,b). The dataset is comprised of 1,024 articles collected from 3 online Irish news services (`rte.ie`, `irishtimes.com` and `independent.ie`), collected from July to October 2009. Thirty one volunteers have annotated an average of 834 of these articles as having either negative, positive or irrelevant impact on the Irish economy at time of press. There are 70,873 word terms appearing in these articles. In order to reduce the impact of words that are too common (such as "at", "the", "and", etc) we eliminate words that appear in more than 1,000 articles. We also eliminate words that appeared in less than 30 articles. To further reduce the dimensionality of the data, we selected the top 300 most negative words (as indicated by a simple majority vote classifier), the top 300 most positive words and the top 300 most irrelevant words only.

Brew et al. (2010b) note that 45 % of the articles do not have consensus annotations and that "there is some evidence that the learning process would be better off without them [articles with low consensus]". The authors of that paper examined k-nearest neighbours (kNN) and support vector machine (SVM) classifiers also but settled on Naive Bayes following an assessment of the performance of the methods under cross-validation.

2 Model

Let $y_a^{(k)} = (y_{a1}^{(k)}, y_{a2}^{(k)}, \ldots, y_{aJ}^{(k)})$ be the annotation of article a by annotator k, where $y_{aj}^{(k)} = 1$ if article a is annotated as being of type j and $y_{aj}^{(k)} = 0$ otherwise. We model the annotator bias as per Dawid and Skene (1979). Error rates, or biases in reporting, are modelled via a matrix of conditional probabilities for each annotator, that is, the probability that annotator k records annotation j given a *true* (but unobserved) type i is denoted by $\pi_{ij}^{(k)}$. These probabilities sum to unity across j for each i and k. The observed annotations are thus a probabilistic (multinomial) function of these π matrices.

If we let the true type of article a be T_a, where $T_{ai} = 1$ if the article is of type i and $T_{ai} = 0$ otherwise. Then, the likelihood for the recorded annotations $y_a = (y_a^{(1)}, y_a^{(2)}, \ldots, y_a^{(K)})$ on article a given a true type T_a is given by

$$\mathscr{L}(\pi | y_a, T_a) \propto \prod_i^J \left\{ \prod_k^K \prod_j^J (\pi_{ij}^{(k)})^{y_{aj}^{(k)}} \right\}^{T_{ai}} \tag{1}$$

where J is three for our sentiment levels (negative, positive and irrelevant).

Hence, the complete-data likelihood of the full annotation dataset (including unobserved true types) across all A articles is

$$\mathscr{L}(\pi, p | y_1, y_2, \ldots, y_A, T_1, T_2, \ldots, T_A) \propto \prod_a^A \prod_i^J \left\{ p_i \prod_k^K \prod_j^J (\pi_{ij}^{(k)})^{y_{aj}^{(k)}} \right\}^{T_{ai}} \tag{2}$$

where p_i is the marginal probability of type i.

Another goal of the sentiment analysis described in Brew et al. (2010b) is to train a classifier to distinguish which word terms appear in which types of article. The trained classifier may then be used to automatically label un-annotated articles. Although word-term frequencies are available in the dataset, we model only the presence or absence of these features (word terms). Let $w_a = (w_{a1}, w_{a2}, \ldots, w_{aN})$ be a binary vector that indicates the presence and absence of words in document a. We employ a Bernoulli likelihood for term w_a given that the article is of type i, that is $T_{ai} = 1$. That is, we use a Naive Bayes classifier to learn the probability of an article type given the words that appear in the article. Although the Naive Bayes assumption is unlikely to hold exactly in practice, there is much evidence to suggest that it can yield excellent classification results (Domingos and Pazzani 1997; Hand and Yu 2001).

The product of Bernoullis likelihood for all N word terms w_a appearing in article a given T_a is then

$$\mathcal{L}(\theta|w_a, T_a) = \prod_{i}^{J} \left\{ \prod_{n}^{N} (\theta_{ni})^{w_{an}} (1-\theta_{ni})^{1-w_{an}} \right\}^{T_{ai}} \tag{3}$$

where θ_{ni} is the probability that word term w_n appears in an article of type i.

The full likelihood for the data is then a product of Eq. (2) and a term in the form of Eq. (3) for each article, yielding Eq. (4),

$$\mathcal{L}(\theta, p, \pi|w, y, T) = \prod_{a}^{A}\prod_{i}^{J} \left\{ p_i \prod_{k}^{K}\prod_{j}^{J} (\pi_{ij}^{(k)})^{y_{aj}^{(k)}} \prod_{n}^{N} (\theta_{ni})^{w_{an}} (1-\theta_{ni})^{1-w_{an}} \right\}^{T_{ai}}. \tag{4}$$

3 EM Algorithm

Since T, p and π are unknown in Eq. (2), we proceed as per Dawid and Skene (1979). We then extend the EM algorithm to yield a joint estimation that learns θ within the same EM iterations as it learns the values of missing data T, the marginal probabilities p and annotator bias matrices π. The algorithm proceeds as follows:

1. **for all articles** a:
2. initialize T using $\hat{T}_{ai} = \mathbf{E}[T_{ai}] = \sum_k y_{ai}^{(k)}/K$
3. initialize p using $\hat{p}_i = \sum_a T_{ai}/A$
4. estimate all π values via maximum likelihood expression

$$\hat{\pi}_{ij}^{(k)} = \frac{\sum_a \hat{T}_{ai} y_{aj}^{(k)}}{\sum_j \sum_a \hat{T}_{ai} y_{aj}^{(k)}}. \tag{5}$$

5. estimate all θ and p via maximum likelihood expressions

$$\hat{\theta}_{ni} = \frac{\sum_a w_{an} \hat{T}_{ai}}{\sum_a \hat{T}_{ai}} \quad \text{and} \quad \hat{p}_i = \frac{\sum_a \hat{T}_{ai}}{A}. \tag{6}$$

6. `re-estimate` T `using`

$$\hat{T}_{ai} = \mathbf{E}[T_{ai}] = \frac{p_i \prod_k^K \prod_j^J (\hat{\pi}_{ij}^{(k)})^{y_{aj}^{(k)}} \prod_n^N (\hat{\theta}_{ni})^{w_{an}} (1 - \hat{\theta}_{ni})^{1-w_{an}}}{\sum_{i'} p_{i'} \prod_k^K \prod_j^J (\hat{\pi}_{i'j}^{(k)})^{y_{aj}^{(k)}} \prod_n^N (\hat{\theta}_{ni'})^{w_{an}} (1 - \hat{\theta}_{ni'})^{1-w_{an}}}. \quad (7)$$

7. **`repeat 4 to 6 until convergence`**

with convergence assumed once the change in log-likelihood fell below 10^{-4}.

In contrast, the decoupled estimator of the above method estimates the biases π, document types T and marginal probabilities p first, as in Dawid and Skene (1979). The Naive Bayes parameters θ are then fitted using the final estimates of the missing data values from the first stage; the decoupled estimation approach is similar to that taken by Smyth et al. (1994).

4 Results

4.1 Simulated Data

To test and compare the algorithm described in Sect. 3 with the decoupled estimator, we simulated data 200 times. For each run, we use the marginal probabilities $p = (0.3, 0.3, 0.4)$ of each of the three types of "article". True types for A "articles" are simulated directly with these marginal probabilities. We then construct K conditional probability matrices $\pi^{(k)}$ of size 3×3, one for each "annotator". The value of $\pi_{ij}^{(k)}$ gives the probability that annotator k annotates an article of type i with label j. Finally, we also simulate observed word terms w for each article using the conditional probabilities of words occurring in each type of article as given in θ.

Two hundred such simulated data sets were analysed and for each data set the biases were randomly sampled uniformly over the range 0.1–0.5 and split evenly between the two wrong types with the balance allocated to the correct type. This was done identically for all simulated annotators which is equivalent to having a single random annotator performing multiple annotations and the number of these annotators was sampled uniformly between 2 and 6. The words were assigned a type according to p and the word-type probabilities θ were 0.1 to appear in an article of opposite type and 0.8 to appear in an article of the same type.

Both models are then evaluated on four performance metrics:

1. The mean error in expectation of type:

$$\sum_a (1 - \mathbf{E}[T_{ai}])/A \quad (8)$$

where the true value of article a is type i.

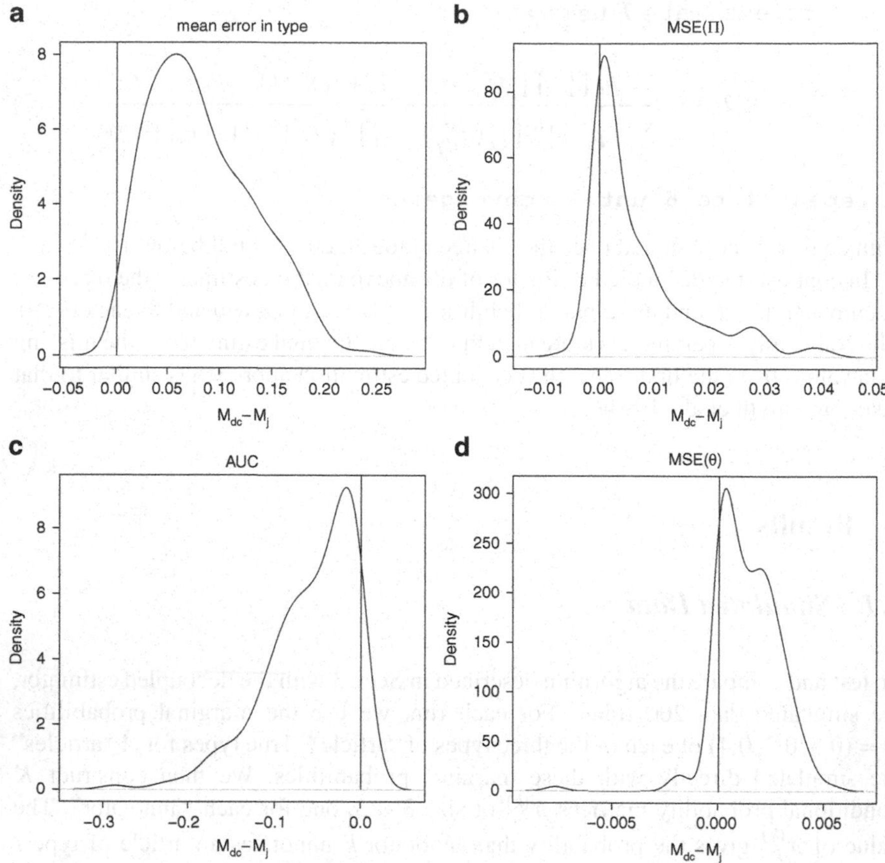

Fig. 1 Kernel density estimates of comparative performance measurements across multiple simulations. Two hundred runs of the simulated dataset analysis were performed and the difference in performance measure is computed for decoupled model (M_{dc}) and the joint model (M_j). **(a)** shows the difference in mean error of type T. **(b)** shows the difference in mean squared error of bias π. **(c)** shows the AUC difference and **(d)** shows the mean squared error difference of word association θ

2. The mean squared error from the π matrix of bias probabilities.
3. The mean area under the ROC curve (AUC) for each of the three possible types.
4. The mean squared error from the θ matrix of word-type probabilities.

We subtracted the above four statistics under the joint estimation model M_j from the decoupled estimation model M_{dc} for repeated simulations. The mean paired difference between the above performance measures were 0.193, 0.009, −0.103 and 0.009, respectively. All four were strongly statistically significantly different from zero under a t-test for the paired differences with p-values all less than 2.2×10^{-16}. Figure 1 shows kernel density estimates of these differences for the above statistics

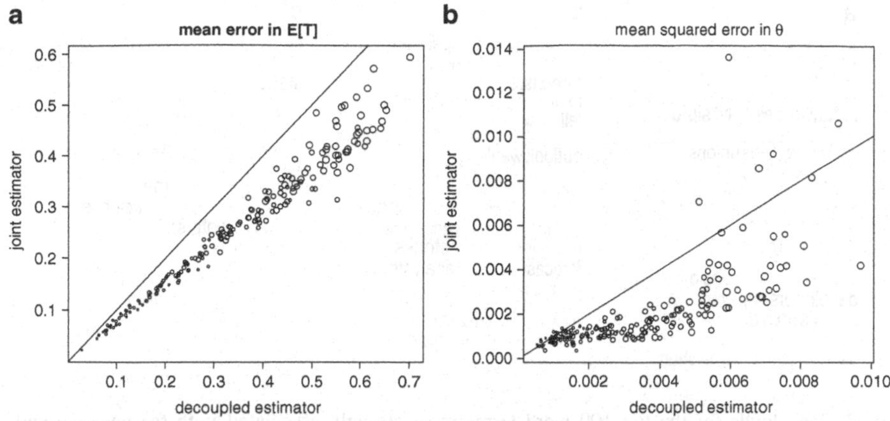

Fig. 2 Comparison of performance across 200 iterations of simulated data. (**a**) shows the mean error in type T (as per Eq. (8)) and (**b**) shows the mean squared error in word-to-type association θ. The size of the circles in the plot is proportional to the bias and each circle represents a single run. *Lines* with unit slope are added for reference

Table 1 Cross-tabulation of article classification and model

	Majority vote	Decoupled estimator	Joint estimator
Negative	540	493	424
Positive	288	289	206
Irrelevant	196	242	394

across the 200 simulation runs. Figure 2 indicates that the joint estimator's increase in performance is greater for higher biases. The size of the circles in the plot is proportional to the sampled bias and each circle represents a single run.

4.2 Sentiment Data

We next present our results on the sentiment dataset. The interquartile range for the bias matrix off diagonal terms is (0.110, 0.517), indicating a level of bias comparable to the simulated dataset. Table 1 shows the breakdown of classification with model for the media sentiment dataset. Figure 3 depicts tag clouds for word terms that have the strongest power for the negative, positive and irrelevant article types, under the joint estimation procedure. These tag clouds appear to show sensible word term associations to both positive and negative sentiment; for example, the names of the finance minister ("Brian", "Lenihan") and the new agency to deal with toxic debt ("NAMA") are included in the negative tag cloud and words like ("Germany", "recovery") are included in the positive tag cloud. The tag cloud

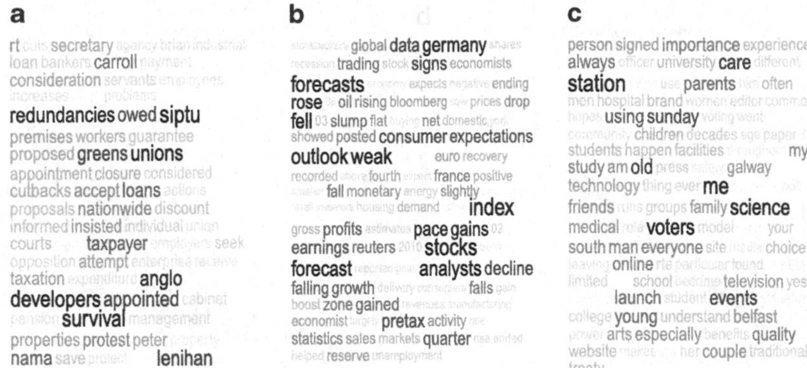

Fig. 3 Tag clouds for the top 100 word terms most strongly associated with (**a**) negative and (**b**) positive and (**c**) irrelevant articles. Most of the words appear to have an intuitively correct association with article type

for the word terms for with the strongest predictive power for the irrelevant article types are given in Fig. 3c. Interestingly, most of the words in this tag cloud are non economic terms.

5 Discussion

We have demonstrated that the joint estimator makes use of the word term association with article type and thus outperforms the decoupled estimator for both bias estimation and classification. This boost in performance is related to the ratio of information in the features to the biases; if the annotators are all in agreement then the word term classifier will contribute little to the model. If there is bias in the annotations and the word terms are influenced by the article type then they will have a larger impact on the model and the joint estimation model will outperform the decoupled estimation model.

The joint estimator can achieve a target level of accuracy in article labelling using fewer biased annotators than the decoupled or majority vote labeling. This suggests that our method could be used to generate savings in the context of crowd-sourcing with inexpert or otherwise biased annotators. There is a computational cost associated with the joint estimation; the time to perform 100 iterations for the decoupled and joint algorithms was approximately 16 and 50 s respectively. The joint algorithm does not seem to take more iterations to converge; for example, using the criterion that a change in log-likelihood of less than 10^{-3} required 38 and 36 iterations respectively. For a change of less than 10^{-2} they took 33 and 35.

The methodology outlined in the paper could be easily adapted to other model-based classifiers where samples are labeled using noisy annotations.

Acknowledgements This work is supported by the Science Foundation Ireland under Grant No. 08/SRC/I1407: Clique: Graph & Network Analysis Cluster.

References

Brew, A., Greene, D., & Cunningham, P. (2010a). The interaction between supervised learning and crowdsourcing. In *NIPS workshop on computational social science and the wisdom of crowds*, Whistler, Canada.

Brew, A., Greene, D., & Cunningham, P. (2010b). Using crowdsourcing and active learning to track sentiment in online media. In H. Coelho, R. Studer, & M. Wooldridge (Eds.), *ECAI 2010 – 19th European conference on artificial intelligence* (pp. 1–11). Berlin: IOS.

Dawid, A., & Skene, A. (1979). Maximum likelihood estimation of observer error-rates using the EM algorithm. *Journal of the Royal Statistical Society. Series C (Applied Statistics), 28*(1), 20–28.

Domingos, P., & Pazzani, M. (1997). On the optimality of the simple Bayesian classifier under zero-one loss. *Machine Learning, 29*, 103–130.

Hand, D. J., & Yu, K. (2001). Idiot's Bayes—not so stupid after all? *International Statistical Review, 69*(3), 385–398.

Pang, B., & Lee, L. (2008). Opinion mining and sentiment analysis. *Foundations and Trends in Information Retrieval, 2*(1–2), 1–135.

Raykar, V., Yu, S., Zhao, L., Valadez, G., Florin, C., Bogoni, L., & Moy, L. (2010). Learning from crowds. *Journal of Machine Learning Research, 11*, 1297–1322.

Rogers, S., Girolami, M., & Polajnar, T. (2010). Semi-parametric analysis of multi-rater data. *Statistics and Computing, 20*, 317–334.

Smyth, P., Fayyad, U. M., Burl, M. C., Perona, P., & Baldi, P. (1994). Inferring ground truth from subjective labelling of venus images. In G. Tesauro, D. S. Touretzky, & T. K. Leen (Eds.), *Advances in neural information processing systems* (Vol. 7, pp. 1085–1092). Cambridge: MIT.

Tetlock, P. C. (2007). Giving content to investor sentiment: The role of media in the stock market. *The Journal of Finance, 62*(3), 1139–1168.

Visualizing Data in Social and Behavioral Sciences: An Application of PARAMAP on Judicial Statistics

Ulas Akkucuk, J. Douglas Carroll*, and Stephen L. France

Abstract In this paper, we describe a technique called PARAMAP for the visualization, scaling, and dimensionality reduction of data in the social and behavioral sciences. PARAMAP uses a criterion of maximizing continuity between higher dimensional data and lower dimensional derived data, rather than the distance based criterion used by standard distance based multidimensional scaling (MDS). We introduce PARAMAP using the example of scaling and visualizing the voting patterns of Justices in the US Supreme Court. We use data on the agreement rates between individual Justices in the US Supreme Court and on the percentage swing votes for Justices over time. We use PARAMAP, metric MDS, and nonmetric MDS approaches to create a voting space representation of the Justices in one and two dimensions. We test the results using a metric that measures neighborhood agreement of points between higher and lower dimensional solutions. PARAMAP produces smooth, easily interpretable, solutions, with no clumping together of points.

---

*We dedicate this paper to our supervisor and mentor Professor J. Douglas Carroll, who sadly passed away in June, 2011.

U. Akkucuk (✉)
Bogazici University, Department of Management, Istanbul, Turkey
e-mail: ulas.akkucuk@boun.edu.tr

J.D. Carroll
Rutgers Business School, Newark and New Brunswick, Newark, NJ, USA

S.L. France
Lubar School of Business, University of Wisconsin-Milwaukee, Milwaukee, WI, USA
e-mail: france@uwm.edu

B. Lausen et al. (eds.), *Algorithms from and for Nature and Life*, Studies in Classification, Data Analysis, and Knowledge Organization, DOI 10.1007/978-3-319-00035-0_14,

1 Introduction

This paper was inspired by a set of New York Times articles (Stevenson 2005; Stevenson et al. 2005) on the retiring centrist Supreme Court Justice Sandra Day O'Connor. The articles describe the pivotal role that Justice O'Connor played on the United States Supreme Court. The articles detail both agreement rates and swing voting patterns for the Justices sitting on the U.S. Supreme Court as of July 2005. The articles use this information to show the influence of Justice O'Connor, given her role as a centrist Justice, between the liberal and conservative blocks on the court.

Each judgment in the Supreme Court is a binary decision, with each Justice voting on one of two outcomes. There are nine Justices in the court, with the overall judgment made by majority voting. The votes cannot be split, and given the odd number of Justices in the court, a majority decision will always be made.

We analyze the Supreme Court decisions using the agreement rate and swing vote tables given in the aforementioned NYT article. Table 1 gives a lower triangular matrix of percentage agreement rates in judgments between pairs of Justices. O'Connor was closest to the center of the court, with the least variation in agreement rates. The Justices most in agreement were Scalia and Thomas, while the Justices least in agreement were Scalia and Stevens. Table 2 details how often each Justice was on the winning side of a 5–4 vote, i.e. how often they were the "swing vote". The data show how O'Connor sided with the winning majority on 77 % of 5–4 votes, more than any other Justice. The second most influential Justice in this respect is Kennedy, who sided with the winning majority on 72 % of 5–4 votes.

In the Political Science literature, both Factor Analysis (MacDonald et al. 1991) and MDS (Brazill and Grofman 2002) have been used to analyze voting patterns and to give spatial representations of voting data. Factor analysis has been used to investigate the relationship between the political party position and the evaluation of parties in a democratic multi-party system (MacDonald et al. 1991). In Brazill and Grofman (2002), it is shown that MDS outperforms Factor Analysis in recovering lower dimensional spatial representations of binary voting data and that when the data fit a unidimensional Guttman scale the MDS recovered solution is perfect with no error.

The purpose of the analyses given in this paper is to provide a visual representation of the voting patterns of the Justices. In order to create lower dimensional representations of proximity, we use the techniques of multidimensional scaling (MDS) and of a procedure called "Parametric mapping" (or "PARAMAP"). The PARAMAP technique was originally proposed by Carroll in Shepard and Carroll (1966). We compare the PARAMAP solutions with those gained from both metric and nonmetric MDS. PARAMAP works by minimizing an index of continuity, which is derived from a metric proposed by von Neumann (1941)). This metric gives an inverse measure of trend based upon the ratio of the mean square successive difference of the data to the variance of the data. The derived measure is called kappa (κ) and is given in 1. Here, for a pair of points i and j, d_{ij} is the input configuration distance, and D_{ij} is the output configuration distance. By default, the distances are calculated with the Euclidean metric. Detailed derivations of (κ) can be found in Akkucuk (2004) and Akkucuk and Carroll (2006). A more general definition of a larger class of measures of continuity or smoothness is given in Shepard and Carroll (1966).

$$\kappa = \sum_i \sum_{j \neq i} \frac{d_{ij}^2}{D_{ij}^4} \Bigg/ \left(\sum_i \sum_{j \neq i} \frac{1}{D_{ij}^2} \right)^{-2} \tag{1}$$

2 Experimental Design

We used both the agreement and swing vote data (Tables 1 and 2) and visualized the data in one and two dimensions. To create the metric and nonmetric MDS lower dimensional solutions, we used the KYST application (Kruskal et al. 1997). To create the PARAMAP solutions we used a combination of the software developed by Akkucuk and the KNITRO solver (Byrd et al. 2006) in conjunction with an AMPL script.

A 9×9 proximity matrix was created for the swing data. Each element of the matrix was the Euclidean distance between the swing votes for each Justice. Each dimension y in the multidimensional space corresponded to one of the years from 1994 to 2004. We define the Euclidean distances between Justices i and j for the swing votes, with p_{iy} and p_{jy} defined as the swing votes for year y.

$$D^2(SW)_{ij} = \frac{1}{11} \left(\sum_{y=1}^{11} \left(p_{iy} - p_{jy}\right)^2 \right) \tag{2}$$

A 9×9 proximity matrix was created for the agreement rates. Each distance between Justices i and j was defined as the Euclidean distance between the Justices' agreement rates with the agreement rates for each individual Justice k. This matrix, derived from what can already be defined as a proximity matrix for the Justices, is a second order proximity matrix.

$$D^2(AG)_{ij} = \frac{1}{9} \left(\sum_{k=1}^{9} \left(a_{ik} - a_{jk}\right)^2 \right) \tag{3}$$

The two 9×9 matrices defined in (2) and (3) were averaged. The resulting matrix was then processed using KYST, for both metric and nonmetric MDS. The resulting KYST solutions were then used as a starting point for subsequent solutions.

A new combined matrix was created from the agreement matrix and the transposed raw swing vote matrix. Both matrices were scaled; each entry in the agreement matrix was multiplied by the square root of $1/9$ $(1/3)$, so that the squared

Table 1 Agreement matrix

	S	B	G	O	C	K	R	L
B	62							
G	66	72						
O	63	71	78					
C	33	55	47	55				
K	36	47	49	50	67			
R	25	43	43	44	71	78		
L	14	25	28	31	44	58	66	
T	15	24	26	29	44	59	68	79

S Stevens; *B* Breyer; *G* Ginsburg; *O* Souter; *C* OConnor; *K* Kennedy; *R* Rehnquist; *L* Scalia; *T* Thomas

Table 2 Swing vote matrix

	C	K	T	R	L	O	S	B	G
94–95	69	81	50	63	56	38	50	44	50
95–96	82	82	64	64	64	36	27	36	45
96–97	72	78	56	67	56	50	50	39	33
97–98	67	87	73	60	47	33	40	53	40
98–99	69	56	63	56	63	50	50	39	44
99–00	83	72	83	72	78	33	28	22	28
00–01	78	78	67	67	67	41	33	33	37
01–02	81	71	67	71	67	33	38	43	29
02–03	93	50	43	57	50	57	50	57	43
03–04	79	63	63	53	53	47	47	42	47
04–05	65	65	53	59	59	59	47	53	47

Codes as per Table 1

distances were effectively multiplied by $1/9$. The swing vote matrix was multiplied by the square root of $1/11$, so that each squared distance was effectively multiplied by $1/11$. Given the rescaled 9×9 agreement matrix A and the rescaled 9×11 swing vote matrix SW, the combined matrix is defined as:

$$\mathbf{Y} = \begin{bmatrix} A_{1,1} \cdots A_{1,9} & SW_{1,1} \cdots SW_{1,11} \\ \cdot \quad \cdot \quad \cdot & \cdot \quad \cdot \quad \cdot \\ \cdot \quad \cdot \quad \cdot & \cdot \quad \cdot \quad \cdot \\ A_{9,1} \cdots A_{9,9} & SW_{9,1} \cdots SW_{9,11} \end{bmatrix} \tag{4}$$

The matrix **Y** was used as input to the PARAMAP and MDS algorithms, with $n = 9$ points embedded in a 20 dimensional space. Each technique was tested with 2,000 random starting configurations. Multiple random starting solutions were used due to the fact that both MDS and PARAMAP functions produce a non-convex solution space. The problem of local minima can be mitigated (if not completely eliminated) by using a large number of different starting configurations.

3 Experimentation and Results

When reporting results, we give the derived solutions for 1 and 2 dimensions. We also include both the criteria for minimizing the algorithm (STRESS for Nonmetric and Metric MDS, or κ for PARAMAP) and the solution agreement rate. For each method, the solution reported in this section is the one that has the best value of the minimization criterion among the 2,000 solutions.

The agreement rate (Akkucuk 2004; Akkucuk and Carroll 2006) is a measure of similarity between solution configurations and is analogous to the Rand index of cluster solution agreement (Hubert and Arabie 1985). For both the input and output configurations, the k nearest neighbors are calculated for each solution point i. Let a_i represent the number of points in both the higher and lower dimensional neighborhoods of point i. The agreement rate A is equal to the sum of a_i across all i divided by $(k * n)$, where k is the size of the neighborhood and n is the number of solution points. For reporting purposes, the value of $k = 3$ was used, as this value of k gave the best spread of results. The one dimensional output solutions are shown in Fig. 1 and the two dimensional output solutions are shown in Fig. 2.

For the one dimensional derived solutions, the PARAMAP and metric MDS solutions have the highest agreement rate (0.8148). Both these solutions have the same ordering of Justices. The nonmetric MDS solution has a lower agreement rate (0.6293), as several of the points have co-located in a degenerate solution. We also ran multiple runs of a combinatorial seriation (or ordering) algorithm (Hubert et al. 1977). The algorithm found the same ordering as the optimal metric MDS and PARAMAP solutions. We can give an interpretation of the solution as the political alignment of the Justices. Each solution has a mirror image solution with symmetry across the 0 value of the axis, but we use the solution with the more "liberal" Justices to the left and the more "conservative" Justices to the right, as per political convention.

The PARAMAP solution gives a smooth continuum of Justices, with O'Connor in the center of the continuum. This fits in with the original NYT article, which had O'Connor as the swing Justice. The nonmetric MDS solution has the Justices in two clumps; we can interpret these clumps as splitting the Justices into liberal and conservative groups. This solution gives less interpretation than the PARAMAP solution; we cannot for example distinguish between O'Connor, who is a centrist Justice, and Thomas, who has the most conservative voting pattern of all the Justices. The metric MDS solution is intermediate to the PARAMAP and nonmetric MDS solutions. The Justices are displayed in a similar order to the PARAMAP solution, but the spacing is more uneven. Nonmetric MDS uses a monotonic regression procedure, which emphasizes order over distance, which can lead to the clumping seen in the solution. Metric MDS uses standard linear regression to fit the parameters, and thus derived solutions recreate distances as accurately as possible. The continuity criterion for PARAMAP leads to smooth, evenly spaced points.

For the two dimensional derived solutions, the PARAMAP solution has the highest agreement rate (0.9630), followed by the metric MDS solution (0.9259), and

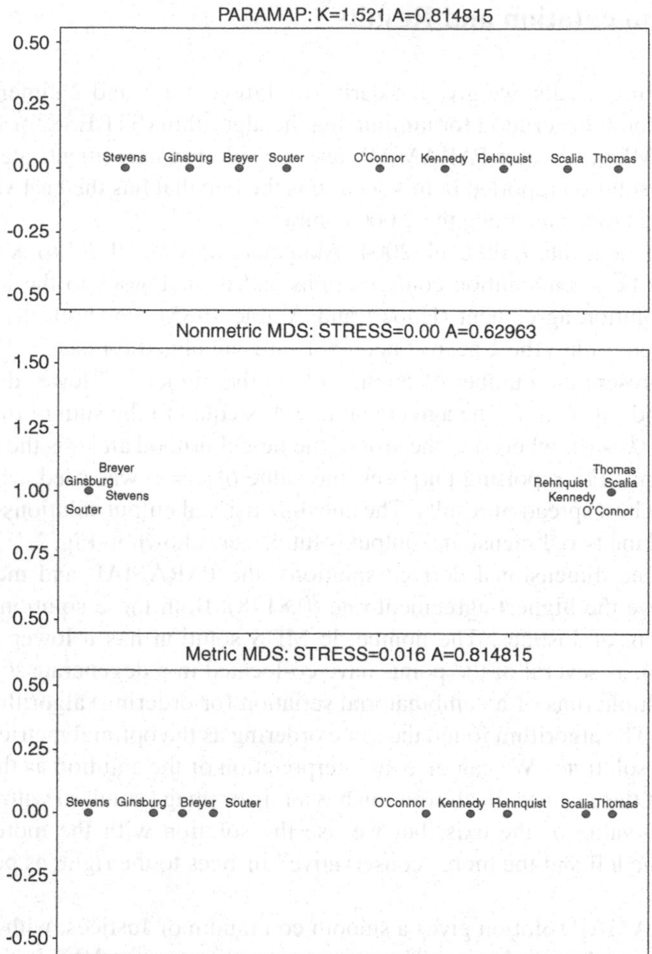

Fig. 1 1D solutions for PARAMAP, nonmetric MDS, and metric MDS

then the nonmetric MDS solution (0.9253). As per the one dimensional solution, the nonmetric MDS solution has some clumping together of points, while the metric MDS solution and the PARAMAP solution are more spread out. The clumping is not as severe for the two dimensional nonmetric MDS solution as for the one dimensional nonmetric MDS solution, but three of the liberal Justices are clumped together (Ginsberg, Breyer, and Souter) and there are two groups of conservative Justices. We can interpret these groups as highly conservative (Scalia and Thomas) and centrist conservative (Rehnquist and Kennedy). In the PARAMAP solution there are two distinct clusters of Justices. The first cluster contains Souter, Stevens, Ginsburg, and Breyer, and the second contains Scalia, Thomas, and Rehnquist. The two outliers are O'Connor and Kennedy, who are closer to the second cluster

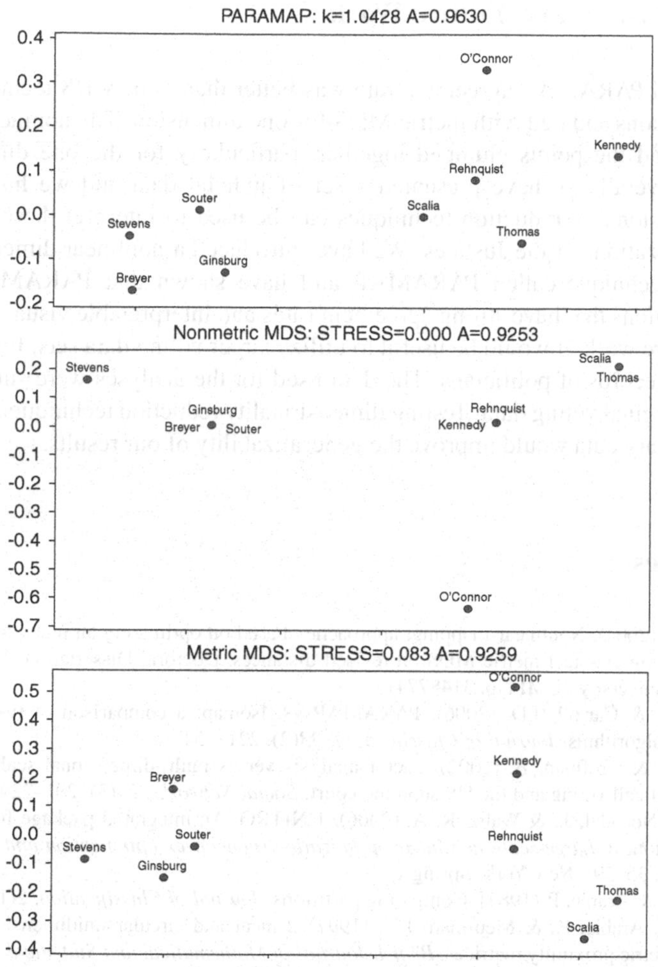

Fig. 2 2D solutions for PARAMAP, nonmetric MDS, and metric MDS

than the first. One could label the first group of four Justices as liberal and the second group of three Justices as conservative. The other two Justices are outliers, closer to the second group than to the first. This configuration implies that these outliers are the swing Justices, deciding votes where the two major voting blocs disagree. Looking back to the original swing vote data in Table 2, both O'Connor and Kennedy have far higher percentages of swing votes than the other Justices. The metric MDS solution is similar to the PARAMAP solution, but gives a slightly different interpretation for the five conservative Justices. Here, the two most conservative Justices (Scalia and Thomas) are grouped together and there is a group of three more centrist Justices, who are oriented vertically, so that the swing Justices (Kennedy and O'Connor) are furthest away from Scalia and Thomas.

4 Conclusions and Future Work

The optimal PARAMAP agreement rate was better than both MDS techniques for two dimensions and tied with metric MDS for one dimension. The nonmetric MDS solutions had the points clumped together, particularly for the one dimensional solution. Overall, we have presented a set of judicial data and we have shown how dimensionality reduction techniques can be used to interpret these data and give visualizations of the Justices. We have introduced a nonlinear dimensionality reduction technique called PARAMAP and have shown that PARAMAP gives output solutions that have strong agreement rates and interpretable visualizations.

For future work, it would be useful to utilize larger voting data sets, for example the voting records of politicians. The data used for the analyses were summarized from the original voting data. Testing dimensionality reduction techniques using the original binary data would improve the generalizability of our results.

References

Akkucuk, U. (2004). Nonlinear mapping: approaches based on optimizing an index of continuity and applying classical metric MDS on revised distances. Doctoral Dissertation. Newark, NJ: Rutgers University. (UMI No. 3148774)

Akkucuk, U., & Carroll, J.D. (2006). PARAMAP vs. Isomap: a comparison of two nonlinear mapping algorithms. *Journal of Classification, 23*(2), 221–254.

Brazill, T. J., & Grofman, B. (2002). Factor analysis versus multi-dimensional scaling: Binary choice roll-call voting and the US supreme court. *Social Networks, 24*(3), 201–229.

Byrd, R. H., Nocedal, J., & Waltz, R. A. (2006). KNITRO: An integrated package for nonlinear optimization. In *Large-scale nonlinear optimization (nonconvex optimization and its applications)* (pp. 35–59). New York: Springer.

Hubert, L. J., & Arabie, P. (1985). Comparing partitions. *Journal of Classification, 2*(1), 193–218.

Hubert, L. J., Arabie, P., & Meulman, J. J. (1997). Linear and circular unidimensional scaling for symmetric proximity matrices. *British Journal of Mathematical and Statistical Psychology, 50*(2), 253–284.

Kruskal, J. B., Young, F. W., & Seery, J. B. (1977). *How to use KYST a very flexible program to do multidimensional scaling and unfolding*. Murray Hill: ATT Bell Labs.

MacDonald, S. E., Listhaug, O., & Rabinowitz, G. (1991). Issues and party support in multiparty systems. *American Political Science Review, 85*(4), 1107–1131.

Shepard, R. N., & Carroll, J. D. (1966). Parametric representation of nonlinear data structures. In *Multivariate Analysis* (pp. 561–592). New York: Academic.

Stevenson, R. W. (2005, July 2). Court in transition: The overview: O'Connor to retire, touching off battle over court. *The New York Times*.

Stevenson, R. W., Greenhouse, L., Kirkpatrick, D. D., Lewis, N. A., & Stolberg, S. G. (2005 July 2). Changes at the U.S. supreme court. *The New York Times*.

von Neumann, J. (1941). Distribution of the ratio of the mean square successive difference to the variance. *The Annals of Mathematical Statistics, 12*(4), 367–395.

Properties of a General Measure of Configuration Agreement

Stephen L. France

Abstract Variants of the Rand index of clustering agreement have been used to measure agreement between spatial configurations of points (Akkucuk 2004; Akkucuk and Carroll 2006; Chen 2006; Chen and Buja 2009). For these measures, the k-nearest neighbors of each point are compared across configurations. The agreement measure can be generalized across multiple values of k (France and Carroll 2007). The generalized agreement metric is denoted as ψ. In this paper, we further generalize ψ to the case of more than two configurations. We develop a partial agreement measure as a neighborhood agreement equivalent of the partial correlation coefficient. We demonstrate the use of ψ and partial ψ using an illustrative example. (MATLAB implementations of routines for calculating ψ and partial ψ are available at https://sites.google.com/site/psychminegroup/.)

1 Introduction

This paper describes a methodology for comparing configurations of points. Consider a set of solution configurations $\mathbf{C}_1, \mathbf{C}_2, \ldots, \mathbf{C}_i, \ldots, \mathbf{C}_n$. Each solution configuration $\mathbf{C}_i$ contains n points embedded in an m_i dimensional space. The similarity or agreement between any two solution configurations i and j can be expressed as a function $s(\mathbf{C}_i, \mathbf{C}_j)$. There are a whole host of applications for which the calculation of a similarity/agreement metric between configurations could be useful. Applications are described in France and Carroll (2007) and Lueks et al. (2011). A summary list is given below.

S.L. France (✉)
Lubar School of Business, University of Wisconsin-Milwaukee, Milwaukee, WI 53201-0742, USA
e-mail: france@uwm.edu

B. Lausen et al. (eds.), *Algorithms from and for Nature and Life*, Studies in Classification, Data Analysis, and Knowledge Organization, DOI 10.1007/978-3-319-00035-0_15,

1. Dimensionality reduction techniques can be used to take data embedded in a higher dimensional space and create an embedding of the data in a lower dimensional space. As the number of degrees of freedom for the data is positively correlated to the data dimensionality, it may not be possible to perfectly embed the data. Different dimensionality reduction techniques may use different quality or optimization criteria, so a suitable agreement metric can be used to provide a neutral measure of solution quality. A dimensionality reduction technique can generate multiple solutions. An agreement metric can be used in several ways.
   (a) Some dimensionality reduction methods are parameterized. For example, Isomap (Joshua et al. 2000) and Local MDS (Chen 2006; Chen and Buja 2009) have an input parameter for neighborhood size. An agreement metric can be used to help find the optimal parameter values.
   (b) Some dimensionality reduction methods minimize a criterion function. The criterion function is not necessarily convex, so solution procedures may not be guaranteed to find a globally optimal solution. For example, distance based MDS (DDMDS) (Kruskal 1964) and PARAMAP (Akkucuk 2004; Akkucuk and Carroll 2006) both have non-convex optimization functions. An agreement metric can be used to help compare solutions and evaluate solution stability (France and Carroll 2009).
2. An agreement metric can be used to test longitudinal data. For example, a marketing manager may wish to test how a perceptual map of product preferences changes over time.
3. One may wish to examine interactions between more than two configurations. For example, given configurations **A**, **B**, and **Z**, one may wish to discount the effect of configuration **Z** on the relationship between configurations **A** and **B**.

In this paper we describe a methodology for testing agreement between item configurations. This methodology utilizes neighborhood agreements or rankings to give a rank order measure of agreement between configurations. We review the current literature and methodology. We extend the methodology by describing a partial agreement metric and we give a short illustrative example to show how both the agreement metric and partial agreement metric can be used in social science applications.

2 Literature Review

The agreement metrics described in this paper are based upon the Rand index for clustering agreement. The Rand index (Rand 1971) was devised to compare clustering configurations and to help test the reliability of cluster analysis techniques. A version of the Rand index, adjusted for random agreement, is described in Hubert and Arabie (1985). Versions of the Rand index to calculate agreement between solutions were developed independently in Akkucuk (2004), Akkucuk and Carroll

(2006), Chen (2006), and Chen and Buja (2009). The basic agreement rate metric (*AR*) is described below.

Consider two solution configurations **A** and **B**. **A** is an $n \times m_1$ matrix and **B** is an $n \times m_2$ matrix. Given a distance metric function *f* that conforms to the distance axioms, let $f(\mathbf{A}) = \mathbf{D}_A$ and $f(\mathbf{B}) = \mathbf{D}_B$. For each item *i*, item *j* is one of the *k* nearest neighbors of *i* if *d(i,j)* is one of *k* smallest values of $d(i,l)$, where $l = 1 \ldots n$ and $l \neq i$. An $n \times k$ matrix of nearest neighbor indexes can be created for each configuration. Let $\mathbf{N}_A$ be the matrix of nearest neighbors for configuration **A** and $\mathbf{N}_B$ be the matrix of nearest neighbors for configuration **B**. Let a_i be the number of indexes in both row *i* of $\mathbf{N}_A$ and row *i* of $\mathbf{N}_B$. The overall agreement is given in (1).

$$AR = \frac{1}{kn} \sum_{i=1}^{n} a_i \tag{1}$$

An adjusted agreement metric (A^*) is described in Akkucuk (2004) and Akkucuk and Carroll (2006). Random agreement is subtracted from the agreement metric by averaging the agreement from multiple empirically generated random samples. In Chen (2006) and Chen and Buja (2009) an adjusted agreement metric is created by assuming a hyper-geometric distribution. The expected agreement is given in (2) and the adjusted agreement is given in (3).

$$E\left[AR\left(k\right)\right] = \frac{1}{kn} \sum_{i=1}^{n} a_i = \frac{1}{kn} \sum_{i=1}^{n} k \cdot p\left(A\right) = \frac{kn}{kn} \left(\frac{k}{n-1}\right) = \frac{k}{n-1} \tag{2}$$

$$AR^* = \frac{1}{kn} \sum_{i=1}^{n} \left[a_i - \frac{k^2}{n-1} \right] \tag{3}$$

A generalized agreement metric (France and Carroll 2007) is given in (4). This agreement metric is denoted as ψ. It is calculated across all *k* and takes account of random agreement.

$$\psi = \frac{\sum\limits_{k=1}^{n-1} \left(AR(k) - E\left[AR\left(k\right)\right]\right)}{\sum\limits_{k=1}^{n-1} \left(1 - E\left[AR\left(k\right)\right]\right)} \tag{4}$$

In (4), equal weights are given for all values of *k*. A further generalization of ψ, given in (5), allows for a weighting function.

$$\psi_{f(k)} = \frac{\sum\limits_{k=1}^{n-1} \left(f\left(k\right)\left(AR(k) - E\left[AR\left(k\right)\right]\right)\right)}{\sum\limits_{k=1}^{n-1} \left(f\left(k\right)\left(1 - E\left[AR\left(k\right)\right]\right)\right)} \tag{5}$$

The function can be restricted to certain values of k. For example, setting $f(k) = 1$ for $k = 1 \ldots 4$ averages evenly over the 4 nearest neighbors. A weighting that is even for values of k from 1 to $n/4$ and then linearly declines to 0 at $n/2$ is given in (6). The weighting scheme used is reliant on the application. For example, a cellphone company marketing manager may wish to examine a perceptual map of brand positions. The manager may want the answer to the question, "Am I competing more closely with Samsung or Nokia?" Thus, a good quality solution would have strong recovery of nearest neighbors. However, for visualization of a large scale nonlinear manifold, the overall global recovery of the manifold shape may be more important than the recovery of nearest neighbors.

$$f(k) = \begin{cases} 1 & 0 \le k \le n/4 \\ 1 - \frac{k-(n/4)}{(n/4)} & n/4 < k \le n/2 \end{cases} \tag{6}$$

Several properties of ψ are given in France and Carroll (2007). These properties are listed below. The proofs are given in France and Carroll (2007).

1. AR is not monotonic with respect to k.
2. $\sup\{\psi_{f(k)}\} = 1$ and is independent of $f(k)$.
3. If $f(k) = c$ for some constant c then $\inf\{\psi_{f(k)}\} = \frac{\Upsilon - n}{(n-2)}$ where Υ is defined in (7).

$$\Upsilon = 2\sum_{i=1}^{\lfloor n/2 \rfloor} \left(\frac{\lfloor \frac{n-1}{2} \rfloor + i}{n - (2 \cdot \lceil (n-1)/2 \rceil) + 2 \cdot i} \right) \Big/ \left(\frac{n}{\lfloor \frac{n-1}{2} \rfloor + i} \right) \tag{7}$$

The ψ metric can be thought of as analogous to a discrete GINI coefficient (Corrodo 1921), but with the "inequality" curve above rather than below the line of equality (or random agreement). Given a line of random agreement over k and the value of AR plotted across k, an unweighted ψ coefficient measures the total proportion of the area above the random agreement line that is below the AR line.

The AR and ψ metrics described in this section measure the proportion of items that are in both configurations. These metrics are symmetric and they are not affected by the order of the configurations. If one was to define a_i as the number of indexes in row i of $\mathbf{N}_A$ but not in row i of $\mathbf{N}_B$ or vice versa then the metrics would be asymmetric. Asymmetric agreement metrics are described in Kaski et al. (2003). A framework for both symmetric and asymmetric agreement metrics is given in Lee and Verleysen (2009). The framework assumes a source high dimensional configuration and a derived low dimensional configuration. The nearest neighbor ranking of item j for item i is $\hat{r}_{ij}$ for the input configuration and r_{ij} for the output configuration. Hard and soft deviations from agreement are listed below.

1. Hard Intrusion: $r_{ij} \le k < \hat{r}_{ij}$
2. Soft Intrusion: $r_{ij} < \hat{r}_{ij} \le k$
3. Hard Extrusion: $r_{ij} \le k < \hat{r}_{ij}$
4. Soft Extrusion: $\hat{r}_{ij} < r_{ij} \le k$

The intrusions and extrusions can be summarized in a "co-ranking" matrix. The AR and ψ agreement metrics measure cases where there is a hard intrusion or extrusion. A method of using a diagonal subset of the co-ranking matrix is described in Lueks et al. (2011). What may be a hard inclusion or exclusion at one value of k may be a soft inclusion or exclusion at a larger value of k. For example, consider a situation where item l is the second nearest neighbor of item i in configuration **A** and the fourth nearest neighbor of item i in configuration **B**. Item l gives a hard intrusion/extrusion for $k = 2$ and $k = 3$, but a soft intrusion/extrusion for $k > 4$. For ψ, item l affects the agreement rate for $k = 2$ and $k = 3$, creating a range of "non-agreement".

3 Extension

We extend previous work by describing a partial agreement metric. The partial agreement metric is analogous to the partial correlation coefficient (Fisher 1924). The rationale behind the partial agreement metric is to discount some configuration **Z** when calculating the agreement between configurations **A** and **B**. A marketing example could be the calculation of the agreement between perceptual product maps for a consumer after two different promotions **A** and **B**. The configuration for the consumer's previous perceptual map **Z** would be discounted from the equation in order to emphasize the differences between configuration **A** and configuration **B**. The equation for partial agreement is given in (8).

$$\psi_{AB \cdot Z} = \frac{\psi_{AB} - \psi_{AZ}\psi_{BZ}}{\sqrt{1 - \psi_{AZ}^2}\sqrt{1 - \psi_{BZ}^2}} \tag{8}$$

As per the properties of the partial correlation coefficient; if $-1 \leq \psi_{AB} \leq 1$, $-1 \leq \psi_{AZ} \leq 1$, and $-1 \leq \psi_{BZ} \leq 1$ then $-1 \leq \psi_{AB \cdot Z} \leq 1$. If $\psi_{AZ} = 1$, $\psi_{AZ} = -1$, $\psi_{BZ} = 1$, or $\psi_{BZ} = -1$ then the partial agreement metric is undefined.

4 Illustrative Example

To illustrate the use of the partial agreement metric (partial ψ), we give an illustrative example. Data were taken from a survey on international educational achievement (Barro and Lee 2001). The data contain details of schooling achievement at various levels (primary, secondary, and tertiary). We used a subset of the data that contains information for all respondents aged 25 plus. Data are included for the years 1965–2000 at 5 year intervals. There are nine continuous variables detailing educational attainment. The data are given at the country and year level and are grouped into six regions. We range scaled the data and calculated Euclidean

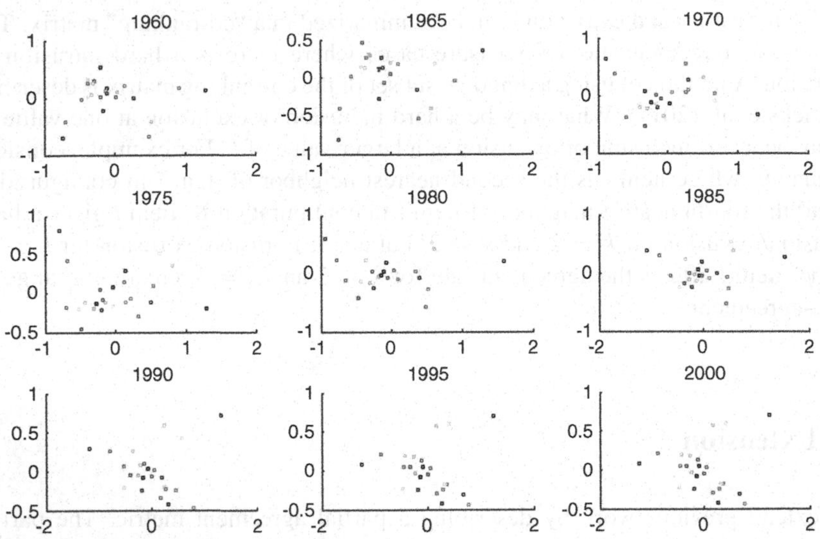

Fig. 1 MDS maps for Europe and North America

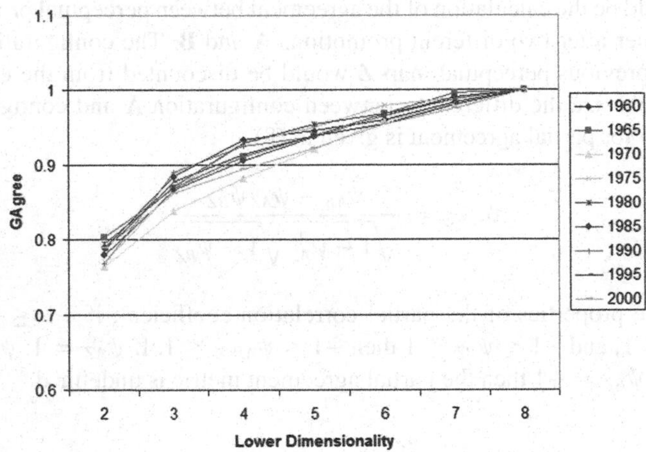

Fig. 2 MDS recovery agreement ($GAAgree = \psi$) for Europe and North America

distances between countries in each region. We created MDS embeddings from 2 to 8 dimensions for each combination of region and year. As an example, the 2 dimensional MDS maps for the Europe and North America region are given in Fig. 1.

We calculated the unweighted ψ value between each lower dimensional embedding and its source higher dimensional configuration. Agreements were calculated for all 6 regions, but for the sake of brevity, we only give agreements for the Europe

Europe and North America

ψ	1960	1965	1970	1975	1980	1985	1990	1995	2000
1960	1								
1965	0.765718	1							
1970	0.46258	0.50754	1						
1975	0.360726	0.40644	0.585421	1					
1980	0.350851	0.376349	0.462979	0.601564	1				
1985	0.378848	0.404712	0.504833	0.570006	0.721457	1			
1990	0.2449	0.255952	0.337062	0.498731	0.546954	0.550954	1		
1995	0.186175	0.212287	0.309803	0.461391	0.497316	0.502662	0.775731	1	
2000	0.187667	0.200313	0.286543	0.43126	0.476317	0.487007	0.726797	[illegible]	1

Partial ψ	1960	1965	1970	1975	1980	1985	1990	1995
1965	1							
1970	0.2689	1						
1975	0.217088	0.506184	1					
1980	0.178811	0.362172	0.543865	1				
1985	0.192564	0.401692	0.502052	0.679112	1			
1990	0.109732	0.260331	0.453834	0.50779	0.510627	1		
1995	0.110344	0.256787	0.430213	0.469531	0.475245	0.76647	1	
2000	0.089614	0.22936	0.39686	0.446267	0.457538	0.714923	[illegible]	1

Middle East and North Africa (MENA)

ψ	1960	1965	1970	1975	1980	1985	1990	1995	2000
1960	1								
1965	[illegible]	1							
1970	0.769857	[illegible]	1						
1975	0.629541	0.709508	0.709388	1					
1980	0.638945	0.666963	0.649287	0.758586	1				
1985	0.59414	0.622158	0.699431	0.685923	0.732027	1			
1990	0.441326	0.523337	0.55501	0.661151	0.580367	0.731818	1		
1995	0.451932	0.495984	0.507488	0.626816	0.549759	0.608706	0.792993	1	
2000	0.347811	0.408337	0.460414	0.547924	0.478924	0.538993	0.654064	0.792047	1

Partial ψ	1960	1965	1970	1975	1980	1985	1990	1995
1965	1							
1970	0.555433	1						
1975	0.438763	0.453202	1					
1980	0.332559	0.320583	0.596208	1				
1985	0.299885	0.471462	0.499049	0.569535	1			
1990	0.315397	0.375855	0.54979	0.432261	0.650616	1		
1995	0.248939	0.280271	0.49388	0.380348	0.474118	0.741485	1	
2000	0.23047	0.321956	0.451588	0.355911	0.440693	0.594977	0.759078	1

Fig. 3 Values of ψ and partial ψ between configurations

and North America region, which are plotted in Fig. 2. For each year, the value of ψ is plotted against output dimensionality. The value of ψ increases with output dimensionality (and available degrees of freedom) and for $k = n - 1$ dimensions, there is perfect agreement.

For each region, we calculated ψ as per (4) between every combination of years. We then calculated partial ψ between every combination of years, but attempted to remove influence of the original 1960 configuration by using 1960 as the "covariate". The rationale behind the use of partial ψ is to discount a region's initial educational attainment. When comparing configurations, the agreement is not a measure of overall change, but of change in the relative positions of the countries. Thus a region with an uneven increase in educational attainment will have lower rates of agreement than a region with a more even increase in educational attainment. Fig. 3 gives tables of ψ and partial ψ between year configurations for the Europe and North America region and for the MENA (Middle East and North Africa) region. Values of ψ greater than 0.5 are shaded in gray, with darker gray shades corresponding to higher levels of agreement. There is very strong agreement for the MENA region, particularly for the years from 1965–1985. The partial ψ values are less than the ψ values, as any changes in the configurations are amplified due to the removal of the starting configuration. The results suggest a slow, even pace of educational development in MENA in these years relative to Europe and North America. One reason for these results is that tertiary eduction expanded rapidly in parts of Europe and North America during these years, but did not expand to such a great degree in the MENA region.

5 Conclusions and Future Work

In this paper, we review work on the generalized agreement metric ψ. We introduce a measure of partial agreement (partial ψ). An illustrative example shows how partial ψ can be used to emphasize changes in agreement relative to a base configuration. There is much scope for future work on ψ and on similar agreement metrics. The interpretation of ψ is currently subjective and there are no rigorous statistical techniques for testing the significance of ψ. Future work could examine the statistical properties of ψ and could explore bootstrapping approaches for calculating confidence intervals for ψ. While the agreement metrics described in this paper utilize ordinal data, they are reliant on the distance metric used to calculate the rank order nearest neighbors representations. Work could be done to test the effects of different distance metrics on the value of ψ. Work could be done to optimize ψ for a given dimensionality reduction problem. This could build on work on optimizing the Rand index (Brusco and Steinley 2008).

References

Akkucuk, U. (2004). *Nonlinear mapping: Approaches based on optimizing an index of continuity and applying classical metric MDS on revised distances*. Ph.D. dissertation, Rutgers.

Akkucuk, U., & Carroll, J. D. (2006). PARAMAP vs. Isomap: a comparison of two nonlinear mapping algorithms. *Journal of Classification, 23*(2), 221–254.

Barro, R. J., & Lee, J. Wha. (2001). International data on educational attainment: updates and implications. *Oxford Economic Papers, 53*(3), 541–563.

Brusco, M., & Steinley, D. (2008). A binary integer program to maximize the agreement between partitions. *Journal of Classification, 25*(2), 185–193.

Chen, L. (2006). *Local multidimensional scaling for nonlinear dimension reduction, graph layout and proximity analysis*. Ph.D. dissertation, University of Pennsylvania.

Chen, L., & Buja, A. (2009). Local multidimensional scaling for nonlinear dimension reduction, graph drawing, and proximity analysis. *Journal of the American Statistical Association, 104*(486), 209–219.

Corrodo, G. (1921). Measurement of inequality and incomes. *The Economic Journal, 31*, 124–126.

Fisher, R. A. (1924). The distribution of the partial correlation coefficient. *Metron, 3*, 332.

France, S. L., & Carroll, J. D. (2007). Development of an agreement metric based upon the Rand index for the evaluation of dimensionality reduction techniques, with applications to mapping customer data. In P. Perner (Ed.), *Machine learning and data mining in pattern recognition* (pp. 499–517). Heidelberg: Springer.

France, S. L. & Carroll, J. D. (2009). DEMScale: Large scale MDS accounting for a ridge operator and demographic variables. In N. M. Adams, C. Robardet, A. Siebes, & J.-F. Boulicaut (Eds.), *Proceedings of the 8th international symposium on intelligent data analysis*. Heidelberg: Springer.

Hubert, L. J., & Arabie, P. (1985). Comparing partitions. *Journal of Classification, 2*(1), 193–218.

Kaski, S., Nikkila, J., Oja, M., Venna, J., Toronen, P., & Castren, E. (2003). Trustworthiness and metrics in visualizing similarity of gene expression. *BMC Bioinformatics, 4*(1), 48. M3: 10.1186/1471-2105-4-48.

Kruskal, J. B. (1964). Multidimensional scaling for optimizing a goodness of fit metric to a nonmetric hypothesis. *Psychometrika, 29*(1), 1–27.

Lee, J. A. & Verleysen, M. (2009). Quality assessment of dimensionality reduction: Rank-based criteria. *Neurocomputing, 72*(7–9), 1431–1443.

Lueks, W., Mokbel, B., Biehl, M., & Hammer, B. (2011). How to evaluate dimensionality reduction? In *Proceedings of the workshop – new challenges in neural computation* (Vol. 5, pp. 29–37). ISSN:1865-3960. http://www.techfak.uni-bielefeld.de/~fschleif/mlr/mlr_05_2011.pdf;MLR0511.

Rand, W. M. (1971). Objective criteria for the evaluation of clustering methods. *Journal of the American Statistical Association, 66*(336), 846–850.

Tenenbaum, J. B., de Silva, V., & Langford, J. C. (2000). A global geometric framework for nonlinear dimensionality reduction. *Science, 290*(5500), 2319–2323.

Convex Optimization as a Tool for Correcting Dissimilarity Matrices for Regular Minimality

Matthias Trendtel and Ali Ünlü

Abstract Fechnerian scaling as developed by Dzhafarov and Colonius (e.g., Dzhafarov and Colonius, J Math Psychol 51:290–304, 2007) aims at imposing a metric on a set of objects based on their pairwise dissimilarities. A necessary condition for this theory is the law of Regular Minimality (e.g., Dzhafarov EN, Colonius H (2006) Regular minimality: a fundamental law of discrimination. In: Colonius H, Dzhafarov EN (eds) Measurement and representation of sensations. Erlbaum, Mahwah, pp. 1–46). In this paper, we solve the problem of correcting a dissimilarity matrix for Regular Minimality by phrasing it as a convex optimization problem in Euclidean metric space. In simulations, we demonstrate the usefulness of this correction procedure.

1 Preliminaries

For a set of stimuli $X = \{x_1, x_2, \ldots, x_n\}$, let $\psi : X \times X \rightarrow \mathbb{R}_+$ be some discriminability measure, mapping pairs of stimuli $x_i \in X$ and $x_j \in X$ into the set of nonnegative reals. For example, a pair of line segments (x_i, x_j) may be repeatedly presented to an observer (or a group of observers), and $\psi(x_i, x_j)$ may be estimated by the frequency of responses "they are different (in length)". Possible examples are numerous, and more can be found in Dzhafarov and Colonius (2006b). In such a pairwise presentation paradigm, even if stimuli x_i and x_j have the same value

M. Trendtel (✉)
Division for Methodology and Statistics, Federal Institute for Educational Research, Innovation & Development of the Austrian School System (BIFIE), Salzburg, Austria
e-mail: m.trendtel@bifie.at

A. Ünlü
Chair for Methods in Empirical Educational Research, TUM School of Education, Technische Universität München, München, Germany
e-mail: ali.uenlue@tum.de

B. Lausen et al. (eds.), *Algorithms from and for Nature and Life*, Studies in Classification, Data Analysis, and Knowledge Organization, DOI 10.1007/978-3-319-00035-0_16,

(say, they are line segments of the same length), they must occupy different spatial and/or temporal positions. This difference in spatial or temporal locations does not enter in the comparison, but it may affect the way people perceive lengths, and this in turn may lead to $\psi(x_i, x_i)$ being larger than $\psi(x_i, x_j)$ for some distinct x_i and x_j, in the same way as it may lead to $\psi(x_i, x_j) \neq \psi(x_j, x_i)$. Therefore the notion of observation area was introduced (Dzhafarov 2002; Dzhafarov and Colonius 2006b). Henceforth we call first observation area the set of stimuli presented, say, first in time or on the left, and second observation area the set of stimuli presented second or on the right.

In the context of pairwise same-different comparisons, it is a well-established empirical fact that $\psi(x_i, x_j)$, however obtained, is not a metric. So the data have to be modified to make such data-analytic procedures as multidimensional scaling (MDS; e.g., Kruskal and Wish 1978) applicable. Also, the class of allowable metrics in MDS is usually a priori restricted to Minkowskian metrics in low-dimensional spaces of real-component vectors. By contrast, Fechnerian scaling (FS) deals directly with ψ-data subject to same-different comparisons, and it imposes no a priori restrictions on the class of metrics computed from ψ. The only property of the ψ-data which is required by FS is Regular Minimality (RM).

This principle postulates the existence of pairs of stimuli that are mutually the most similar ones to each other. In regard to discrimination probability matrices (discrimination probabilities being presented as an $n \times n$ matrix $\Psi = (\psi_{ij})_{i,j=1,\ldots,n}$), this means that a matrix Ψ satisfies RM, iff every row has a unique minimum entry which is also the unique minimum entry in its column. For instance, the matrix of ψ-data

$$\begin{pmatrix} & x_1 & x_2 & x_3 \\ x_1 & 0.2 & 0.1 & 0.5 \\ x_2 & 0.7 & 0.3 & 0.2 \\ x_3 & 0.1 & 0.6 & 0.3 \end{pmatrix}$$

satisfies RM, with (x_1, x_2), (x_2, x_3), and (x_3, x_1) being pairs of mutually most similar stimuli. Here, the first symbol in every pair refers to a row object (all row objects belonging to one, the "first", observation area) and the second symbol refers to a column object (in the "second" observation area).

FS imposes a metric G, if RM is satisfied. For every pair of objects (x_i, x_j) we consider all possible chains of objects $(x_i, x_{k_1}, \ldots, x_{k_r}, x_j)$. Presupposing RM, for each such a chain we compute what is called its psychometric length. Then we find a chain with minimal psychometric length, and take this minimal value for the quasidistance from x_i to x_j (referred to as the oriented Fechnerian distance). Quasidistance is a pairwise measure which satisfies all metric properties except for symmetry. In FS we symmetrize this quasimetric and transform it into a metric, taking it for the "true" or "overall" Fechnerian distance $G(x_i, x_j)$ between x_i and x_j. (For a detailed discussion refer to Dzhafarov (2002), Dzhafarov and Colonius (2006a), Dzhafarov and Colonius (2006b), Dzhafarov and Colonius (2007), and Dzhafarov et al. (2011).)

RM can be generalized to nonnegative reals:

Definition 1. Let $S_n = \{\sigma : \{1, \ldots, n\} \to \{1, \ldots, n\} : \sigma \text{ a permutation}\}$ be the group of permutations on $\{1, \ldots, n\}$. A matrix $\Psi = (\psi_{ij})_{i,j=1,\ldots,n} \in \mathbb{R}_+^{n^2}$, viewed as the vector $(\psi_{11}, \psi_{12}, \ldots, \psi_{1n}, \psi_{21}, \psi_{22}, \ldots, \psi_{2n}, \ldots, \psi_{n1}, \psi_{n2}, \ldots, \psi_{nn})^T$, of discriminability measures satisfies RM, iff there exists an $\sigma \in S_n$ such that, for any $i = 1, \ldots, n$,

$$\psi_{i\sigma(i)} < \psi_{ij} \text{ for } j \neq \sigma(i), \text{ and}$$
$$\psi_{i\sigma(i)} < \psi_{j\sigma(i)} \text{ for } j \neq i.$$

Then, σ is also uniquely determined, and we say that Ψ satisfies RM in σ-form. The set of all $n \times n$ matrices satisfying RM in σ-form (for one given permutation σ) is denoted by RM_σ^n.

The square matrix Ψ is the matrix of true (unknown) discrimination probabilities ψ_{ij} in the population of reference. Same-different comparisons can be modeled by a Bernoulli random variable: 1 if the response is "different", with "success" probability ψ_{ij}, and 0 if the response is "same", with probability $1 - \psi_{ij}$. The relative success counts $\hat{\psi}_{ij}$ from independent samples of independent responses are the maximum likelihood estimators (MLEs) for the ψ_{ij}'s. The population matrix Ψ is unknown and estimated from the data using its MLE $\hat{\Psi} = \left(\hat{\psi}_{ij}\right)_{i,j=1,\ldots,n}$.

The observed data matrix $\hat{\Psi}$ may not satisfy RM, although the underlying population matrix Ψ may do. In other words, the compliance of a matrix of discrimination probabilities with RM must be tested statistically. In the literature, a parametric hypothesis test based on a measure was proposed by Ünlü et al. (2010) and a nonparametric test based on permutations was derived in Dzhafarov et al. (2011). However, these tests do not allow correcting the data for RM.

In this paper, a method is proposed for correcting a dissimilarity matrix for RM in an "optimal" way, with respect to the Euclidean metric. We interpret Ψ and $\hat{\Psi}$ as points in the n^2-dimensional nonnegative orthant and propose finding that RM-compliant point M of the orthant which minimizes the Euclidean norm $\|\hat{\Psi} - M\|$ (up to arbitrary $\epsilon > 0$; see Sect. 3). Stated in terms of convex optimization, this problem is solved by expressing it as an equivalent convex optimization problem:

$$\text{minimize } g(M)$$
$$\text{subject to } M \in \mathcal{D},$$

where $g : \mathbb{R}_+^{n^2} \to \mathbb{R}_+$ is a convex function and $\mathcal{D} \subset \mathbb{R}_+^{n^2}$ is a convex feasible set.

2 Convexity and Regular Minimality

We apply convex geometry (e.g., Dattorro 2009; Ekeland and Temam 1999) to discriminability measure matrices. We prove the convexity of the set RM_σ^n of all $n \times n$ matrices satisfying RM in a specified σ-form. This allows us to abstract the constraints of the optimization problem.

Lemma 1. *Let* $M = (m_{ij})_{i,j=1,\ldots,n}$ *and* $M' = (m'_{ij})_{i,j=1,\ldots,n}$ *in* RM_σ^n *be matrices that satisfy RM in* σ*-form. Then any convex combination of* M *and* M' *also satisfies RM in* σ*-form. In particular, the set* RM_σ^n *of all matrices satisfying RM in* σ*-form is convex.*

Proof. Let $0 \leq \lambda \leq 1$. Obviously, the convex combination $\lambda M + (1-\lambda)M'$ defines a discriminability measure matrix, with entries in $\mathbb{R}_+ = [0, \infty)$. Let $i, k, l \in \{1, \ldots, n\}$ such that $k \neq \sigma(i)$ and $l \neq i$. According to Definition 1,

$$m_{i\sigma(i)} < m_{ik},\ m_{i\sigma(i)} < m_{l\sigma(i)} \text{ and } m'_{i\sigma(i)} < m'_{ik},\ m'_{i\sigma(i)} < m'_{l\sigma(i)}.$$

Therefore,

$$\lambda m_{i\sigma(i)} + (1-\lambda)m'_{i\sigma(i)} < \lambda m_{ik} + (1-\lambda)m'_{ik} \text{ and}$$
$$\lambda m_{i\sigma(i)} + (1-\lambda)m'_{i\sigma(i)} < \lambda m_{l\sigma(i)} + (1-\lambda)m'_{l\sigma(i)}.$$

Since the inequalities in Definition 1 are strict, the set RM_σ^n is not closed. To apply fundamental results in convex geometry, however, it is necessary to consider its topological closure $\overline{RM_\sigma^n}$. This means, inter alia, that a matrix which is an element of $\overline{RM_\sigma^n} \setminus RM_\sigma^n$ violates RM by ties only. Since RM_σ^n is convex, so is its closure $\overline{RM_\sigma^n}$. As a consequence of the Hahn-Banach separation theorem (e.g., Hiriart-Urruty and Lemaréchal 2001) a closed convex set in $\mathbb{R}^{n^2}$ is the intersection of all halfspaces that contain it. In our case, every inequality in Definition 1 represents a halfspace bounded by a hyperplane, which can be represented using a matrix variable.

As an example, consider an 3×3 matrix $M \in \overline{RM_{\sigma=\text{id}}^3}$

$$M = \begin{pmatrix} m_{11} & m_{12} & m_{13} \\ m_{21} & m_{22} & m_{23} \\ m_{31} & m_{32} & m_{33} \end{pmatrix}$$

with nonnegative entries m_{ij}, and with $\sigma = \text{id}$ the identity. Since M satisfies RM in ($\sigma = \text{id}$)-form except for ties, the diagonal entries m_{11}, m_{22}, and m_{33} are minimal in the rows and columns, and, for instance,

$$m_{11} \leq m_{12}.$$

Reading M as the vector $\mathbf{M} = (m_{11}, m_{12}, m_{13}, m_{21}, m_{22}, m_{23}, m_{31}, m_{32}, m_{33})^T$ this inequality is equivalent to

$$b \cdot \mathbf{M} \leq 0,$$

where $b = (1, -1, 0, 0, 0, 0, 0, 0, 0)$ and "$\cdot$" is the matrix product. In this case, the set $\mathscr{H}_- = \{\mathbf{M} \in \mathbb{R}^{3^2} | b \cdot \mathbf{M} \leq 0\}$ forms a halfspace, which is bounded by the hyperplane $\partial\mathscr{H} = \{\mathbf{M} \in \mathbb{R}^{3^2} | b \cdot \mathbf{M} = 0\}$. If we construct such vectors b for every inequality according to Definition 1 and merge them row-wise into a matrix B, we obtain

$$B = \begin{pmatrix} 1 & -1 & 0 & 0 & 0 & 0 & 0 & 0 & 0 \\ 1 & 0 & -1 & 0 & 0 & 0 & 0 & 0 & 0 \\ 0 & 0 & 0 & -1 & 1 & 0 & 0 & 0 & 0 \\ 0 & 0 & 0 & 0 & 1 & -1 & 0 & 0 & 0 \\ 0 & 0 & 0 & 0 & 0 & 0 & -1 & 0 & 1 \\ 0 & 0 & 0 & 0 & 0 & 0 & 0 & -1 & 1 \\ 1 & 0 & 0 & -1 & 0 & 0 & 0 & 0 & 0 \\ 1 & 0 & 0 & 0 & 0 & 0 & -1 & 0 & 0 \\ 0 & -1 & 0 & 0 & 1 & 0 & 0 & 0 & 0 \\ 0 & 0 & 0 & 0 & 1 & 0 & 0 & -1 & 0 \\ 0 & 0 & -1 & 0 & 0 & 0 & 0 & 0 & 1 \\ 0 & 0 & 0 & 0 & 0 & -1 & 0 & 0 & 1 \end{pmatrix},$$

and then the following equivalence holds

$$M \in \overline{RM^3_{\sigma=\text{id}}} \iff B \cdot \mathbf{M} \leq 0.$$

This example is now generalized. The matrix B can be constructed systematically for a given dimension n and a specified σ-form using the following procedure: For $l = 1, \ldots, n$, let

$$B_l^1(n, \sigma) = \left(b_l^1(n, \sigma)_{ij}\right) \in \{-1, 0, 1\}^{n-1 \times n}$$

and

$$B_l^2(n, \sigma) = \left(b_l^2(n, \sigma)_{ij}\right) \in \{-1, 0, 1\}^{n-1 \times n^2}$$

with

$$b_l^1(n, \sigma)_{ij} = \begin{cases} 1 & \text{for } j = \sigma(l), \\ -1 & \text{for } i = j \text{ and } i < \sigma(l), \\ -1 & \text{for } i = j - 1 \text{ and } \sigma(l) \leq i, \\ 0 & \text{else,} \end{cases}$$

and

$$b_l^2(n,\sigma)_{ij} = \begin{cases} 1 & \text{for } j = (l-1)n + \sigma(l), \\ -1 & \text{for } j = (i-1)n + l \text{ and } i < \sigma(l), \\ -1 & \text{for } j = i \cdot n + l \text{ and } \sigma(l) \leq i, \\ 0 & \text{else.} \end{cases}$$

Define

$$B(n,\sigma) = \begin{pmatrix} B_1^1(n,\sigma) & 0 & \dots & 0 & 0 \\ 0 & B_2^1(n,\sigma) & 0 & \dots & 0 \\ \vdots & & \ddots & & \vdots \\ 0 & & \dots & 0 & B_n^1(n,\sigma) \\ & & B_1^2(n,\sigma) & & \\ & & \vdots & & \\ & & B_n^2(n,\sigma) & & \end{pmatrix},$$

which is a matrix with $2n(n-1)$ rows and n^2 columns (and entries in $\{-1, 0, 1\}$).

Now, for a matrix $M = (m_{ij})_{i,j=1,\dots,n} \in \mathbb{R}_+^{n^2}$,

$$M \in \overline{RM_\sigma^n} \iff B(n,\sigma) \cdot \mathbf{M} \leq 0. \quad (1)$$

This equivalence gives a matrix representation for the closed convex set of all $n \times n$ matrices satisfying RM in σ-form except for ties.

3 Convex Optimization and Regular Minimality

Setting up the convex optimization problem is now straightforward. Assume we have observed a data matrix $\hat{\Psi} \in \mathbb{R}_+^{n^2}$ which violates RM in σ-form. The question posed is this: can one correct the data for RM in a principled way?

As a possible answer, we propose to consider the following convex optimization problem:

$$\begin{aligned} &\text{minimize} \quad \|\hat{\Psi} - M\| \\ &\text{subject to} \quad M \in \overline{RM_\sigma^n}, \end{aligned}$$

where $\|.\|$ is the Euclidean norm. Using the matrix representation in Eq. 1, this problem can be expressed as

$$\begin{aligned}\text{minimize} \quad & \sqrt{\sum_{i=1}^{n}\sum_{j=1}^{n}(\hat{\psi}_{ij}^2 - 2\hat{\psi}_{ij}m_{ij} + m_{ij}^2)} \\ \text{subject to} \quad & B(n,\sigma)\cdot\mathbf{M} \leq 0, \\ & \mathbf{M} \in \mathbb{R}_+^{n^2}.\end{aligned}$$

The data $\hat{\psi}_{ij}$ are constants, and $\sqrt{.}$, $x \mapsto x + c$, and $x \mapsto 2x$ are strictly increasing monotone functions. Therefore, an equivalent formulation of this program is

$$\begin{aligned}\text{minimize} \quad & \tfrac{1}{2}\sum_{i=1}^{n}\sum_{j=1}^{n} m_{ij}^2 - \sum_{i=1}^{n}\sum_{j=1}^{n}\hat{\psi}_{ij}m_{ij} \\ \text{subject to} \quad & B(n,\sigma)\cdot\mathbf{M} \leq 0, \\ & \mathbf{M} \in \mathbb{R}_+^{n^2}.\end{aligned}$$

This problem is a quadratic program (QP). The constraints describe a polyhedron, the optimality conditions for convex optimization are satisfied, and there exists a unique global optimum for this problem. For details, see, for instance, Boyd and Vandenberghe (2009) and Roberts and Varberg (1973).

The solution M^{opt} of this QP comes with two caveat:

Caveat 1. If $\hat{\Psi} \notin \overline{RM_\sigma^n}$ is observed, M^{opt} obtained from the QP by such popular algorithms as Goldfarb and Idnani (1983) will be an element of the boundary of $\overline{RM_\sigma^n}$. That is, M^{opt} will violate RM by at least one tie. Yet, to obtain a "solution" M^* satisfying RM, geometrically speaking, we "proceed a tiny bit further into" the interior of $\overline{RM_\sigma^n}$.

More precisely, for some feasible small $\epsilon > 0$ (e.g., machine accuracy), we take

$$M^* = \hat{\Psi} + (1+\epsilon)(M^{opt} - \hat{\Psi}) \in RM_\sigma^n$$

as the final solution, that is, the "optimal" RM-compliant approximation to $\hat{\Psi}$. This correction procedure was used in the simulation study in the next section.

Caveat 2. If $\hat{\Psi} \in \overline{RM_\sigma^n} \setminus RM_\sigma^n$, that is, if $\hat{\Psi}$ lies on the boundary of $\overline{RM_\sigma^n}$, then $\hat{\Psi}$ violates RM in σ-form by ties only, and $\hat{\Psi}$ will be the solution of the QP. In this case, $\hat{\Psi}$ belongs to at least one hyperplane bounding the polyhedron $\overline{RM_\sigma^n}$, and we would have to choose a direction (e.g., a combination of inward normal vectors) along which "to proceed a tiny bit further into" the interior of $\overline{RM_\sigma^n}$. This procedure is not discussed in this paper.

4 Simulation Study

We present the results of a simulation study. To investigate the usefulness of the presented correction procedure, we considered dimensions n of the stimulus space ranging from 5 to 8, and sample sizes of $m = 10, 50, 100, 150$. We set

Table 1 Results from the simulation study. n and m represent the dimension of the stimulus space and the sample size, respectively. If the Euclidean distance $\|\Psi - M^*\|$ was smaller or greater than $\|\Psi - \hat{\Psi}\|$, this was counted as an "Improvement" or "Worsening", respectively. Columns 5 and 11 ("Violated by ties only") represent the number of cases when a simulated matrix $\hat{\Psi}$ violated RM by ties only, and Columns 6 and 12 ("No violation") when $\hat{\Psi}$ satisfied RM

n	m	Improvement	Worsening	Violated by ties only	No violation	n	m	Improvement	Worsening	Violated by ties only	No violation
5	10	676,968	0	230,367	92,665	7	10	845,537	0	142,434	12,029
5	50	470,358	0	116,571	413,071	7	50	669,026	0	134,099	196,875
5	100	375,825	0	72,685	551,490	7	100	560,690	0	99,143	340,167
5	150	325,268	0	52,864	621,868	7	150	495,047	0	79,779	425,174
6	10	777,770	0	186,955	35,275	8	10	888,833	0	107,356	3,811
6	50	582,230	0	129,701	288,069	8	50	739,053	0	128,299	132,648
6	100	477,019	0	87,940	435,041	8	100	633,212	0	104,015	262,773
6	150	417,806	0	67,185	515,009	8	150	565,612	0	85,897	348,491

$\epsilon = 10^{-6}$ (see Caveat 1). For each stimulus space dimension n, 1,000 population matrices $\Psi = (\psi_{ij})_{i,j=1,\ldots,n}$ of discrimination probabilities satisfying RM (σ = id) were drawn. For each Ψ and for any sample size m, 1,000 sample matrices $\hat{\Psi} = (\hat{\psi}_{ij})_{i,j=1,\ldots,n}$ were simulated, using binomial distributions $Bi(\psi_{ij}, m)$. The correction procedure was performed to obtain M^*. The Euclidean distance between Ψ and M^* was compared to the distance between Ψ and $\hat{\Psi}$. All computations were done in the software environment R (www.r-project.org), and the algorithm by Goldfarb and Idnani (1983) was used.

Table 1 reports the results of the simulation study.

For any setting of n and m, there was not a single case out of 1,000,000 simulations that produced a worsening. Whenever a solution $M^* \neq \hat{\Psi}$ of the optimization procedure was returned, an improvement in the Euclidean distance was observed. Taking into account the fact that the set of all matrices satisfying RM becomes very (very) small in relation to all possible matrices in the unit hypercube (see Trendtel et al. 2010), this result is not very surprising. If the true discrimination probability matrix Ψ satisfies RM and the observed matrix $\hat{\Psi}$ does not, it is very likely that almost every matrix M satisfying RM is closer to Ψ than $\hat{\Psi}$ is, with respect to the Euclidean distance.

5 Discussion

RM is an important property in psychophysics. In this paper we have proposed a correction procedure for RM based on convex optimization.

The relative frequencies of violations by ties only are high enough to justify further research on the problem described in Caveat 2. For instance, the geometrical properties of $\overline{RM_\sigma^n}$ may be investigated in order to justify an "optimal" direction into the interior of this set. Or, the interior of the feasible region may be traversed to reach an optimal solution. In future research, the discussion must be extended incorporating such additional constraints as ones that represent confidence intervals. Such an endeavor may allow, both, testing and correcting for RM simultaneously.

Acknowledgements We are deeply indebted to Professor Ehtibar N. Dzhafarov for introducing us to this topic.

References

Boyd, S., & Vandenberghe, L. (2009). *Convex optimization*. New York: Cambridge University Press.

Dattorro, J. (2009). *Convex optimization & euclidean distance geometry*. Palo Alto: Meboo.

Dzhafarov, E. N. (2002). Multidimensional Fechnerian scaling: pairwise comparisons, regular minimality, and nonconstant self-similarity. *Journal of Mathematical Psychology, 46*, 583–608.

Dzhafarov, E. N., & Colonius, H. (2006a). Reconstructing distances among objects from their discriminability. *Psychometrika, 71*, 365–386.
Dzhafarov, E. N., & Colonius, H. (2006b). Regular minimality: a fundamental law of discrimination. In: H. Colonius & E. N. Dzhafarov (Eds.), *Measurement and representation of sensations* (pp. 1–46). Mahwah: Erlbaum.
Dzhafarov, E. N., & Colonius, H. (2007). Dissimilarity cumulation theory and subjective metrics. *Journal of Mathematical Psychology, 51*, 290–304.
Dzhafarov, E. N., Ünlü, A., Trendtel, M., & Colonius, H. (2011). Matrices with a given number of violations of regular minimality. *Journal of Mathematical Psychology, 55*, 240–250.
Ekeland, I., & Temam, R. (1999). *Convex analysis and variational problems*. Philadelphia: SIAM.
Goldfarb, D., & Idnani, A. (1983). A numerically stable dual method for solving strictly convex quadratic programs. *Mathematical Programming, 27*, 1–33
Hiriart-Urruty, J., & Lemaréchal, C. (2001). *Fundamentals of convex analysis*. Berlin: Springer.
Kruskal, J. B., & Wish, M. (1978). *Multidimensional scaling*. Beverly Hills: Sage.
Roberts, A. W., & Varberg, D. E. (1973). *Convex functions*. New York: Academic.
Trendtel, M., Ünlü, A., & Dzhafarov, E. N. (2010). Matrices satisfying Regular Minimality. *Frontiers in Quantitative Psychology and Measurement, 1*, 1–6.
Ünlü, A., & Trendtel, M. (2010). Testing for regular minimality. In A. Bastianelli & G. Vidotto (Eds.), *Fechner Day 2010* (pp. 51–56). Padua: The International Society for Psychophysics.

Principal Components Analysis for a Gaussian Mixture

Carlos Cuevas-Covarrubias

Abstract Given a p-dimensional random variable $\mathbf{X}$, Principal Components Analysis defines its optimal representation in a lower dimensional space. In this article we assume that $\mathbf{X}$ is distributed according to a Mixture of two Multivariate Normal Distributions and we project it onto an optimal vector space. We propose an original combination of principal components and linear discriminant analysis where the area under the ROC curve appears as the link between both methods. We represent $\mathbf{X}$ in terms of a small number of independent factors with maximum contribution to the area under the ROC curve of an optimal linear discriminant function. A practical example illustrates how these factors describe the differences between two categories in a simple classification problem.

1 Introduction

Principal Components Analysis (PCA) and Linear Discriminant Analysis (LDA) are very important methods of Multivariate Statistics. Given a p-dimensional random variable $\mathbf{X}$, PCA defines its optimal representation in a lower dimensional space; this representation is usually assessed in terms of a percentage of total variation expressed as a function of the eigenvalues of the covariance matrix (Izenman 2008). LDA assumes that Ω, the sample space of $\mathbf{X}$, is partitioned into two different categories: Ω_0 and Ω_1. Given $\mathbf{x}$, a particular realization of $\mathbf{X}$, LDA is used to infer whether $\mathbf{x}$ corresponds to an observation from Ω_0 or Ω_1 (Izenman 2008). The area under the ROC curve is one many criteria to assess the global quality of this clasification procedure. Probably, the earliest discussion about PCA for a Gaussian Mixture is the one given in Kullback (1968), where PCA and LDA are discussed

C. Cuevas-Covarrubias (✉)
Anahuac University, Huixquilucan, State of Mexico, Mexico
e-mail: ccuevas@anahuac.mx

B. Lausen et al. (eds.), *Algorithms from and for Nature and Life*, Studies in Classification, Data Analysis, and Knowledge Organization, DOI 10.1007/978-3-319-00035-0_17,

from an information theoretic point of view. Chang (1983) presents an interesting discussion about PCA for Gaussian Mixtures and proves that those components associated to the largest eigenvalues do not necessarily contain the largest amount of information. More recently, Caprihan (2008) present an interesting application of the proposal in Chang (1983) in a medical diagnosis context. They calculate the eigenvectors of the total sum of squares matrix and select those factors that maximize the Mahalanobis distance between both considered categories. In this article, we assume that **X** is distributed according to a mixture of two multivariate normal distributions, and we explore an original combination of PCA and LDA where the area under the ROC curve appears as the link between both methods. Our objective is to represent **X** in terms of a small number of factors. In a similar way to Chang (1983) and Caprihan (2008) we miximize the separability between Ω_0 and Ω_1. However, we select those eigenvectors with maximum contribution to the area under the ROC curve of an optimal dicriminant function. Contrasting with Chang (1983) and Caprihan (2008) we diagonalize both covariance matrices in the probability model; therefore, the final result is a set of factors simultaneously independent in both elements[1] of the Gaussian mixture. We call this idea "Mutual Principal Components" (MPCA).

2 Risk Scores and ROC Curves

In its simplest form, discriminant analysis is used to summarize a vector of covariates **X** into a univariate risk score S, and to discriminate between two groups or populations (Ω_0,Ω_1) according to the following decision rule:

$$\text{Classify in } \begin{cases} \Omega_1 \textit{ if } \; S > t \\ \Omega_0 \textit{ if } \; S \leq t, \end{cases} \tag{1}$$

where t is a decision threshold. Given a fixed t in (1), the classification rule can be assessed in terms of its error rates: $\beta(t) = Pr[S \leq t|\Omega_1]$ (false negative rate) and $\alpha(t) = Pr[S > t|\Omega_0]$ (false positive rate). Some times the assessment is based on $1 - \beta(t)$ and $1 - \alpha(t)$. Given a score S with class conditional distribution functions $F_0(t) = \Pr[S \leq t|\Omega_0]$ and $F_1(t) = \Pr[S \leq t|\Omega_1]$, its *ROC*[2] curve is the following set

$$ROC = \{(x, y)|x = 1 - F_0(t), y = 1 - F_1(t), -\infty < t < \infty\}. \tag{2}$$

The ROC curve is a plot of the sensitivity $(1 - \beta(t))$ expressed as a function of the false positive rate $(\alpha(t))$ for every possible t (Krzanowski and Hand 2008).

[1]Class conditional distributions

[2]Receiver Operating Characteristic

The area covered by the ROC curve (denoted as A) is a global measure of the performance of S (Bamber 1975). In general terms, the closer A is to 1, the better the performance of S. When both class conditional distributions are equal, then S is not informative; in this case $A = \frac{1}{2}$.

ROC curves are invariant to monotonous transformations of the score (Bamber 1975). Therefore, if there is a monotonous transformation $T(S)$ of S such that its class conditional distributions are both normal, then its *ROC* curve is given by

$$ROC = \{(u,v)|u = 1 - \Phi(t), v = 1 - \Phi\left(\frac{t-d}{r}\right), -\infty < t < \infty\}, \quad (3)$$

where Φ denotes a standard normal distribution function; $d = \frac{E(T(S)|\Omega_1) - E(T(S)|\Omega_0)}{\sqrt{Var(T(S)|\Omega_0)}}$ and $r^2 = \frac{Var(T(S)|\Omega_1)}{Var(T(S)|\Omega_0)}$. Any variable S with this property is said to be a binormal score with parameters (d, r^2) (Krzanowski and Hand 2008; Metz and Xiaochuan 1999). The area under the ROC curve of such a score is

$$A = \Phi\left(\frac{d}{\sqrt{1+r^2}}\right). \quad (4)$$

3 Linear Discriminant Analysis

Let $\mathbf{X}$ be a p-dimensional vector of covariates such that its class conditional distributions are both multivariate normal. Thus, its class conditional density given Ω_i is:

$$f_i(\mathbf{x}) = \frac{1}{\|2\pi\Sigma_i\|^{1/2}} \exp -\frac{1}{2}(\mathbf{x} - \boldsymbol{\mu}_i)^t \Sigma_i^{-1}(\mathbf{x} - \boldsymbol{\mu}_i), \quad (5)$$

with $i = 0, 1$. The parameters $\boldsymbol{\mu}_i$ and Σ_i represent the class conditional expectation and class conditional covariance matrix of $\mathbf{X}$ given Ω_i respectively. When a vector $\mathbf{X}$ is normally distributed in both classes we refer to it as a *Multivariate Normal Score* (MNS). Let $\mathbf{X}$ be a MNS and let θ be a constant vector. In principle, any linear combination $S = \theta^t(\mathbf{X} - \boldsymbol{\mu}_0)$ could be used in a classification rule like the one given in Eq. (1). Thus, it is important to find that linear score with the best global performance. Given the normality assumption on $\mathbf{X}$, any linear combination of its components is a binormal score. The area under its ROC curve, given by Eq. (4), takes the following form: $A_S(\theta) = \Phi\left[\frac{\theta^t(\boldsymbol{\mu}_1 - \boldsymbol{\mu}_0)}{\sqrt{\theta^t(\Sigma_0+\Sigma_1)\theta}}\right]$ (Su and Liu 1993). We are interested on finding θ_* such that $A_S(\theta_*)$ is maximum; this optimal score is given in Theorem 1 (see demonstration in Su and Liu (1993)).

Theorem 1. *Let $\mathbf{X}$ be a MNS and let $\mathbf{A} = (\boldsymbol{\mu}_1 - \boldsymbol{\mu}_0)(\boldsymbol{\mu}_1 - \boldsymbol{\mu}_0)^T$ and $\mathbf{B} = (\Sigma_0 + \Sigma_1)$. Then, no linear combination of the elements of $\mathbf{X}$ has an area under its ROC curve larger than $A_S(\theta_*) = \Phi(\sqrt{\varphi})$, where φ is the only positive eigenvalue of $\mathbf{B}^{-1}\mathbf{A}$ and $\boldsymbol{\theta}_*$ its corresponding normalized eigenvector.*

It is possible to show that in Theorem 1

$$\varphi = (\boldsymbol{\mu}_1 - \boldsymbol{\mu}_0)^t B^{-1}(\boldsymbol{\mu}_1 - \boldsymbol{\mu}_0) \tag{6}$$

and

$$\theta_* = \frac{1}{\sqrt{\varphi}} B^{-1}(\boldsymbol{\mu}_1 - \boldsymbol{\mu}_0). \tag{7}$$

4 Mutual Principal Components

A MNS **X** is expressed in its canonical form (Kullback 1968) when $(\mathbf{X}|\Omega_0) \sim N(\mathbf{0}, \mathbf{I})$ and $(\mathbf{X}|\Omega_1) \sim N(\boldsymbol{\delta}, \Lambda)$, where **I** is the identity matrix and Λ is diagonal. The canonical form of **X** implies that its class conditional covariance matrices are both diagonal; therefore, its components are independent in both classes. Every MNS can be transformed to its canonical form with just a linear transformation.

Definition 1. Let **X** be a multivariate score such that its class conditional covariance matrices (Σ_0, Σ_1) are both positive definite and with all their eigenvalues with multiplicity one; also let (μ_0, μ_1) be its class conditional expectations. The Principal Components Vector *(PCV)* of **X** is given by the following linear transformation

$$\mathbf{Z} = \Gamma^T \Sigma_0^{-1/2}(\mathbf{X} - \mu_0), \tag{8}$$

where $\Gamma\Lambda\Gamma^T$ is the spectral decomposition of $\Sigma_0^{-1/2}\Sigma_1\Sigma_0^{-1/2}$, and $\Sigma_0^{1/2}$ is any square root of Σ_0. We refer to $\mathbf{T} = \Sigma_0^{-1/2}\Gamma$ as the *Principal Components Transformation Matrix.*

When **X** is a MNS, **Z** is just its canonical form. Therefore, $(\mathbf{Z}|\Omega_1) \sim N(\boldsymbol{\delta}, \Lambda)$, where $\delta = \Gamma^T\Sigma_0^{-1/2}(\mu_1 - \mu_0)$ and $\Lambda = \text{diag}\{\lambda_i\}_{i=1}^p$ is the matrix of eigenvalues of $\Sigma_0^{-1/2}\Sigma_1\Sigma_0^{-1/2}$. Thus, according to Eqs. (6) and (7) , the linear combination of the elements of **Z** with maximum area under its ROC curve is given in terms of the following vector of coefficients:

$$\xi_* = \left[\frac{\delta_1}{1+\lambda_1}, \frac{\delta_2}{1+\lambda_2}, \ldots, \frac{\delta_p}{1+\lambda_p}\right]^t. \tag{9}$$

The area under its ROC curve is

$$A_Z(\xi_*) = \Phi\left[\sqrt{\sum_{i=1}^p \frac{\delta_i^2}{1+\lambda_i}}\right]. \tag{10}$$

Something interesting about this principal components transformation is that

$$\theta_* = \Sigma_0^{-\frac{1}{2}} \Gamma \xi_*. \tag{11}$$

The columns of $\mathbf{T}$ form a set of linearly independent[3] eigenvectors of $\mathbf{R} = \Sigma_0^{-1}\Sigma_1$ and the diagonal of Λ contains their respective eigenvalues (see Harville 1997, p. 562).

Each principal component Z_i is a linear score itself and the area under its ROC curve is

$$A_i = \Phi\left[\sqrt{\frac{\delta_i^2}{1+\lambda_i}}\right]. \tag{12}$$

The main objective of PCA is to represent random vectors in a linear space of lower dimension. $A_Z(\xi_*)$ can be used as a criterion to asses and control this reduction of dimensionality (see Eqs. (10) and (14)). Once $\mathbf{X}$ is transformed into its PCV $\mathbf{Z}$, its p components can be ordered as $Z_{(1)}, Z_{(2)}, \ldots, Z_{(p)}$ where

$$\frac{\delta_{(i)}^2}{1+\lambda_{(i)}} \geq \frac{\delta_{(i+1)}^2}{1+\lambda_{(i+1)}} \tag{13}$$

with $i = 1, 2, \ldots p-1$. After ordering the components of $\mathbf{Z}$, the sequence $\{\rho_k\}_{i=1}^{p}$ of log odds ratios $\rho_k = \frac{\log \frac{A_Z|k}{1-A_Z|k}}{\log \frac{A_Z(\xi_*)}{1-A_Z(\xi_*)}}$ is computed. In this ratio

$$A_Z|k = \Phi\left[\sqrt{\sum_{i=1}^{k} \frac{\delta_i^2}{1+\lambda_i}}\right] \tag{14}$$

is the maximum area under the ROC curve that can be obtained with a linear combination of the first k principal components $(Z_{(1)}, Z_{(2)}, \ldots, Z_{(k)})$. Any dimension reduction can imply a smaller area under the ROC curve of the final linear score. Therefore, if the minimum log odds ratio that can be afforded is the $100(1-p)\%$ of $\log \frac{A_Z(\xi_*)}{1-A_Z(\xi_*)}$, the new multivariate score in a lower dimension is obtained by selecting the first k components, where k is the minimum k such that $\rho_k \geq p$.

As we mentioned before, the columns of our transformation matrix $\mathbf{T}$ are linearly independent eigenvectors of $\Sigma_0^{-1}\Sigma_1$. If the inner product $\langle \mathbf{u}, \mathbf{v} \rangle_{\mathbf{B}} = \mathbf{u}^t \mathbf{B} \mathbf{v}$ is now considered[4]; the columns of $\mathbf{T}$ are not only linearly independent, but actually orthogonal. Therefore, θ_* can be expressed as a linear combination of the columns of $\mathbf{T}$: *i.e.* $\theta_* = a_1\mathbf{t}_1 + a_2\mathbf{t}_2 + \ldots + a_p\mathbf{t}_p$ with $a_i = \frac{\langle \theta_*, t_i \rangle_{\mathbf{B}}}{\|\mathbf{t}_i\|_{\mathbf{B}}^2}$. However, we know that $\mathbf{T}^T\mathbf{B}\mathbf{T} = \mathbf{I} + \Lambda$, therefore $\|t_i\|_{\mathbf{B}}^2 = 1 + \lambda_i$. On the other hand,

[3] But not necessarily orthogonal

[4] remember that $\mathbf{B} = (\Sigma_0 + \Sigma_1)$

$$\langle \theta_*, \mathbf{t}_i \rangle_{\mathbf{B}} = \frac{1}{\sqrt{\varphi}} [\mathbf{B}^{-1}(\mu_1 - \mu_0)]^T \mathbf{B} \mathbf{t}_i = \frac{\delta_i}{\sqrt{\varphi}} \tag{15}$$

where $\varphi = \sum_{i=1}^{p} \frac{\delta_i^2}{1+\lambda_i}$. Therefore,

$$a_i = \frac{1}{\sqrt{\varphi}} \cdot \frac{\delta_i}{1+\lambda_i} \tag{16}$$

and Eq. (11) takes the form

$$\theta_* \sqrt{\varphi} = \frac{\delta_1}{1+\lambda_1}\mathbf{t}_1 + \frac{\delta_2}{1+\lambda_2}\mathbf{t}_2 + \ldots + \frac{\delta_p}{1+\lambda_p}\mathbf{t}_p. \tag{17}$$

The fact that $\mathbf{T}$ is $\mathbf{B}$-orthogonal can be used in a reduction of dimensionality process. In order to identify those components with the largest contribution to S_* we can use the squared cosine of the angle between θ_* and each column of $\mathbf{T}$. If we denote this angle as π_i, then

$$\cos \pi_i = \frac{\langle \theta_*, \mathbf{t}_i \rangle}{\|\theta_*\|_{\mathbf{B}} \|\mathbf{t}_i\|_{\mathbf{B}}} = \frac{1}{\sqrt{\varphi}} \frac{\delta_i}{\sqrt{1+\lambda_i}}. \tag{18}$$

Thus, each squared cosine measures the contribution of its corresponding principal component to φ, the squared Mahalanobis distance between the centers of both class conditional distributions.[5] Again, the principal components can be ordered as $\{\mathbf{Z}_{(1)}, \mathbf{Z}_{(2)}, \ldots, \mathbf{Z}_{(p)}\}$ according to their contribution to φ. To reduce the dimensionality, we select just enough components to get a minimum proportion of φ (or a maximum angle with respect to $\theta*$). This approach is analogous to the one based on the contributions to $A_Z(\xi_*)$. Given Theorem 1 and Eq. (18), we can see that

$$A_i = \Phi\left[\sqrt{\varphi} \cos \pi_i\right] \text{ and } A_{\mathbf{Z}|k} = \Phi\left[\sqrt{\varphi \sum_{i=1}^{k} (\cos \pi_i)^2}\right]. \tag{19}$$

Showing that $A_i = \frac{1}{2}$ for any principal component such that $\langle \mathbf{t_i}, \theta_* \rangle_{\mathbf{B}} = 0$.

5 A Practical Example

Consider the species $\Omega_0 =$ *Versicolour* and $\Omega_0 =$ *Virginica* from the Fisher's Irises data set. Each sample unit is represented in terms of a vector of four variables $\mathbf{X} = (X_1 = \text{sepal length},\ X_2 = \text{sepal width},\ X_3 = \text{petal length},\ X_4 = \text{petal width})$.

[5]This makes our proposal similar to the methodology applied in Caprihan (2008).

Table 1 Fisher's Irises: Principal components analysis

Z_i	A_i	Rank	$A_{S\|i}$	$\log \frac{A_{S\|i}}{1-A_{S\|i}}$	$100 * \rho_i\%$	$\sum \frac{\delta_i^2}{1+\lambda_i}$	$(\cos \pi_i)^2$	π_i (°)
1	0.780	(3)	0.960	3.1911	57.33	3.0887	0.4344	48.76
2	0.950	(2)	0.993	4.9840	89.91	6.1139	0.8599	21.98
3	0.730	(4)	0.995	5.3343	96.23	6.7303	0.9466	13.36
4	0.960	(1)	0.996	5.5428	100.00	7.1094	1.0000	0.00

This data set is frequently analyzed in the literature; its class conditional distributions are approximately multivariate normal. Thus, the optimal linear score S_* is assumed to be binormal with $d = 3.93$ and $r^2 = 1.17$. The principal components transformation matrix is

$$\mathbf{T} = \begin{pmatrix} 0.43 & 1.57 & 1.12 & 2.35 \\ -1.13 & -2.49 & 3.26 & -1.23 \\ 3.30 & 0.33 & -1.25 & -2.67 \\ 8.92 & 2.58 & -0.96 & -0.18 \end{pmatrix} \tag{20}$$

The canonical form of **X** is given as in def. (1) with $\delta = (2.02, 2.76, -0.89, -2.30)^T$ and $\Lambda = diag[5.65, 1.53, 1.13, 0.71]$ Table (1) shows the principal components analysis of **X**. The first set of columns shows, for each component, the area under its ROC curve and its corresponding rank. The second set of columns contains the sequence of cumulative areas for the first k components and its corresponding log odds ratio. The final set shows the angle between θ^* and the coefficients for the best linear combination of $(Z_{(1)}, Z_{(2)}, \ldots Z_{(k)})$; its squared cosine is also given.

According to Table 1, **X** can be represented by its first two principal components keeping almost the 90% of the total information in **X**. The angle between θ_* and the linear space generated by the coefficients of ($Z_{(1)}$ and $Z_{(2)}$) is less than 22 degrees.[6] We conclude that **X** can be satisfactorily represented in a bidimensional space. Figure 1 shows the plot of $Z_{(1)}$ vs $Z_{(2)}$; it is evident how both groups can be separated by a straight line. The coefficients of $Z_{(1)}$ and $Z_{(2)}$ are given by the forth and second columns of **T** respectively. In a simplistic interpretation, we could say that $Z_{(1)}$ mainly indicates the difference of the sepal and petal lengths and $Z_{(2)}$ indicates the difference of the petal and sepal widths. Taking this into account, Fig. 1 would suggest that the Virginica irises tend to have longer and wider petals with shorter and narrower sepals than the Versicolour ones.

[6]This is less than angle formed by the hands of a clock at five past twelve.

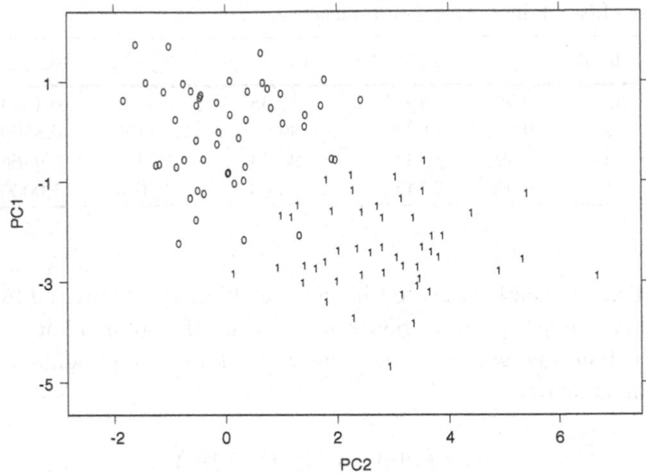

Fig. 1 First vs second discriminant components: Versicoulor = 0, Virginica = 1

6 Conclusion

We have explored an original application of ROC curves to Principal Components Analysis. Given a random sample coming from a mixture of two normal distributions, we diagonalize both covariance matrices simultaneously. The area under the ROC curve helps to identify those components with the largest contribution to an optimal discriminant function. Our discussion concentrates on the two categories problem. However, this method can be applied in a three categories context as long as the covariance matrices involved are proportional to each other (see Harville 1997). Mutual Principal Components is similar to other methods reported in the literature. As in Chang (1983) and Caprihan (2008), it identifies those principal components that maximize the Mahalanobis distance between both class conditional distributions. However, our proposal produces independent factors.

References

Bamber, D. (1975). The area above the ordinal dominance graph and the area bellow the receiver operating characteristic graph. *Journal of Mathematical Psychology, 12*, 387–415.

Caprihan, A., Pearlson, G. D., & Calhoun, V. D. (2008). Aplication of principal component analysis to distinguish patients with schizophrenia from healthy controls based on fractional anisotropy measurements. *Neuroimage, 42*(2), 675–682.

Chang, W. C. (1983). Using principal components before separating a mixture of two multivariate normal distributions. *Applied Statistics, 32*(3), 267–275.

Harville, D. A. (1997). *Matrix algebra from a statistitian's perspective*. New York: Springer.

Izenman, A. J. (2008). *Modern multivariate statistical techniques: regression, classification and manifold learning*. New York: Springer.
Krzanowski, W. J., & Hand, D. J. (2008). *ROC curves for continuous data*. Boca Raton, FL: CRC/Taylor and Francis/Chapman & Hall.
Kullback, S. (1968). *Information theory and statistics*. Mineola, NY: Dover.
Metz, C. E., & Xiaochuan, P. (1999). "Proper" binormal ROC curves: Theory and maximum likelihood estimation. *Journal of Mathematical Psychology, 43*, 1–33.
Su, J. Q., & Liu, J. S. (1993). Linear combinations of multiple diagnostic markers. *Journal of the American Statistical Association, 88*(424), 1350–1355.

Interactive Principal Components Analysis: A New Technological Resource in the Classroom

Carmen Villar-Patiño, Miguel Angel Mendez-Mendez, and Carlos Cuevas-Covarrubias

Abstract Principal Components Analysis (PCA) is a mathematical technique widely used in multivariate statistics and pattern recognition. From a statistical point of view, PCA is an optimal linear transformation that eliminates the covariance structure of the data. From a geometrical point of view, it is simply a convenient axes rotation. A successful PCA application depends, at a certain point, on the comprehension of this geometrical concept; however, to visualize these axes rotation can be an important challenge for many students. At the present time, undergraduate students are immersed in a social environment with an increasing amount of collaborative and interactive elements. This situation gives us the opportunity to incorporate new and creative alternatives of knowledge transmission. We present an interactive educational software called Mi-iPCA, that helps students understand geometrical foundations of Principal Components Analysis. Based on the Nintendo's Wiimote students manipulate axes rotation interactively in order to get a diagonal covariance matrix. The graphical environment shows different projections of the data, as well as several statistics like the percentage of variance explained by each component. Previous applications of this new pedagogical tool suggest that it constitutes an important didactic support in the classroom.

C. Villar-Patiño (✉) · M.A. Mendez-Mendez
Faculty of Engineering, Anahuac University, Edo. de Mexico, Mexico
e-mail: maria.villar@anahuac.mx; mmendez@anahuac.mx

C. Cuevas-Covarrubias
Actuarial Sciences School, Anahuac University, Edo. de Mexico, Mexico
e-mail: ccuevas@anahuac.mx

B. Lausen et al. (eds.), *Algorithms from and for Nature and Life*, Studies in Classification, Data Analysis, and Knowledge Organization, DOI 10.1007/978-3-319-00035-0_18,

1 Introduction

Principal Component Analysis (PCA) is a mathematical technique widely used in multivariate statistics for finding patterns in data, of high dimension spaces. The mathematical procedure to get the principal components transformation is defined in terms of the correlation structure between the variables considered. From a geometrical point of view, this transformation is an axes rotation. For many students, to visualize this axes rotation and understand its meaning may be a difficult task.

Technological advances and better understanding of psychological and social aspects of Human Computer Interaction (HCI) have lead to a recent explosion of new interaction forms. The time where human computer interfaces was limited to desktop computers using mouse and keyboard for interaction is passing away. Novel input devices like Wiimote and multi-touch surfaces are increasing in popularity (Shaer 2009). Some theoretical foundations of HCI , suggest that embodiment and physical engagement are becoming a key issue (Holmquist et al. 2010). With 87.4 million of Nintendo's Wii consoles sold by August 2011 (VGChartz Network 2011), the Wiimote became a familiar device for most of young people, who find it enjoyable. According to O'Malley and Fraser (2004):

> It is commonly believed that physical action is important in learning, and there is a good deal of research evidence in psychology to support this. Piaget and Bruner showed that children can often solve problems when given concrete materials to work with before they can solve them symbolically … So evidence suggests that young children (and adults as well) can in some senses 'know' things without being able to express their understanding through verbal language or without being able to reflect on what they know in an explicit sense.

In our project, students use Mi-iPCA in order to manipulate axes rotations interactively. They control this device using a simple twist of their wrist. They rotate axes to diagonalize the covariance matrix of the actual coordinates representing data. There are three different graphical projections to visualize the data set. Mi-iPCA was developed using free GNU/GPL libraries and was evaluated in two ways: subjective using an opinion questionnaire and objective with a learning experiment that includes a comprehension test.

2 Previous Work

There are several educational tools using the Wiimote as an interaction device. There are also commercial and open source software to visualize PCA. However, it seems that Mi-iPCA is the first educational tool specifically designed to facilitate the understanding on how PCA works through an ingenious application of the Wiimote.

Shaffer et al. (2005), discuss how video games have the potential to change the nowadays landscape of education. They describe an approach where they design learning environments built on the educational properties of games. They promote a new model of learning "through meaningful activity in virtual worlds as a preparation for meaningful activity in our post-industrial, technology-rich, real world". Mi-iPCA is not a key stone in a new model of learning, but we realize that undergraduate students are immersed in a social environment with an increasing amount of collaborative and interactive elements. This situation brings us the opportunity to incorporate new and creative alternatives of knowledge transmission. Many examples of the usage of Wii related technology can be found in the literature. Pearson and Bailey (2007) evaluate and identify the accessibility of the Nintendo Wii console for supporting disabled learners in an educational context. Holmquist et al. (2010) describe a collaborative project between Standford University, Sdertn University and two primary schools (one from Sweden and one from US) to introduce science education through the use of computers and video games; the Wiimote was selected as the interaction device for this project. They believe that by integrating elements of video game technologies into their educational curricula, they can enhance students' interest in their educational activities, materials and technologies. Daniels (2009) reported that since 2008, undergraduate students from Iowa State University participate in a laboratory based Wiimotes interaction, as part of their course: "Introduction to Computer Engineering and Problem Solving I". Promising results were obtained.

Jeong et al. (2009) developed an interactive system for PCA called iPCA. The objective is to help users to get a better understanding of PCA applications. It was designed as a statistical analysis tool rather than a pedagogical device. This system allows users to visualize data in four views: projection view, data view, eigenvector view and correlation view. In order to demonstrate the usefulness of their system, they conducted a comparative evaluation in relation with SAS/INSIGTH's; a well-known commercial tool. Their evaluation consisted on performing four analysis tasks in high dimensional data sets. This tasks were solved by 12 undergraduate students. The time needed to solve the problem with each system was measured. They applied a subjective questionnaire asking the participant to feedback about the system. They found that the use of iPCA for analysis increased the performance of the students because the interface design and the set of interactions facilitated the discovering of the relation between coordinate spaces and data dimensions.

3 Mi-iPCA and PCA

PCA is an exploratory tool designed by Karl Pearson in 1901 to identify unknown trends in a multidimensional data. The algorithm was introduced to psychologist's in 1933 by Harold Hotelling. Today we know that implementing PCA is equivalent to obtain the Spectral Decomposition of the covariance matrix of the data set (Garcia 2008). PCA is a useful statistical technique that is widely used for applications such as dimensionality reduction, feature extraction and data visualization (Bishop 2006).

Any real symmetric $m \times m$ matrix A has a spectral decomposition of the form:

$$A = U \Lambda U^T \tag{1}$$

where U is an orthonormal matrix and Λ is a diagonal matrix. The columns of U are the eigenvectors of A and the diagonal elements of Λ are their corresponding eigenvalues. The eigenvalue equation can be expressed as:

$$Ax = \lambda x \tag{2}$$

In Mi-iPCA we incorporate two commonly PCA definitions: PCA as the orthogonal projection of data onto a lower dimensional space where the variance of the projected data is maximized (the principal subspace), and PCA as the linear projection that minimizes the mean squared distances between the data points and their projections (just as originally suggested by Karl Pearson). Because it is designed with a pedagogical objective, Mi-iPCA works on a three dimensional space only. The original data must be standardized (subtract mean and divide by the standard deviation). The covariance matrix is continuously computed and displayed. The definition for the covariance matrix for 3 dimensional data set, represented by x, y and z coordinates is Smith (2002):

$$\mathbf{A} = \begin{pmatrix} cov(x,x) & cov(x,y) & cov(x,z) \\ cov(y,x) & cov(y,y) & cov(y,z) \\ cov(z,x) & cov(z,y) & cov(z,z) \end{pmatrix} \tag{3}$$

The actual PCA is determined by the eigenvectors and eigenvalues of the covariance matrix A which are calculated numerically. Each position of the Wiimote defines an orthonormal basis for the three dimensional Euclidean space. Thus, every element of the sample is projected to this basis and displayed on the screen. This task is performed continuously as long as the student keeps on moving the Wiimote. The projection on the computer screen is updated automatically, giving a clear sensation of real movement. The PCA problem is solved once the Wiimote projection space coincides with the actual eigenvectors of the covariance matrix. Students can compare the quality of the solution obtained by direct comparison of the orthonormal basis with these eigenvectors. The covariance matrix of the projected data is also displayed. Thus, students can constantly analyze it and stop once it is diagonal.

4 Software Description and Implementation

Mi-iPCA uses free GNU/GPL libraries and was developed and tested in various Linux distributions with the 2.6 kernel. GNU Scientific Library (GSL) (GNU Scientific Library 2011) was used for the statistics and mathematical procedures. OpenGL and GLUT libraries were applied for the graphical interface. Cwiid was employed to establish the Wiimote remote communication through Bluetooth.

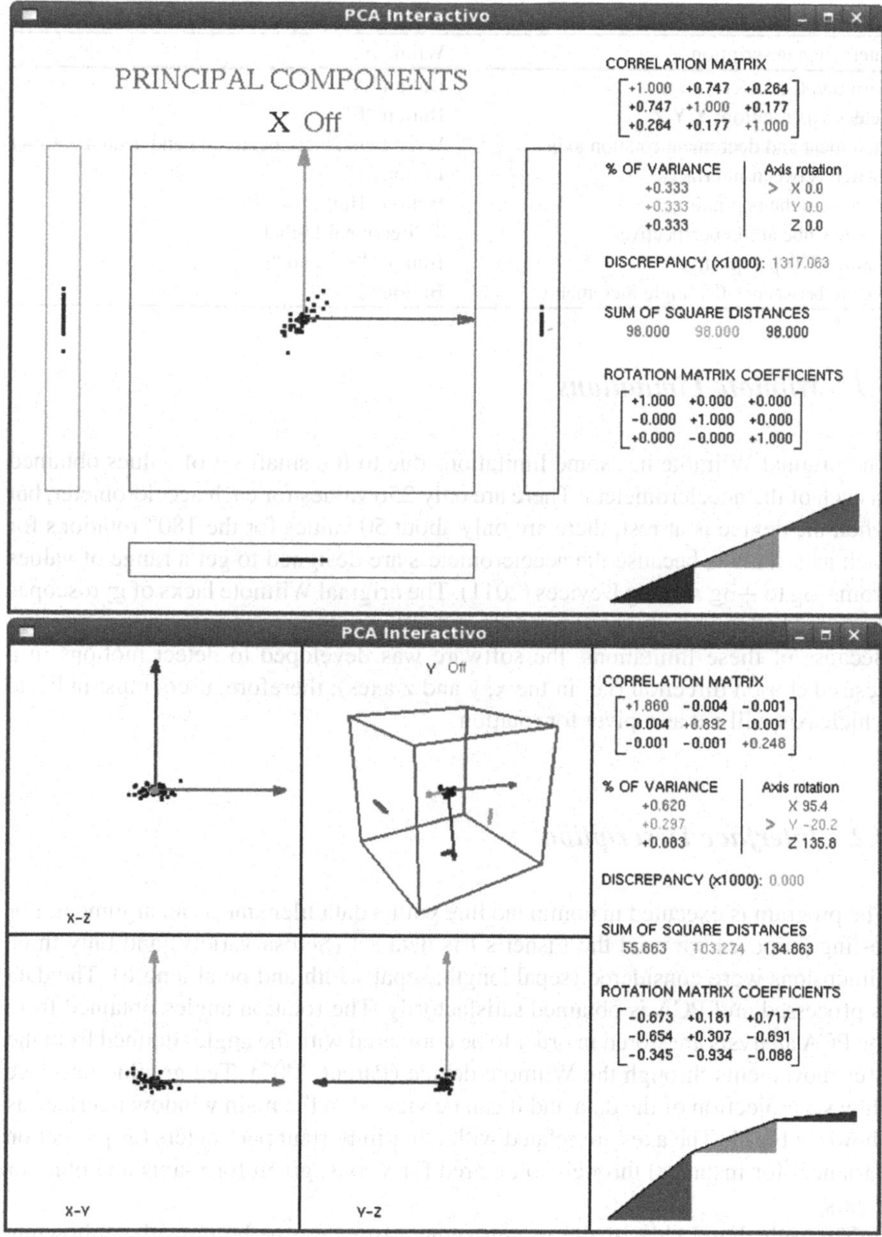

Fig. 1 Initial scatter plot interface with Setosa original data (*top*) and the solution in a four view with 2D scatter plots and a 3D isometric perspective (*down*)

Table 1 Wiimote interaction

Interaction description	Wiimote
Turn on/off accelerometer	Button "A"
Select axis rotation: X Y Z	Button "B"
Increment and decrement rotation axis	Wrist twist, increment(ccw) and decrement(cw)
Switch between interfaces	Button "1"
Start from the beginning	Button "Home"
Rotate Cube at 3D perspective	4-directional D-Pad
Zoom at 3D perspective	Buttons "+" and "−"
Toggle between 1-0.1 angle increments	Button "2"

4.1 Wiimote Limitations

The original Wiimote has some limitations due to the small set of values obtained in each of the accelerometers. There are only 256 values for each accelerometer, but when the device is at rest, there are only about 50 values for the 180° rotations for each axis. This is because the accelerometers are designed to get a range of values from −3g to +3g Analog Devices (2011). The original Wiimote lacks of gyroscopes that would allow us getting the position of the Wiimote in a three-dimensional space. Because of these limitations, the software was developed to detect motions in a desired chosen direction (i.e. in the x, y and z axes); therefore, users must indicate which axis will act as a pivot for rotation.

4.2 Interface Description

The program is executed in command line with a data filename as an argument. For testing purposes we used the Fisher's Iris data set (Setosa variety) and only three dimensions were considered (sepal length, sepal width and petal length). The data is processed and PCA is obtained satisfactorily. The rotation angles obtained from the PCA analysis are stored in order to be compared with the angles defined from the user movements through the Wiimote device (Parent 2002). The graphic interface shows a projection of the data and it can be viewed in the main window interface as shown in Fig. 1. The axes are related with other important parameters (as projection variances for instance) through colors: red for x-axis, green for y-axis and blue for z-axis.

The right hand side of the screen shows, from top to bottom, the following: Correlation matrix (remember that data are standardized), percentage of variance explained by each component and rotation angle, discrepancy of the solution obtained with respect to the actual PCA solution, sum of the squared distance between each data points and the line of best fit (Pearson's criteria), rotation matrix. At the bottom the percentage of variance explained by each component is

Table 2 Groups schedule

Group	Stage 1	Stage2	Stage3
Red	Theory	Test	Mi-iPCA
Green	Theory	Mi-iPCA	Test
Blue	Mi-iPCA	Theory	Test

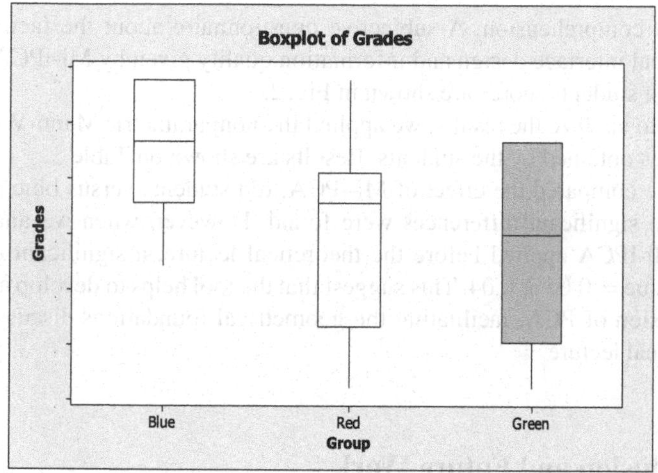

Fig. 2 Students grades boxplot

graphically reported on a scree plot. Student can choose between three ways to visualize data: scatter plot of the projected data and its projection over the axes; a 3D isometric perspective; or three separate 2D scatter plots plus an isometric representation. We show an example of these three interfaces in Fig. 1.

The Wiimote has a set of buttons, that can be used besides the accelerometers. We provide the student with the interaction possibilities described in Table 1:

5 Testing Mi-iPCA

The application was tested by 24 undergraduate students selected at random, 16 from the Actuarial Sciences Programme and 8 Engineering Students. The students attended a theoretical lecture on PCA and a computer practice using Mi-iPCA. Then, a test with 9 questions about geometrical foundations of PCA was applied to all of them. This group of students was divided at random into three groups (green with 6 students, blue with 5 students and red whit 13 students); proportions between actuaries and engineers were preserved within each group. The order between theory and practice was controlled, and applied according to Table 2:

The theoretical lecture had a 20 min duration. During the computer practice, students worked in pairs for 20 min. The test was designed to measure the PCA

Table 3 Mann-Whitney test applied to students grades

	Blue + green	Red	Blue	Green	Blue	Red
Median	6.66	5.55	6.667	5	6.667	5.556
Mann-Whitney	154.0		39.5		65.0	
P Value		0.17		0.05		0.04

geometrical comprehension. A subjective questionnaire about the facility on the use, graphical interface design and information quality given by Mi-iPCA was also applied. Test students scores are shown in Fig. 2.

In order to analyze the results, we applied the nonparametric Mann-Withney test on the scores obtained by the students. Results are shown on Table 3.

When we compared the effect of Mi-iPCA, red students versus blue plus green students, no significant differences were found. However, when we analyzed the effect of Mi-iPCA applied before the theoretical lecture, a significant effect was found: p-value = 0.05 & 0.04. This suggest that the tool helps to develop an intuitive comprehension of PCA, facilitating the geometrical foundations discussed during the theoretical lecture.

6 Conclusion and Future Work

Using the Wiimote, the user can change the visualized projection, the rotation of axes and the 3D projection. Students can evaluate how close their empirical rotation is to the actual PCA solution. The results suggest that the use of Mi-iPCA is effective to develop intuitive comprehension of principal components analysis. The responses to the subjective questionnaire, indicate that students consider Mi-iPCA useful. They liked the interface and enjoyed the axes rotation interaction. When the user is near from the analytical solution, the Wiimote interaction may be too sensitive, making harder to reach the actual analytical solution. To incorporate the Wiimotion plus, with a gyroscope integrated could give us a new interesting interaction environment. The results are encouraging and new experiments will take place in order to confirm the effect before/after theory. The authors have special interest in studying the effect of Mi-iPCA in different statistical courses for non technical and non mathematician students, like pedagogy, psychology and medicine.

References

Shaer, O. (2009). Tangible user interfaces: past, present, and future directions. *Foundations and Trends in Human – Computer Interaction, 3*(1–2), 1–137. Mike Casey.

Holmquist, L. E., Ju, W., Jonsson, M., Tholander, J., Ahmet, Z., Sumon, S. I., Acholonu, U., & Winograd, T. (2010). Wii science: teaching the laws of nature with physically engaging video game technologies. *CHI 2010 workshop: video games as research instruments,* Altlanta.

VGChartz Network. (2011, August). *Video games, charts, articles, news, reviews, community, forums at the VGChartz Network* [Internet]. Available from http://www.vgchartz.com/#Worldwide%20Totals. Accessed August 19, 2011.

O'Malley, C., & Fraser, D. S. (2004). *Literature review in learning with tangible technologies*. Bristol, UK: NESTA Futurelab.

Shaffer, D. W., Squire, K. R., Halverson, R., Gee, J. P., et al. (2005). Video games and the future of learning. *Phi delta kappan, 87*(2), 104.

Pearson, E., & Bailey, C. (2007). Evaluating the potential of the Nintendo Wii to support disabled students in education. *ICT: providing choices for learning: proceedings ascilite*, Singapore.

Daniels, T. E. (2009). Integrating engagement and first year problem solving using game controller technology. In *Proceedinging of the 39th IEEE annual frontiers in education conference, 2009*, FIE'09 (pp. 1–6).

Jeong, D. H., Ziemkiewicz, C., Fisher, B., Ribarsky, W., & Chang, R. (2009). iPCA: an interactive system for PCA-based visual analytics [Wiley Online Library]. *Computer Graphics Forum, 28*(3), 267–774.

Garcia, E., PCA and SPCA Tutorial Personal Publication. (2008, March). Available at http://www.miislita.com/information-retrieval-tutorial/pca-spca-tutorial.pdf

Bishop, C. M. (2006). *Pattern recognition and machine learning*. New York: Spinger.

Smith, L. I. (2002). *A tutorial on principal components analysis*. Ithaca: Cornell University.

GNU Scientific Library GSL – GNU Scientific Library – GNU Project – Free Software Foundation (FSF) [Internet]. (2011, March). Available from http://www.gnu.org/s/gsl/. Accessed 29 March, 2011.

Analog Devices Inc. (2011, March). *ADXL330: small, low power, 3-axis 3g imems accelerometer* [Internet]. Available from http://www.analog.com/en/mems-sensors/inertial-sensors/adxl330/products/product.html. Accessed 29 March, 2011.

Parent, R. (2002). *Computer animation: algorithms and techniques*. San Francisco: Morgan Kaufmann.

One-Mode Three-Way Analysis Based on Result of One-Mode Two-Way Analysis

Satoru Yokoyama and Akinori Okada

Abstract Several analysis models for proximity data have been introduced. While most of them were for one-mode two-way data, some analysis models which are able to analyze one-mode three-way data have been suggested in recent years. One-mode three-way multidimensional scaling and overlapping cluster analysis models were suggested. Furthermore, several studies were done for comparison between the results of analyses by one-mode two-way and by one-mode three-way data, where both of them are generated from the same source of data. In the present study, the authors suggest the analysis of one-mode three-way data based on one-mode two-way analysis for overlapping clusters. To evaluate the necessity of one-mode three-way analysis, firstly one-mode three-way data are reconstructed from the clusters and weights obtained by one-mode two-way overlapping cluster analysis. Secondly the reconstructed one-mode three-way data are subtracted from original one-mode three-way data. The subtracted data were analyzed by one-mode three-way overlapping cluster analysis model. The result of analysis discloses the components of proximities which can be expressed by one-mode three-way analysis but not by one-mode two-way analysis.

1 Background

Various analysis models for proximity data have been introduced. For example, multidimensional scaling (MDS) and cluster analysis models are suggested. Most of them were for one-mode two-way data. In recent years, some analysis models

S. Yokoyama (✉)
Department of Business Administration, Faculty of Economics, Teikyo University, Tokyo, Japan
e-mail: satoru@main.teikyo-u.ac.jp

A. Okada
Graduate School of Management and Information Sciences, Tama University, Tama, Tokyo, Japan

B. Lausen et al. (eds.), *Algorithms from and for Nature and Life*, Studies in Classification, Data Analysis, and Knowledge Organization, DOI 10.1007/978-3-319-00035-0_19,

which are able to analyze one-mode three-way data have been suggested. De Rooij and Heiser (2000) and Nakayama (2005) suggested MDS models. Yokoyama et al. (2009) suggested an overlapping cluster analysis model. The purpose of one-mode three-way analysis is to obtain the results which cannot be obtained in one-mode two-way analysis.

In Yokoyama et al. (2009) and Yokoyama and Okada (2010a), one-mode two-way and one-mode three-way data were generated from the same source, such as POS (Point Of Sale) data, and were analyzed by using overlapping cluster analysis models. The results of one-mode two-way and of one-mode three-way analyses were compared. However, the research on differences between one-mode two-way results and one-mode three-way results or the necessity of one-mode three-way analysis are not performed. Yokoyama and Okada (2010a,b) investigated the necessity of one-mode three-way analysis to use the correlation coefficient between one-mode three-way data and the index calculated from the triplet of one-mode two-way data which were obtained from the three objects corresponding to the one-mode three-way data, and Yokoyama and Okada (2010c, 2011a) applied one-mode two-way result to one-mode three-way data as the external analysis.

In the present study,to evaluate the necessity of one-mode three-way analysis, or, to investigate the effect of the one-mode two-way results on the one-mode three-way data, the authors introduce the procedure descripted below. To associate with authours' earlier works, they decided to used overlapping cluster analysis models in the present study. One-mode three-way data are reconstructed from the clusters and weights which are obtained by one-mode two-way overlapping cluster analysis. Then, the reconstructed one-mode three-way data are subtracted from original one-mode three-way data. The subtracted one-mode three-way data are analyzed by overlapping cluster analysis model. The analysis discloses the components of data which can be expressed by one-mode three-way analysis but not by one-mode two-way analysis. The present procedure was applied to several data to evaluate it.

2 Overlapping Cluster Analysis Models and Principle of the Present Study

Overlapping cluster analysis model was suggested by Shepard and Arabie (1979). In this model called ADCLUS (ADditive CLUStering), similarity between objects i and j, s_{ij}, is predicted as

$$s_{ij} \cong \sum_{r=1}^{R} w_{r(2)} p_{ir(2)} p_{jr(2)} + c_{(2)} = \sum_{r=1}^{R+1} w_{r(2)} p_{ir(2)} p_{jr(2)}, \quad (1)$$

where $w_{r(2)}$ is the nonnegative weight of the r-th cluster, $c_{(2)}$ is the additive constant, the $(R+1)$-th weight is the additive constant. $p_{ir(2)}$ is binary; if object i belongs to cluster r, $p_{ir(2)}$ is 1, otherwise it is 0.

Yokoyama et al. (2009) suggested a one-mode three-way overlapping clustering model, and proposed an algorithm fitting that model for one-mode three-way similarities. In this model, one-mode three-way similarity among objects i, j, and k, s_{ijk}, is predicted as

$$s_{ijk} \cong \sum_{r=1}^{R+1} w_{r(3)} p_{ir(3)} p_{jr(3)} p_{kr(3)}. \tag{2}$$

Moreover, Yokoyama et al. (2010) have improved the algorithm to eliminate negative weights.

In the present study, one-mode three-way data $\hat{s}_{ijk}$ which were reconstructed from the one-mode two-way result are calculated by the following equation:

$$\hat{s}_{ijk} = \sum_{r=1}^{R+1} w_{r(2)} p_{ir(2)} p_{jr(2)} p_{kr(2)}. \tag{3}$$

One-mode three-way reconstructed data $\hat{s}_{ijk}$ are subtracted from original one-mode three-way data:

$$\tilde{s}_{ijk} = s_{ijk} - \hat{s}_{ijk}. \tag{4}$$

Finally, the subtracted one-mode three-way data $\tilde{s}_{ijk}$ are analyzed by one-mode tree-way overlapping cluster analysis, then the results are compared with one-mode two-way and one-mode three-way results.

By the present procedure, we can say the following. If $\tilde{s}_{ijk} \simeq s_{ijk}$, the results of the analyses of original and subtracted one-mode three-way data are similar. The two VAF (Variance Accounted For) values which was defined by Arabie and Carroll (1980) should be almost equal. The effect of one-mode two-way result is very small. If $\tilde{s}_{ijk} \simeq 0$, the effect of one-mode two-way results is very large. The result of the analysis of subtracted one-mode three-way data does not have any meaning. Thus, it is not necessary to analyze one-mode three-way data.

3 Applications

In the present study, to evaluate the necessity of one-mode three-way analysis, the procedure which introduced in Sect. 2 is applied to two kinds of data. One is the POS data of convenience stores, and the other is the access log data.

3.1 The Analysis of POS Data

From the POS data of convenience stores in residential areas frequencies of joint purchases were derived. The data were drawn from the Nikkei POS Data. In the data, all customers bought at least Beer. The items in the data were classified into 14 categories: Liquors, Soft drinks, Tabaco, Ice creams, Frozen foods, Cooked foods, Bread, Sweets, Snacks, Dessert, Processed foods, Instant foods, Dairy products, and Daily necessities. One-mode two-way and one-mode three-way data were calculated from the frequency of the joint purchases among two or three categories.

In the one-mode two-way analysis, the resulting maximum VAFs in seven through three clusters were 0.944, 0.931, 0.911, 0.890, and 0.860, respectively. Because of the VAF values and the interpretation of the results, the six cluster result is adopted as the solution in the present analysis. The result is shown in the left side of Table 1. Liquors belong to all clusters, Soft drinks and Sweets belong to all clusters except cluster 4. Tabaco, Cooked foods, Processed foods, and Dairy products belong more than three clusters. These seven categories were jointly purchased. In addition, the combination of Sweets and Snacks were appeared in clusters 2 and 5, and Cooked foods, Bread, and Sweets were appeared in clusters 3 and 6, these combinations of categories were also often jointly purchased.

Then, one-mode three-way reconstructed data were calculated by this result and subtracted data were calculated, and the subtracted data were analyzed by one-mode three-way overlapping cluster analysis under the same condition of one-mode two-way analysis. We set the maximum number of clusters to six and the minimum number of clusters to three. The largest VAF for each number of clusters determined the maximum VAF at that number of clusters. In the analysis of the subtracted data, the resulting maximum VAFs in six through three clusters were 0.300, 0.293, 0.261, and 0.259. To simplify the comparison, the six cluster result was adopted as the solution. The result is shown in the right side of Table 1. It is thought that the VAF was quite small and additive constant c was large in the present result, the subtracted data seem not have the structures.

Moreover, original one-mode three-way data were analyzed. The VAFs in seven through three clusters were 0.703, 0.707, 0.700, 0.710, and 0.670. The result of six clusters is shown in Table 2.

By the comparison between the results of one-mode two-way and original one-mode three-way data analyses, we can see that the cluster structures are not completely same but similar. For example, Liquors, Soft drinks, and Sweets belong to many clusters, and clusters 4 and 5 are similar to clusters 2 and 3 in the result of one-mode two-way analysis (right side of Table 1).

In the present POS data analyses,original one-mode three-way data analysis does not have new interpretation of the analysis,and from the result of the analysis for subtracted data, it is concluded that the necessity of the one-mode three-way data analysis is low.

Table 1 Results of one-mode two-way and subtracted data analyses of POS data

	One-mode two-way analysis cluster						Subtracted data analysis cluster					
Category	1	2	3	4	5	6	1	2	3	4	5	6
Liquors	1	1	1	1	1	1	1	0	1	0	1	0
Soft drinks	1	1	1	0	1	1	0	1	1	1	0	1
Tabaco	1	0	1	1	0	0	0	0	0	0	0	1
Ice creams	0	0	0	0	1	0	0	0	0	1	0	0
Frozen foods	0	0	0	0	0	0	0	0	0	0	0	0
Cooked foods	0	1	1	1	0	1	0	0	0	0	0	1
Bread	0	0	1	0	0	1	0	0	0	0	0	0
Sweets	1	1	1	0	1	1	1	1	0	1	1	1
Snacks	0	1	0	0	1	0	1	1	0	0	0	0
Dessert	0	0	0	0	0	1	0	0	0	0	0	0
Processed foods	0	1	0	1	0	1	0	0	0	0	0	1
Instant foods	0	1	1	0	0	0	0	0	0	0	1	0
Dairy products	0	0	1	0	1	1	0	0	0	0	0	1
Daily necessities	0	0	0	0	1	0	0	0	1	0	1	1
Weights	.340	.183	.171	.145	.122	.109	.691	.367	.332	.251	.211	.097
Additive constant	.031						.304					

Table 2 Results of original one-mode three-way data analysis of POS data

	One-mode three-way analysis cluster					
Category	1	2	3	4	5	6
Liquors	1	1	1	1	1	0
Soft drinks	0	1	1	1	1	0
Tabaco	0	0	0	0	1	1
Ice creams	0	0	0	0	0	0
Frozen foods	0	0	0	0	0	0
Cooked foods	0	0	0	1	1	1
Bread	0	0	0	0	1	0
Sweets	1	1	1	1	1	0
Snacks	1	0	0	1	0	0
Dessert	0	0	0	0	0	0
Processed foods	0	0	0	1	0	1
Instant foods	0	0	0	1	0	0
Dairy products	0	0	0	0	1	0
Daily necessities	0	0	1	0	0	0
Weights	.435	.279	.259	.204	.194	.130
Additive constant	.051					

Table 3 Results of one-mode two-way and subtracted data analyses of access log data

	One-mode two-way analysis cluster					Subtracted data analysis cluster				
Page	1	2	3	4	5	1	2	3	4	5
TOP	1	1	0	0	1	0	0	0	0	1
Info.	1	0	1	0	0	0	0	0	0	1
Feature	0	0	1	0	0	0	0	0	0	1
BF1	0	0	0	1	1	1	0	0	0	1
BF2	0	0	0	1	1	1	1	0	0	0
BF3	0	0	0	0	1	1	1	0	1	0
BF4	0	0	0	0	0	0	1	1	1	0
SF1	0	0	0	0	0	0	0	1	1	0
SF2	0	0	0	0	0	0	0	1	0	0
Fee	0	1	0	0	0	0	0	0	0	0
Weight	.840	.815	.726	.558	.319	.702	.552	.454	.433	.276
Additive constant	.132					.298				

3.2 *The Analysis of Access Log Data*

The data were access logs of a web site of a company which supplies access log analysis services. The data were used and analyzed in Yokoyama and Okada (2011b). As shown in Yokoyama and Okada (2011b), the web site mainly consists of pages which introduce several functions and the list of the fee of the services. We focus our attention on 10 pages which are linked to the global navigation, Top page (TOP), Information page (Info.), the page to introduce the feature (Feature), the page to introduce of the first basic function (BF1), the second basic function (BF2), the third basic function (BF3), the fourth basic function (BF4), the first special function (SF1), the second special function (SF2), and information of the fee (Fee). The link buttons of these 10 pages were represented sequentially from the top to the bottom. One-mode two-way and one-mode three-way data are generated from the transition among the 10 pages.

In the present analysis, one-mode two-way and one-mode three-way data were symmetrized and analyzed by overlapping cluster analysis models. The symmetrized data were transformed so that transformed data are in the range of $[0, 1]$, we set the maximum number of clusters to six and the minimum number of clusters to three. The largest VAF for each number of clusters determined the maximum VAF at that number of clusters.

The resulting maximum VAFs in six through three clusters were 0.783, 0.722, 0.653, and 0.583 in the one-mode two-way analysis, and 0.815, 0.784, 0.719, and 0.626 in the one-mode three-way analysis. Because of the VAF values and the interpretation of the results, the five cluster result was adopted as the solution in the present analysis. The results are shown in the left side of Table 3, and in Table 4. Adjacent pages belong to mainly the same clusters in both results, it is interpreted

Table 4 Results of original one-mode three-way data analysis of access log data

	One-mode three-way analysis cluster				
Page	1	2	3	4	5
TOP	0	0	0	0	1
Info.	0	0	0	0	1
Feature	0	0	0	0	1
BF1	1	0	0	0	1
BF2	1	1	0	0	0
BF3	1	1	0	1	0
BF4	0	1	1	1	0
SF1	0	0	1	1	0
SF2	0	0	1	0	0
Fee	0	0	0	0	0
Weight	.953	.516	.423	.403	.254
Additive constant	.047				

that these pages were viewed sequentially. However,most of clusters in one-mode two-way results consist of two pages, and most of clusters in one-mode three-way analysis consist of three pages. It shows that one-mode three-way analysis is needed.

Along the present procedure, one-mode three-way reconstructed data were calculated by this results and subtracted data were calculated. The subtracted data were analyzed by one-mode three-way overlapping cluster analysis. The resulting maximum VAFs in six through three clusters were 0.777, 0.701, 0.624, and 0.508, the five cluster result is shown in the right side of Table 3.

The three analyses resulted in similar VAFs for each number of clusters. The cluster structures of the subtracted and original data analyses are the same. This shows that the result of one-mode three-way data is not affected from one-mode two-way results. Thus, the present procedure shows that one-mode three-way analysis is also needed.

4 Conclusion and Future Study

In the present study,we introduced the procedure to evaluate the necessity of one-mode three-way analysis and to investigate the effect of the one-mode two-way analysis. The present procedure was applied to the POS data and the access log data.The characteristic and interesting results were obtained from the present analyses to evaluate VAF values and cluster structures. Based on the present study, we think the simulation study of the procedure is needed to examine the procedure. In addition, one-mode two-way data, subtracted data, and one-mode three-way data were transformed as described earlier in Sect. 3.2, in the analyses of the present procedure, we think that a detailed study for the standardization is necessary.

Moreover, in Yokoyama and Okada (2010c, 2011a), the cluster structure of one-mode two-way result $p_{ir(2)}$ were adapted to one-mode three-way data s_{ijk}, the weights $\boldsymbol{w}_{(32)} = [w_{r(32)}]$ were calculated as

$$\boldsymbol{w}_{(32)} = \left(\boldsymbol{Q}'_{(2)}\boldsymbol{Q}_{(2)}\right)^{+} \boldsymbol{Q}_{(2)}\boldsymbol{s}_{(3)}, \tag{5}$$

where $\boldsymbol{s}_{(3)}$ is the M-dimensional column vector with s_{ijk} as element (here $M = {}_n\mathrm{C}_3$ and n is the number of objects), $\boldsymbol{Q}_{(2)} = [p_{ir(2)}p_{jr(2)}]$ is the $M \times r$ matrix, $\boldsymbol{w}_{(32)}$ is the $(R+1)$-dimensional column vector, and $\left(\boldsymbol{Q}'_{(2)}\boldsymbol{Q}_{(2)}\right)^{+}$ is the Moore-Penrose generalized inverse of $\left(\boldsymbol{Q}'_{(2)}\boldsymbol{Q}_{(2)}\right)$. Then, the weights which are obtained by two analyses $\boldsymbol{w}_{(2)}$, and $\boldsymbol{w}_{(3)}$ can be compared. We can apply this approach to the present procedure, the reconstructed one-mode three-way data are calculated by the following equations instead of Eq. (3),

$$\hat{s}^{*}_{ijk} = \sum_{r=1}^{R+1} w_{r(3)}\, p_{ir(2)}\, p_{jr(2)}\, p_{kr(2)}. \tag{6}$$

We think that this procedure is worth to investigate.

Finally, it is obvious that the present procedure is applicable not only to one-mode three-way data analysis but also to one-mode more higher way data analysis. Moreover, the present procedure is applicable to another analysis models suche as MDS models. Further improvements should be possible.

References

Arabie, P., & Carroll, J. D. (1980). Mapclus: A mathematical programming approach to fitting the ADCLUS model. *Psychometrika, 45*, 211–235.

De Rooij, M., & Heiser, W. J. (2000). Triadic distance models for the analysis of asymmetric three-way proximity data. *British Journal of Mathematical and Statistical Psychology, 53*, 99–119.

Nakayama, A. (2005). A multidimensional scaling model for three-way data analysis. *Behaviormetrika, 32*, 95–110.

Shepard, R. N., & Arabie, P. (1979). Additive clustering: Representation of similarities as combinations of discrete overlapping properties. *Psychological Review, 86*, 87–123.

Yokoyama, S., Nakayama, A., & Okada, A. (2009). One-mode three-way overlapping cluster analysis. *Computational Statistics, 24*, 165–179.

Yokoyama, S., Nakayama, A., & Okada, A. (2010). An application of one-mode three-way overlapping cluster analysis. In L. J. Hermann & W. Claus (Eds.), *Classification as a tool for research* (pp. 193–200). Berlin: Springer.

Yokoyama, S., & Okada, A. (2010a). Consideration about the necessity of one-mode three-way data analysis. In *Proceedings of the 26th meeting of the Japanese classification society* (in Japanese), Fukuoka (pp. 13–14).

Yokoyama, S., & Okada, A. (2010b). A study on the necessity of analysis in one-mode proximity data. *Proceedings of the 34th annual conference of the German classification society*, Karlsruhe (p. 197).

Yokoyama, S., & Okada, A. (2010c): External analysis of overlapping cluster analysis. *Proceedings of the 28th meeting of the Japanese classification society* (in Japanese), Tokyo (pp. 13–14).

Yokoyama, S., & Okada, A. (2011a): External analysis of POS data. *Proceedings of the 29th Meeting of the Japanese Classification Society* (in Japanese), Kyoto (pp. 11–12).

Yokoyama, S., & Okada, A. (2011b). A study on the necessity of one-mode three-way analysis. *The Teikyo University Economic Review, 45*, 143–150.

Latent Class Models of Time Series Data: An Entropic-Based Uncertainty Measure

José G. Dias

Abstract Latent class modeling has proven to be a powerful tool for identifying regimes in time series. Here, we focus on the classification uncertainty in latent class modeling of time series data with emphasis on entropy-based measures of uncertainty. Results are illustrated with an example.

1 Introduction

Latent class models are a powerful tool for capturing unobserved heterogeneity in a wide range of social and behavioral science data (see, for example, McLachlan and Peel 2000 or Ramos et al. 2011).

Let $\mathbf{y}_i$ denote a T-dimensional observation and $D = \{\mathbf{y}_1, \ldots, \mathbf{y}_n\}$ a sample of size n. Each data point is assumed to be a realization of the random variable $\mathbf{Y}$ coming from an S-component mixture probability density function (p.d.f.)

$$f(\mathbf{y}_i; \boldsymbol{\varphi}) = \sum_{w=1}^{S} \pi_w f_w(\mathbf{y}_i; \boldsymbol{\theta}_w), \tag{1}$$

where π_w are positive mixing proportions that sum to one, $\boldsymbol{\theta}_w$ are the parameters defining the conditional distribution $f_w(\mathbf{y}_i; \boldsymbol{\theta}_w)$ for latent class w, and $\boldsymbol{\varphi} = \{\pi_1, \ldots, \pi_{S-1}, \boldsymbol{\theta}_1, \ldots, \boldsymbol{\theta}_S\}$. Note that $\pi_S = 1 - \sum_{w=1}^{S-1} \pi_w$. The log-likelihood function for an LC model – assuming i.i.d. observations – has the form $\ell(\boldsymbol{\varphi}; \mathbf{y}) =$

J.G. Dias (✉)
UNIDE, ISCTE – Instituto Universitário de Lisboa, BRU – Business Research Centre, Lisbon, Portugal

Edifício ISCTE, Av. das Forças Armadas, 1649–026 Lisboa, Portugal
e-mail: jose.dias@iscte.pt

B. Lausen et al. (eds.), *Algorithms from and for Nature and Life*, Studies in Classification, Data Analysis, and Knowledge Organization, DOI 10.1007/978-3-319-00035-0_20,

$\sum_{i=1}^{n} \log f(\mathbf{y}_i; \boldsymbol{\varphi})$, which is straightforward to maximize (yielding the MLE - maximum likelihood estimator) by the EM algorithm (Dempster et al. 1977).

From parameter estimates of the LC model one can derive the posterior probability that an observation belongs to a certain latent class conditional on its response pattern. From the ML parameter estimates, Bayes' theorem gives the posterior probability that observation i was generated by latent class w:

$$\hat{\alpha}_{iw} = \frac{\hat{\pi}_w f_w(\mathbf{y}_i; \hat{\boldsymbol{\theta}}_w)}{\sum_{v=1}^{S} \hat{\pi}_v f_v(\mathbf{y}_i; \hat{\boldsymbol{\theta}}_v)}, \tag{2}$$

where $\hat{\pi}_w$ and $\hat{\boldsymbol{\theta}}_w$ are the ML estimates of π_w and $\boldsymbol{\theta}_w$, respectively.

The $\hat{\alpha}_{iw}$ values define a soft partitioning/clustering of the data set, since $\sum_{w=1}^{S} \hat{\alpha}_{iw} = 1$ and $\hat{\alpha}_{iw} \in [0, 1]$. Let c_i represent the true cluster membership (the missing data) of observation i. Then, the optimal Bayes rule assigning observation i to the class with maximum posterior probability can be defined as follows:

$$\hat{c}_i = \arg\max_{w} \hat{\alpha}_{iw}, i = 1, \ldots, n, \tag{3}$$

where $\hat{c}_i$ is the estimate of c_i.

In this paper, we address the following question: How can we measure the level of uncertainty in the mapping from the [0, 1] soft partition to the {0,1} hard partition obtained by applying the optimal Bayes rule in time series analysis?

The paper is organized as follows: Sect. 2 presents the full mixture hidden Markov model used to obtain the posterior probabilities of the regimes; Sect. 3 introduces a measure of classification uncertainty; Sect. 4 studies its behavior using a synthetic example; and Sect. 5 illustrates the application of the procedure to a panel data set of twenty European stock markets. Section 6 gives concluding remarks.

2 Latent Class Modeling of Time Series Data

The mixture of hidden Markov models (MHMM-S) (Dias et al. 2008; Ramos et al. 2011) is defined by the density:

$$f(\mathbf{y}_i; \boldsymbol{\varphi}) = \sum_{w_i=1}^{S} \pi_w \sum_{z_{i1}=1}^{K} \cdots \sum_{z_{iT}=1}^{K} f(z_{i1}|w_i) \prod_{t=2}^{T} f(z_{it}|z_{i,t-1}, w_i) \prod_{t=1}^{T} f(y_{it}|z_{it}), \tag{4}$$

where the conditional distribution within each latent class is given by a hidden Markov model with K regimes. $\boldsymbol{\varphi}$ is the vector containing all parameters in the model. Thus, we assume that within latent class w the sequence $\{z_{i1}, \ldots, z_{iT}\}$ is in agreement with a first-order Markov chain. Moreover, we assume that the observed value y_{it} at a particular time point depends only on the regime at this time point;

i.e, conditionally on the regime z_{it}, the response y_{it} is independent of other time points, which is often referred to as the local independence assumption. As far as the first-order Markov assumption for the latent regime switching conditional on latent class membership w is concerned, it is important to note that this assumption is not as restrictive as one may initially think. It does clearly not imply a first-order Markov structure for the responses y_{it}. The standard hidden Markov model (HMM) (Baum et al. 1970) is a special case of the MHMM-S that is obtained by eliminating the time-constant latent variable w from the model, that is, by assuming that there is no unobserved heterogeneity across time series.

The characterization of the MHMM is provided by:

- π_w is the prior probability of belonging to the latent class w;
- $f(z_{i1}|w_i)$ is the initial-regime probability; that is, the probability of having a particular initial regime conditional on belonging to latent class w with multinomial parameter $\lambda_{kw} = P(Z_{i1} = k|W_i = w)$;
- $f(z_{it}|z_{i,t-1}, w_i)$ is a latent transition probability; that is, the probability of being in a particular regime at time point t conditional on the regime at time point $t - 1$ and latent class membership; assuming a time-homogeneous transition process, we have $p_{jkw} = P(Z_{it} = k|Z_{i,t-1} = j, W_i = w)$ as the relevant multinomial parameter. Note that the MHMM-S allows that each latent class has its specific transition or regime-switching dynamics, whereas in a standard HMM it is assumed that all cases have the same transition probabilities;
- $f(y_{it}|z_{it})$, the probability density of having a particular value y_{it}, conditional on the regime occupied at time point t, is assumed to have the form of a univariate normal (or Gaussian) density function. This distribution is characterized by the parameter vector (μ_k, σ_k^2) containing the mean (μ_k) and variance (σ_k^2) for regime k. Note that these parameters are assumed invariant across latent classes, an assumption that may, however, be relaxed.

Since $f(\mathbf{y}_i; \varphi)$, defined by Eq. (4), is a mixture of densities across S latent classes and K regimes, it defines a flexible Gaussian mixture model that can accommodate deviations from normality in terms of skewness and kurtosis. For example, for two regimes ($K = 2$), the MHMM-S has $4S + 3$ free parameters to be estimated, including $S-1$ class sizes, S initial-regime probabilities, $2S$ transition probabilities, 2 conditional means, and 2 conditional variances.

Maximum likelihood (ML) estimation of the parameters of the MHMM-S involves maximizing the log-likelihood function: $\ell(\varphi; \mathbf{y}) = \sum_{i=1}^{n} \log f(\mathbf{y}_i; \varphi)$, a problem that can be tackled by the Expectation-Maximization (EM) algorithm (Dempster et al. 1977). The E step computes the joint conditional distribution of the $T + 1$ latent variables given the data and the current provisional estimates of the model parameters. In the M step, standard complete data ML methods are used to update the unknown model parameters using an expanded data matrix with the estimated densities of the latent variables as weights. Since the EM algorithm requires the computation of $S \cdot 2^T$ entries in the E step, which makes this algorithm impractical or even impossible to apply with more than a few time points. However, for hidden Markov models, a special variant of the EM algorithm was proposed that is usually referred to as the forward-backward or Baum-Welch algorithm (Baum et al. 1970). The Baum-Welch algorithm circumvents the computation of this joint posterior distribution making use of the conditional independencies implied by the model.

An important modeling issue is the setting of S and K, the number of latent classes and regimes needed to capture the unobserved heterogeneity across time series. The selection of S and K is typically based on information statistics such as the Bayesian Information Criterion (BIC) (Schwarz 1978) defined as:

$$BIC_{S,K} = -2\ell_{S,K}(\hat{\varphi}; \mathbf{y}) + N_{S,K} \log n, \tag{5}$$

where $N_{S,K}$ is the number of free parameters of the model and n is the sample size.

3 An Entropic-Based Uncertainty Measure

Classification uncertainty can be measured by the posterior probabilities $\hat{\alpha}_{is}$. An aggregate measure of classification uncertainty is the entropy. For LC models, the entropy is

$$EN(\boldsymbol{\alpha}) = -\sum_{i=1}^{n}\sum_{s=1}^{S} \alpha_{is} \log \alpha_{is}. \tag{6}$$

Its normalized version has been used as a model selection criterion, indicating the level of separation of latent classes (Dias and Vermunt 2006, 2008). The relative entropy that scales the entropy to the interval [0,1] is given by

$$E = 1 - EN(\boldsymbol{\alpha})/(n \log S). \tag{7}$$

For well-separated latent classes, $E \approx 1$; for ill-separated latent classes, $E \approx 0$. This provides a method for assessing the "fuzzyness" of the partition of the data under the hypothesized model. The ML estimate of E – $\hat{E}$ – can be obtained using the MLE ($\hat{\alpha}_{is}$) of α_{is} in Eq. (7).

We propose an extension of the relative entropy to panel data, that we call Entropy Regime Classification Measure (ERCM). For the time series $\mathbf{y}_i$, the ERCM is given by

$$ERCM_i = 1 + \frac{1}{T \log K} \sum_{t=1}^{T}\sum_{k=1}^{K} \alpha_{itk} \log(\alpha_{itk}), \tag{8}$$

where $\alpha_{itk} = P(Z_{it} = k|\mathbf{y}_i)$ is the probability that time series i is in regime k at time t conditional on the observed data.

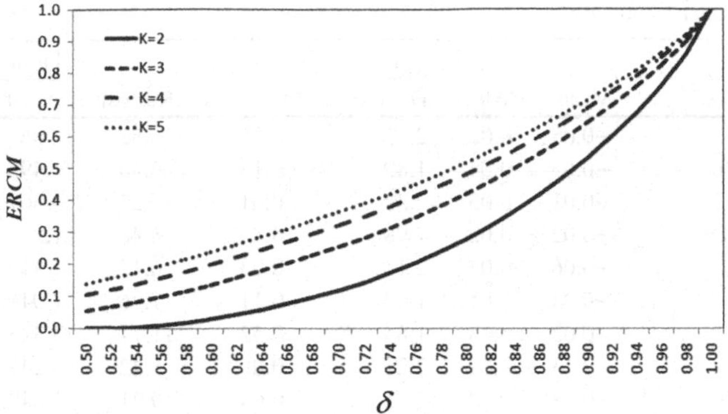

Fig. 1 ERCM as function of K and δ

4 A Synthetic Example

To understand the behavior of ERCM, let us assume that the posterior probabilities for each time series i and time t – $\boldsymbol{\alpha}_{it} = (\alpha_{it1}, \ldots, \alpha_{itK})$ – are:

$$\boldsymbol{\alpha}_{it} = \left(\delta, \frac{1-\delta}{K-1}, \ldots, \frac{1-\delta}{K-1}\right), \tag{9}$$

i.e., it represents a regime with probability δ and the remaining $K-1$ regimes with identical probability: $(1-\delta)/(K-1)$. Replacing vector $\boldsymbol{\alpha}_{it}$, in Eq. (8), we obtain the expression below for ERCM:

$$ERCM_{K,\delta} = 1 + \frac{1}{\log K}\left[\delta \log \delta + (1-\delta)\log\left(\frac{1-\delta}{K-1}\right)\right].$$

Figure 1 depicts the ERCM as function of K and δ with values $\{2, 3, 4, 5\}$, and $[0.5, 1.0]$, respectively. For example, for $K = 2$ and $\delta = 0.5$, the classification uncertainty is maximum and then ERCM $= 0$.

As expected, the relation between δ and ERCM is nonlinear. For the same value of δ, increasing K leads to an increase of ERCM as the value of $(1-\delta)/(K-1)$ decreases with an increase in K, and it becomes clearer the 'right' regime. In the opposite case, with $\delta = 1$, then ERCM $= 1$, with $0 \cdot \log 0 = 0$.

5 Application

Modeling the dynamics of stock market returns has been an important challenge in modern financial econometrics. The statistics and dynamics of correctly specified distributions provide more accurate and detailed input for financial asset pricing

Table 1 Summary statistics

Stock market	Mean	Median	Std. Deviation	Skewness	Kurtosis	Jarque-Bera test statistics	Jarque-Bera test p-value
Austria (OE)	−0.05	0.02	2.12	−0.21	6.62	564.41	0.000
Belgium (BG)	−0.05	0.04	1.89	−0.18	6.40	495.51	0.000
Czech Rep. (CZ)	−0.01	0.05	2.32	0.01	15.23	6,367.55	0.000
Denmark (DK)	−0.02	0.02	1.98	−0.23	8.46	1,275.01	0.000
Finland (FN)	−0.06	−0.04	2.14	0.08	6.12	413.55	0.000
France (FR)	−0.04	0.01	1.97	0.11	7.96	1,046.75	0.000
Germany (BD)	−0.02	0.08	1.84	0.52	11.77	3,323.29	0.000
Greece (GR)	−0.13	−0.02	2.26	0.00	5.42	247.02	0.000
Hungary (HN)	−0.05	0.00	2.81	0.04	9.21	1,640.73	0.000
Ireland (IR)	−0.11	−0.02	2.27	−0.46	6.65	599.30	0.000
Italy (IT)	−0.07	0.04	2.05	0.07	7.51	865.40	0.000
Netherlands (NL)	−0.05	0.02	2.02	−0.16	8.17	1,140.84	0.000
Norway (NW)	−0.03	0.12	2.72	−0.32	6.52	542.43	0.000
Poland (PO)	−0.04	0.04	2.39	−0.17	6.71	588.35	0.000
Portugal (PT)	−0.06	0.02	1.85	−0.05	9.03	1,545.13	0.000
Russia (RS)	−0.01	0.09	2.88	−0.17	15.06	6,197.63	0.000
Spain (ES)	−0.05	0.04	2.03	0.13	7.80	980.94	0.000
Sweden (SD)	−0.02	0.03	2.47	0.20	6.04	398.23	0.000
Switzerland (SW)	0.00	0.05	1.47	0.11	7.62	907.38	0.000
United Kingdom (UK)	−0.03	0.07	1.98	−0.04	8.61	1,338.27	0.000

and risk management. For example, investors buy or sell securities according to their expectation of the market regime. In addition, portfolio risk reduction might be achieved by procedures that take into account the synchronization of market regimes. Therefore, regime switching uncertainty is key in financial modeling.

The data set used in this article are daily closing prices from 4 July 2007 (the start of the subprime crisis) to 11 July 2011 for twenty European stock market indexes drawn from Datastream database and listed in Table 1. The series are expressed in US dollars. In total, we have 1,038 end-of-the-day observations per country. Let P_{it} be the observed daily closing price of market i on day t, $i = 1, \ldots, 20$ and $t = 0, \ldots, 1{,}037$. The daily rates of return are defined as the log-returns multiplied by 100: $y_{it} = 100 \times \log(P_{it}/P_{i,t-1})$, $t = 1, \ldots, T$, with $T = 1{,}037$.

This period was a very harsh one for the European stock markets. Table 1 provides descriptive statistics of the time series, while Fig. 2 depicts the log-returns time series. It can be seen that the mean is not positive for all markets in this period, however only for three markets – Finland, Greece, and Ireland – the median is negative. Stock markets show, instead, very different patterns of dispersion (Fig. 2); the largest standard deviation is found in Russia, Hungary, and Norway, while the smallest is in Switzerland (1.47). Return rate distributions are diverse in terms of skewness and the kurtosis (which equals 0 for normal distributions) shows high positive values, indicating heavier tails and more peakness than the normal distribution. The Jarque-Bera test rejects the null hypothesis of normality for all twenty stock markets.

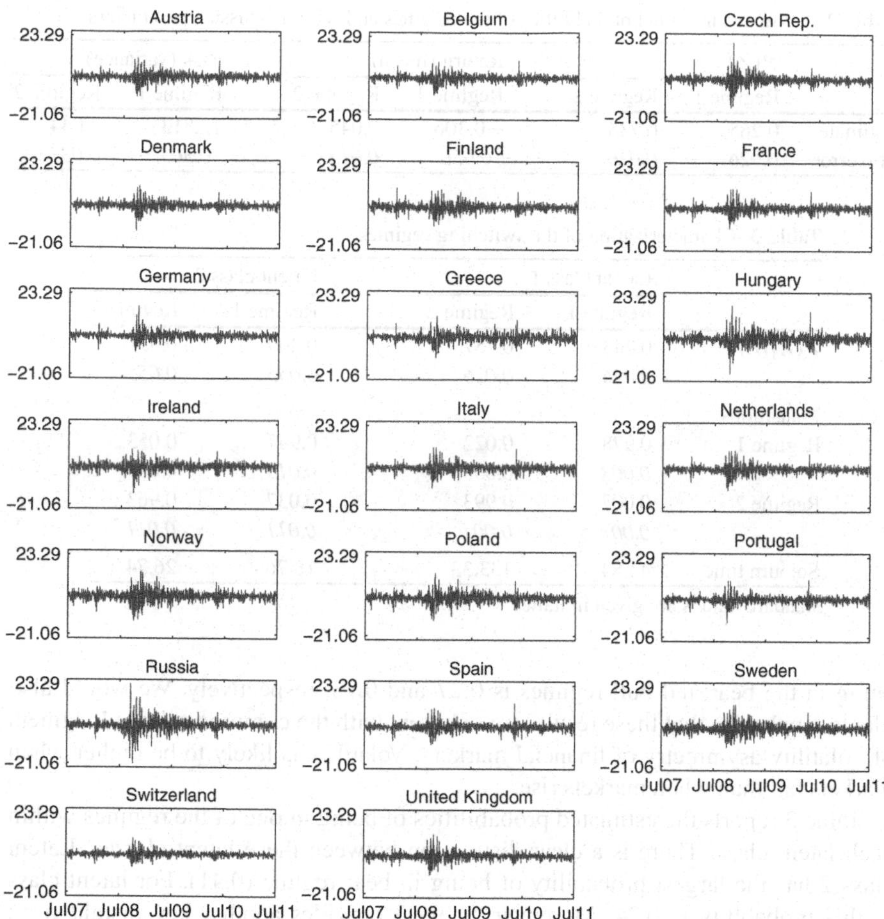

Fig. 2 Log-return time series for 20 European stock markets

Overall, these stock market features seem well suited to be modeled using MHMMs as we want to model simultaneously the 20 time series with typical cluster volatility. Given the traditional dichotomization of financial markets into "bull" and "bear" markets, we assume $K = 2$. We estimated models characterized by different number of latent classes ($S = 1, \ldots, 5$). To minimize the impact of local maxima, 300 different starting values for the parameters are used for each model. The model with two latent classes ($S = 2$) yielded the lowest BIC value ($\ell_2(\hat{\varphi}; \mathbf{y}) = -42{,}479.61$, $N_2 = 11$ and $BIC_2 = 84{,}992.17$).

Table 2 provides information on the two regimes that were identified ($K = 2$); that is, the average proportion of markets in regime k over time and the mean and variance of the returns in regime k. The result is in line with the common dichotomization of financial markets into "bull" and "bear" markets. Consistently, the reported means show that one of the regimes is associated with positive returns (bull market) and the other with negative returns (bear market). The probability of

Table 2 Estimated marginal probabilities of the regimes and within Gaussian parameters

	$P(Z)$		Return (mean)		Risk (variance)	
	Regime 1	Regime 2	Regime 1	Regime 2	Regime 1	Regime 2
Estimate	0.265	0.735	−0.306	0.045	12.813	2.044
Std. error	*0.026*	*0.026*	*0.050*	*0.012*	*0.305*	*0.031*

Table 3 Characterization of the switching regimes

	Latent class 1		Latent class 2	
	Regime 1	Regime 2	Regime 1	Regime 2
$P(Z\|W)$	0.243	0.757	0.409	0.591
	0.026	*0.026*	*0.055*	*0.055*
Transitions				
Regime 1	0.978	0.022	0.947	0.053
	0.003	*0.003*	*0.015*	*0.015*
Regime 2	0.007	0.993	0.037	0.963
	0.001	*0.001*	*0.011*	*0.011*
Sojourn time	44.84	133.33	18.76	26.74

Standard values are given in italics

being in the bear and bull regimes is 0.27 and 0.73, respectively. We would also like to emphasize that these results are coherent with the common acknowledgment of volatility asymmetry of financial markets. Volatility is likely to be higher when markets fall than when markets rise.

Table 3 reports the estimated probabilities of being in one of the regimes within each latent class. There is a clear distinction between these latent classes. Latent class 2 has the largest probability of being in bear regime (0.41). For latent class 1, this probability is 0.24. Moreover, Table 3 provides another key insight from our analysis. It gives the transition probabilities between the two regimes for both latent classes. First, notice that both latent classes show regime persistence. Once a stock market jumps into a regime, it is likely to remain within the same regime for a while, which is coherent with stylized facts in financial markets. Second, latent class 1 shows lower propensity to move from a bull regime to a bear regime (0.007) than latent class 2. Third, latent class 1 shows higher probability to jump from a bear to a bull regime than latent class 2.

The sojourn time is the expected number of days that a stock market stays in a given regime. For regime k and latent class w it can be obtained by $1/(1-p_{kkw})$. As reported in Table 3, stock markets in latent class 2 stay the shortest number of days in both bear and bull markets, and consequently being the less stable group.

Table 4 summarizes the results related to the distribution of stock market across latent classes. From the posterior class membership probabilities, the probability of belonging to each of the latent classes conditional on the observed data, we found only two stock markets are more likely to belong to latent class two: Greece (1.00) and Hungary (0.99). For most of the stock markets the posterior probability is

Table 4 Estimated posterior probabilities and ERCM

	Posterior probability			
Markets	Latent class 1	Latent class 2	Modal	ERCM
Austria (OE)	1.00	0.00	1	0.876
Belgium (BG)	1.00	0.00	1	0.922
Czech Rep. (CZ)	1.00	0.00	1	0.910
Denmark (DK)	1.00	0.00	1	0.911
Finland (FN)	1.00	0.00	1	0.854
France (FR)	1.00	0.00	1	0.905
Germany (BD)	1.00	0.00	1	0.923
Greece (GR)	0.00	1.00	2	0.605
Hungary (HN)	0.01	0.99	2	0.672
Ireland (IR)	0.84	0.16	1	0.777
Italy (IT)	1.00	0.00	1	0.862
Netherlands (NL)	1.00	0.00	1	0.914
Norway (NW)	0.99	0.01	1	0.801
Poland (PO)	1.00	0.00	1	0.846
Portugal (PT)	1.00	0.00	1	0.888
Russia (RS)	1.00	0.00	1	0.893
Spain (ES)	1.00	0.00	1	0.852
Sweden (SD)	0.98	0.02	1	0.809
Switzerland (SW)	1.00	0.00	1	0.936
United Kingdom (UK)	1.00	0.00	1	0.892

precise (the probability of the most likely latent class is always one or very close to one), the exception being Ireland with probability 0.16 of belonging to latent class 2. By combining the classification information with the descriptive statistics in Table 1, latent class 1 tends to contain countries with lower volatility than latent class 2.

Based on the posterior probabilities from the estimated model, Table 4 also reports the estimate of the ERCM for each stock market. Above 0.9 we have Switzerland (0.936), Germany (0.923), Belgium (0.922), Netherlands (0.914), Denmark (0.9111), Czech Republic (0.910), and France (0.905) as the least uncertain stock markets. On the other hand, Greece (0.605) and Hungary (0.672) are the most uncertain stock markets with ERCM below 0.7. The third most uncertain market is Ireland (0.777). Thus, the ERCM complements the values of posterior probabilities providing a more detailed indicator of stock market uncertainty. These results are consistent with financial market stylized facts.

6 Conclusions

This paper provides an extension to the MHMM model (Dias et al. 2008; Ramos et al. 2011) as a tool for measuring classification uncertainty in financial time series analysis. The proposed measure of uncertainty – Entropy Regime Classification

Measure (ERCM) – reveals the amount of uncertainty in a given time series. In the analysis of a sample of twenty stock markets for a period of 1,038 days, the best model was the one with two latent classes, with distinct types of regime switching. We conclude that the European market uncertainty in this period of analysis ranged from a minimum for Switzerland (ERCM = 0.936) and a maximum for Greece (ERCM = 0.605).

This approach should be further explored namely in applications where more than two regimes are needed. For instance, in modeling electricity prices it is standard to use at least three regimes as a result of abnormality in the markets and spike prices.

Acknowledgements The author would like to thank the Fundação para a Ciência e a Tecnologia (Portugal) for its financial support (PTDC/EGE-GES/103223/2008 and PEst-OE/EGE/UI0315/2011) and the three referees for their very valuable comments.

References

Baum, L. E., Petrie, T., Soules, G., & Weiss, N. (1970). A maximization technique occurring in the statistical analysis of probabilistic functions of Markov chains. *Annals of Mathematical Statistics, 41*, 164–171.

Dempster, A. P., Laird, N. M., & Rubin, D. B. (1977). Maximum likelihood from incomplete data via the EM algorithm (with discussion). *Journal of the Royal Statistical Society B, 39*, 1–38.

Dias, J. G., & Vermunt, J. K. (2006). Bootstrap methods for measuring classification uncertainty in latent class analysis. In A. Rizzi & M. Vichi (Eds.), *COMPSTAT 2006: proceedings in computational statistics*, Rome (pp. 31–41). Heidelberg: Physica-Verlag.

Dias, J. G., & Vermunt, J. K. (2008). A bootstrap-based aggregate classifier for model-based clustering. *Computational Statistics, 23*(4), 643–659.

Dias, J. G., Vermunt, J. K., & Ramos, S. B. (2008). Heterogeneous hidden Markov models. In P. Brito (Ed.), *COMPSTAT 2008: proceedings in computational statistics*, Porto (pp. 373–380). Heidelberg: Physica-Verlag.

McLachlan, G. J., & Peel, D. (2000). *Finite mixture models*. New York: Wiley.

Ramos, S. B., Vermunt, J. K., & Dias, J. G. (2011). When markets fall down: Are emerging markets all the same? *International Journal of Finance and Economics, 16*(4), 324-338.

Schwarz, G. (1978). Estimating the dimension of a model. *Annals of Statistics, 6*, 461–464.

Regularization and Model Selection with Categorical Covariates

Jan Gertheiss, Veronika Stelz, and Gerhard Tutz

Abstract The challenge in regression problems with categorical covariates is the high number of parameters involved. Common regularization methods like the Lasso, which allow for selection of predictors, are typically designed for metric predictors. If independent variables are categorical, selection strategies should be based on modified penalties. For categorical predictor variables with many categories a useful strategy is to search for clusters of categories with similar effects. We focus on generalized linear models and present L_1-penalty approaches for factor selection and clustering of categories. The methods proposed are investigated in simulation studies and applied to a real world classification problem.

1 Introduction

When selecting a regression model with categorical predictors two questions should be answered: (1) Which categorical predictors should be included in the model? and (2) Which categories within one categorical predictor should be distinguished? The latter question poses the question of which categories differ from one another with respect to the dependent variable.

In this paper, we assume a generalized linear model (GLM; McCullagh and Nelder 1989). That means, given (potentially vector-valued) explanatory variables $x_i, i = 1, \ldots, n$, response values y_i are (conditionally) independent, and the (conditional) distribution of y_i belongs to a simple exponential family with (conditional) expectation $E(y_i|x_i) = \mu_i$. This expectation μ_i is related to the so-called linear predictor $\eta_i = z_i^T\beta$ by $\mu_i = h(\eta_i) = h(z_i^T\beta)$, resp., $\eta_i = g(\mu_i)$, where $h(\cdot)$ is a known one-to-one, sufficiently smooth response function, and $g(\cdot)$ is the link

J. Gertheiss (✉) · V. Stelz · G. Tutz
Department of Statistics, LMU Munich, Akademiestr. 1, 80799 Munich, Germany
e-mail: jan.gertheiss@stat.uni-muenchen.de

B. Lausen et al. (eds.), *Algorithms from and for Nature and Life*, Studies in Classification, Data Analysis, and Knowledge Organization, DOI 10.1007/978-3-319-00035-0_21,

function, that is, the inverse of $h(\cdot)$. The parameter vector β denotes the vector of (unknown) regression coefficients (including a constant α), and z_i is a design vector in a general form (and therefore denoted as z_i instead of x_i), which is determined as an appropriate function $z_i = z(x_i)$ of the covariates, cf. Fahrmeir and Tutz (2001). A typical example of z_i is (one observation of) a dummy-coded categorical predictor. More precisely, if a categorical predictor x_j has potential levels $0, \ldots, K_j$, we have dummy variables z_{jk}, with $z_{jk} = 1$ if $x_j = k$ and 0 otherwise, $k = 0, \ldots, K_j$. For identifiability of β-coefficients from above, we specify reference category $k = 0$, such that $\beta_{j0} = 0$ for all $j = 1, \ldots, p$, where p denotes the number of categorical predictors considered.

2 L1-Regularization for Categorical Predictors

One problem with dummy coding is the high number of parameters involved, which makes pure maximum likelihood estimation unstable when the sample size is only modest, and interpretation of results may be difficult. Therefore we prefer penalized likelihood estimation, where instead of the usual unpenalized (log-)likelihood $l(\beta) = \sum_i \log(f(y_i|\theta_i, \phi))$, the penalized likelihood

$$l_p(\beta) = l(\beta) - \lambda J(\beta) \quad (1)$$

is maximized. The strength of penalty $J(\beta)$ is controlled by λ. The decisive point is a suitable choice of $J(\beta)$.

Under the assumption of a classical linear model (with approx. normally distributed response values), Bondell and Reich (2009) and Gertheiss and Tutz (2010) presented penalties for nominal and ordinal covariates that make factor selection and level clustering possible. For unordered categories, the penalty has the form

$$J(\beta) = \sum_{j=1}^{p} \sum_{l>k} w_{jkl} |\beta_{jl} - \beta_{jk}|, \quad (2)$$

with β_{jk} denoting the coefficient of dummy z_{jk}. If predictor levels are ordered, we only consider differences of adjacent coefficients β_{jk} and $\beta_{j,k-1}$ in penalty (2). The L_1-norm/Lasso-type penalty (see Tibshirani 1996) effects that coefficients may be set equal, which results in fusion/clustering of categories. If all coefficients of a factor are set equal to the dummy coefficient of the reference category (which is 0), the respective factor is excluded from the model. Weights w_{jkl} are incorporated to account for unbalanced designs. In addition, if weights are chosen as proportional to $|\hat{\beta}_{jl}^{(\text{ml})} - \hat{\beta}_{jk}^{(\text{ml})}|^{-1}$, where $\hat{\beta}_{jl}^{(\text{ml})}$ denotes the usual maximum likelihood (ml) estimate, an adaptive version (following Zou 2006) with nice asymptotic properties (as selection and fusion consistency) is obtained; see Bondell and Reich (2009) and Gertheiss and Tutz (2010) for details. In this paper, we use the proposed penalties for generalized linear models and corresponding penalized likelihood estimation (1).

Table 1 Dummy coefficients β_j (excluding the reference category) and marginal class probabilities ϑ_j (including the reference category) used for data generation in simulation study (3)

Nominal predictors	Ordinal predictors
$\beta_1 = (0, 0, 0)^T$	$\beta_5 = (0, 0, 0)^T$
$\vartheta_1 = (0.5, 0.1, 0.25, 0.15)^T$	$\vartheta_5 = (0.3, 0.2, 0.35, 0.15)^T$
$\beta_2 = (0, 0, 0, 0, 0, 0, 0)^T$	$\beta_6 = (0, 0, 0, 0, 0, 0, 0)^T$
$\vartheta_2 = (0.05, 0.15, 0.15, 0.1, 0.1, 0.05, 0.2, 0.2)^T$	$\vartheta_6 = (0.05, 0.15, 0.15, 0.2, 0.2, 0.05, 0.1, 0.1)^T$
$\beta_3 = (0, 1, 1)^T$	$\beta_7 = (-0.5, -0.5, -1)^T$
$\vartheta_3 = (0.15, 0.1, 0.25, 0.5)^T$	$\vartheta_7 = (0.4, 0.3, 0.2, 0.1)^T$
$\beta_4 = (0, 0, -1, -1, -1, 1.5, 1.5)^T$	$\beta_8 = (0, 1, 1, 2)^T$
$\vartheta_4 = (0.05, 0.1, 0.1, 0.2, 0.25, 0.05, 0.1, 0.15)^T$	$\vartheta_8 = (0.2, 0.25, 0.1, 0.25, 0.2)^T$

3 Numerical Experiments

Given categorical covariates $x_1, \ldots, x_8$, we generate (conditionally) independent response values y_i, $i = 1, \ldots, n$, each following a Poisson distribution with conditional expectation

$$E(y_i|x_{i1}, \ldots, x_{i8}) = E(y_i|z_i) = \mu_i = \exp(z_i^T \beta), \tag{3}$$

with parameter vector $\beta = (\alpha, \beta_1^T, \ldots, \beta_8^T)^T$ containing dummy coefficients, and design vectors z_i consisting of dummy-coded predictors. We assume four nominal $(x_1, \ldots, x_4)$ and four ordinal $(x_5, \ldots, x_8)$ predictors. True dummy coefficients used in the simulation are given in Table 1 (excluding coefficients $\beta_{j0} = 0$, $j = 1, \ldots, 8$, of the respective reference category). As the constant, we use $\alpha = -1$. We consider different sample sizes $n = 100, 200, 400, 600$, and generate 100 training data sets for each case. Categorical predictors x_j are assumed as being mutually independent, and class labels are randomly drawn with marginal class probabilities $\vartheta_j = (\vartheta_{j0}, \vartheta_{j1}, \ldots)^T$, $j = 1, \ldots, 8$, as given in Table 1.

On each simulated data set dummy coefficients are fitted employing the categorical L_1-penalty (2), which is able to select variables and to cluster categories. We consider the standard (non-adaptive) approach as well as the adaptive version (with weights depending on ml estimates). For computation, we use the R package `lqa` (Ulbricht 2010) where local quadratic approximation as proposed by Fan and Li (2001) is generalized. Penalty parameters are determined by cross-validation.

We investigate errors of parameter estimates and prediction accuracies. Figure 1 shows boxplots of the MSE of parameter estimates observed in the simulation study. It is seen that for small sample sizes the standard approach is superior to the adaptive version. This is apparently caused by the fact that pure ml estimates, where the adaptive version is based on, are quite unstable when sample sizes n are small (with respect to the number (38) of unknown parameters). If n increases, maximum likelihood estimates become more accurate and the adaptive version outperforms the standard approach.

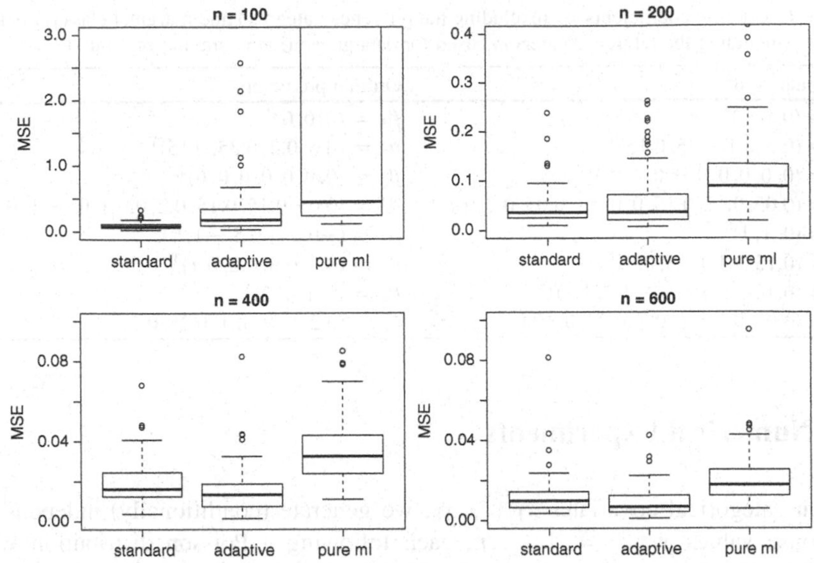

Fig. 1 Empirical mean squared errors of parameter estimates observed after 100 runs of simulation scenario (3) with different sample sizes n; considered are the standard and adaptive version of categorical L_1-penalty (2) as well as pure maximum likelihood estimation of dummy coefficients. For $n = 100$, the boxplot for pure ml is not completely shown because the range of ml MSEs is too large

For evaluating prediction accuracies, a test set of size 10,000 is generated for each simulation scenario (i.e., for each training sample size), and regression parameters estimated in each simulation run are used to predict the respective test response values. Resulting errors are measured in terms of test set deviances and illustrated in Fig. 2. Results are similar to those found in Fig. 1 (MSEs). If n is low compared to the number of parameters which are to be estimated, the standard approach should be preferred over the adaptive version, whereas the latter is a good choice when many observations are available. In any case, pure ml estimates are outperformed.

A problem found in the simulation is the high number of cases where truly zero coefficients (or differences thereof) are fitted as nonzero. Even when the adaptive penalty is used corresponding false positive rates are sometimes around 60 %. By contrast, false negatives (i.e., truly nonzero coefficients or differences thereof which are fitted as zero) are hardly observed. A possible explanation of this finding is that the algorithm implemented in the `lqa` package stops before an acceptable approximation of the 'exact' solution is reached. If only ordinal predictors are given, R package `glmpath` (Park and Hastie 2007) can be used instead of `lqa` (see also Sect. 4). With `glmpath`, selection and fusion performance is much better.

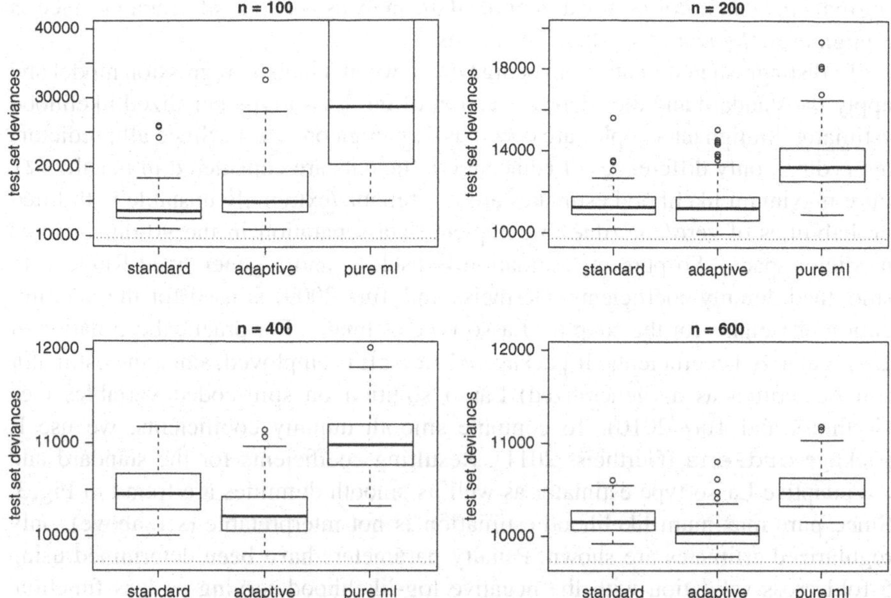

Fig. 2 Empirical test set deviances observed after 100 runs of simulation scenario (3) with different sample sizes n; considered are the standard and adaptive version of categorical L_1-penalty (2) as well as pure maximum likelihood estimation of dummy coefficients; the *dotted line* indicates the test set error if the true underlying model is used for prediction. For $n = 100$, the boxplot for pure ml is not completely shown because the range of ml deviances is too large

4 Wisconsin Breast Cancer Database

We apply the proposed penalized likelihood approach to data from the Wisconsin breast cancer database. The considered data were originally obtained from the University of Wisconsin Hospitals, Madison from Dr. William H. Wolberg, who periodically reported his clinical cases (see also Wolberg and Mangasarian 1990). The data are also available from the UCI Repository of machine learning databases (Newman et al 1998), and are part of the R package `mlbench` (Leisch and Dimitriadou 2010). The dataset consists of 699 samples[1]; 16 samples, however, have been removed because of missing values. Available covariates are nine cytological characteristics graded on a 1 to 10 scale at the time of sample collection, with 1 being the closest to normal tissue and 10 the most anaplastic (cf. Wolberg and Mangasarian 1990). In detail, we have *clump thickness, uniformity of cell size, uniformity of cell shape, marginal adhesion, single epithelial cell size, bare nuclei, bland chromatin,*

[1]The original dataset (Wolberg and Mangasarian 1990) was of size 369 (reported January 1989). Two instances were removed later and additional groups of all in all 332 samples were collected (between October 1989 and November 1991).

normal nucleoli, mitoses. The aim of the data analysis is to classify each instance as *benign* or *malignant* using these covariates.

To distinguish malignant from benign cells, we fit a logistic regression model and apply the standard and the adaptive version of our Lasso-type penalized likelihood estimator. Malignant samples are coded as 1, benign ones as 0. Since all predictors are ordinal, only differences of adjacent coefficients are considered in penalty (2). Pure maximum likelihood estimates are not reliable/extremely unstable with fitted probabilities of zero/one (due to complete data separation in the 90-dimensional predictor space). So pure ml estimation is useless, and a generalized Ridge with smoothed dummy coefficients (Gertheiss and Tutz 2009) is used for the determination of weights for the adaptive Lasso-type estimator. For practical estimation of L_1-regularized coefficients, R package `glmpath` is employed, since the estimator can be written as a (generalized) Lasso solution on split-coded variables (see Gertheiss and Tutz 2010). To compute smooth dummy coefficients, we use R package `ordPens` (Gertheiss 2011). Resulting coefficients for the standard and the adaptive Lasso-type estimator as well as smooth dummies are found in Fig. 3. Since pure maximum likelihood estimation is not interpretable (see above), only regularized estimates are shown. Penalty parameters have been determined using 5-fold cross-validation with the negative log-likelihood serving as loss function. Constant α is fitted as -4.75, -5.00 and -5.35 for the standard L_1-type approach, the adaptive version and the quadratic Ridge penalization (the 'smooth dummies'), respectively. Since for all covariates grade 1 is taken as reference category, the fitted probability that a sample is malignant, given all nine cytological characteristics are graded as 1, is $\exp(\hat{\alpha})/(1 + \exp(\hat{\alpha}))$, and hence below 1 % for each method. If the grade increases, the probability of malignancy increases, too, because coefficient functions in Fig. 3 are increasing. In the case of the generalized Ridge estimator (dotted lines), however, a few inconsistencies are observed. If the Lasso-type estimator is applied, coefficient curves are (monotone increasing) step functions. That means, categories are clustered and some relevant jumps between categories are selected. In the case of *marginal adhesion*, for example, such a jump is located between grade 5 and 6. By clustering, the degrees of freedom of the model are distinctly reduced to 20 (standard approach), resp. 15 (adaptive weights). If adaptive weights are used, the predictor *mitoses* is even completely removed from the model, since all corresponding dummy coefficients are fitted as zero.

For evaluating the performance of the considered methods, we randomly draw a test set of size 100 from the data at hand. On the remaining data the methods are trained and used to predict the test samples, and test set deviances as well as misclassification rates are computed. This procedure is independently repeated 100 times. The performance of the adaptive L_1- and the quadratic regularization is very similar (with small advantages for the generalized Ridge), and superior to the standard Lasso-type estimator (not shown, see Stelz 2010). On average (median), the observed misclassification rates are 4 % for the adaptive L_1- and the Ridge-type regularization (see also Stelz 2010), and thus similar to the values obtained by Wolberg and Mangasarian (1990), who treated grades as real numbers

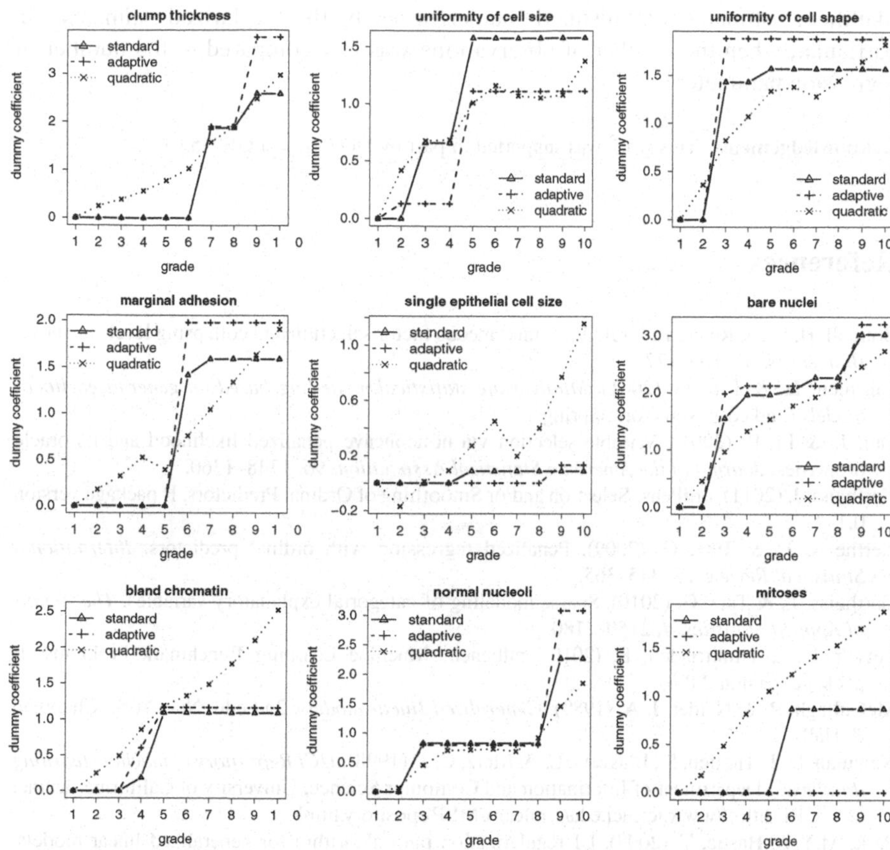

Fig. 3 Fitted dummy coefficients for the Wisconsin breast cancer database, using Lasso-type (standard/adaptive) regularization of categorical predictors, or quadratic first-order difference penalization of dummy coefficients (Gertheiss and Tutz 2009)

and used sets of separating hyper-planes to classify the data. The advantage of our logit model is that results, i.e., fitted coefficient curves, can be interpreted easily. Moreover, for each instance (and future samples as well), the (estimated) probability of malignancy is obtained directly. The advantage of L_1-regularization over a quadratic penalty is that predictor levels can be fused.

5 Summary

We showed that regularization techniques proposed for fusion of levels of categorical predictors in the classical linear model can also be applied in the GLM framework. In simulation studies and real world data evaluation, pure maximum likelihood estimates were distinctly outperformed by the regularized estimates – in particular, when the number of observations was low compared to the number of regression parameters.

Acknowledgements This work was supported in part by DFG project GE2353/1-1.

References

Bondell, H. D., & Reich, B. J. (2009). Simultaneous factor selection and collapsing levels in anova. *Biometrics, 65*, 169–177.

Fahrmeir, L., & Tutz, G. (2001). *Multivariate statistical modelling based on generalized linear models* (2nd ed.). New York: Springer.

Fan, J., & Li, R. (2001). Variable selection via nonconcave penalized likelihood and its oracle properties. *Journal of the American Statistical Association, 96*, 1348–1360.

Gertheiss, J. (2011). ordPens: Selection and/or Smoothing of Ordinal Predictors. R package version 0.1–7

Gertheiss, J., & Tutz, G. (2009). Penalized regression with ordinal predictors. *International Statistical Review, 77*, 345–365.

Gertheiss, J., & Tutz, G. (2010). Sparse modeling of categorial explanatory variables. *The Annals of Applied Statistics, 4*, 2150–2180.

Leisch, F., & Dimitriadou, E. (2010). mlbench: Machine Learning Benchmark Problems. R package version 2.0-0

McCullagh, P., & Nelder, J. A. (1989). *Generalized linear models*, 2nd edn. New York: Chapman & Hall

Newman, D. J., Hettich, S., Blake, C. L., & Merz, C. J. (1998). *UCI Repository of machine learning databases*. Department of Information and Computer Science, University of California, Irvine, CA, URL http://www.ics.uci.edu/~mlearn/MLRepository.html

Park, M.Y, & Hastie, T. (2007). L1 regularization-path algorithm for generalized linear models. *Journal of the Royal Statistical Society, B 69*, 659–677.

Stelz, V. (2010). *L1-Regularisierung bei kategorialen Prädiktoren in generalisierten linearen modellen*. Master thesis, Ludwig-Maximilians-University Munich

Tibshirani, R. (1996). Regression shrinkage and selection via the lasso. *Journal of the Royal Statistical Society B, 58*, 267–288.

Ulbricht, J. (2010). lqa: Penalized Likelihood Inference for GLMs. R package version 1.0–3

Wolberg, W. H., & Mangasarian, O. L. (1990). Multisurface method of pattern separation for medical diagnosis applied to breast cytology. *Proceedings of the National Academy of Sciences, 87*, 9193–9196.

Zou, H. (2006). The adaptive lasso and its oracle properties. *Journal of the American Statistical Association, 101*, 1418–1429.

Factor Preselection and Multiple Measures of Dependence

Nina Büchel, Kay F. Hildebrand, and Ulrich Müller-Funk

Abstract Factor selection or factor reduction is carried out to reduce the complexity of a data analysis problems (classification, regression) or to improve the fit of a model (via parameter estimation). In data mining there are special needs for a process by which relevant factors of influence are identified in order to achieve a balance between bias and noise. Insurance companies, for example, face data sets that contain hundreds of attributes or factors per object. With a large number of factors, the selection procedure requires a suitable process model. A process like that becomes compelling once data analysis is to be (semi) automated.

We suggest an approach that proceeds in two phases: In the first one, we cluster attributes that are highly correlated in order to identify factor combinations that—statistically speaking—are near duplicates. In the second phase, we choose factors from each cluster that are highly associated with a target variable. The implementation requires some form of non-linear canonical correlation analysis. We define a correlation measure for two blocks of factors that will be employed as a measure of similarity within the clustering process. Such measures, in turn, are based on multiple indices of dependence. Few indices have been introduced cf. Wolff (Stochastica 4(3):175–188, 1980), 'Few indices have been introduced in the literature'. All of them, however, are hard to interpret if the number of dimensions considerably exceeds two. For that reason we come up with signed measures that can be interpreted in the usual way.

N. Büchel (✉) · K.F. Hildebrand · U. Müller-Funk
European Research Center for Information Systems (ERCIS), University of Münster, Münster, Germany
e-mail: buechel@ercis.de; hildebrand@ercis.de; funk@ercis.de

B. Lausen et al. (eds.), *Algorithms from and for Nature and Life*, Studies in Classification, Data Analysis, and Knowledge Organization, DOI 10.1007/978-3-319-00035-0_22,

1 Approaches to Factor Selection

Factor selection, typically, is but factor reduction because in most cases we are restricted to a given pool of factors available. The problem, obviously, only occurs in a non-trivial way if that pool is large—as it is the case in data mining. Data mining—in contrast to ordinary data analysis—is characterized by 1. a large number of data sets coming from a heterogeneous universe and 2. a comparatively large number of factors measured on different scales. Both features together imply that the data has to be modeled by a general unidentifiable finite mixture model. Accordingly, procedures which are linear or likelihood-based become obsolete. In that context, factor selection has been tackled by regularizing relevant black box techniques, e.g. EBPN or SVM. Corresponding procedures first take all factors into account and only subsequently discard part of them by means of thresholds. Note, however, that the thresholding device is applied to quantities that are typically biased and loaded with noise. Moreover, that method leads to rather complex optimization problems. A more formal post-analysis approach—popular in classical data analysis—is to employ formal tests for that purpose. However, it seems impossible to complement black box procedures with fixed sample size testing procedures replacing the F-test even if we add heroic distributional assumptions such as normality. In principle it is possible to rely on asymptotic tests like the LR-test which in the presence of many factors is not trustworthy. Trees, on the other hand, sequentially select relevant factors from the pool at hand, but can only successfully deal with a limited number of features. Therefore, it is conclusive to incorporate factor selection into data preprocessing.

From a statistical point of view, factor selection can be done in different ways. Each alternative has a set of methods associated with it. The approaches and the corresponding tools are:

- *Thinning approach*: "group and select"
  - Correlation indices
  - Simultaneous testing

The problem with simultaneous testing is that—apart from distributional assumptions—it requires some sort of Bonferroni adjustment depending on the number of tests to be applied. Here, that number is not known and its upper bound (worst case) is exceedingly large. Furthermore, there is the

- *Formative approach*: "a few new factors from many old ones"
  - (Kernelized) linear projection methods, e.g. principal component analysis, canonical correlation analysis, projection pursuit and exploratory partial least squares (PLS)
  - Matching: multidimensional scaling, homogeneity analysis, classical factor analysis, correspondence analysis

All these tools share one unpleasant feature: Their results are not interpretable in the original problem domain as the new factors are typically rather artificial constructs. This problem is circumvented by the

- *Reflective approach*: "limited number of prescribed latent variables expressed by a much larger set of manifest variables"
  - Linear structural equations
  - PLS path modeling

In the present context we are not interested in that sort of "inverse" factor selection problem or, more precisely, factor analysis.

For classification problems, there are further approaches in the machine learning literature: filters and wrappers. Both techniques are boolean in character and work in a brute force manner. Cf. Hall (1999) for an overview.

In what follows, we shall concentrate on thinning within the preselection context.

2 A Process Model for Factor Selection

In order to structure the process of handling data that requires factor selection, we propose a process model. The main process can be seen in Fig. 1.[1] It starts with a sequence of functions, correlating the all factors with the target variable and discarding the tailing factors. A threshold has to be defined in advance and reflect the amount of factors. The sub process *Generate factor clusters* will be explained later. For now, let us assume that it produces a set of clusters containing factors that are in some way similar. Eventually, the goal is to create mixed-scale clusters. Once these clusters of factors have been generated, it is reasonable to determine a representative for each cluster. You can think of this representative as an actual factor that came to lie in the center of that cluster but it may also be an artificially created factor that did not exist in the data to begin with. The representatives can then be correlated with the target variable again in order to further reduce their number. The step of removing non-promising representatives is optional, in case there are none. At last, the model fit is determined. Should it not be satisfactory, previously discarded factors can be included to the factor set. Once the desired model fit is reached, the process ends.

The crucial part happens in the sub process *Generate Factor Clusters* (see Fig. 2). Of course, there should be an optional way of handling factors according to their scale. This represented by the left path in the diagram. Here, well-known clustering methods apply. However, we will also introduce measures that operate scale-independently. Thus, the function *Apply scale-independent measures* (see Fig. 2) is explained in Sect. 3 in more detail.

---

[1]We use Event-driven Process Chains (EPC) as a modeling language. Details can be found in Becker and Schütte (1996).

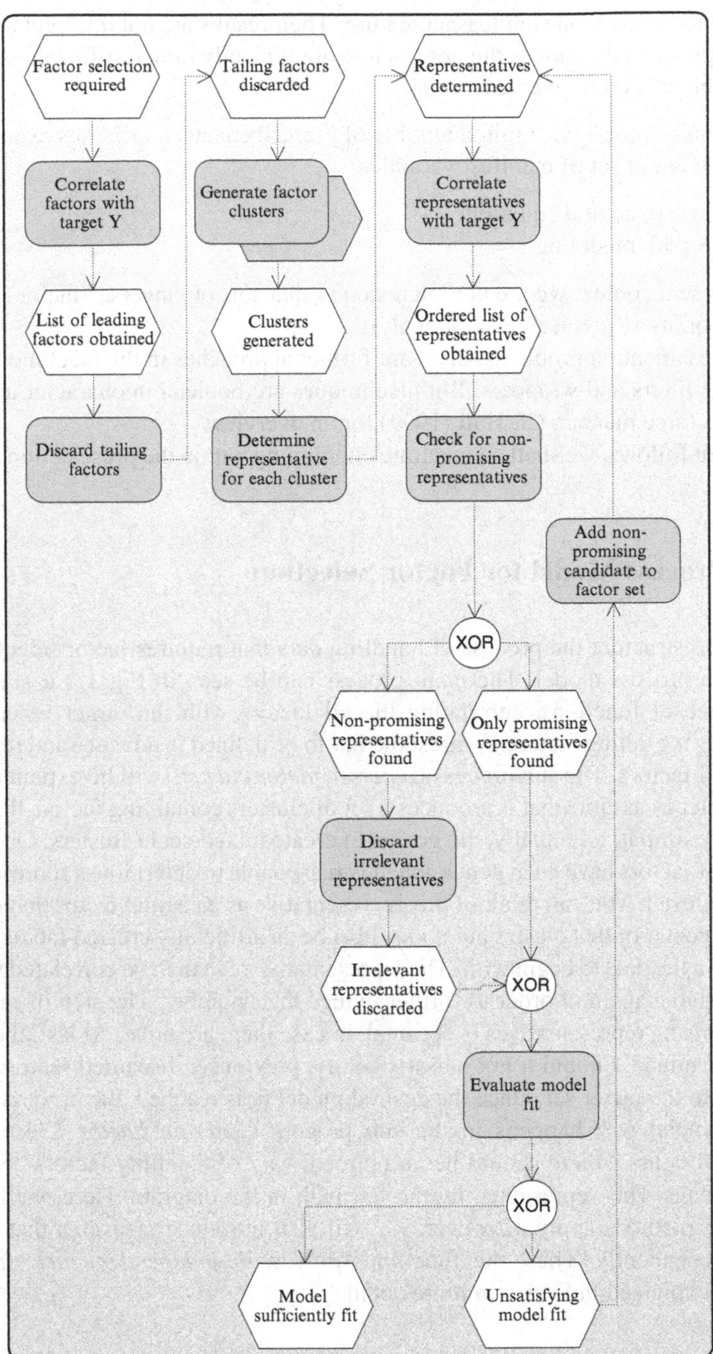

Fig. 1 Main process for factor selection

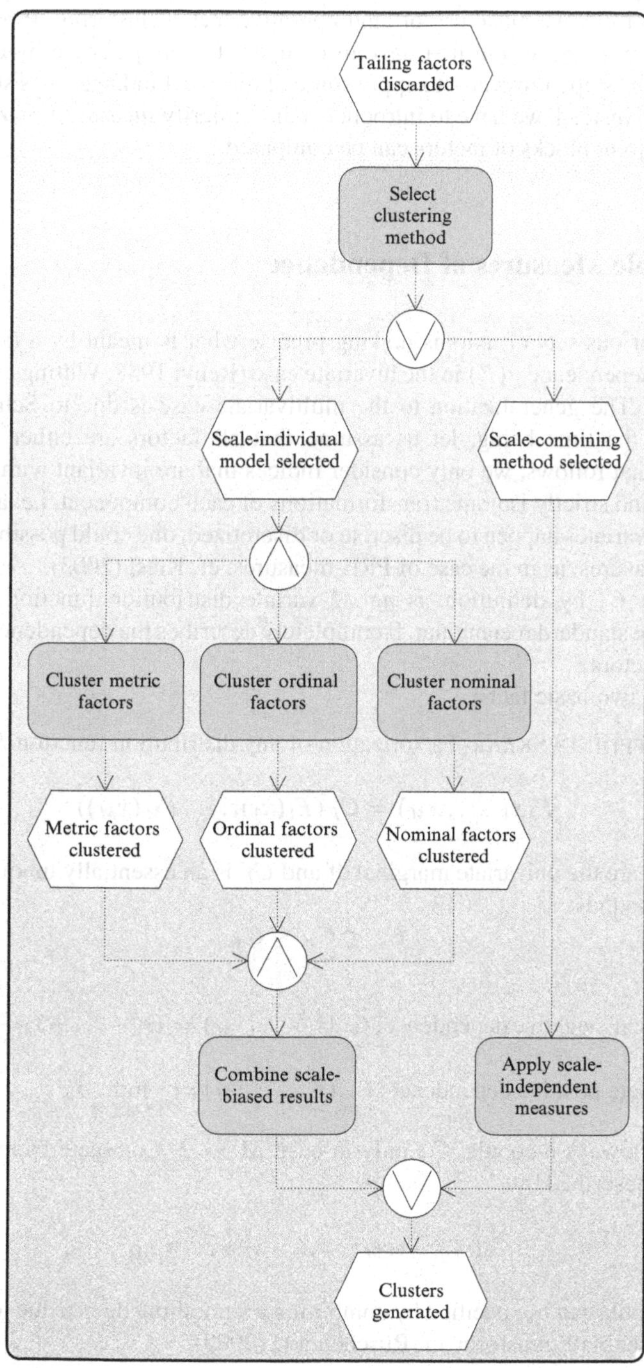

Fig. 2 Sub process *Generate Factor Clusters*

More precisely, we make use of agglomerative factor clustering. Proceeding in a traditional way, we would first have to compare factors pairwise by correlating them. The next step, however—to apply some of the usual linkages— is statistically meaningless. Instead, we have to introduce a dissimilarity measure $1 - \alpha$ by means of which disjoint blocks of factors can be compared.

3 Multiple Measures of Dependence

There are various sets of axioms making precise what is meant by a (symmetric) measure of dependence $\gamma(F)$ in the bivariate case (Renyi 1958; Witting and Müller-Funk 1995). The generalization to the multivariate case is due to Schmid et al. (2010). For the time being, let us assume that all factors are either ordinal or metric. In what follows, we only consider indices that are invariant with respect to continuous and strictly isotonic transformations of each component, i.e. are copula-based. If all variates happen to be discrete or discretized, one could possibly develop different measures, as in the case of PRE-measures, cf. Kiesl (2003).

A copula C, by definition, is an M-variate distribution function where all marginals are standard rectangular. It completely describes the dependence structure among all factors.

We recall two basic facts:

- The HOEFFDING–SKLAR–Factorization of any distribution function F:

$$F(x_1, \ldots, x_M) = C_F(F_1(x_1), \ldots, F_M(x_M))$$

  where F_i are the univariate marginal df and C_F is an essentially unique copula.
- Frechet Bounds:

$$C_- \leq C_F \leq C_+$$

  - Complete negative dependence: $C_-(s_1, \ldots, s_M) = (s_1 + \ldots + s_M - M + 1)^+$
  - Complete positive dependence: $C_+(s_1, \ldots, s_M) = \min_{1 \leq m \leq M} s_m$

  C_+ is always a copula, C_- only in case $M = 2$. Complete factor independence is described by

$$C_0(s_1, \ldots, s_M) = s_1 * s_2 * \ldots * s_M.$$

A unique copula can be specified by combining a smoothing device due to Ferguson with the Rosenblatt transform, cf. Rüschendorf (2009).

In order to design multiple measures of dependence, a reference copula C_R is chosen, typically, $C_R = C_0$. The difference between the true and the reference copula is assessed by means of some "loss function" L. The resulting quantity is condensed into a real-valued measure by means of averaging. Subsequently, it is normed by a constant. That way we arrive at measures

$$\gamma(F) = \gamma(C_F) = const * \int_{[0,1]^M} L(C_F - C_R)du.$$

Sometimes, a copula different from C_0 is used for averaging. Examples include the following indices, cf. Schmid et al. (2010)

- Signed measures with $C_R = C_0$: Spearman, Fechner-Kendall,
- Unsigned measures with $C_R = C_0$: Hoeffding, Schweizer-Wolff,
- Measures with $C_R = \frac{1}{2}(C_+ + C_-)$: Gini-Cograduation, Spearman's Footrule.

The first two groups of indices suffer from the fact that C_0—on the average—is "no longer the midpoint" of C_+ and C_- for $M > 2$ but is "wandering" towards the lower bound, cf. Wolff (1980). This makes it hard to give a meaningful interpretation to indices of the first and second categories. The reference "copula" chosen for the third category has no intuitive appeal and is but a numerical quantity.

In the context of factor selection, we are primarily interested in detecting high correlation not just the deviation from the independence case. Taking additionally into account the midpoint problem, it becomes natural to choose C_+ as a reference. This leads to a new class of indices: First, we define a preliminary index

$$\delta(F) = c_M * \int_{[0,1]^M} (C_+(u) - C_F(u))du \in [0,1],$$

$$= c_M * \int_{R^M} (C_+(F_1(x_1), \ldots, F_M(x_M)) - F(x_1, \ldots, x_M))F_1(dx_1) \ldots F_M(dx_M),$$

where c_M is norming constant ensuring $\delta(C_-) = 1$,

$$c_M = (M+1)/(1 - \frac{1}{M!}).$$

Note, that

$$\delta(C_F) = 1 - c_M \int_{[0,1]^M} (C_F(u) - C_-)du,$$

i.e., we could have defined δ equivalently with the "pseudo"–copula C_-.

In order to create a measure γ which can be interpreted in the usual way, i.e.

$$\gamma(C_+) = 1, \gamma(C_-) = -1, \gamma(C_0) = 0,$$

we employ suitable transformations. For that purpose, we locate the value of independence: For $C_F = C_0$ we get

$$v_M = \delta(C_0) = c_M(\frac{1}{M+1} - \frac{1}{2^M}).$$

Now, we choose any $q_M : [0,1] \to [-1,1]$ that is strictly antitonic, concave and continuous and moreover satisfies the norming conditions

$$q_M(0) = 1, q_M(v_M) = 0, q_M(1) = -1.$$

Define

$$\gamma(F) = q_M(\delta(F)).$$

Based on the quantity $\gamma(F)$ we now derive similarity measures for blocks of factors (allowing for different block sizes). Let $\mathscr{I}$ and $\mathscr{J}$ be disjoint sets of factors. The pertaining copulas are denoted by $C_F(\cdot|\mathscr{I})$, $C_F(\cdot|\mathscr{J})$, respectively. The argument underlying the FRECHET bounds and the monotonicity of γ provides us with the inequality:

$$\gamma(C_F(\cdot|\mathscr{I})+C_F(\cdot|\mathscr{J})-1)^+) \leq \gamma(C_F(\cdot|\mathscr{I}\cup\mathscr{J})) \leq \gamma(\min(C_F(\cdot|\mathscr{I}), C_F(\cdot|\mathscr{J})))$$

The bounds, again, correspond to cases where $\mathscr{I}$ completely determines $\mathscr{J}$ or vice versa. The inequality expresses the fact that enlarging a set of factors corresponds to a weakening of the inner dependence among its elements.

Motivated by that inequality, we put

$$\alpha(\mathscr{I},\mathscr{J}) = \max(\frac{\gamma(C_F(\cdot|\mathscr{I})+C_F(\cdot|\mathscr{J})-1)^+)}{\gamma(C_F(\cdot|\mathscr{I}\cup\mathscr{J}))}, \frac{\gamma(C_F(\cdot|\mathscr{I}\cup\mathscr{J}))}{\gamma(\min(C_F(\cdot|\mathscr{I}), C_F(\cdot|\mathscr{J})))})$$

Of course, $1-\alpha(\mathscr{I},\mathscr{J})$ is a measure of dissimilarity—which are typically used in agglomerative clustering. Such a procedure based on α is meant to detect factor sets, comprising statistical quasi doubles. To complete our thinning procedure we have to select from each cluster one or a few representatives that are highly predictive with suspect to a target variable. This can be achieved by means of α, where now $\mathscr{I}$ corresponds to the target variable and $\mathscr{J}$ to a set of predictors. Alternatively, one can make use of a generalized version of the FECHNER-KENDALL correlation coefficient (to be dealt with in a forthcoming paper).

Up to now all factors were assumed to be at least ordinal. In order to include nominal factors into the analysis, there are essentially two alternatives:

- All factors are discretized and a normed version of the χ^2 statistic is employed. The obvious drawback of that device is the enormous number of cells and the sparsity of the corresponding frequency tables. Moreover, ordinal variables are not dealt with in a proper way.

- Enumerate the levels of each nominal factor in an arbitrary way and treat them as ordinal variates. Proceed as before, i.e. compute γ, depending on the class labels. Repeat that device with any permutation of the class labels, separately for each nominal factor. Let γ be the maximum of all those values.

The empirical analysis is based on an M-dimensional i.i.d. sample with parent df F, emp. df $\hat{F}_N$ and empirical copula $\hat{C}_N$. The empirical measure of dependence is labeled $\hat{\gamma}_N = \gamma(\hat{F}_N)$. The question of consistency of $\hat{\gamma}_N$ and the distributional convergence of the error terms are settled by the following two assertions:

- Strong Consistency:
$$\hat{\gamma}_N \to \gamma(\hat{F}_N) F a.s.,$$
- Central Limit Theorem:
$$\sqrt{N}(\hat{\gamma}_N - \gamma(F)) \longrightarrow_w N(0, \tau^2))$$
$$\tau^2 = \sigma^2(F) \cdot q'^2(\delta(C_F)),$$
where, in principle, $\sigma^2(F)$ can be worked out by integration. As it is unknown in practice, it has to be estimated as well.

The first statement follows from the Glivenko-Cantelli theorem and the SLLN. The second one is based on a functional limit theorem, cf. Rüschendorf (1976).

$$\sqrt{N}(\hat{C}_N(u) - C_F(u)) \longrightarrow_w B_C(u) - \sum_{m=1}^{M} D_m C_F(u) B_C^{(m)}(u_m).$$

B_C is a tied down Brownian sheet with intensity C—i.e. the multivariate version of a tied down Brownian motion. The partial derivatives $D_m C_F$ are assumed to be continuous. Assuming furthermore that q_M is differentiable at $\delta(C_F)$, the result follows from CRAMERS theorem (Δ-method).

In order to work out the limiting distribution of $\hat{\alpha}_N$, we shall have to determine the limiting normal distribution of the vectors of empirical quantities involved in its definition.

4 Conclusion

We have given an introduction to the approaches for factor selection in data mining. In Sect. 2, we have introduced a process model for factor selection that facilitates (semi-) automation and structures the task at hand. In Sect. 3 we introduced a measure of dependence γ that can handle both metric and ordinal factors. This measure only partially fills the void of scale-independent measures since nominal factors remain a problem.

Concerning the empirical analysis, the next thing to do is to apply the process model—including the similarity measure α—to real life data provided by an insurance company. We refrained from testing the approach on the basis of synthetic data as so many arbitrary specifications would have been required that a representative evaluation as well as an assessment of the simulation result would have been rendered impossible.

With regards to the tools, future work will include

- Studying variants of γ by evaluating different choices of transformations q_M and
- The effect of squared loss,
- The search for PRE measures applicable to metric variables beyond the classical R^2-measure based on linear models,
- A treatment of multiple dependencies that allows for nominal variables as well.

As for the foundation of the empirical analysis, it would be desirable to have a basic limit theorem that does without the technical assumption of continuous partial derivatives. Such a variant requires a different topological setting and a more direct method of proof.

References

Becker, J. & Schütte, R. (1996). *Handelsinformationssysteme*. Verl. Moderne Industrie, Landsberg/Lech.

Hall, M. A. (1999). *Correlation-based feature subset selection for machine learning*. PhD thesis, Department of Computer Science, University of Waikato, Hamilton, New Zealand.

Kiesl, H. (2003). *Ordinale Streuungsmasse: Theoretische Fundierung und statistische Anwendung*. PhD thesis, Universität Bamberg.

Renyi, A. (1958). On measures of dependence. *Acta mathematica hungarica, 9*, 441–451.

Rüschendorf, L. (1976). Asymptotic distributions of multivariate rank order statistics. *The Annals of Statistics, 4*, 912–923.

Rüschendorf, L. (2009). On the distributional transform, sklar's theorem, and the empirical copula process. *Journal of Statistical Planning and Inference, 139*, 3921–3927.

Schmid, F., Blumentritt, T., Gaißer, S., Ruppert, M., & Schmidt, R. (2010). Copula-based measures of multivariate association. In F. Durante, W. Härdle, P. Jaworski, & T. Rychlik (Eds.), *Workshop on copula theory and its applications*, Warsaw. Berlin Heidelberg: Springer-Verlag.

Witting, H., & Müller-Funk, U. (1995). *Mathematische Statistik II*. Stuttgart: Teubner Verlag.

Wolff, E. F. (1980). N-dimensional measures of dependence. *Stochastica, 4*(3), 175–188.

Intrablocks Correspondence Analysis

Campo Elías Pardo and Jorge Eduardo Ortiz

Abstract We propose a new method to describe contingency tables with double partition structures in columns and rows. Furthermore, we propose new superimposed representations, based on the introduction of variable dilations for the partial clouds associated with the partitions of the columns and the rows. We illustrate our contributions with the analysis of some mortality data in Spanish regions.

1 Introduction

Many applications lead to build up contingency tables (CT) with double partition structures in columns and rows as showed in Fig. 1. To deal with this kind of tables, Cazes et al. (1988) developed the Internal Correspondence Analysis (ICA). We propose an alternative analysis, named Intrablocks Correspondence Analysis (IBCA), as the Correspondence Analysis (CA) of a CT with respect to its Intrablocks Independence Model, using the methodology proposed by Escofier (1984). Furthermore, we introduce variable dilations to the partial points in the superimposed representations. Some advantages of the new method with respect to ICA, and of the use of the variable dilations instead of a constant one, are shown at the end of the paper.

C.E. Pardo (✉)
Departamento de Estadística, Universidad Nacional de Colombia, Bogotá, Colombia
e-mail: cepardot@unal.edu.co

J.E. Ortiz
Facultad de Estadística, Universidad Santo Tomás, Bogotá, Colombia
e-mail: jorgeortiz@usantotomas.edu.co

B. Lausen et al. (eds.), *Algorithms from and for Nature and Life*, Studies in Classification, Data Analysis, and Knowledge Organization, DOI 10.1007/978-3-319-00035-0_23,

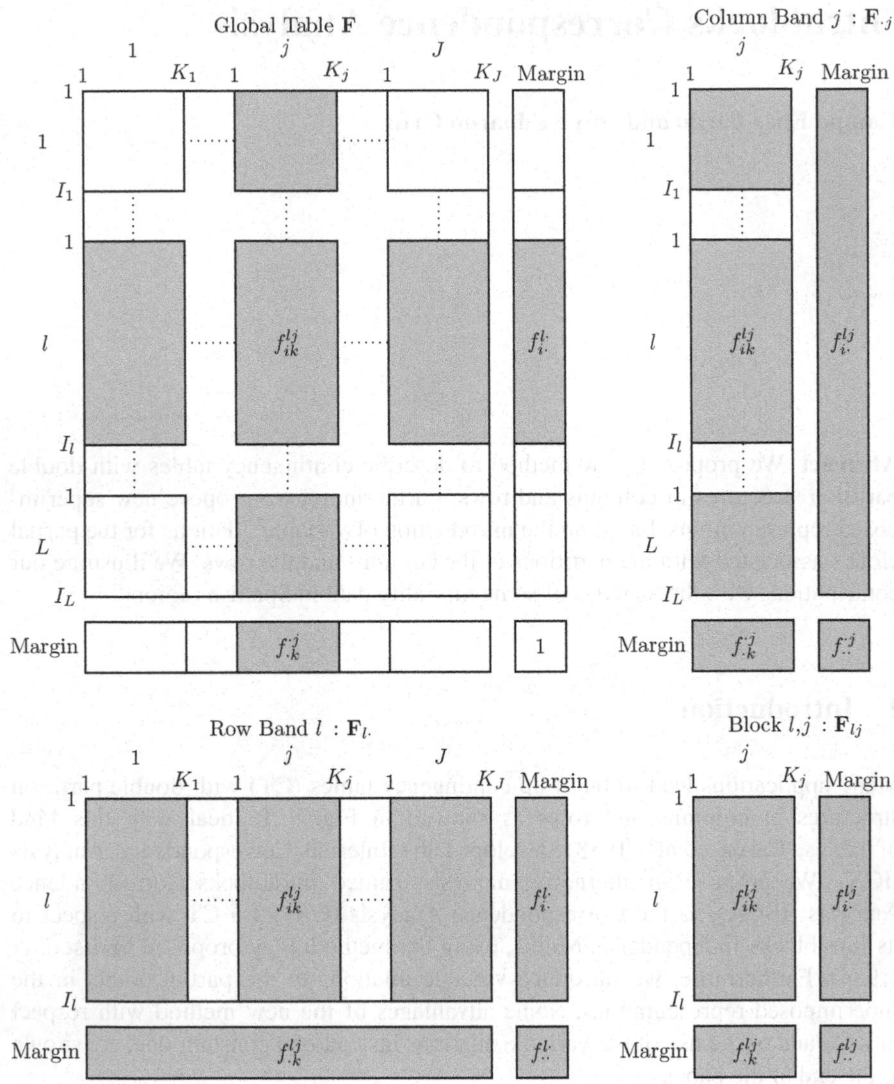

Fig. 1 The global table **F** of proportions, partitioned into rows and columns groups

2 Notation

We adopt the notation used by Bécue et al. (2005). Let $\mathbf{F}$ be an $I \times K$ table of proportions, i.e., the sum of its elements is equal to 1. The I rows are grouped in L row bands with, respectively, $I_1, I_2, \ldots, I_L$ rows ($\sum_{l=1}^{L} I_l = I$), and the K columns are structured in J column bands with, respectively, $K_1, K_2, \ldots, K_J$ columns ($\sum_{j=1}^{J} K_j = K$). This induces a partition of $\mathbf{F}$ into $L \times J$ subtables, named blocks (Fig. 1). The block (l, j) has I_l rows and K_j columns. A symbol used to denote the cardinality of a set is also used to denote the set itself.

Let us consider the block $\mathbf{F}_{lj}$. Its margins and totals are shown in Fig. 1. Let $\mathbf{B}_{lj}$ be the block obtained from the independence model associated to $\mathbf{F}_{lj}$. Thus, the general term of $\mathbf{B}_{lj}$ is $b_{ik}^{lj} = \frac{f_{i\cdot}^{lj} f_{\cdot k}^{lj}}{f_{\cdot\cdot}^{lj}}$. The $L \times J$ blocks $\mathbf{B}_{lj}$ conform the matrix model $\mathbf{B}$, which inherits the same double partition structure of $\mathbf{F}$. Moreover, the two tables have the same margins and totals.

Let $\mathbf{X}$ be the matrix with the same dimensions of $\mathbf{F}$ and $\mathbf{B}$ and general term $x_{ik}^{lj} = \frac{f_{ik}^{lj} - b_{ik}^{lj}}{f_{i\cdot}^{l\cdot} f_{\cdot k}^{\cdot j}}$. $PCA(\mathbf{X}, \mathbf{M}, \mathbf{D})$ is used to denote the Principal Components Analysis of $\mathbf{X}$ with metric and weight matrices given by $\mathbf{M}$ and $\mathbf{D}$. This is performed by calculating the eigenvalues and eigenvectors of the matrix $\mathbf{X}'\mathbf{D}\mathbf{X}\mathbf{M}$ (See Pagès 2004).

3 Intrablocks Correspondence Analysis (IBCA)

CA is a graphical method used to describe the associations between row and column categories of a CT and can be seen as the description of de differences between the observed counts and the expected values using the independence model $\mathbf{H}$. The $CA(\mathbf{F})$, with skeleton showed in the Fig. 1, is the $PCA(\mathbf{D}_I^{-1}(\mathbf{F} - \mathbf{H})\mathbf{D}_K^{-1}, \mathbf{D}_K, \mathbf{D}_I)$, where $\mathbf{D}_I = diag(f_{i\cdot}^{l\cdot})$, $\mathbf{D}_K = diag(f_{\cdot k}^{\cdot j})$ and the general term of $\mathbf{H}$: $h_{ik}^{lj} = f_{i\cdot}^{l\cdot} f_{\cdot k}^{\cdot j}$. In the generalisation of the CA proposed by Escofier (1984), any model is used instead of $\mathbf{H}$.

IBCA is defined as the CA of $\mathbf{F}$ with respect to the Intrablock Independence Model $\mathbf{B}$. IBCA uses the same metrics and weights of the CA of $\mathbf{F}$ and it is denoted here by $CA(\mathbf{F}, \mathbf{B})$. IBCA is equivalent to $PCA(\mathbf{X}, \mathbf{D}_K, \mathbf{D}_I)$, where $\mathbf{X} = \mathbf{D}_I^{-1}(\mathbf{F} - \mathbf{B})\mathbf{D}_K^{-1}$ with general term as defined in Sect. 2.

3.1 IBCA and Log-Linear Models

In some situations, the contingency table with double partition structure can be obtained by "flattening" a four way contingency table. Let us consider the four factors A, B, C, D. The flattening of the table is performed by considering a new categorical variable in the rows whose levels are defined by crossing the categories of A and B and doing the same for C and D in the columns.

The intrablocks independence model $\mathbf{B}$ is also the estimation of the log-linear model $[ABC]\,[ACD]$, which adds the four main effects, the first order interactions AB, AC, AD, BC, CD and the second order interactions ABC and ACD. Thus, the $CA(\mathbf{F}, \mathbf{B})$ is the analysis of the intrablock interactions BD and the higher order interactions containing BD (Van der Heijden 1987).

3.2 Comparison of the Partial and Global Structures

IBCA offers a representation of the global structure of the rows and columns on principal planes in a CA-like way and the rows (the columns) as described separately by every column band (row band) are projected as illustrative partial points. In that representation the partial points are amplified by the number of column bands J (row bands L) (Bécue et al. 2005). We introduce another amplification as shown below.

Let us associate to each column band matrix j of $\mathbf{X}$ as defined in IBCA applied to $\mathbf{F}$, the cloud N_I^j of the rows as described by only the columns of $\mathbf{X}_{*j}$. The coordinates of the N_I^j points in $\mathbb{R}^K$ with variable dilation are:

$$\tilde{\mathbf{X}}_{*j} = \begin{array}{|c|c|c|c|} \hline \mathbf{0} & \mathbf{0} & \frac{1}{f_{..}^{.j}}\mathbf{X}_{*j} & \mathbf{0} \\ \hline \end{array}$$

The centroid of the cloud $N_{(l,i)}^J$ with J partial points, each one with weight $f_{..}^{.j}$ is the global point $\mathbf{x}_{(l,i)}$. The projections have the same property. The *constant dilation* by J, used in previous work, is a particular case of our proposal if we take $f_{..}^{.j} = \frac{1}{J}$ for all j. With the variable dilations, partial points belonging to low weight bands are highlighted and the more weighted partial points are closer to their mean points.

The superimposed representations of the partial and global column clouds are obtained in a symmetric way.

4 Comparison Between IBCA and ICA

ICA is the $CA(\mathbf{F}, \mathbf{C})$. The model $\mathbf{C}$ takes into account the principal effects: total, row bands and column bands, in the context of linear models. The model $\mathbf{B}$ of IBCA takes into account the principal effects and the interactions $\mathbf{BD}$, as a log-linear model.

In both methods, the cloud of the rows N_I are divided in L subclouds N_{I_l}, each one with centroid in the origin (weights: $f_{i.}^{l.} / f_{..}^{l.}, i \in I_l$); and the cloud of the columns N_K are divided in J subclouds N_{K_j}, each one with centroid in the origin (weights: $f_{.k}^{.j} / f_{..}^{.j}, k \in K_j$).

The IBCA superimposed representations have an important advantage with respect to those of the ICA: in the IBCA, if all the values of a row i corresponding to a block $\mathbf{F}_{lj}$ are zero, the associated partial point j is located at the origin. In the ICA representations this is not always true. The same situation may occur when there is a column of zeros within a block $\mathbf{F}_{lj}$.

5 Example: Mortality in Spain

Eurostat (2008) reports the crude mortality rates, by 100,000 inhabitants, classified in 65 causes of death, sex and 5-years age ranges and aggregated in three consecutive years. The comparison between regional mortalities is made using standardised mortality rates, i.e. the weighted average with the reference population as defined in European-Communities (2009). In this work, the standardised rates for two age ranges: 35–64 (premature mortality) and 65 and over (non premature mortality) were computed using the crude mortality rates averaged for three consecutive years, i.e. 1994, 1995 and 1996 for 1995, and 2004, 2005 and 2006 for 2005.

From the standardised rates we build up the table shown in Fig. 2, consisting of eight blocks, each one crossing the 17 autonomous regions of Spain with the causes of death responsible for at least 2 % of the $sex \times age$ groups in one of the 2 years. To help the interpretation of the factorial planes we include three development indicators of the autonomous regions: GDP per capita, illiteracy rate and unemployment rate, obtained from online datasets of INE of Spain (INE-España 2010).

5.1 Global Representation Using the IBCA

The IBCA provides a common framework for comparing, in the row space the mortalities of the communities in the 2 years, and in the column space the causes of death in the four $sex \times age$ groups.

The plot of eigenvalues suggests keeping the first three axes representing 62.7 % of the total inertia. The weights of the column bands associated to non premature male (61.4 %) and female (30.1 %) mortalities are higher than the weights of the column bands associated to premature mortalities (male 6.4 %, female 2.2 %). The contributions of subclouds to the inertia have similar behaviour (non premature: males 54.3 %, females: 32.5 %; premature: males 9.0 %, females 4.2 %). The weights and the inertias of the two subclouds of the row profiles are balanced.

The first factorial axis highlights the separation of Canary Islands from the rest of the communities (the contributions to the inertia are 30.8 % for 1995 and 35.6 % for 2005). This occurs mainly because the two profiles of the Canary Islands mortality rates are higher than the average mortality due to diabetes and cardiac ischemia in elderly men and women (the contributions to inertia are: *f2Diabetes* 17.0 %, *m2Diabetes* 16.0 %, *m2IscHeart* 13.9 % and *f2IscHeart* 11.7 %). In contrast, male and female mortalities for malignant neoplasm of stomach in adults are lower than the mean mortalities for these causes. The same occurs in the mortality due to transportation accidents, especially in adult women.

The factorial plane 2–3 (Fig. 3) is a good summary of mortality profiles in the other regions. On the second (horizontal) axis, the communities are sorted by their degree of development, where more developed regions lie on the left

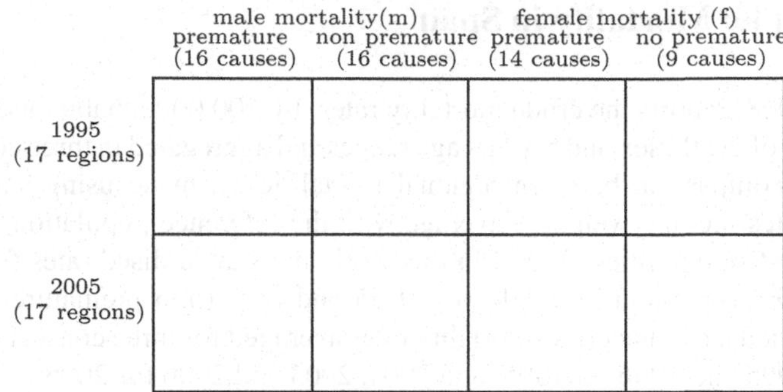

Fig. 2 CT scheme (year × region) × cause (sex × age): juxtaposition of 8 tables. In this table: $L = 2$, $I = 34$, $J = 4$ and $K = 55$; $I_1 = 17$ and $I_2 = 17$; $K_1 = 16$, $K_2 = 16$, $K_3 = 14$ and $K_4 = 9$ (See Fig. 1)

(correlations: GDP per capita −0.289, illiteracy rate 0.327 and unemployment 0.502). Thus, mortality profiles are partially explained by the degree of development of the regions. Death causes such as mental and behavioural disorders are associated with more developed communities, and transportation accidents and cerebrovascular diseases with less developed regions.

5.2 Superimposed Representations

Autonomous communities have two points in the global representations (1995 and 2005) showing the stability or the instability of their mortality profiles. The differences due to $sex \times age$ groups can be appreciated in the superimposed representations. Figure 4, is an extract of the superimposed representation of some communities. In this figure, it can be seen, for example that, from a global point of view, the profiles of Navarre are almost identical. However, from the point of view of each of the $sex \times age$ groups, they are a bit different. Madrid shows differences in both global and partial profiles between the 2 years. The superimposed representations allow the comparison between profiles of different regions from the point of view of each $sex \times age$ group. For example, male premature mortalities under Murcia and Asturias are very similar in 1995. Non premature mortalities have more weights than premature ones; and consequently, these partial points are closer to their global ones.

The superimposed representations of columns allow to see the stability or instability of the mortality cause profiles and the comparison between them, from partial (each year) and global (both years) points of view. The reader can use the package *pamctdp* to obtain some superimposed representations of the causes of death and other outputs.

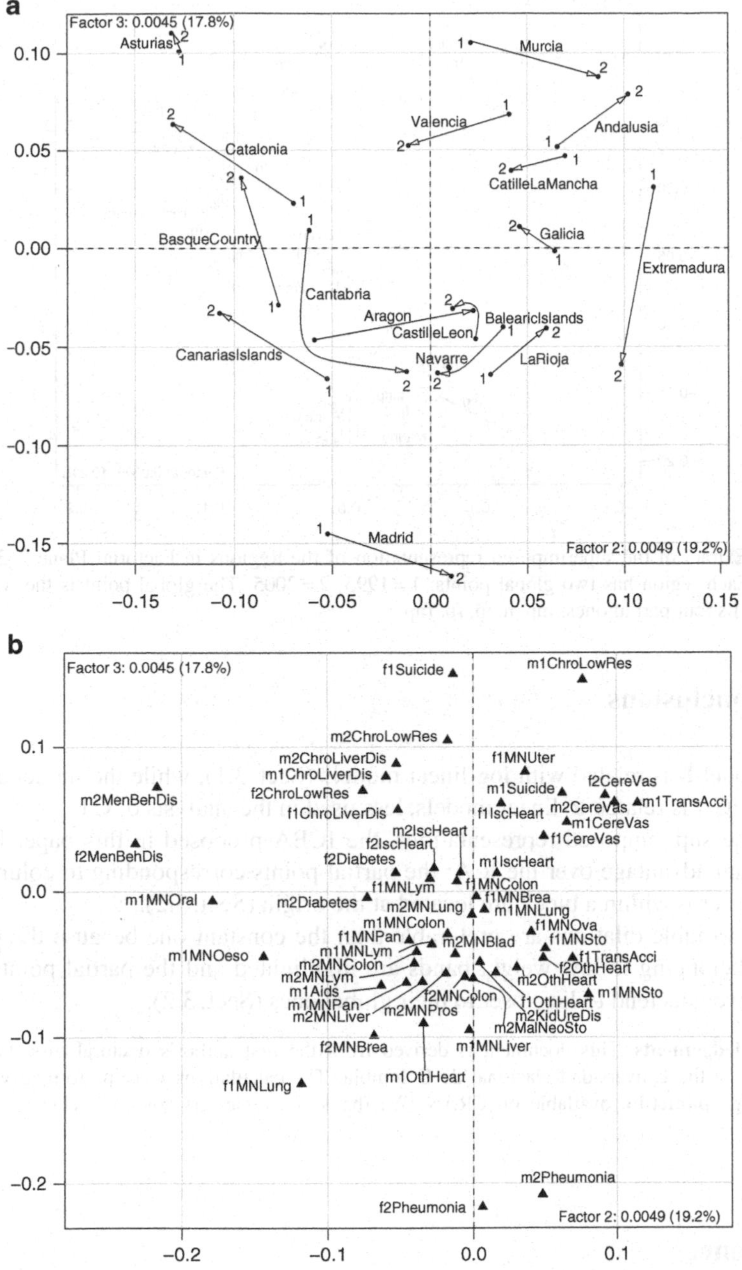

Fig. 3 Factorial Plane 2–3 of the IBCA for Spain mortality. (**a**) Autonomous Communities. The arrows go from 1995 (1) to 2005 (2) (**b**) Causes: m1: premature male, m2: non premature male, f1: premature female, f2: non prematura female

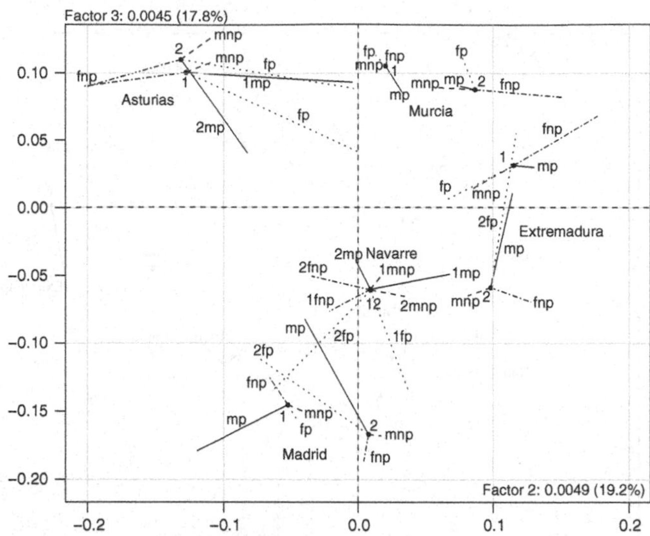

Fig. 4 Extract of the superimposed representation of the Regions in Factorial Plane 2–3 of the IBCA. Each region has two global points: 1 = 1995, 2 = 2005. The global point is the weighted mean of its four partial ones: mp, mnp, fp, fnp

6 Conclusions

The model **B** is related with log-linear models (Sect. 3.1), while the model associated to ICA is related to linear models, less used in the analysis of CT.

In the superimposed representations, the ICBA proposed in this paper has an important advantage over the ICA: the partial points corresponding to columns or rows of zeros within a block are located at the origin (Sect. 3.2).

The variable dilations are preferable than the constant one because the partial points belonging to low weight bands are highlighted and the partial points with higher weights tend to lie closer to their global ones (Sect. 3.2).

Acknowledgements This document is derived from the first author's doctoral dissertation in Statistics at the Universidad Nacional de Colombia. The calculations were performed with the R package **pamctdp**, available on CRAN. We thank the reviewers for their suggestions and corrections.

References

Bécue, M., Pagès, J., & Pardo, C. (2005). Contingency table with a double partition on rows and columns. Visualization and comparison of the partial and global structures. In: Janssen, J. & Lenca, P. (Eds.), *Proceedings ASMDA 2005*, ENST Bretagne (pp. 355–364).

Cazes, P., Chessel, D., & Dolédec, S. (1988). L'analyse des correspondances internes d'un tableau partitionné. Son usage en hydrobiologie. *Revue de Statistique Appliquee, 36*(1), 39–54.

Escofier, B. (1984). Analyse factorielle en référence à un modèle. Application à l'analyse de tableaux d'échanges. *Revue de Statistique Appliquee, 32*(4), 25–36.

European-Communities (2009). *Health statistics – Atlas on mortality in the European Union*. Statistical books, Office for Official Publications of the European Communities, Luxembourg.

Eurostat (2008). Eurostat-population and social conditions. Web, http://epp.eurostat.ec.europa.eu.

INE-España (2010). Inebase. Base de datos de INE, Instituto Nacional de Estadística de España, http://www.ine.es/inebmenu/indice.htm.

Pagès, J. (2004). Multiple factor analysis: main features and application to sensory data. *Revista Colombiana de Estadística, 27*(1), 1–26.

Van der Heijden, P. (1987). *Correspondence analysis of longitudinal categorical data*. Leiden: DSWO Press.

Determining the Similarity Between US Cities Using a Gravity Model for Search Engine Query Data

Paul Hofmarcher, Bettina Grün, Kurt Hornik, and Patrick Mair

Abstract In this paper we use the gravity model to estimate the similarity of US cities based on data provided by Google Trends (GT). GT allows to look up search terms and to obtain ranked lists of US cities according to the relative frequencies of requests for each term. The occurences of the US cities on these ranked lists are used to determine the similarities with the gravity model. As search terms for GT serve dictionaries derived from the General Inquirer (GI), containing the categories *Economy* and *Politics/Legal*. The estimated similarity scores are visualized with multidimensional scaling (MDS).

1 Introduction

Undoubtedly, the internet has ushered a new era and changed our lives. According to http://www.internetworldstats.com/ 245,000,000 people are using the World Wide Web in the United States (US) of America. These are approximately 78.2 % of the population. Within the internet usage search engines play a fundamental role when it comes to managing and searching information in the web. One of the most prominent web search engines is Google with a global market share of 82.8 % in May 2011 (see http://en.wikipedia.org/wiki/Web_search_engine).

In this work we aim at eliciting the similarity of US cities based on millions of queries performed with Google. Data are available through Google Trends (GT; see

P. Hofmarcher (✉) · K. Hornik · P. Mair
Institute for Statistics and Mathematics, WU (Vienna University of Economics and Business), Augasse 2–6, 1090 Wien, Austria
e-mail: paul.hofmarcher@wu.ac.at; kurt.hornik@wu.ac.at; patrick.mair@wu.ac.at

B. Grün
Department of Applied Statistics, Johannes Kepler University Linz, Altenbergerstraße 69, 4040 Linz, Austria
e-mail: bettina.gruen@jku.at

B. Lausen et al. (eds.), *Algorithms from and for Nature and Life*, Studies in Classification, Data Analysis, and Knowledge Organization, DOI 10.1007/978-3-319-00035-0_24,

http://www.google.com/trends; Google 2010). GT allows to look up how often a certain search term was queried and to determine a relative ranking of the regions and cities where internet users have requested this term using Google. Getting insight into the interests of cities and their differences based on the search engine entries is not only per se an interesting question, but may be useful for e.g., tourism analysis, political analysis, or discussing economic questions.

Previous uses of GT data include the analysis of economic questions, like forecasting unemployment or economic activity (see Askitas and Zimmermann 2009; Choi and Yi 2009). Ginsberg et al. (2009) use search engine query data to detect influenza epidemics. We illustrate the value of such data, by using GT data to investigate the similarities between US cities with respect to interest in economical and political/legal issues. Similarities between US cities are estimated based on the frequency of occurrence on the ranked lists provided by GT for search terms for two categories: *Economic* and *Political/Legal*. The dictionaries for these two categories are taken from the General Inquirer (GI) (see http://www.wjh.harvard.edu/~inquirer/; Stone et al. 1966; Wilson et al. 2005). To get a measure of similarity which allows for easy and intuitive interpretation we perform a *gravity model* (Haynes and Fotheringham 1988). Multidimensional scaling (MDS) is used to visualize the estimated similarities.

This article is organized as follows: In the next section we will give an overview of the data used. Section 3 describes the methods used in our application. Section 4 presents the results. Finally, Sect. 5 concludes and discusses further possible research on this topic.

2 Data

Google Trends (GT)

GT allows the user to investigate how often the search term of interest was queried during a given time period. It also has the ability to split up the data, with special attention given to the breakdown of information by countries and cities.

All results from GT are normalized, which means that Google divides the data sets by a common variable to cancel out the variable's effect on the data. Normalization of the data also allows to compare ranking of cities, because otherwise densely populated areas would be at the top for many search queries just because there are a lot more searches in absolute terms in densely populated areas (Google 2010).

To rank cities, GT calculates the ratio of searches of the considered term coming from a city (based on IP address information) divided by total searches from the same city. Afterwards, these ratios are scaled in a way that the top city has score 1. Usually GT shows the "Top 10" ranked cities. The terms' ranks used in this work are based on their search traffic between January 2004 and November 2010.

General Inquirer Lexica

Economic and political/legal search terms of interest are taken from the General Inquirer (GI; see Stone et al. 1966; Wilson et al. 2005) using the tag categories "Economy", "Politics" and "Legal". GI is a collection of the following sources: the Harvard IV-4 dictionary, the Lasswell value dictionary, several categories recently constructed, and "marker" categories primarily developed as a resource for disambiguation. The GI tag categories reflect the language of a particular "institution". In more detail, we make use of the *Econ*, *Polit* and *Legal* subcategories. *Econ* contains 455 words of an economic, commercial, industrial, or business orientation, including roles, collectivities, acts, abstract ideas, and symbols, including references to money. It includes names of common commodities in business. *Polit* includes 241 words having a clear political character, including political roles, collectivities, acts, ideas, ideologies, and symbols. *Legal* contains 173 words relating to legal, judicial, or police matters.

Further Information

Due to the smaller sample sizes of the politics and the legal categories, we merge the latter two to one category "Politics & Legal". Not all words contained in our categories "Economy" and "Politics & Legal" were available on GT. Some words were not available, e.g, due to a too small number of search entries containing them. In total we collected results for 438 words for the economic category and 407 for the political/legal category. The intersection of these two categories contains 30 words. These are words like *capital, welfare, exchange, equity*, but also unexpected terms like *auditor, run, blue (chip)*.

The set of cities considered is restricted to the 20 largest US cities plus Washington DC for the economic terms and to the US state capitals plus Washington DC for the political/legal terms.

3 Methodology

A Gravity Model to Measure Similarity

There is a wide spectrum of research areas using the gravity model, like economics, social sciences, transportation sciences, and consumer behavior. As a comprehensive, general reference for gravity models we refer to Haynes and Fotheringham (1988).

In the traditional gravity model, the interaction I between two entities (cities) is assumed to be proportional to the masses (i.e., the occurrences of a city in the ranks of the search terms) and the distance between these two entities. Formally, for cities i, j we get,

$$I_{ij} = \frac{P_i P_j}{d_{ij}^2}, \tag{1}$$

where P_i is the appearance frequency of city i in the search queries and d_{ij} denotes a distance measure between i and j. I_{ij} stands for the interaction between i and j. The interaction I_{ij} is measured by the number of co-occurrences of cities i and j in the search queries. In this work we want to get a "gravity–distance" between the cities, i.e., we resolve Eq. 1 to

$$d_{ij} = \sqrt{\frac{P_i P_j}{I_{ij}}}. \tag{2}$$

A "gravity–distance"[1] is used for measuring the similarity of cities instead of an Euclidean or Manhattan norm because of its easy and intuitive interpretation.[2] Firstly, this distance is decreasing in interaction, i.e., the number of common occurrences of cities i and j. Thus if both cities i and j appear in the ranks of a given search term, I_{ij} increases by 1. Secondly, Eq. 2 is weighted by the frequency of the appearance of city i in all considered search terms. Intuitively, a high probability of appearing in a search term increases the distance between two objects if the interaction I does not equally increase. The main advantage of the gravity model compared to any other distance measure is that it captures a balance between the general search interest P_i and the "common interests" between the cities I_{ij}. The gravity model is widely used in economics and the social sciences to model the "trade" between countries (see http://en.wikipedia.org/wiki/Gravity_model_of_trade).

Multidimensional Scaling

In order to visualize our distance matrix $d_{ij} \quad i, j = 1, \ldots, n$, we make use of multidimensional scaling (MDS; Borg and Groenen 2005), a technique for projecting our high dimensional distance matrix d_{ij} to a lower dimensional space (here 2-dim), in which the results can be visualized in an appropriate way. Formally one aims at minimizing

$$S_D(z_1, \ldots, z_N) = \sqrt{\sum_{i \neq j} w_{ij}(d_{ij} - \|z_i - z_j\|)^2} \tag{3}$$

[1]Note that the gravity distance is not a distance measure in the strict mathematical sense, since it neither fulfills $d(x, x) = 0$ nor the triangle inequality.

[2]We also performed our estimation for the Euclidean norm which gave roughly the same results.

with lower dimensional coordinates (2-dim) z_i for city i. The weights w_{ij} are set to 1 if d_{ij} is known and zero otherwise. A detailed description of MDS can be found in Borg and Groenen (2005). The advantage of this approach is that we can look at the MDS plot and immediately see how closely related the cities are. MDS is performed using the R extension package **smacof** (de Leeuw and Mair 2009).

4 Results

Table 1 summarizes the appearance frequencies of the cities for the different dictionaries. Washington DC ranks in both categories on the first position, which might not be unexpected for the political terms category. Seventy percent appearance frequency for political/legal terms and 52.1 % for economic terms paint a clear picture. A possible interpretation may be found in the unique economic constitution of Washington DC. The federal government accounted for about 27 % of the jobs in Washington DC (see http://en.wikipedia.org/wiki/Washington,_D.C). Washington DC has an economy with a high percentage of professional and business service jobs. Such regions therefore seem to be hot spots for search queries containing the terms in the considered dictionaries.

For political terms Atlanta, the capital of Georgia, appears for every third term, followed by Boston, Massachusetts, with 26 %. For the economic key terms, next to Washington DC we find New York (the largest US city) and Chicago (3rd largest city) on the subsequent ranks. Both have an appearance frequency above 40 %. Then there is a gap of about 10 % and we find the west coast cities Los Angeles (2nd largest US city) and San Francisco (rank 13 for total population). Out of the 20 largest cities plus Washington DC three do not appear at all in our GT ranks for economic terms: Memphis, El Paso, and San Jose. These are mainly cities in the south of the US. In total we find for both dictionaries a vague pattern of dominance by cities from the geographic north-east.

Political/Legal Terms

Figure 1 illustrates the similarity of the US state capitals plus Washington DC based on political/legal keywords. In total 23 state capitals and Washington DC appear within the data. Thus under half of the US state capitals appear within the GT ranks when entering political/legal key terms. Cities with a very small appearance frequency can be found on the left of Fig. 1. These are state capitals like Olympia (state: Washington) and Salem (Oregon). Tallahassee (Florida) and Little Rock (Arkansas) appear only in the ranks of the term *democrat* and are thus projected to the same coordinates.[3] The similarity of the most frequently occurring cities

[3] In order to get the figures readable, identical projections are slightly shifted.

Table 1 Appearance frequency of the cities for economic and political/legal dictionaries

Political/legal terms		Economic terms	
City	Frequency in %	City	Frequency in %
Washington	70.5	Washington	52.1
Atlanta	35.9	New York	42.0
Boston	26.0	Chicago	40.9
Phoenix	21.1	Los Angeles	29.5
Denver	12.0	San Francisco	25.8

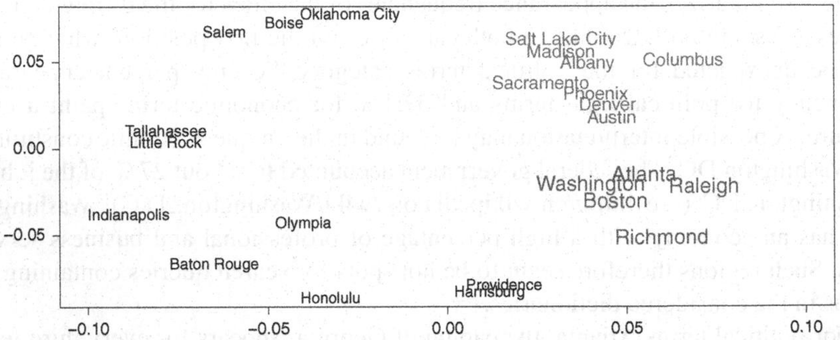

Fig. 1 MDS for the political/legal keywords. Font sizes and gray shades are intended to indicate the "grouping" of cities

Boston (Massachusetts), Atlanta (Georgia) and Washington DC can be found in a cluster at the bottom right. Based on the appearance in the GT ranks the gravity model and MDS estimate a small distance of Richmond (Virginia) and Raleigh (North Carolina) to those cities. Again we observe a geographic proximity and an average appearance rate of 109 terms (out of 438) is observed within this group. At least three of those cities appear in the ranks of *congress, conference, presidential, voter, governor, administration*. Next we observe within a "group" the state capitals of California (Sacramento), New York (Albany), Utah (Salt Lake City), Texas (Austin), Arizona (Phoenix), Colorado (Denver), Wisconsin (Madison). Common terms within this group are *legislative, legislator, capitol*.

Economic Terms

For the economic search terms, Fig. 2 displays the similarity of 18 out of the 20 largest US cities plus Washington DC. On the right hand side of Fig. 2 we find those cities which appear very rarely within the GT ranks of economic terms. Within these cities we find, e.g., MotorCity Detroit – which ranks at the 11th most populous city in the US. Detroit appears only within the GT ranks for the term *wealthy*. Next to those cities we find a "cluster" containing the cities Phoenix, San Diego, Austin and

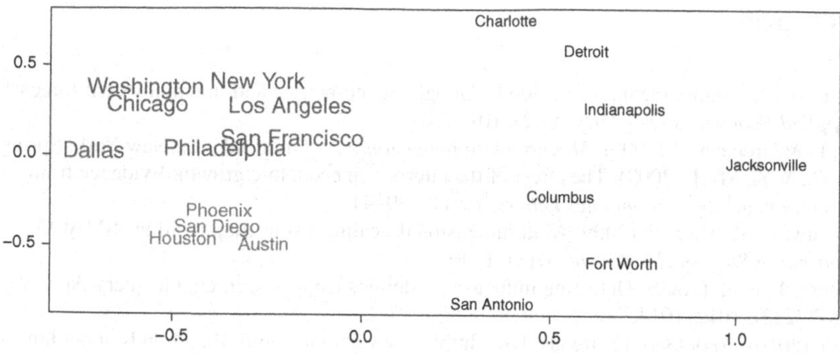

Fig. 2 MDS for the economic keywords

Houston. Due to the geographic location of these cities we call it the south-cluster. All 4 cities appear within the ranks of the terms like *discount* and *ranch*. Three out of four cities additionally appear within the ranks of *debt, enrollment, hanger, seller* and *technician*. Finally, we find again the group of the most prominent US cities, Washington, Chicago, New York, Los Angeles, San Francisco, but also Dallas and Philadelphia. This is the most powerful cluster for the economic terms. Those 7 cities have in common that all appear within the ranks of *endow* and *inherit*. 6 out of 7 cities appear in the ranks of *coverage, default, gross* and *rich*.

5 Conclusions and Discussion

The possibility to get data from the web which account for the "world's" interest in the entered search terms offers a wide range of applications. To the best of our knowledge this work is the first which aims at determining the similarity of cities based on the internet search terms entered by people living there. In total we find a pattern which nourishes the hypothesis that GT ranks are dominated by cities in the east and north. For both categories (economic, political/legal) cities from the north and east dominate in terms of appearance frequency. Additionally we find for the economic terms that high similarity for cities in the south (Houston, Austin, San Diego, Phoenix) is estimated.

The possibility to get the data from GT allows to model for arbitrary keywords of interest the interaction between cities and regions. Following the lines of Michel et al. (2011), introducing the concept of *Culturomics*, it would be of interest to extend our approach to larger data sets and estimate a regional similarity of used words or estimate how certain words spread over regions. Our results set the scene for further research, explaining the patterns in greater detail and, e.g., performing regressions methods which allow for a deeper understanding of the interaction between web search and socio-economic factors.

References

Askitas, N., & Zimmermann, K. F. (2009). Google econometrics and unemployment forecasting. *Applied Economics Quarterly, 55*(2), 107–120.

Borg, I., & Groenen, P. (2005). *Modern multidimensional scaling* (2nd ed.). New York: Springer.

Choi, C., & Yi, M. H. (2009). The effect of the internet on economic growth: Evidence from cross-country panel data. *Economics Letters, 105*(1), 39–41.

de Leeuw, J., & Mair, P. (2009) Multidimensional scaling using majorization: SMACOF in R. *Journal of Statistical Software, 31*(3), 1–30.

Ginsberg, J., et al. (2009). Detecting influenza epidemics using search engine query data. *Nature, 457*(7232), 1012–1014.

Google (2010). About Google trends. URL http://www.google.com/intl/en/trends/about.html

Haynes, K. E., & Fotheringham, A. S. (1988). *Gravity and spatial interaction models*. Beverly Hills: Sage.

Michel, J. B., et al. (2011). Quantitative analysis of culture using millions of digitized books. *Science, 331*(6014), 176–182.

Stone, P. J., Dunphy, D. C., & Smith, M. S. (1966) *The general inquirer: A computer approach to content analysis*. Cambridge: MIT.

Wilson, T., Wiebe, J., & Hoffmann, P. (2005). Recognizing contextual polarity in phrase-level sentiment analysis. In *Proceedings of Human Language Technology Conference and Conference on Empirical Methods in Natural Language Processing*, Vancouver (pp. 347–354).

Part IV
Bioinformatics and Biostatistics

An Efficient Algorithm for the Detection and Classification of Horizontal Gene Transfer Events and Identification of Mosaic Genes

Alix Boc, Pierre Legendre, and Vladimir Makarenkov

Abstract In this article we present a new algorithm for detecting partial and complete horizontal gene transfer (HGT) events which may give rise to the formation of mosaic genes. The algorithm uses a sliding window procedure that analyses sequence fragments along a given multiple sequence alignment (MSA). The size of the sliding window changes during the scanning process to better identify the blocks of transferred sequences. A bootstrap validation procedure incorporated in the algorithm is used to assess the bootstrap support of each predicted partial or complete HGT. The proposed technique can be also used to refine the results obtained by any traditional algorithm for inferring complete HGTs, and thus to classify the detected gene transfers as partial or complete. The new algorithm will be applied to study the evolution of the gene *rpl12e* as well as the evolution of a complete set of 53 archaeal MSA (i.e., 53 different ribosomal proteins) originally considered in Matte-Tailliez et al. (Mol Biol Evol 19:631–639, 2002).

1 Introduction

Bacteria and viruses adapt to changing environmental conditions via horizontal gene transfer (HGT) and intragenic recombination leading to the formation of mosaic genes, which are composed of alternating sequence parts belonging either to the original host gene or stemming from the integrated donor sequence (Doolittle 1999; Zhaxybayeva et al. 2004). An accurate identification and classification of

---

A. Boc · P. Legendre
Université de Montréal, C.P. 6128, succursale Centre-ville, Montréal, QC H3C 3J7, Canada
e-mail: alix.boc@umontreal.ca; pierre.legendre@umontreal.ca

V. Makarenkov (✉)
Département d'Informatique, Université du Québec à Montréal, C.P.8888, succursale Centre Ville, Montreal, QC H3C 3P8, Canada
e-mail: makarenkov.vladimir@uqam.ca

B. Lausen et al. (eds.), *Algorithms from and for Nature and Life*, Studies in Classification, Data Analysis, and Knowledge Organization, DOI 10.1007/978-3-319-00035-0_25,

mosaic genes as well as the detection of the related gene transfers are among the most important challenges posed by modern computational biology (Koonin 2003; Zheng et al. 2004). Partial HGT model assumes that any part of a gene can be transferred among the organisms under study, whereas traditional (complete) HGT model assumes that only an entire gene, or a group of complete genes, can be transferred (Makarenkov et al. 2006a,b). Mosaic genes can pose several risks to humans including cancer onset or formation of antibiotic-resistant genes spreading among pathogenic bacteria (Nakhleh et al. 2005). The term "mosaic" stems from the pattern of interspersed blocks of sequences having different evolutionary histories, but being combined in the resulting allele subsequent to recombination events. The recombined segments can derive from other strains in the same species or from other more distant bacterial or viral relatives (Gogarten et al. 2002; Hollingshead et al. 2000). Mosaic genes are constantly generated in populations of transformable organisms, and probably in all genes (Maiden 1998).

Many methods have been proposed to address the problem of the identification and validation of complete HGT events (e.g., Boc et al. 2010, Hallett and Lagergren 2001, and Nakhleh et al. 2005), but only a few methods treat the much more challenging problem of inferring partial HGTs and predicting the origins of mosaic genes (Denamur et al. 2000; Makarenkov et al. 2006b). We have recently proposed (Boc and Makarenkov 2011) a new method allowing for detection and statistical validation of partial HGT events using a sliding window approach.

In this article we describe an extension of the algorithm presented in Boc and Makarenkov (2011), considering sliding windows of variable size. We will show how the new algorithm can be used: (1) to estimate the robustness of the obtained HGT events; (2) to classify the obtained transfers as partial or complete; (3) to classify the species under study as potential donors or receivers of genetic material.

2 Algorithm

Here we present the new algorithm for inferring partial horizontal gene transfers using a sliding window of adjustable size. The idea of the method is to provide the most probable partial HGT scenario characterizing the evolution of the given gene. It takes as input a species phylogenetic tree representing the traditional evolution of the group of species under study and a multiple sequence alignment (MSA) representing the evolution of the gene of interest for the same group of species. A sliding window procedure, with a variable window size, is carried out to scan the fragments of the given MSA (see Fig. 1). In the algorithm Boc and Makarenkov (2011), the sliding window size was constant, thus preventing the method from detecting accurately the exact lengths of the transferred sequences (i.e. only an approximate length of the transferred sequence blocks was provided). In this study, the most appropriate size of the sliding window is selected with respect to the significance of the gene transfers inferred for different overlapping MSA intervals. The HGT significance is computed as the average HGT bootstrap support (Boc et al. 2010) obtained for the corresponding fixed MSA interval.

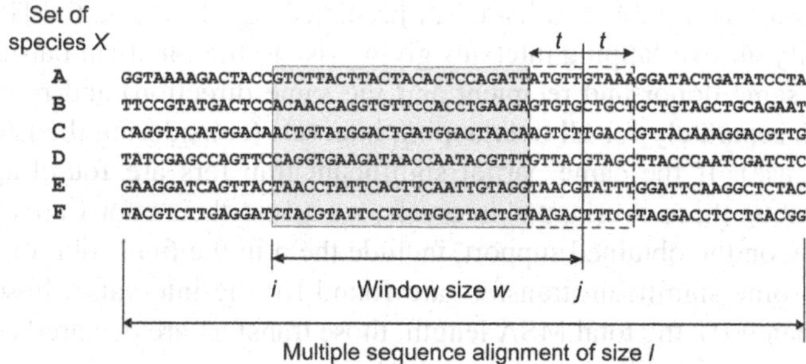

Fig. 1 New algorithm uses a sliding window of variable size. If the transfers obtained for the original window position $[i\,; i+w-1]$ are significant, we refine the obtained results by searching in all the intervals of types $[i\,; i+w-1-t+k]$ and $[i\,; i+w-1+t-k]$, where $k=0,\ldots,t$ and t is a fixed window contraction/extension parameter

The algorithm includes the three following main steps:

***Step* 1.** Let X be a set of species and l is the length of the given MSA. We first define the initial sliding window size w ($w = j-i+1$, see Fig. 1) and the window progress step s. The species tree, denoted T, characterizing the evolution of the species in X can be either inferred from the available taxonomic or morphological data, or can be given. T must be rooted to take into account the evolutionary time-constraints that should be satisfied when inferring HGTs (Boc et al. 2010; Hallett and Lagergren 2001).

***Step* 2.** For i varying from 1 to $|l-w+1|$, we first infer (e.g., using PhyML, Guindon and Gascuel (2003)) a partial gene tree T' from the subsequences located within the interval $[i\,; i+w-1]$ of the given MSA. If the average bootstrap support of the edges of T' constructed for this interval is significant (i.e. >60 % in this study), then we apply a standard HGT detection algorithm (e.g., Boc et al. 2010) using as input species phylogenetic tree T and partial gene tree T'. If the transfers obtained for this interval are significant, then we perform the algorithm for all the intervals of types $[i\,; i+w-1-t+k]$ and $[i\,; i+w-1+t-k]$, where $k=0,\ldots,t$ (see Fig. 1) and t is a fixed window contraction/extension parameter (in our study the value of t equal to $w/2$ was used). If for some of these intervals the average HGT significance is greater than or equal to the HGT significance of the original interval $[i\,; i+w-1]$, then we adjust the sliding window size w to the length of the interval providing the greatest significance, which may be typical for the dataset being analyzed. If the transfers corresponding to the latter interval have an average bootstrap score greater than a pre-defined threshold (e.g. when the average HGT bootstrap score of the interval is >50 %), we add them to the list of predicted partial HGT events and advance along the given MSA with the progress step s. The bootstrapping procedure for HGT is presented in Boc et al. (2010).

***Step* 3.** Using the established list of all predicted significant partial HGT events, we identify all overlapping intervals giving rise to the identical partial transfers (i.e., the same donor and recipient and the same direction) and re-execute the algorithm separately for all overlapping intervals (considering their total length in each case). If the same partial significant transfers are found again when concatenating these overlapped intervals, we assess their bootstrap support and, depending on the obtained support, include them in the final solution or discard them. If some significant transfers are found for the intervals whose length is greater than 90 % the total MSA length, those transfers are declared complete.

The time complexity of the described algorithm is as follows:

$$O(r \times (\frac{t \times (l - w)}{s} \times (C(Phylo\_Inf) + C(HGT\_Inf)))), \quad (1)$$

where $C(Phylo\_Inf)$ is the time complexity of the tree inferring method used to infer partial gene phylogenetic trees, $C(HGT\_Inf)$ is the time complexity of the HGT detection method used to infer complete transfers and r is the number of replicates in the HGT bootstrapping. The simulations carried out with the new algorithm (due the lack of space, the simulation results are not presented here) showed that it outperformed the algorithm described in Boc and Makarenkov (2011) in terms of HGT prediction accuracy, but was slower than the latter algorithm, especially in the situations when large values of the t parameter were considered.

3 Application Example

We first applied the new algorithm to analyze the evolution of the gene *rpl12e* for the group of 14 organisms of Archaea originally considered in Matte-Tailliez et al. (2002). The latter authors discussed the problems encountered when reconstructing some parts of the species phylogeny for these organisms and indicated the evidence of HGT events influencing the evolution of the gene *rpl12e* (MSA size for this gene was 89 sites). In Boc et al. (2010), we examined this dataset using an algorithm for predicting complete HGTs and found five complete transfers that were necessary to reconcile the reconstructed species and gene *rpl12e* phylogenetic trees (see Fig. 2a). These results confirm the hypothesis formulated in Matte-Tailliez et al. (2002). For instance, HGT 1 between the cluster (*Halobacterium sp.*, *Haloarcula mar.*) and *Methanobacterium therm.* as well as HGTs 4 and 5 between the clade of *Crenarchaeota* and the organisms *Thermoplasma ac.* and *Ferroplasma ac.* have been characterized in Matte-Tailliez et al. (2002) as the most likely HGT events occurred during the evolution of this group of species. In this study, we first applied the new algorithm allowing for prediction of partial and complete HGT event to confirm or discard complete horizontal gene transfers presented in Fig. 2a,

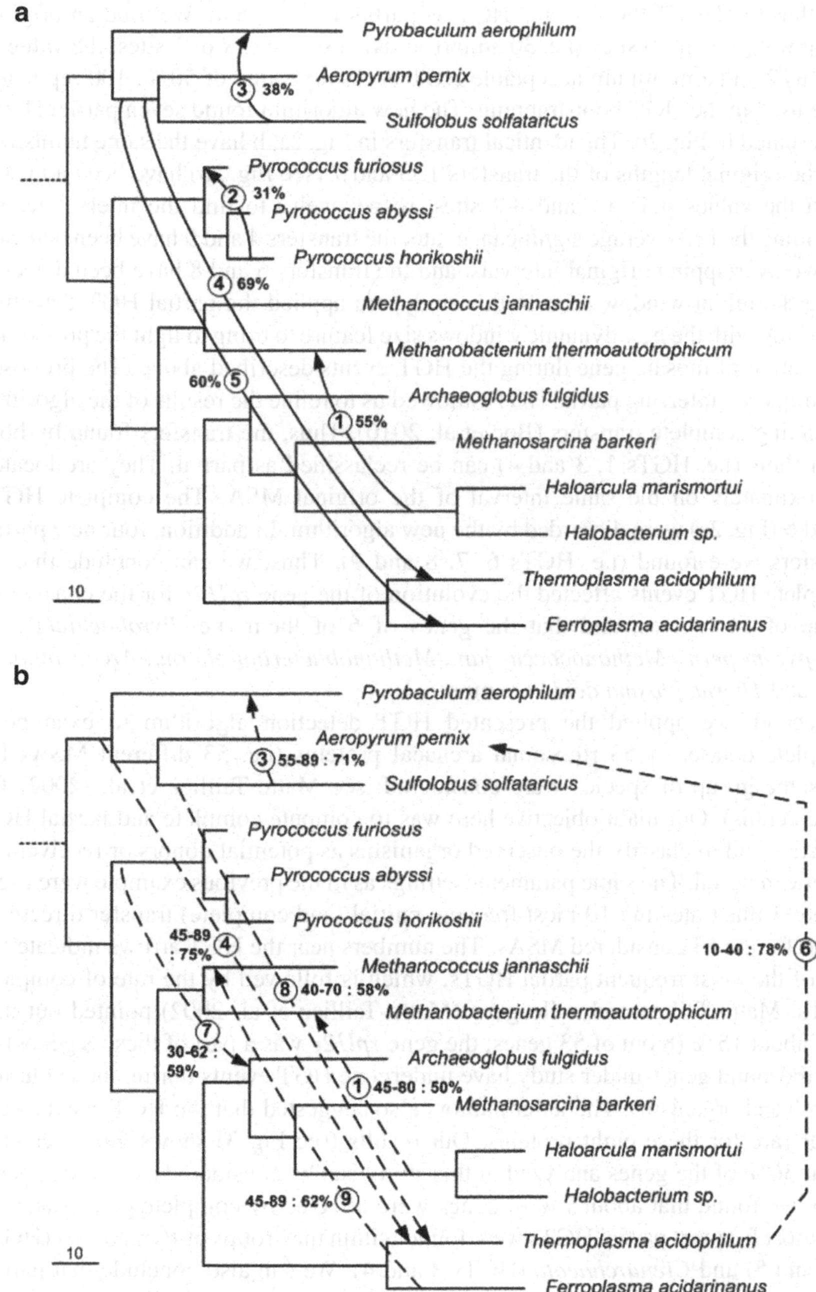

Fig. 2 Species tree (Matte-Tailliez et al. 2002, Fig. 1a) encompassing: (**a**) five complete horizontal gene transfers, found by the algorithm described in Boc et al. (2010), indicated by *arrows*; numbers on HGTs indicate their order of inference; HGT bootstrap scores are indicated near each HGT *arrow*; and (**b**) seven partial HGTs detected by the new algorithm; the identical transfers have the same numbers in the positions A and B of the figure; the interval for which the transfer was detected and the corresponding bootstrap score are indicated near each HGT *arrow*

and thus to classify the detected HGT as partial or complete. We used an original window size w of 30 sites (i.e. 30 amino acids), a step size s of 5 sites, the value of $t = w/2$ and a minimum acceptable HGT bootstrap value of 50 %; 100 replicates were used in the HGT bootstrapping. The new algorithm found seven partial HGTs represented in Fig. 2b. The identical transfers in Fig. 2a, b have the same numbers.

The original lengths of the transfers 1, 3 and 7 (see Fig. 2b) have been adjusted (with the values +4, +5 and +2 sites, respectively) to find the interval length providing the best average significance rate, the transfers 4 and 9 have been detected on two overlapping original intervals, and the transfers 6 and 8 have been detected using the initial window size. In this study, we applied the partial HGT detection algorithm with the new dynamic windows size feature to bring to light the possibility of creation of mosaic gene during the HGT events described above. The proposed technique for inferring partial HGTs allowed us to refine the results of the algorithm predicting complete transfers (Boc et al. 2010). Thus, the transfers found by both algorithms (i.e. HGTs 1, 3 and 4) can be reclassified as partial. They are located approximately on the same interval of the original MSA. The complete HGTs 2 and 5 (Fig. 2a) were discarded by the new algorithm. In addition, four new partial transfers were found (i.e. HGTs 6, 7, 8 and 9). Thus, we can conclude that no complete HGT events affected the evolution of the gene *rpl12e* for the considered group of 14 species, and that the genes of 6 of them (i.e. *Pyrobaculum aer.*, *Aeropyrum pern.*, *Methanococcus jan.*, *Methanobacterium therm.*, *Archaeoglobus fulg.* and *Thermoplasma acid.*) are mosaic.

Second, we applied the presented HGT detection algorithm to examine a complete dataset of 53 ribosomal archaeal proteins (i.e. 53 different MSAs for the same group of species were considered; see Matte-Tailliez et al. (2002) for more details). Our main objective here was to compute complete and partial HGT statistics and to classify the observed organisms as potential donors or receivers of genetic material. The same parameter settings as in the previous example were used. Figure 3 illustrates the 10 most frequent partial (and complete) transfer directions found for the 53 considered MSAs. The numbers near the HGT arrows indicate the rate of the most frequent partial HGTs, which is followed by the rate of complete HGTs. Matte-Taillez and colleagues (Matte-Tailliez et al. 2002) pointed out that only about 15 % (8 out of 53 genes; the gene *rpl12e* was a part of these 8 genes) of the ribosomal genes under study have undergone HGT events during the evolution of archaeal organisms. The latter authors also suggested that the HGT events were rather rare for these eight proteins. Our results (see Fig. 3) shows, however, that about 36 % of the genes analyzed in this study can be considered as mosaic genes. Also, we found that about 7 % of genes were affected by complete gene transfers. The most frequent partial HGTs were found within the groups of *Pyrococcus* (HGTs 1, 2 and 5) and *Crenarchaeota* (HGTs 3 and 4). We can also conclude that partial gene transfers were about five times more frequent than partial HGT events.

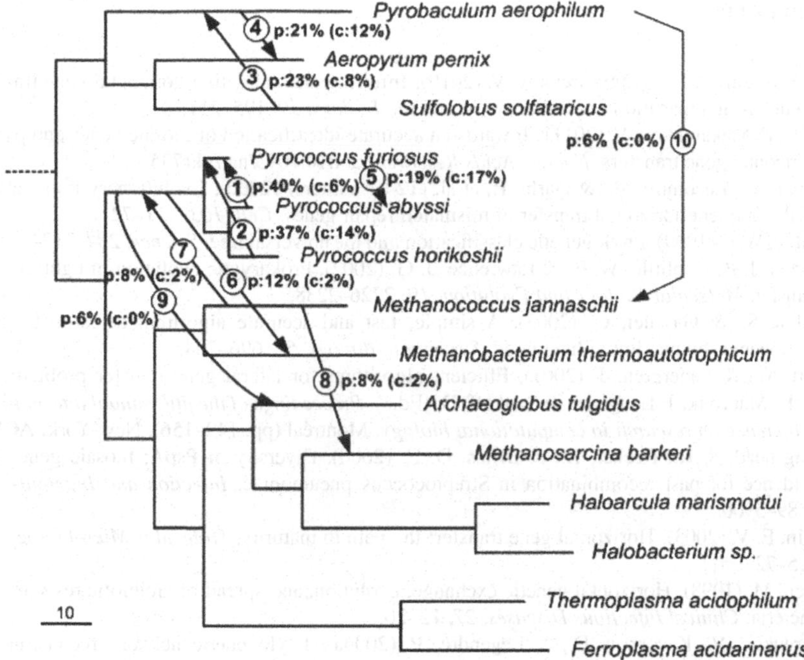

Fig. 3 Species tree (Matte-Tailliez et al. 2002, Fig. 1a) with the 10 most frequent HGT events obtained by the new algorithm when analyzing separately the MSAs of 53 archaeal proteins. The first value near each HGT arrow indicates the rate of partial HGT detection (p) and the second value indicates the rate of complete HGT detection (c)

4 Conclusion

In this article we described a new algorithm for inferring partial and complete horizontal gene transfer events using a sliding window approach in which the size of the sliding window is adjusted dynamically to fit the nature of the sequences under study. Such an approach aids to identify and validate mosaic genes with a better precision. The main advantage of the presented algorithm over the methods used to detect recombination in sequence data is that it allows one to determine the source (i.e. putative donor) of the subsequence being incorporated in the host gene. The discussed algorithm was applied to study the evolution of the gene *rpl12e* and that of a group of 53 ribosomal proteins in order to estimate the pro-portion of mosaic genes as well as the rates of partial and complete gene transfers characterizing the considered group of 14 archaeal species. In the future, this algorithm could be adapted to compute several relevant statistics regarding the functionality of genetic fragments affected by horizontal gene transfer as well as to estimate the rates of intraspecies (i.e. transfers between strains of the same species) and interspecies (i.e. transfers between distinct species) HGT.

References

Boc, A., Philippe, H., & Makarenkov, V. (2010). Inferring and validating horizontal gene transfer events using bipartition dissimilarity. *Systematic Biology, 59*, 195–211.
Boc, A., & Makarenkov, V. (2011). Towards an accurate identification of mosaic genes and partial horizontal gene transfers. *Nucleic Acids Research*. doi:10.1093/nar/gkr735.
Denamur, E., Lecointre, G., & Darlu, P., et al. (12 co-authors) (2000). Evolutionary implications of the frequent horizontal transfer of mismatch repair genes. *Cell, 103*, 711–721.
Doolittle, W. F. (1999). Phylogenetic classification and the universal tree. *Science, 284*, 2124–2129.
Gogarten, J. P., Doolittle, W. F., & Lawrence, J. G. (2002). Prokaryotic evolution in light of gene transfer. *Molecular Biology and Evolution, 19*, 2226–2238.
Guindon, S., & Gascuel, O. (2003). A simple, fast and accurate algorithm to estimate large phylogenies by maximum likelihood. *Systematic Biology, 52*, 696–704.
Hallett, M., & Lagergren, J. (2001). Efficient algorithms for lateral gene transfer problems. In N. El-Mabrouk, T. Lengauer, & Sankoff, D. (Eds.), *Proceedings of the fifth annual international conference on research in computational biology*, Montréal (pp. 149–156). New-York: ACM.
Hollingshead, S. K., Becker, R., & Briles, D. E. (2000). Diversity of PspA: mosaic genes and evidence for past recombination in Streptococcus pneumoniae. *Infection and Immunity, 68*, 5889–5900.
Koonin, E. V. (2003). Horizontal gene transfer: the path to maturity. *Molecular Microbiology, 50*, 725–727.
Maiden, M. (1998). Horizontal genetic exchange, evolution, and spread of antibiotic resistance in bacteria. *Clinical Infectious Diseases, 27*, 12–20.
Makarenkov, V., Kevorkov, D., & Legendre, P. (2006a). Phylogenetic network reconstruction approaches. In *Applied mycology and biotechnology* (International Elsevier series: Bioinformatics, Vol. 6, pp. 61–97). Amsterdam: Elsevier.
Makarenkov, V., Boc, A., Delwiche, C. F., Diallo, A. B., & Philippe, H. (2006b). New efficient algorithm for modeling partial and complete gene transfer scenarios. In V. Batagelj, H. H. Bock, A. Ferligoj, A. Ziberna, (Eds.), *Data science and classification* (pp. 341–349). Berlin: Springer.
Matte-Tailliez, O., Brochier, C., Forterre, P., & Philippe, H. (2002). Archaeal phylogeny based on ribosomal proteins. *Molecular Biology and Evolution, 19*, 631–639.
Nakhleh, L., Ruths, D., & Wang, L. S. (2005). RIATA-HGT: A fast and accurate heuristic for reconstructing horizontal gene transfer. In L. Wang, (Ed.), *Lecture notes in computer science* (pp. 84–93). Kunming: Springer.
Zhaxybayeva, O., Lapierre, P., & Gogarten, J. P. (2004). Genome mosaicism and organismal lineages. *Trends in Genetics, 20*, 254–260.
Zheng, Y., Roberts, R. J., & Kasif, S. (2004). Segmentally variable genes: a new perspective on adaptation. *PLoS Biology, 2*, 452–464.

Complexity Selection with Cross-validation for Lasso and Sparse Partial Least Squares Using High-Dimensional Data

Anne-Laure Boulesteix, Adrian Richter, and Christoph Bernau

Abstract Sparse regression and classification methods are commonly applied to high-dimensional data to simultaneously build a prediction rule and select relevant predictors. The well-known lasso regression and the more recent sparse partial least squares (SPLS) approach are important examples. In such procedures, the number of identified relevant predictors typically depends on a complexity parameter that has to be adequately tuned. Most often, parameter tuning is performed via cross validation (CV). In the context of lasso penalized logistic regression and SPLS classification, this paper addresses three important questions related to complexity selection: (1) Does the number of folds in CV affect the results of the tuning procedure? (2) Should CV be repeated several times to yield less variable tuning results?, and (3) Is complexity selection robust against resampling?

1 Background

The most straightforward approach to build a prediction rule based on a large number of predictors is to first select a subset of "relevant predictors" and then use a standard classifier such as discriminant analysis or logistic regression. However, in this two-step approach the subset of predictors selected in the first step may not be optimal for the second step. Sparse regression methods can be seen as a solution to this problem because they select predictors and build a prediction rule simultaneously. Lasso regression (Tibshirani 1996) or its more recent variants elastic nets (Zou 2005) or SCAD (Fan and Li 2001) select predictors by setting some of the regression coefficients to zero through the application of an L_1-penalty.

A.-L. Boulesteix (✉) · A. Richter · C. Bernau
Institut für Medizinische Informationsverarbeitung, Biometrie und Epidemiologie, Universität München (LMU), Munich, Germany
e-mail: boulesteix@ibe.med.uni-muenchen.de

B. Lausen et al. (eds.), *Algorithms from and for Nature and Life*, Studies in Classification, Data Analysis, and Knowledge Organization, DOI 10.1007/978-3-319-00035-0_26,

The so-called sparse partial least squares (SPLS) approach recently suggested by Chun and Keles (2010) and generalized to class prediction by Chung and Keles (2010) embeds an L_1-penalty into PLS dimension. This results in a sparse prediction rule in the sense that the PLS components are built as linear combinations of a subset of predictors instead of using all predictors as in standard PLS.

Sparse regression methods in general and the two examples considered here in particular involve one or several tuning parameter(s) determining the number of predictors used to construct the prediction rule. In practice, this parameter is almost always chosen by internal cross-validation: the prediction error is estimated for different candidate values of the parameter(s) via cross-validation and the value yielding the smallest error is selected. However, there are no widely established rules with respect to the type of cross-validation to be used. Does the number of folds in CV affect the results of the tuning procedure? Should CV be repeated several times to yield less variable tuning results? Are variable selection and complexity selection robust against resampling? In this paper, we address these questions in an extensive empirical study using five real high-dimensional gene expression data sets with binary response class. The outcome we consider in this study is the number of predictors selected for building the prediction rule, not the prediction accuracy. We perform only one CV-loop – for tuning purposes.

2 Methods

2.1 *Lasso Regression*

Lasso regression is similar to standard logistic regression except that a penalty of the form $\lambda \sum_{j=1}^{p} |\beta_j|$ is added to the optimality criterion to be minimized, where λ is the *penalty parameter* and $\beta_1, \dots, \beta_p$ denote the regression coefficients of the p predictors. This penalty is termed L_1-penalty in contrast to the more widely used L_2 penalty of the form $\lambda \sum_{j=1}^{p} \beta_j^2$.

2.2 *Sparse Partial Least Squares (SPLS)*

Similarly to standard PLS regression, SPLS constructs a small number c of orthogonal latent components such that they have maximal covariance with the response. In standard PLS each latent component t_i $(i = 1, \dots, c)$ is a linear combination $t_i = Xw_i$ of all predictors, where w_i is usually denoted as weight vector. Each weight vector w_i $(i = 1, \dots, c)$ is constructed to maximize the squared sample covariance of t_i with the response: $w_i = \text{argmax}_w\, w^T X^T Y Y^T X w$ s.t.: $w_i^T w_i = 1$ and $t_i^T t_j = 0$ $(j = 1, \dots, i-1)$. Similarly to the L_1-penalty in lasso, Chun and Keles (2010) add a L_1-penalty term to the objective function. The strength of the penalty is determined by a sparsity parameter denoted as $\eta \in [0, 1)$ and affects the number of predictors selected to construct the PLS components. For $\eta = 0$ all predictors contribute to the PLS components while the number of predictors decreases considerably for $\eta \to 1$. We refer to the original publications and to our codes for more details.

2.3 *Cross-validation for Parameter Tuning*

The parameters λ (for lasso) and c and η (for sparse PLS) are almost always chosen by cross-validation (CV) in practical data analysis. The basic idea is to split the available data set D into K approximately equally sized and non–overlapping folds: $D = D_1 \cup \ldots \cup D_K$ where $D_i \cap D_j = \emptyset$ for $i \neq j$. In the first CV iteration the first learning set $\mathcal{L}_1 = D_1 \cup \ldots \cup D_{K-1}$ is used for learning a prediction rule, which is subsequently applied to the first test set $\mathcal{T}_1 = D_K$ and evaluated by computing its error rate. This process is repeated for all K folds. The error rate is then averaged over K folds. The results can be made more stable by performing CV repeatedly with different random partitions of the data D and averaging the resulting CV errors. When used for tuning purposes, CV is applied with different values of the parameter successively. The parameter value yielding the smallest CV error is then selected. The number K of folds as well as the number of random partitions (if repeated CV is performed) are important parameters of the tuning procedure.

3 Empirical Study

3.1 *Study Design*

The empirical study is based on five microarray gene expression data sets with different samples sizes n, different (large) numbers of predictors p and a binary class outcome (e.g. diseased versus healthy). The characteristics of the five data sets are summarized in Table 1. R codes (http://www.r-project.org) and preprocessed data sets for reproducing our analyses are available from: www.ibe.med.uni-muenchen.de/organisation/mitarbeiter/020_professuren/boulesteix/cvcomplexity

Study 1 is designed to assess the variability of the number of selected predictors across different random CV partitions, the effect of the number of CV folds K (here $K = 3, 5, 10, 20$) and the effect of repetitions of CV (here 1, 5, 10) on the number of selected predictors. For each of the five data sets, we perform 2,000 runs of cross-validation successively, i.e. we consider 2,000 different random partitions. For each of the 2,000 runs, we store the number of selected predictors corresponding to the selected parameter value. This study is performed for lasso as implemented in the function `cv.glmnet` from the `R` package `glmnet` and

Table 1 Data sets used in the empirical study

Disease	Size (size of classes 0/1)	Number of predictors	Reference
Colon cancer	$n = 47$ (22/25)	$p = 22{,}283$	Ancona et al. (2006)
Sepsis	$n = 70$ (16/54)	$p = 22{,}283$	Tang et al. (2009)
Parkinson	$n = 105$ (55/50)	$p = 22{,}283$	Scherzer et al. (2007)
Prostate cancer	$n = 102$ (50/52)	$p = 12{,}625$	Singh et al. (2002)
Breast cancer	$n = 250$ (179/107)	$p = 22{,}283$	Wang et al. (2005)

for sparse PLS as implemented in the function `cv.spls` from the `R` package `spls`. The considered candidate values of the parameters are those that are automatically determined by these functions.

Study 2 is designed to assess the effect of small changes of the data sets on the number of predictors determined by leave-one-out CV. The number of folds is thus fixed to $K = n$, leading to a unique possible partition. In study 2, the source of variability is now the elimination of a randomly selected subset of observations. In other words, leave-one-out CV is run on subsamples. The size of the subsamples is fixed to $0.95n$, $0.9n$ and $0.8n$ successively.

Study 3 is designed exactly as study 1, except that a preliminary univariate variable selection is now performed before the analysis. We thus consider reduced data sets instead of the whole data sets whose dimensions are displayed in Table 1. Variable selection is performed based on the p-value of the two-sample t-test, i.e. the p^* predictors with smallest p-values are selected, where p^* successively takes the values $p^* = 100$ and $p^* = 500$.

3.2 Results

The results of study 1 suggest that the number of selected predictors highly depends on the partition chosen to perform CV, as illustrated in Fig. 1 (top row). For all investigated data sets, the interquartile range (i.e. the difference between the upper and lower quartiles) approximately equals the median for lasso regression. Figure 2 displays the boxplots of the number of selected predictors for different CV partitions with sparse PLS: obviously, the number of selected predictors is much more variable than for lasso regression. While some CV partitions lead to the selection of a few tens of predictors, the number of predictors reaches several thousands for many of them. On the whole, these results suggest that the number of selected predictors selected by sparse methods with CV-tuned complexity parameter should *not* be extensively interpreted, because it is highly variable.

An other conclusion of study 1 is that it is useful to run CV several times, i.e. with different random partitions successively, and to select the parameter value that minimizes the average error. By average error, we mean here the average over several random partitions. As can be seen from the middle row (average over 5 CV partitions) and from the bottom row (average over 10 CV partitions), the number of

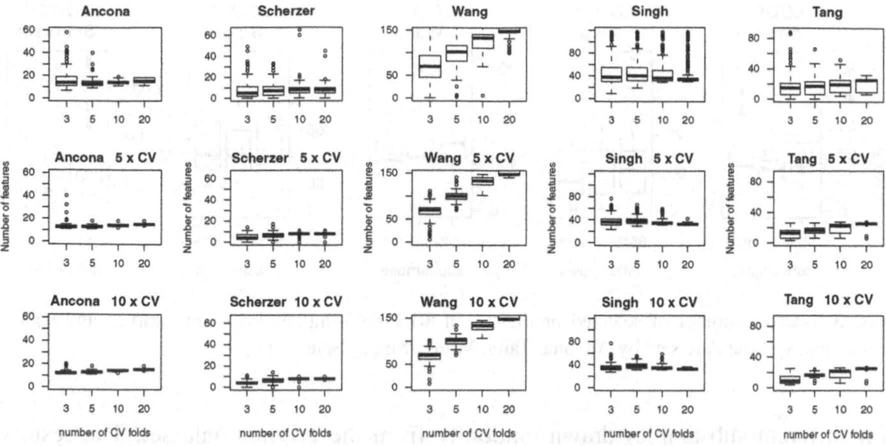

Fig. 1 Lasso. Number of selected predictors in the data sets by Ancona, Tang, Wang, Singh, Scherzer with a single CV run (*top row*), 5 averaged CV runs (*middle row*), and 10 averaged CV runs (*bottom row*)

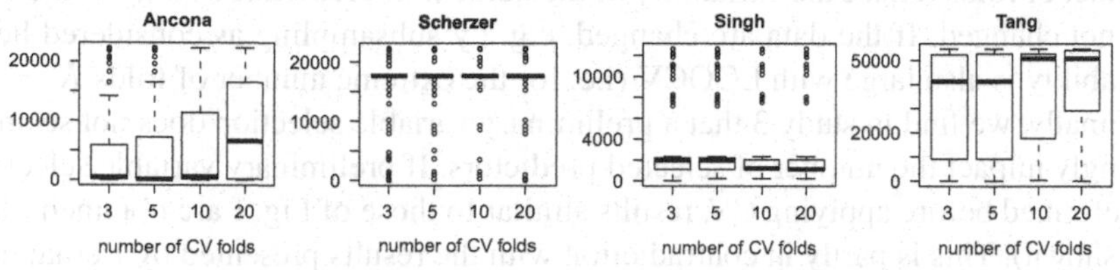

Fig. 2 Sparse PLS. Number of selected predictors in the data sets by Ancona, Tang, Singh, Scherzer with a single CV run. Analyses with the Wang data set were computationally unfeasible

selected predictors is noticeably less variable than the number of selected predictors chosen based on a single CV partition. However, the variability remains high for most data sets even after averaging.

Further, it can be clearly seen from the bottom row of Fig. 1 that the number of CV folds K also affects the number of selected predictors. Higher numbers of folds lead to the selection of more predictors in average, especially for the Wang data set. Moreover, the variability of the number of selected predictors seems to decrease with increasing number of folds. This would be an argument in favor of large numbers of folds (e.g. $K = 20$). Another argument is that we are interested in the performance of the different parameter values for a data set of size n rather than for a data set of size $2n/3$ ($2n/3$ is the size of learnings sets when $K = 3$). In the extreme case of leave-one-out (LOO) CV, there would not be any variability at all, since there is only one possible partition: the partition where each single observation forms its own fold.

However, LOOCV is generally known to have a high unconditional variance (Braga-Neto and Dougherty 2004), i.e. to yield very different results if the data set is changed. Study 2 is performed to address this problem: LOOCV is performed

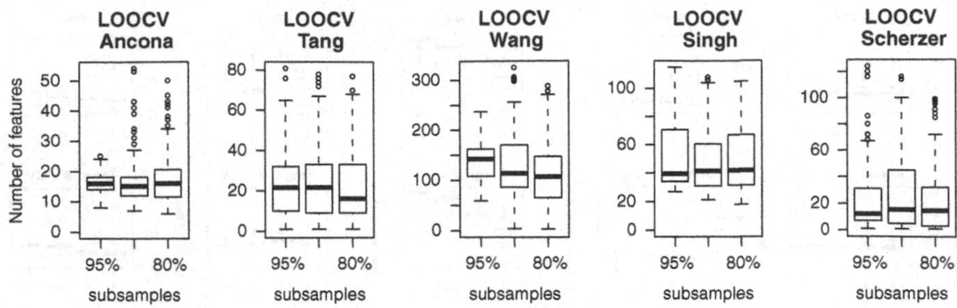

Fig. 3 **Lasso**. Number of selected predictors in 80 %-subsamples, 90 %-subsamples and 95 %-subsamples of the data sets by Ancona, Tang, Wang, Singh, Scherzer

for different subsamples drawn randomly from the original data set. The results displayed in Fig. 3 show that the number of selected predictors highly depends on the considered subsample, even if we consider large subsamples containing 95 % of the original data set. All in all, the results of studies 1 and 2 suggest that higher number of folds reduce the variability of the number of predictors as long as the data are not changed. If the data are changed, e.g. by subsampling as considered here, variability is also large with LOOCV (i.e. for the extreme number of folds $K = n$).

Finally, we find in study 3 that a preliminary variable selection does not seem to strongly impact the number of selected predictors. If preliminary variable selection is peformed before applying CV, results similar to those of Fig. 1 are obtained (data not shown). This is partly in contradiction with the results presented by Bernau and Boulesteix (2010). However, we also observe big differences between data sets: the number of selected predictors after variable selection is higher than without variable selection for some data sets, but lower for other data sets. It is thus difficult to draw general conclusions based on the five considered data sets.

4 Conclusions

The number of selected predictors is sometimes considered as a meaningful output of the lasso method in the literature. The problem is that it directly depends on the penalty parameter λ that is usually chosen by CV – a procedure known to be highly unreliable in the "$n \ll p$" setting (Dougherty et al. 2011; Hanczar et al. 2007). The aim of our study was to investigate the behaviour of the number of selected predictors in different CV settings. Our results suggest that the number of selected predictors highly depends on the particular random partition of the data used to perform CV. Two different partitions may yield two completely different numbers of selected predictors. Note that two models with completely different numbers of predictors may lead to similar prediction accuracies. Furthermore, we do not claim that CV is useless for parameter tuning. Choosing parameters by CV is better than trying only one value or trying several values to finally report the best results only. The number of selected predictors itself, however, should not be given much attention, since it may be different with another random partition.

This variability if even more pronounced in the case of sparse PLS, where a partition may lead to the selection of a handful of predictors, while the next one selects several thousands of predictors. Let us try to give an intuitive explanation for the difference between the behaviours of lasso and sparse PLS with respect to the variability of the number of selected predictors. PLS regression is essentially a method that can handle a high number of predictors: it is designed such that all predictors have a modest but non-zero contribution to the PLS components and thus to the prediction rule. Selecting a sparse PLS rule is difficult because most contributions are very small anyway: there is thus no big difference between a sparse PLS prediction involving 20 predictors and a "sparse" PLS prediction rule involving 1,000 predictors. In contrast, lasso regression is essentially similar to ML regression, that can only handle a very small number of predictors. Such a method is expected to yield a clearer difference between complex and sparse prediction rules, and thus a less ambiguous decision in favor of one of the candidate parameters values.

Another interesting result is that CV with many folds (i.e. with large training sets) leads to the selection of more predictors than CV with few folds. A potential intuitive explanation is that in a small training data set a complex model is more likely to overfit the data and perform badly. CV with large training sets (i.e. large numbers of folds) is more representative of what would happen with the whole data set. In this perspective, large numbers of folds should be preferred to small numbers of folds. However, one should not think that making the training sets larger solves all problems. By choosing a large K, one also makes the results highly specific of the particular data set at hand. A corrolar is that changes of the data set (for instance exclusion of a certain proportion of the observations) usually lead to important changes in the results, as illustrated by study 2 in our paper in the case of leave-one-out CV. Quite generally, leave-one-out CV is known to have a large *unconditional* variance over data sets generated from a particular data generating process (Braga-Neto and Dougherty 2004).

To conclude, let us point out that the high variability of the number of selected predictors does not necessarily lead to bad prediction accuracy. Prediction accuracy may be excellent both with 10 and 120 selected predictors. Moreover, the high variability does not mean that the selected predictors are the wrong ones or that one should not perform variable selection at all. It simply means that the number of selected predictors should not be considered as an interpretable output of lasso or sparse PLS. While it makes sense to fit a sparse model with a penalty parameter chosen by CV, it certainly does not make sense to, say, compare the numbers of selected predictors for two different response variables or consider that the non-selected predictors are completely irrelevant. With a different CV partition the results may have looked completely different. This variability partly remains but is noticeably reduced when a repeated CV is performed instead of a single CV run. It is also expected to decrease noticeably when the total number of predictors decreases. In future research one could investigate the variability of the tuning procedure in connection to the dimensionality and relate the empirical results of our study to theoretical results on the properties of CV-based complexity selection in easier settings.

References

Ancona, N., Maglietta, R., & Piepoli, A., et al. (2006). On the statistical assessment of classifiers using DNA microarray data. *BMC Bioinformatics, 7*, 387.

Bernau, C., & Boulesteix, A. L. (2010). Variable selection and parameter tuning in high-dimensional prediction. In electronic *COMPSTAT Proceedings*, Paris.

Braga-Neto, U., & Dougherty, E. R. (2004). Is cross-validation valid for small-sample microarray classification? *Bioinformatics, 20*, 374–380.

Chun, D., & Keles, S. (2010). Sparse partial least squares regression for simultaneous dimension reduction and variable selection. *Journal of the Royal Statistical Society, 72*, 3–25.

Chung, D., & Keles, S. (2010). Sparse Partial Least Squares Classification for High Dimensional Data. *Statistical Applications in Genetics and Molecular Biology, 9, 17*.

Dougherty, E. R., Zollanvari, A., & Braga-Neto, U. M. (2011). The illusion of distribution-free small-sample classification in genomics. *Current Genomics, 12*, 333–341.

Fan, J., & Li, R. (2001). Variable selection via nonconcave penalized likelihood and its oracle properties. *Journal of the American Statistical Association, 96*, 1348–1360.

Hanczar, B., Hua, J., & Dougherty, E. R. (2007). Decorrelation of the true and estimated classifier errors in high-dimensional settings. *EURASIP Journal on Bioinformatics and Systems Biology, 2007*, 38473.

Scherzer, C. R., Eklund, A. C., & Morse, L. J. et al. (2007). Molecular markers of early Parkinson's disease based on gene expression in blood. *Proceedings of the National Academy of Science, 104*, 955–960.

Singh, D., Febbo, P. G., & Ross, K. (2002). Gene expression correlates of clinical prostate cancer behavior. *Cancer Cell, 1*, 203–209.

Tang, B. M., McLean, A. S., & Dawes, I. W. et al. (2009). Gene-expression profiling of peripheral blood mononuclear cells in sepsis. *Critical Care Medicine, 37*, 882–888.

Tibshirani, R. (1996). Regression shrinkage and selection via the Lasso. *Journal of the Royal Statistical Society, Series B, 58*, 267–288.

Wang, Y., Klijn, J. G., & Zhang, Y. et al. (2005). Gene-expression profiles to predict distant metastasis of lymph-node-negative primary breast cancer. *Lancet, 365*, 671–679.

Zou, H. (2005). Regularization and variable selection via the elastic net. *Journal of the Royal Statistical Society B, 67*, 301–320.

A New Effective Method for Elimination of Systematic Error in Experimental High-Throughput Screening

Vladimir Makarenkov, Plamen Dragiev, and Robert Nadon

Abstract High-throughput screening (HTS) is a critical step of the drug discovery process. It involves measuring the activity levels of thousands of chemical compounds. Several technical and environmental factors can affect an experimental HTS campaign and thus cause systematic deviations from correct results. A number of error correction methods have been designed to address this issue in the context of experimental HTS (Brideau et al., J Biomol Screen 8:634–647, 2003; Kevorkov and Makarenkov, J Biomol Screen 10:557–567, 2005; Makarenkov et al., Bioinformatics 23:1648–1657, 2007; Malo et al., Nat Biotechnol 24:167–175, 2006). Despite their power to reduce the impact of systematic noise, all these methods introduce a bias when applied to data not containing any systematic error. We will present a new method, proceeding by finding an approximate solution of an overdetermined system of linear equations, for eliminating systematic error from HTS screens by using a prior knowledge on its exact location. This is an important improvement over the popular B-score method designed by Merck Frosst researchers (Brideau et al., J Biomol Screen 8:634–647, 2003) and widely used in the modern HTS.

V. Makarenkov (✉)
Département d'informatique, Université du Québec à Montréal, c.p. 8888 succ. Centre-Ville, Montreal, QC, H3C-3P8, Canada
e-mail: makarenkov.vladimir@uqam.ca

P. Dragiev
Département d'informatique, Université du Québec à Montréal, c.p. 8888 succ. Centre-Ville, Montreal, QC, H3C-3P8, Canada

Department of Human Genetics, McGill University, 1205 Dr. Penfield Ave. Montreal, QC, H3A-1B1, Canada

R. Nadon
Department of Human Genetics, McGill University, 1205 Dr. Penfield Ave. Montreal, QC, H3A-1B1, Canada

B. Lausen et al. (eds.), *Algorithms from and for Nature and Life*, Studies in Classification, Data Analysis, and Knowledge Organization, DOI 10.1007/978-3-319-00035-0_27,

1 Introduction

Contemporary drug development practice comprises an initial step of testing a large number of chemical compounds in order to identify those that show promising activity against a given biological target (Brideau et al. 2003; Makarenkov et al. 2006, 2007; Malo et al. 2006). This step, known as high-throughput screening (HTS), is fulfilled by employing robotic equipment which takes precise measurements of compounds activity levels. The obtained measurements are then compared and the entities with 'the best' activity levels are selected as *hits* for further clinical trials (Malo et al. 2006). The hit selection process assumes that the measurements provided by HTS equipment correctly represent the activity levels of tested compounds as well as that all compounds are assessed at absolutely the same experimental conditions (Dragiev et al. 2011; Malo et al. 2006). In practice, inconsistency in the environment conditions, such as temperature, humidity and lighting, can disturb highly sensitive HTS readers (Heyse 2002). Procedural factors may as well have a significant systematic impact. For instance, differences in the incubation time for certain groups of compounds may cause variance in the solution concentrations (because of solvent evaporation). Hit selection carried out with measurements that differ from real activity levels results in the cases when some compounds are incorrectly selected as hits – false positives (FP), and other mistakenly overlooked – false negatives (FN).

Quality control and preprocessing normalization techniques have been employed to eliminate or reduce the effect of systematic error in experimental HTS (Makarenkov et al. 2007; Malo et al. 2006). Positive and negative controls are often used throughout the assay in order to assess the plates background levels. *Percent of control* normalization can be applied to the plates to compensate for the plate-to-plate background differences:

$$x'_{ij} = \frac{x_{ij}}{\mu_{pos}} \times 100\%, \tag{1}$$

where μ_{pos} is the mean of the positive controls and x'_{ij} is the normalized value of the raw measurement x_{ij} of the compound in well (i, j) located in the intersection of line i and column j of the given plate. *Normalized percent inhibition* is another control based method:

$$x'_{ij} = \frac{\mu_{pos} - x_{ij}}{\mu_{pos} - \mu_{neg}}, \tag{2}$$

where x_{ij} is the raw measurement of the compound in well (i, j), μ_{pos} is the mean of the positive controls of the plate, μ_{neg} is the mean of the negative controls of the plate and x'_{ij} is the normalized value. *Z-score* is a simple and well-known normalization procedure carried out using the following formula:

$$x'_{ij} = \frac{x_{ij} - \mu}{SD}, \tag{3}$$

where x_{ij} is the raw measurement of the compound in well (i, j), μ and SD are the mean and the standard deviation of the measurements within the given plate and x'_{ij} is the normalized value. *B-score* (Brideau et al. 2003), is a more robust analogue of z-score widely used to correct experimental HTS data. B-score replaces the raw data with the corresponding residuals adjusted by the plate's *median absolute deviation (MAD)*:

$$x'_{ij} = \frac{r_{ij}}{MAD}, \qquad MAD = median\{|r_{ij} - median(r_{ij})|\}, \tag{4}$$

where x'_{ij} is the normalized output value, MAD is the median absolute deviation of all residuals, r_{ij} is the residual for well (i, j) calculated as the difference between the raw measurement and its corresponding fitted value after running the two-way median polish procedure (Tukey 1977) to account for the plate's row and column effects. *Well correction* (Makarenkov et al. 2006, 2007) is another advanced correction method that can remove systematic error appearing along all assay plates and affecting rows, columns or separate wells. It is performed in two steps – least-squares approximation of the measurements carried out for each well location separately across all plates of the assay is followed by the z-score normalization within each well location.

The final step of HTS, hit selection, is usually carried out by identifying the compounds whose activity levels exceed a predefined threshold. Hit selection threshold is typically defined using the mean μ and the standard deviation SD of the measurements with the most used threshold of $\mu - 3SD$ (i.e. the measurements whose values are lower than $\mu - 3SD$ are selected as hits in inhibition assays). In our previous works, we showed that systematic error correction methods should be applied very delicately because the use of any systematic error correction method on error-free HTS data introduces a bias, sometimes very significant, that affects very negatively the accuracy of the hit selection process (Makarenkov et al. 2006, 2007). In our recent work, we described a method for detecting the presence of systematic error in HTS data and thus allowing one to decide whether systematic error correction is needed or not (Dragiev et al. 2011).

2 Matrix Error Amendment Method

In this article, we present a new method, called *Matrix Error Amendment (MEA)*, for systematic error correction of experimental HTS data. It relies on a prior information that systematic error is present in HTS data and that it is row or/and column-located (i.e. the measurements in certain rows and columns are systematically under- or over-estimated). We also assume that the location of the rows and columns affected by systematic noise is known. Such information can either be available in advance or be obtained using the t-test or χ^2 goodness-of-fit test (see Dragiev et al. 2011 for more details).

Let X be a plate of experimental HTS measurements with m rows and n columns. Let x_{ij} be the measurement of the compound located in well (i, j) and μ be the mean value of all measurements on the plate that *are not affected by systematic error*. For a plate of an assay that is not affected by systematic error, we can expect that the mean of the measurements in a given row i is close to μ, which is, in this case, the mean of all measurements on the plate: $\sum_{j=1}^{n} x_{ij} \approx n\mu$. And similarly, for a given column j, we can expect that: $\sum_{i=1}^{m} x_{ij} \approx m\mu$.

Assume that the plate X is affected by systematic error and denote by $r_1, r_2, \ldots, r_p$ $(p < m)$ the rows of X and by $c_1, c_2, \ldots, c_s$ $(s < n)$ the columns of X where the presence of systematic error was confirmed. Let e_{r_i} be the unknown value of systematic error affecting row r_i and e_{c_j} be the unknown value of the systematic error affecting column c_j. The following four sets of equations can be composed:

$$\sum_{j=1}^{n} x_{r_i j} - n e_{r_i} - \sum_{j=1}^{s} e_{c_j} = n\mu, \tag{5}$$

$$\sum_{i=1}^{m} x_{i c_j} - n e_{c_j} - \sum_{i=1}^{p} e_{r_i} = m\mu, \tag{6}$$

$$\sum_{j=1}^{n} x_{ij} - \sum_{j=1}^{s} e_{c_j} = n\mu, \tag{7}$$

$$\sum_{i=1}^{m} x_{ij} - \sum_{i=1}^{p} e_{r_i} = m\mu, \tag{8}$$

where Eq. (5) correspond to rows $r_1, r_2, \ldots, r_p$ affected by row systematic error, (6) to columns $c_1, c_2, \ldots, c_s$ affected by column systematic error, (7) to rows not affected by row systematic error, and (8) to columns not affected by column systematic error.

Typically, systematic error in HTS affects only a few rows or columns of a plate (usually those located on the plate edges Brideau et al. (2003) and Kevorkov and Makarenkov (2005)). Having compounds not affected by systematic error, allows us to estimate μ and leaves e_{r_i} and e_{c_j} the only unknowns in the equation system (5)–(8) above. Thus, in practice, we have a system with $m + n$ linear equations and less than $m + n$ unknowns.

We tested three different approaches to find the most appropriate approximate solution of this system. First, we combined the expressions (5) and (6) in a linear system (9) having $m + n$ equations and $m + n$ unknowns. For plates with at least 3 rows and 3 columns the system (9) always has a unique solution. This was the first way of computing the approximate solution of the system (5) to (8).

Second, by combining all the Eqs. (5)–(8), we composed an overdetermined system of linear equations $Ax = b$ with $m + n$ equations and less than $m + n$ unknowns. We found that in all cases the matrix $A^T A$ was singular what rendered the standard least-square approximation approach inapplicable. Still, we were able to find an approximate solution of the latter system by using the singular value decomposition method. However, the performances of this approach were worse than those attained using the first one.

Third, we attempted to improve our first approach by using the additional information provided by the Eqs. (7) and (8). The set of equations (7) offers estimations for the sum of column errors $\sum_{j=1}^{s} e_{c_j}$ affecting the plate. We used the mean, and in a separate test the median, of those estimations in order to calculate the sum of column errors of the plate. Then, we replaced the corresponding terms in the Eq. (5) by their estimates. In the same way, we used the Eq. (8) in order to estimate the sum of all row errors $\sum_{i=1}^{p} e_{r_i}$ and then substituted the corresponding terms in the Eq. (6). This approach produced results that were usually equivalent to those provided by our first approach, but it appeared to be more sensitive to the location of hits within plates. If most of the hits were among the compounds not affected by systematic error, then the hit selection accuracy was decreasing. In the next section, we will present the results obtained using the first approach as it proved to be the fastest and the steadiest one among the three competing strategies.

$$\left(\begin{array}{ccccc|ccccc} n & 0 & \dots & 0 & 0 & 1 & 1 & \dots & 1 & 1 \\ 0 & n & \dots & 0 & 0 & 1 & 1 & \dots & 1 & 1 \\ \vdots & \vdots & \ddots & \vdots & \vdots & \vdots & \vdots & \ddots & \vdots & \vdots \\ 0 & 0 & \dots & n & 0 & 1 & 1 & \dots & 1 & 1 \\ 0 & 0 & \dots & 0 & n & 1 & 1 & \dots & 1 & 1 \\ \hline 1 & 1 & \dots & 1 & 1 & m & 0 & \dots & 0 & 0 \\ 1 & 1 & \dots & 1 & 1 & 0 & m & \dots & 0 & 0 \\ \vdots & \vdots & \ddots & \vdots & \vdots & \vdots & \vdots & \ddots & \vdots & \vdots \\ 1 & 1 & \dots & 1 & 1 & 0 & 0 & \dots & m & 0 \\ 1 & 1 & \dots & 1 & 1 & 0 & 0 & \dots & 0 & m \end{array}\right) \left(\begin{array}{c} e_{r_1} \\ e_{r_2} \\ \vdots \\ e_{r_{p-1}} \\ e_{r_p} \\ \hline e_{c_1} \\ e_{c_2} \\ \vdots \\ e_{c_{s-1}} \\ e_{c_s} \end{array}\right) = \left(\begin{array}{c} b_{r_1} \\ b_{r_2} \\ \vdots \\ b_{r_{p-1}} \\ b_{r_p} \\ \hline b_{c_1} \\ b_{c_2} \\ \vdots \\ b_{c_{s-1}} \\ b_{c_s} \end{array}\right), \tag{9}$$

where $b_{r_i} = \sum_{j=1}^{n} x_{r_ij} - n\mu$ and $b_{c_j} = \sum_{j=1}^{m} x_{ic_j} - m\mu$.

The final step in our method is the removal of systematic error from the plate once the values of e_{r_i} and e_{c_j} are determined. These values are subtracted from the measurements in the corresponding rows r_i:

$$x'_{r_ij} = x_{r_ij} - e_{r_i}, \quad \text{for all } i, 1 \le i \le p \text{ and for all } j, 1 \le j \le n, \tag{10}$$

and columns c_j:

$$x'_{ic_j} = x_{ic_j} - e_{c_j}, \quad \text{for all } j, 1 \le j \le s \text{ and for all } i, 1 \le i \le m. \tag{11}$$

3 Simulation Study

To evaluate the performances of the MEA method, we carried out simulations with artificially generated data typical for HTS. Assays with 1,250 plates and three different plate sizes were considered: 96-well plates – 8 rows × 12 columns, 384-well plates – 16 rows × 24 columns, and 1,536-well plates – 32 rows × 48 columns. The generated measurement data followed standard normal distribution. The hits were added randomly in such a way that the probability of a given well to contain a hit was the same, fixed a priori, for all wells in the assay regardless their location. All the hit values followed a normal distribution with parameters $\sim N(\mu - 5SD, SD)$, where μ and SD were the mean and standard deviation of the original dataset before the addition of hits. Systematic row and column errors were added to the data. The rows and columns affected by systematic error were chosen randomly and separately for each plate. Thus, the positions of rows and columns affected by systematic noise varied from plate to plate. A small random error was also added to all generated measurements. Formula (12) specifies how the error-affected measurement of the compound in well (i, j) on plate p was calculated:

$$x'_{ijp} = x_{ijp} + er_{ip} + ec_{jp} + rand_{ijp}, \tag{12}$$

where x'_{ijp} is the resulting measurement value, x_{ijp} is the original error-free value, er_{ip} is the row error affecting row i of plate p, ec_{jp} is the column error affecting column j of plate p and $rand_{ijp}$ is the random error in well (i, j) of plate p. Systematic error er_{ip} and ec_{jp} followed a normal distribution with parameters $\sim N(0, C)$. Different values of C were tested: 0, 0.6*SD*, 1.2*SD*, 1.8*SD*, and 2.4*SD*. The random error $rand_{ijp}$ in all datasets followed normal distribution with parameters $\sim N(0, 0.6SD)$. Four systematic error correction methods were tested. Each method comprised the identical hit selection step using the same threshold of $\mu - 3SD$, but the raw data were preprocessed in four different ways prior to hit selection. The B-score correction method (Brideau et al. 2003), the introduced herein MEA method carried out under the assumption that the exact locations of rows and columns affected by systemic error were known (i.e. ideal situation), and second when those locations were experimentally determined by the t-test (Dragiev et al. 2011) as well as the traditional hit selection procedure without any data correction were considered. In each experiment, we measured the total number of false positives and false negatives, and the hit detection rate (i. e. true positive rate).

Two groups of experiments were conducted. The first group used datasets with fixed hit percentage (of 1 %) and different amplitudes of systematic error, ranging from 0 to 2.4*SD*. The second group of experiments considered datasets with the fixed level of systematic error (of 1.2*SD*) and the hit percentage rate varying from 0.5 to 5 % (x-axis). Figures 1 and 2 show the average results for the two groups of experiments. The results, obtained from 500 datasets for each parameters combination, suggest that both variants of MEA outperformed

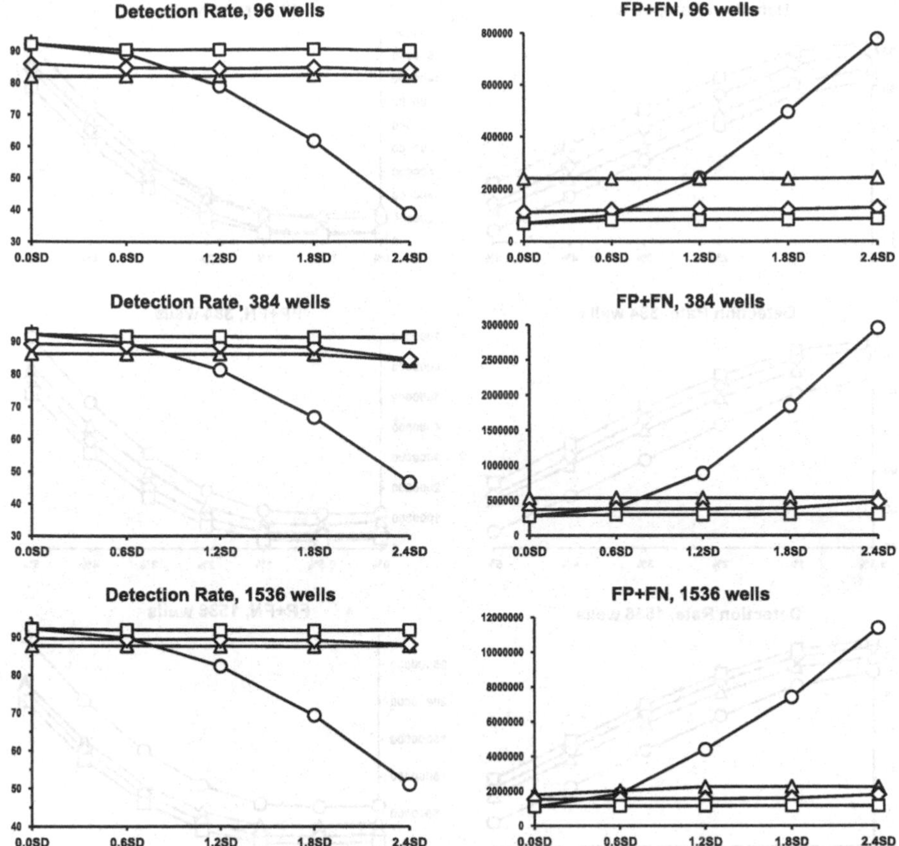

Fig. 1 Hit detection rate and total number of of false positives and false negatives for datasets with fixed hit percentage of 1 %, systematic error was added to 2 rows and 2 columns (96-well plates); 4 rows and 4 columns (384-well plates); 8 rows and 8 columns (1,536-well plates). Compared methods: hit selection without error correction (○), B-score (△), MEA (□), t-test and MEA (◇)

B-score and the traditional hit selection procedure. MEA demonstrated a robust and almost independent of the error amplitude behavior when the exact location of the systematic error was provided. When paired with the t-test, the performance of MEA, decreased for plates affected by large systematic error as well as for plates with high hit percentages, mainly because of the deteriorating accuracy of the t-test in those situations (Dragiev et al. 2011).

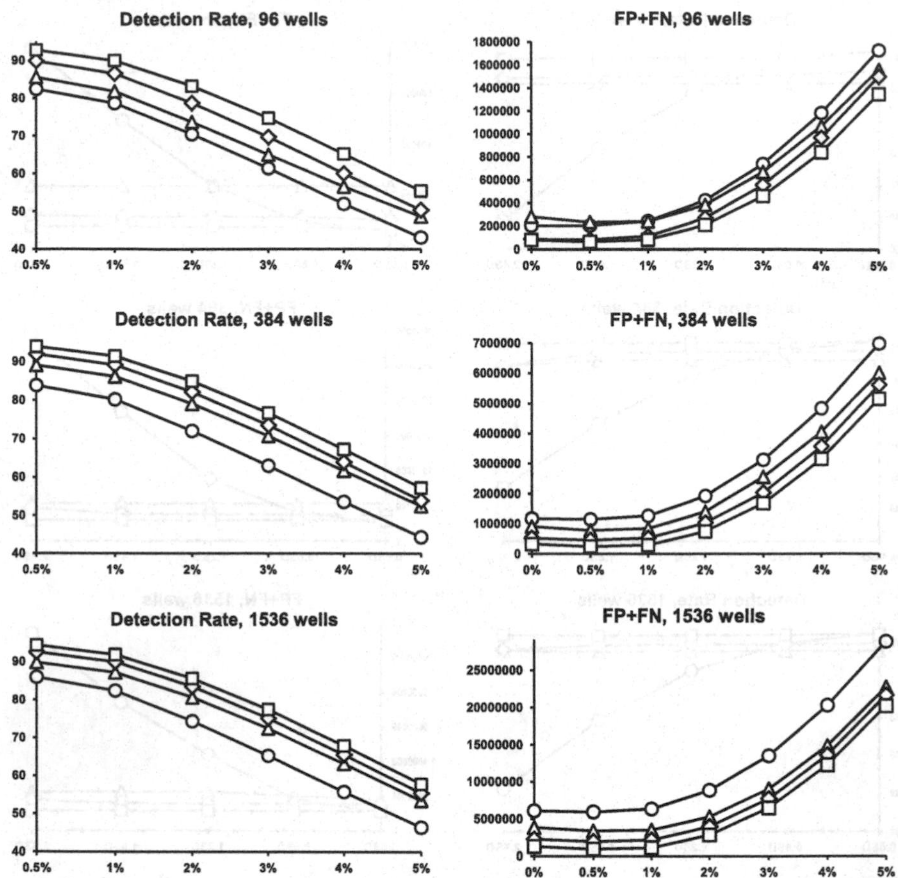

Fig. 2 Hit detection rate and total number of false positives and false negatives for datasets with fixed systematic error amplitude of 1.2*SD* added to 2 rows and 2 columns (96-well plates); 4 rows and 4 columns (384-well plates); 8 rows and 8 columns (1,536-well plates). Compared methods: hit selection without error correction (○), B-score (△), MEA (□), t-test and MEA (◇)

4 Conclusion

We described a novel method, called *Matrix Error Amendment (MEA)*, for eliminating systematic error from experimental HTS data. MEA assumes that the exact location of the rows and columns of the plates affected by systematic error is known. The presence and location of systematic error can be detected using the methodology described in Dragiev et al. (2011). Unlike the popular B-score method which transforms the original dataset into a set of residuals, MEA eliminates systematic error from the plate by adjusting only the measurements affected by the error and leaving all error-free measurements unchanged. Simulations were carried out to evaluate the performance of the new method using datasets with different plate sizes, different amplitudes of systematic error and hit percentages. MEA outperformed both B-score and the traditional uncorrected hits selection procedure in all our experiments. The results provided by the new method were especially encouraging for small plates, low systematic error levels and low hit percentages (see Figs. 1 and 2).

References

Brideau, C., Gunter, B., Pikounis, W., Pajni, N., & Liaw, A. (2003). Improved statistical methods for hit selection in HTS. *Journal of Biomolecular Screening, 8*, 634–647.

Dragiev, P., Nadon, R., & Makarenkov, V. (2011). Systematic error detection in experimental high-throughput screening. *BMC Bioinformatics, 12*, 25.

Heyse, S. (2002). Comprehensive analysis of HTS data. In *Proceedings of SPIE 2002*, Bellingham (Vol. 4626, pp. 535–547).

Kevorkov, D., & Makarenkov, V. (2005). An efficient method for the detection and elimination of systematic error in high-throughput screening. *Journal of Biomolecular Screening, 10*, 557–567.

Makarenkov, V., Kevorkov, D., Zentilli, P., Gagarin, A., Malo, N., & Nadon, R. (2006). HTS-Corrector: new application for statistical analysis and correction of experimental data. *Bioinformatics, 22*, 1408–1409.

Makarenkov, V., Zentilli, P., Kevorkov, D., Gagarin, A., Malo, N., & Nadon, R. (2007). Statistical analysis of systematic errors in HTS. *Bioinformatics, 23*, 1648–1657.

Malo, N., Hanley, J. A., Cerquozzi, S., Pelletier, J., & Nadon, R. (2006). Statistical practice in high-throughput screening data analysis. *Nature Biotechnology, 24*, 167–175.

Tukey, J. W. (1977). *Exploratory data analysis*. Reading: Addison-Wesley.

Local Clique Merging: An Extension of the Maximum Common Subgraph Measure with Applications in Structural Bioinformatics

Thomas Fober, Gerhard Klebe, and Eyke Hüllermeier

Abstract We develop a novel similarity measure for node-labeled and edge-weighted graphs, which is an extension of the well-known maximum common subgraph (MCS) measure. Despite its common usage and appealing properties, the MCS also exhibits some disadvantages, notably a lack of flexibility and tolerance toward structural variation. In order to address these issues, we propose a generalization which is based on so-called quasi-cliques. A quasi-clique is a relaxation of a clique in the sense of being an "almost" complete subgraph. Thus, it increases flexibility and robustness toward structural variation. To construct a quasi-clique, we make use of a heuristic approach, in which so-called local cliques are determined first and combined into larger (quasi-)cliques afterward. We also present applications of our novel similarity measure to the retrieval and classification of protein binding sites.

1 Introduction

Many methods for data analysis are based on a measure of similarity between data objects. In this paper, we are specifically interested in objects represented in the form of a graph, since representations of that kind are commonly used in many application domains, including web mining, image processing, and bioinformatics, just to name a few.

T. Fober (✉) · E. Hüllermeier
Department of Mathematics and Computer Science, Philipps-Universität, 35032 Marburg, Germany
e-mail: thomas@mathematik.uni-marburg.de; eyke@mathematik.uni-marburg.de

G. Klebe
Department of Pharmaceutical Chemistry, Philipps-Universität, 35032 Marburg, Germany

B. Lausen et al. (eds.), *Algorithms from and for Nature and Life*, Studies in Classification, Data Analysis, and Knowledge Organization, DOI 10.1007/978-3-319-00035-0_28,

We will focus on the *maximum common subgraph* (MCS) or, more specifically, the size of this subgraph, as an important and widely used similarity measure on graphs (Bunke and Shearer 1998). The MCS exhibits a number of appealing properties, notably in terms of interpretability. In fact, it does not only produce a numerical degree of similarity, but also an "explanation" of this similarity, namely the MCS itself. Roughly speaking, the MCS of two graphs corresponds to the largest substructure that both of them have in common. In application domains like bioinformatics, where one is typically not only interested in *how* similar two objects are, but also *why* they are similar, an explanation of that kind is arguably important.

A drawback of the MCS is its sensitivity toward errors and small deviations. This becomes especially obvious in application domains like structural bioinformatics, where measurements are noisy and imprecise, and data objects exhibit a high level of variability. As an example, consider edge-weighted graphs representing protein structures. Due to molecular flexibility and noise in the measurement of atomic positions, one cannot expect to find exact matches between two such structures. Instead, one will normally end up with very small common subgraphs that fail to capture the main structural similarities in a proper way. Besides, the MCS is also critical from a complexity point of view, since its computation is an NP-hard problem.

To overcome these problems, we follow up on our idea (Boukhris et al. 2009) of relaxing the condition of exact matches and using a method for detecting maximum "approximate" common subgraphs (MACS). To this end, we employ the concept of a so-called *quasi-clique* of a graph that has recently been studied in the literature (Liu and Wong 2008). The resulting similarity measure is tolerant toward noise and structural deformation.

While the computation of the MCS of two graphs can be done by searching for a maximum clique in the corresponding product graph, the computation of the MACS requires the search for quasi-cliques. To circumvent the problem of NP-hardness, we make use of a heuristic approach, in which so-called local cliques are iteratively merged into larger graphs, preserving a constraint on the degree of connectedness. Eventually, an approximation of the largest quasi-clique in the product graph is obtained, which corresponds to an approximate match between the two original graphs.

The remainder of the paper is organized as follows. Subsequent to recalling the original MCS measure in Sect. 2, its extension is introduced in Sect. 3. In that section, we also present an algorithm for constructing a MACS. Section 4 is devoted to an experimental study. Finally, Sect. 5 concludes the paper.

2 Similarity Based on the Maximum Common Subgraph

A node-labeled and edge-weighted graph G is a 4-tuple (V, E, ℓ_V, ℓ_E), where V is the set of nodes and E the set of edges; moreover, node-labels and edge weights are defined by the functions $\ell_V : V \to \mathcal{L}$ and $\ell_E : E \to \mathbb{R}$, respectively. A widely accepted measure of similarity between two graphs G and G' is the (normalized) size of their MCS:

$$\text{sim}(G, G') = \frac{\text{mcs}(G, G')}{\max\{|G|, |G'|\}}, \tag{1}$$

where $\text{mcs}(\cdot, \cdot)$ is a function computing the maximum common subgraph and $|\cdot|$ returns the size of a graph in terms of the number of nodes. Obviously, (1) assumes values in the unit interval, with values closer to 1 indicating higher similarity between G and G'.

The algorithmic challenge is to realize the function mcs efficiently. This function can be implemented in an indirect way by using the so-called product graph. Here, the categorical product graph will be used, which has the important property that each clique in this graph corresponds to a common subgraph in the two input graphs. Consequently, the MCS is given by the maximum clique in the product graph (Levi 1973). In other words, the problem of finding a maximum common subgraph can be reduced to the problem of clique detection.

Formally, the product graph is $G_\otimes = (V_\otimes, E_\otimes)$ of two node-labeled and edge-weighted graphs $G = (V, E, \ell_V, \ell_E)$ and $G' = (V', E', \ell'_V, \ell'_E)$ is defined by

$$V_\otimes = \{(v_i, v'_j) \in V \times V' : \ell_V(v_i) = \ell'_V(v'_j)\},$$

$$E_\otimes = \left\{((v_i, v'_j), (v_k, v'_l)) \in V_\otimes^2 : \|\ell_E(v_i, v_k) - \ell'_E(v'_j, v'_l)\| \le \epsilon\right\},$$

where ϵ is a threshold for comparing the length of edges.

Cliques are the densest form of subgraphs, since each pair of nodes must be connected by an edge. Considering the retrieval of the MCS by searching for cliques in the product graph $G_\otimes$, this means that all node and edge labels must be equal. As mentioned previously, this requirement is overly restrictive in the context of biological data analysis, especially in the case of structure analysis where edges are labeled with real-valued distances. In the above definition of the product graph, the condition of equal edge weights is already relaxed, since edges are allowed to differ by at most a constant ϵ. Yet, looking for cliques in $G_\otimes$ still means that this condition must hold for *all* pairs of edges in the MCS. Roughly speaking, this approach is tolerant toward possibly numerous though small (measurement) errors but not toward single though exceptionally large deviations. To become flexible in this regard, too, our idea is to replace the detection of cliques in $G_\otimes$ by the detection of *quasi-cliques*.

3 Maximum Approximate Common Subgraphs

A similarity measure on graphs using the MACS instead of the MCS can be defined analogously to (1). Likewise, the computation of the MACS can be reduced to the search for a quasi-clique in the product graph, since each such quasi-clique corresponds to an approximate common subgraph.

Roughly speaking, quasi-cliques are "almost complete" graphs $G = (V, E)$. In the literature, different definitions of quasi-cliques have been proposed. Most of them are based on the degree of the nodes (Liu and Wong 2008), calling G a quasi-clique if every node in V is adjacent to at least $\gamma \cdot (|V| - 1)$ other nodes, where $deg(v)$ is the number of nodes adjacent to v and $\gamma \in]0, 1]$ a relaxation parameter. This is the definition that we shall adopt in this paper. Note that the concept of a γ-quasi-clique is a proper generalization of the concept of a clique, since each clique is a 1-quasi-clique.

Since clique detection is an NP-complete problem (Karp 1972), exact algorithms are feasible only for very small graphs. Therefore, practically relevant problems are usually solved in an approximate way by means of heuristic algorithms, which typically exploit a downward-closure property, namely that a supergraph of a non-clique cannot be a clique either.

Since quasi-cliques are a generalization of cliques, it immediately follows that finding a maximum γ-quasi-clique is an NP-complete problem, too. Unfortunately, the downward-closure property does not hold for quasi-cliques, as one can easily show by counter-examples. Instead, any subset of the set of nodes V in a graph $G = (V, E)$ may form a γ-quasi-clique. Nevertheless, alternative heuristic methods for quasi-clique detection have been developed (e.g. Liu and Wong 2008). However, for large graphs, such algorithms become very inefficient in both time and space. Another approach was proposed in Li et al. (2005), where an efficient algorithm based on local clique detection for interaction graph mining in protein complexes was developed. This approach is modified technically and it is applied to find approximate common subgraphs, hence, it is operating on the product graph $G_{\otimes} = (V_{\otimes}, E_{\otimes})$.

3.1 Detection of Quasi-Cliques by Local Clique Merging

The approach to quasi-clique detection outlined in this section is a heuristic one and, therefore, does not guarantee the optimality of a solution. Practically, however, it turned out to be a viable alternative to exact methods like (Liu and Wong 2008), since the solution quality is at least close to optimal and the runtime acceptable.

Our algorithm requires the specification of three parameters: First of all, γ defines the density of the maximum quasi-clique we are looking for. Since the downward-closure property does not hold for quasi-cliques, intermediate solutions having density below γ cannot be discarded. However, an extension to a γ-quasi-clique becomes unlikely for intermediate solutions whose density is significantly smaller than γ; we test this condition using a second ("cautious") threshold $\gamma' < \gamma$. Finally, a parameter ω is used to control the number of merge operations, since the intermediate solutions that are merged must overlap to a degree of at least ω.

Our heuristic essentially consists of two steps, namely the detection of local cliques and a merging procedure. A local clique in a graph G is a complete subgraph that contains a certain node of G. To detect a local clique in the product graph $G_\otimes = (V_\otimes, E_\otimes)$, a node $v \in V_\otimes$ is selected and a neighborhood graph $G_\otimes^{(v)} = (V_v, E_v)$ is derived, which is defined by $V_v = \{v\} \cup \{v_i \in V_\otimes : (v, v_i) \in E_\otimes\}$ and $E_v = E_\otimes \cap V_v \times V_v$. Unless the clique property is satisfied for $G_\otimes^{(v)}$, the node with the smallest degree in V_v is iteratively removed from $G_\otimes^{(v)}$, together with all adjacent edges; if the smallest degree is shared by more than one node, one of them is chosen at random. Once the clique property is fulfilled, a local clique containing the node v is formed.

Obviously, the set of cliques found in this step will not contain the maximum clique, and in general, these cliques will not even be the maximal clique containing v. Instead, this approach is merely used to efficiently generate a set of cliques to be used in the second step of our procedure, in which "small" cliques are merged into larger quasi-cliques. For this purpose, each node $v \in G_\otimes$ is considered once and used to create a local clique. These local cliques are stored in a set LC (provided they have at least 3 nodes).

In the second step, the cliques in LC are merged iteratively, using a modified version of beam search (Norvig 1991) on the search space of local cliques. To this end, a beam C is defined containing the best solutions. The beam is initialized with the set LC containing all local cliques detected in the first step. While the beam is not empty, all pairs (G_Q, G_L) with $G_Q \in C$ and $G_L \in LC$ are considered. If their overlap $|V_Q \cap V_L|/\min\{|V_Q|, |V_L|\}$ is above the threshold ω, the pair is merged and inserted into a temporal set. Having considered all pairs, the beam C is replaced by the temporal set and the loop is continued. To avoid the risk of loosing the best solution[1] found so far, it is stored in a variable which is returned upon termination of the algorithm.

More concretely, two graphs G and G' are merged by defining the union of their respective node sets, and connecting a pair in the aggregation $\tilde{V} = V \cup V'$ by an edge if this edge also exists in the graph $G_\otimes$; the set $\tilde{E}$ is hence given by $\tilde{V}^2 \cap E_\otimes$. However, in the case of merging two product graphs, the nodes must be considered more carefully: Since product nodes correspond to unique pairs $(v, v') \in V \times V'$, different product nodes can still correspond to the same nodes in G or G'. This many-to-one relationship must be taken into consideration to ensure that valid approximate common subgraphs are produced. Here, a very simple though efficient procedure is applied, which adds a product node into the quasi-clique only if the nodes it represents are not yet contained.

3.1.1 Complexity

Finding all local cliques in the product graph of the graphs $G = (V, E)$ and $G' = (V', E')$, where $n = \max\{|V|, |V'|\}$, takes time $\mathcal{O}(n^4)$. For each of the $\mathcal{O}(n^2)$ nodes of the product graph, the neighborhood graph is constructed in time $\mathcal{O}(1)$ by

[1] The best solution is the largest quasi-clique whose density is at least γ.

making use of adjacency lists. The size of the neighborhood graph is $\mathcal{O}(n^2)$, hence for the removal of nodes violating the clique property, time $\mathcal{O}(n^4)$ is required: $O(n^2)$ times the node with smallest degree is searched, a procedure which requires time $O(n^2)$ per node.

Merging of local cliques has unfortunately exponential complexity, since the number of graphs in the set C can become very large, theoretically up to $\sum_{i=3}^{n}\binom{n}{i} = 2^n - \sum_{i=0}^{2}\binom{n}{i} = \mathcal{O}(2^n)$. However, thanks to the thresholds ω and γ', the true number is typically much smaller than this theoretical bound, and most of the time, the set C is already empty after a few iterations of the merging step.

4 Experimental Study

We applied our method to the comparison of protein structures or, more specifically, protein binding sites. This is an important problem in structural bioinformatics, where graphs are often used for representing the structure and specific properties of biomolecules. Here, we make use of the representation proposed in Schmitt et al. (2002): A protein binding site is modeled as a graph whose nodes and edges represent, respectively, physico-chemical properties (7 such properties are distinguished) associated with representative points on the surface of the binding site and the Euclidean distances between these points.

In a first experiment, we consider the complete CavBase (Kuhn et al. 2006), a database currently storing almost 200,000 protein binding sites. Using 4 different query structures `1EAG.1` (*asparyl protease*), `2EU2.1` (*carbonic anhydrase*), `2OQ5.1` (*serine protease*) and `3HEC.3` (*kinase*), the goal is to retrieve structures from CavBase belonging to the same type of protein. The second study is a classification experiment. We use a dataset consisting of 355 protein binding sites, 214 that bind ATP ligands and 141 that bind NADH ligands. Thus, a dataset defining a binary classification problem is obtained (Fober et al. 2011). Since ATP is a substructure of NADH, and hence the former can possibly bind the same ligands as the latter, this dataset is especially challenging. The binary classification problem is tackled using a simple nearest neighbor classifier in a leave-one-out cross-validation setting. As a baseline to compare with, we use the MCS measure which is implemented in CavBase. Due to computational reasons, however, CavBase only computes an approximation of the MCS: Instead of enumerating all maximal cliques, it only generates the first 100 candidates and selects the largest among these.

The results on the retrieval experiment are summarized in Fig. 1, in which the 25 most similar structures are ordered on the x-axis according to the obtained score. In the case of a positive hit, which means that a protein was found belonging to the same class as the query, a counter giving the position on the y-axis in incremented, which means that the identity (as obtained for the query `2EU2.1`) is the optimal solution.

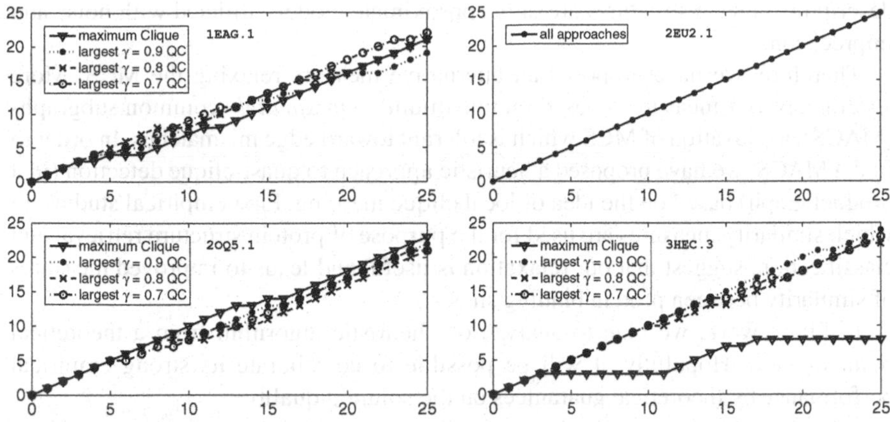

Fig. 1 Results on the retrieval experiment on the four queries

Table 1 Results on the classification experiment (percent correct classification)

Method	MCS	$\mathrm{MACS}_{\gamma'=0.7}^{\gamma=0.9}$	$\mathrm{MACS}_{\gamma'=0.6}^{\gamma=0.8}$	$\mathrm{MACS}_{\gamma'=0.5}^{\gamma=0.7}$	$\mathrm{MACS}_{\gamma'=0.4}^{\gamma=0.6}$
Accuracy	0.769	0.851	0.839	0.862	0.682

As can be seen, the performance depends on the query. For 2EU2.1, all methods perform perfectly. However, this class of proteins is known to be very rigid, hence the use of error-tolerant measures is not necessary. This is also the case for 2OQ5.1, where the error-tolerance even seems to have a detrimental effect. In the case of protein-classes which are known to be more flexible (1EAG.1 and 3HEC.3), on the other hand, error-tolerance clearly improves the results. However, the poor performance of MCS in the case of 3HEC.3 cannot be explained by a lack of error-tolerance alone. Here, a drawback of the CavBase heuristic becomes obvious: Very large structures produce very large product graphs with many maximal (i.e., not further extensible) cliques, so that the maximum clique is probably missed by the first 100 candidates.

The results on the classification experiment given in Table 1 show a similar picture. It clearly sticks out that a higher degree of error-tolerance can increase the classification rate, an indirect proof for an improved similarity measure. Beyond a certain degree of error-tolerance, however, the classification rate drops strongly. Here, the error-tolerance of the similarity measure becomes too high, so that even dissimilar graphs are considered as similar.

5 Conclusions

Maximum common subgraphs have been used successfully as similarity measures for graphs. In this paper, however, we have argued that this measure is overly stringent in the context of protein structure comparison, mainly since graph descriptors of such structures are only approximate models afflicted with noise and imprecision.

Therefore, we have proposed an alternative measure relaxing the MCS. More specifically, our measure is based on maximum *approximately* common subgraphs (MACS), a relaxation of MCS which is tolerant toward edge mismatches. In order to find a MACS, we have proposed a heuristic approach to quasi-clique detection (in a product graph) based on the idea of local clique merging. First empirical studies, in which similarity measures are used for the purpose of protein structure retrieval and classification, suggest that our relaxation is useful and leads to improved measures of similarity between protein binding sites.

In future work, we plan to analyze our heuristic algorithm from a theoretical point of view. Hopefully, it will be possible to corroborate its strong empirical performance by theoretical guarantees on the solution quality.

References

Boukhris, I., Elouedi, Z., Fober, T., Mernberger, M., & Hüllermeier, E. (2009). Similarity analysis of protein binding sites: A generalization of the maximum common subgraph measure based on quasi-clique detection. In *International Conference on Intelligent Systems Design and Applications*, Pisa, Italy (pp. 1245–1250).

Bunke, H., & Shearer, K. (1998). A graph distance metric based on the maximal common subgraph. *Pattern Recognition Letters, 19*(3–4), 255–259.

Fober, T., Glinca, S., Klebe, G., & Hüllermeier, E. (2011). Superposition and alignment of labeled point clouds. *IEEE/ACM Transactions on Computational Biology and Bioinformatics, 8*(6), 1653–1666.

Karp, R. M. (1972). Reducibility among combinatorial problems. In *Complexity of Computer Computations* (pp. 85–103). New York: Plenum Press

Kuhn, D., Weskamp, N., Schmitt, S., Hüllermeier, E., & Klebe, G. (2006). From the similarity analysis of protein cavities to the functional classification of protein families using cavbase. *Journal of Molecular Biology, 359*(4), 1023–1044.

Levi, G. (1973) A note on the derivation of maximal common subgraphs of two directed or undirected graphs. *Calcolo, 9*(1972), 341–352.

Li, X.-L., Tan, S.-H., Foo, C.-S., & Ng, S.-K. (2005) Interaction graph mining for protein complexes using local clique merging. *Genome Informatics, 16*(2), 260–269.

Liu, G., & Wong, L. (2008). Effective pruning techniques for mining quasi-cliques. In *European conference on machine learning and principles and practice of knowledge discovery in databases, part II*, Antwerp, Belgium (pp. 33–49).

Norvig, P. (1991) *Paradigms of artificial intelligence programming*. Burlington: Morgan Kaufmann

Schmitt, S., Kuhn, D., & Klebe, G. (2002) A new method to detect related function among proteins independent of sequence and fold homology. *Journal of Molecular Biology, 323*(2), 387–406.

Identification of Risk Factors in Coronary Bypass Surgery

Julia Schiffner, Erhard Godehardt, Stefanie Hillebrand, Alexander Albert, Artur Lichtenberg, and Claus Weihs

Abstract In quality improvement in medical care one important aim is to prevent complications after a surgery and, particularly, keep the mortality rate as small as possible. Therefore it is of great importance to identify which factors increase the risk to die in the aftermath of a surgery. Based on data of 1,163 patients who underwent an isolated coronary bypass surgery in 2007 or 2008 at the Clinic of Cardiovascular Surgery in Düsseldorf, Germany, we select predictors that affect the in-hospital-mortality. A forward search using the wrapper approach in conjunction with simple linear and also more complex classification methods such as gradient boosting and support vector machines is performed. Since the classification problem is highly imbalanced with certainly unequal but unknown misclassification costs the area under ROC curve (AUC) is used as performance criterion for hyperparameter tuning as well as for variable selection. In order to get stable results and to obtain estimates of the AUC the variable selection is repeated 25 times on different subsamples of the data set. It turns out that simple linear classification methods (linear discriminant analysis and logistic regression) are suitable for this problem since the AUC cannot be considerably increased by more complex methods. We identify the three most important predictors as the severity of cardiac insufficiency, the patient's age as well as pulmonary hypertension. A comparison with full models trained on the same 25 subsamples shows that the classification performance in terms of AUC is not affected or only slightly decreased by variable selection.

J. Schiffner (✉) · S. Hillebrand · C. Weihs
Faculty of Statistics, TU Dortmund, 44221 Dortmund, Germany
e-mail: schiffner@statistik.tu-dortmund.de

E. Godehardt · A. Albert · A. Lichtenberg
Clinic of Cardiovascular Surgery, Heinrich-Heine University, 40225 Düsseldorf, Germany
e-mail: godehard@uni-duesseldorf.de

B. Lausen et al. (eds.), *Algorithms from and for Nature and Life*, Studies in Classification, Data Analysis, and Knowledge Organization, DOI 10.1007/978-3-319-00035-0_29,

1 Introduction

In medical care one important indicator to asses the quality of e.g. a certain type of surgery is the occurrence of complications. In order to help hospitals to develop strategies for quality improvement one important aim is to identify the factors that increase the risk of complications. Hospitals can then particularly take measures to improve care of patients with increased risk. In this paper we are concerned with coronary bypass surgery at the Clinic of Cardiovascular Surgery in Düsseldorf, Germany. We aim to identify factors that affect the in-hospital-mortality, i.e. the risk to die in the aftermath of the surgery. Our analysis is based on data from 2007 and 2008 that were collected at the clinic in Düsseldorf in the course of a nationwide initiative for quality improvement in inpatient care. This paper is organized as follows. In Sect. 2 basic information about quality improvement in inpatient care and particularly in coronary bypass surgery is provided. In Sect. 3 we describe the data in more detail. In Sect. 4 we give a short introduction to variable selection and explain the wrapper approach that is applied in our analysis in detail. Finally, in Sect. 5 the results and our conclusions are presented and an outlook to future work is given.

2 Quality Improvement in Coronary Bypass Surgery

In Germany there exists a nationwide system for external quality improvement of medical care in hospitals. In several service areas, particularly various types of surgeries, several quality indicators concerning the compliance with certain standards as well as postoperative complications are considered. The hospitals document the treatment of patients in the selected service areas and the collected data are submitted to an external agency (currently the AQUA Institute for Applied Quality Improvement and Research in Health Care, http://www.aqua-institut.de). Here, the data are checked and evaluated. Based on the results the hospitals can take measures for quality improvement. In case of abnormal results like an unusually large number of complications further discussions with the respective hospitals are initiated in order to identify the causes and take countermeasures. The results are annually published in form of so called structured quality reports (available at http://www.bqs-qualitaetsreport.de for 2001–2008 and http://www.sqg.de/themen/qualitaetsreport for 2009–2010). One service area under consideration is isolated coronary bypass surgery. Isolated means that no other surgery is done simultaneously. Among the corresponding quality indicators are various postoperative complications (http://www.sqg.de/themen/leistungsbereiche/koronarchirurgie-isoliert.html). Moreover, two types of mortality are distinguished. The so called *in-hospital-mortality* that is in the focus of this paper refers to patients who die during the same inpatient stay regardless of the time of death. Since this indicator does not comprise patients that are transferred to other hospitals at an early stage the 30-days-mortality is usually considered, too.The observed mortality rates for several hospitals are hardly comparable. A clinic with many high-risk patients will have an increased mortality rate but need not be inferior regarding the quality of medical care. For this reason a risk adjustment is done. For each clinic an expected mortality rate for the patient mix at hand is calculated and the ratio of observed and expected mortality is used to assess the quality. The expected mortality rate is computed by means of a logistic regression model that is built on data of all participating hospitals in Germany. The predictor variables contain information regarding the preoperative state of the patients. Since the logistic model is used for risk adjustment information related to the surgery like surgeon, anesthesia and to the postoperative treatment like the intensive care unit is not taken into account. The binary target variable describes the post-operative recovery state of the patient, i.e. if the patient died in hospital in the aftermath of the surgery. Predictors without significant influence on the target variable are removed from the model. The selected logistic model is referred to as the logistic KCH Score where KCH is a German abbreviation for coronary surgery. Until now there are three versions of the KCH Score due to updates of the logistic model: KCH Score 1.0: built on data from 2004, KCH Score 2.0: built on data from 2007 and KCH Score 3.0: built on data from 2008. The KCH Score 3.0 is still in use. The reason for the two updates is that the hospitals were obligated to collect additional preoperative variables that had to be integrated into the logistic model. Moreover, in 2007 the definitions of some variables were changed.

3 Data

In this paper we deal with data collected by the Clinic of Cardiovascular Surgery in Düsseldorf for the quality reports in 2007 and 2008. The data sets at hand contain information regarding the preoperative state of 1,163 (604 in 2007 and 559 in 2008) patients who underwent an isolated coronary bypass surgery. The following preprocessing steps are performed. First, the raw data sets from 2007 and 2008 are merged. Certainly it is also possible to build individual models for the 2 years. But changes over time should not be too severe (note that the KCH Score 3.0 is still in use). Moreover, building two models would mean halving the number of training observations and therefore lead to less reliable results of the variable selection. Based on the raw data from 2007 and 2008 all variables that are used in at least one of the KCH Scores 1.0, 2.0 or 3.0 are calculated. Thus 19 candidate predictor variables are obtained. Besides basic information like sex and age they describe pre-existing conditions and results of medical examinations done before the surgery. If the meaning of a variable has changed over time we take the recent definition. For categorical predictor variables we use the maximum number of categories, even though some categories were not deemed significant in the KCH Scores. The two continuous variables age and body mass index are scaled to zero mean and unit

Table 1 Predictor variables. For categorical predictors the number of values and the KCH Score where this variable is defined are given

(Preoperative) variable	# values	KCH score	(Preoperative) variable	# values	KCH score
Age			nyha (severity of cardiac insufficiency)	3	2.0
Angiography findings	2	3.0	Preoperative creatinine level	2	1.0
bmi (body mass index)					
Critical preoperative state	2	3.0			
Diabetes	2	3.0	Preoperative renal Replacement therapy	2	3.0
Emergency	2	3.0			
Extracardiac arteriopathy	2	3.0	Pulmonary disease	3	3.0
Heart rhythm	3	2.0	Pulmonary hypertension	2	3.0
Left-ventricular dysfunction	3	3.0	Reoperation	2	3.0
Myocardial infarction	2	3.0	Sex	2	3.0
Neurological dysfunction	2	3.0	Troponin	2	2.0

variance. Table 1 shows all predictors. Due to lack of space we cannot give the exact definitions of all variables. Instead, we point to the KCH Score where the variable is used (the exact definitions can be found at http://www.bqs-qualitaetsreport.de).

4 Identification of Risk Factors: Variable Selection and Ranking

Our aim is the identification of risk factors in coronary bypass surgery based on data of the Clinic of Cardiovascular Surgery in Düsseldorf since this may be helpful in quality improvement. Two issues are worth discussing. First, predictors related to the surgical procedure and postoperative medical treatment would be also helpful for quality improvement but unfortunately are not available. Second, there is already a global model based on data of all participating hospitals and the importance of predictors can be seen from the results of the Wald tests that are given in the quality reports. Nevertheless, it makes sense to build an individual model for a single hospital since for a single clinic other factors than those most important in the global logistic model may be relevant. Also predictors like troponin which are not included in the currently used KCH Score 3.0 may be important for a single hospital. Generally, the aim of variable selection is at least three-fold: understanding the data, improving prediction performance and reducing data in order to provide faster and more cost-effective predictors (cp. Guyon et al. 2006, for an overview of variable selection approaches). Our aim clearly lies in data understanding. But prediction accuracy is also an important aspect since an interpretable model with no predictive power is only of little use. Usually three different approaches to variable selection are distinguished, namely filter, embedded and wrapper methods. Filter methods such as RELIEF are applied as part of the data preprocessing and do not require building a predictor. Filter methods often (but not necessarily) provide a variable ranking, i.e. assign an importance value to each variable. The user has to choose a threshold in order to obtain a variable subset. In our case a variable ranking is sufficient since we particularly aim to identify the most important variables. In order to apply embedded or wrapper methods a predictor has to be built. In case of embedded methods like decision trees, forests or LASSO-based methods variable selection is part of the training process. Wrapper approaches use the classifier as a "black box" to assess the goodness of a variable set. We tried simple things first, i.e. we used several filter and embedded methods. But this did not lead to clear results as completely different variable rankings or subsets were obtained by means of distinct methods. We finally tried the wrapper approach which led to satisfactory results and is described in detail in the following. In order to use this approach three important choices have to be made: the selection criterion, the classification method(s) and the search strategy. The standard performance criterion in classification is the error rate. Fortunately the mortality rate is very low and thus the classification problem under consideration is highly imbalanced. Moreover, the misclassification costs in this problem are certainly unequal, but unknown. For these reasons the area under the ROC curve, in short AUC (cp. e.g. Fawcett 2006), is used as selection criterion. Since we do not know in advance which classification method to apply we try seven methods of different complexity. As linear methods we use logistic regression (logreg) and linear discriminant analysis (lda). Moreover, we employ kernel k nearest neighbors (kknn), support vector machines with polynomial and radial kernels (svm.poly and svm.radial), random forests (rF) and gradient boosting machines (gbm). For our analysis we use the software R (R Development Core Team 2011). The classification methods are implemented by the R-packages `stats`, `MASS` (Venables and Ripley 2002), `kknn` (Schliep and Hechenbichler 2010), `kernlab` (Karatzoglou et al. 2004), `randomForest` (Liaw and Wiener 2002) and `gbm` (Ridgeway 2010). All classification methods, except logreg and lda, possess hyperparameters that have to be tuned. Moreover, we would like to assess the effect of variable selection on the prediction performance. Hence an appropriate resampling approach and search strategies for tuning and variable selection are required. In the literature there are only few approaches that account for the mutual dependence of hyperparameters and selected variables and the corresponding software is not readily available. For this reason we apply a two-step procedure: In the first step parameter tuning is done by means of grid search on the whole data set. As performance measure we use the five-fold stratified cross-validated AUC. In Table 2 an overview of the tuned parameters and their values is given. The selected values are printed in bold. In the second step variable selection is done and the performance of all classifiers on the selected and the full variable sets is assessed. For variable selection we apply a nested resampling strategy, again on the whole data set. In the outer loop the variable selection is repeated 25 times for each classification method on different subsamples of size $4/5 \cdot 1{,}163 \approx 929$. In the inner loop the goodness of a variable set is assessed by means of the three-fold stratified cross-validated AUC.

Table 2 Classification methods and corresponding tuning parameters. The values that are tried during the grid search are given in the fourth column. The selected hyperparameter values are printed in bold

Method	Parameter	Definition	Range
kknn	k	Number of nearest neighbors	5, 10, 20, 50, 100, 200, 500, **600**
	distance	Parameter of Minkowski distance	1, **2**, 3
	kernel	Window function	**rectangular**, gaussian
svm.poly	degree	Degree of polynomial	**1**, 2, 3
	C	Regularization constant in Lagrange formulation	$2^{-5}, \mathbf{2^{-4}}, \ldots, 2^4, 2^5$
svm.radial	sigma	Parameter of RBF-function $(\exp(-\sigma \cdot \|u - v\|^2))$	$2^{-5}, 2^{-4}, \ldots, \mathbf{2^{-1}}, \ldots, 2^4, 2^5$
	C	Regularization constant in Lagrange formulation	$2^{-5}, 2^{-4}, \mathbf{2^{-3}}, \ldots, 2^4, 2^5$
rF	ntree	Number of trees	500, 1,000, **2,000**, 3,000
	mtry	Number of candidate variables randomly sampled per split	**3**, 5, 10, 15
	sampsize	Size of sample randomly drawn per tree	$\lceil \mathbf{0.25n} \rceil$, $\lceil 0.5n \rceil$, $\lceil 0.75n \rceil$ with $n = 929 \approx 4/5 \cdot 1{,}163$
gbm	n.trees	Number of iterations	500, **1,000**, 2,000, 3,000
	interaction.depth	Maximum degree of interactions between variables	1, **2**, 3
	distribution	Loss functions	Bernoulli, **adaboost**
	shrinkage	Learning rate	0.001, **0.01**, 0.05
	bag.fraction	Proportion of training observations randomly drawn per tree	**0.25**, 0.5, 0.75

We apply a forward search and stop if the AUC cannot be improved by at least 0.001. The prediction accuracy of all classification methods on the full variable set is assessed on the same 25 subsamples. Besides obtaining performance estimates repeating the variable selection on 25 subsamples has several advantages (cp. Bi et al. 2003). The results are stabilized and the selection frequency across the 25 iterations can be regarded as variable importance measure. Moreover, analyzing the behavior of variables across the different subsamples may provide further insight. However, the fact that hyperparameter tuning was done in advance on the whole data set may lead to optimistically biased AUC values. This is not too severe as the AUC values are only computed to assess if variable selection causes a degradation in performance. Another possible disadvantage is that the hyperparameter values found in advance may not be suitable anymore if a smaller number of features is used. The resampling strategies and variable selection methods are implemented by the R-package `mlr` (Bischl 2010). For calculation of the AUC the R-package `ROCR` (Sing et al. 2005) is used. The source code of the analysis is available at http://www.statistik.tu-dortmund.de/publications_schiffner.html.

5 Results

Table 3 shows the AUC values obtained with all predictor variables and with the selected variables. First of all we can observe that linear classification methods work well on this problem. Using more complex methods like svm, kknn, random forest and gbm results in no or only small improvements. Moreover, variable selection does not lead to much smaller AUC values than using all predictors except for support vector machines. The reason is that the hyperparameter settings found by means of the grid search are not suitable anymore if only few predictors are used. As a result in many of the 25 resampling iterations (7 for svm.poly and 6 for svm.radial) the forward search stopped after the first step since none of the variables led to any improvements in AUC over 0.5. It is possible to adapt the hyperparameters during the variable selection. But this is time-consuming, hence we decided to exclude the support vector machines from the further analysis. The mean number of selected variables per iteration ranges from 3 (for the two svms), 4.5 (for kknn) to 7.7 (for rF). For the reason explained above the support vector machines exhibit the lowest numbers of selected variables with a rather large standard deviation. Fig. 1 shows the selection frequency for all variables and all classification methods except for the support vector machines. For every classification method we counted how often a variable was selected in the 25 subsampling iterations. As the bars are stacked the maximum attainable value is 125. The most important variables for the clinic in Düsseldorf are the severity of cardiac insufficiency (nyha), age, pulmonary hypertension and preoperative renal replacement therapy. The abbreviation nyha stands for New York Heart Association that has developed a functional classification system to assess the stage of heart failure. The nyha variable indicates if a patient is classified as category III or IV where III means symptoms of cardiac insufficiency during less-than-ordinary activity and IV means symptoms already at rest. Rather unimportant variables are angiography findings and reoperation. Concerning the four most important variables the results of all classification methods are similar except for kknn. According to kknn nyha and age are the most important predictors, but pulmonary hypertension is rather unimportant and preoperative renal replacement therapy is not even selected once. Compared to the logistic KCH Score 3.0 we get a rather different variable ranking. According to the p-values of the Wald test statistics the most important variables in the KCH Score 3.0 are critical preoperative state, age, emergency and renal replacement therapy. The other two variables that turned out to be important for the clinic in Düsseldorf, nyha and pulmonary hypertension, are on places 6 and 15 (where the total number of predictors is 17). Future work is mainly gaining deeper understanding of the results. This on the one hand regards differences

Table 3 AUC values and standard deviations obtained with all predictors and with the selected variables. Moreover, the average numbers of selected variables across the 25 resampling iterations and the corresponding standard deviations are shown

	All predictors		Selected variables			
Method	AUC	stand. dev.	AUC	stand. dev.	# selected variables	stand. dev.
logreg	0.81	0.07	0.82	0.07	6.8	1.3
lda	0.80	0.08	0.81	0.07	6.7	1.3
kknn	0.81	0.06	0.78	0.06	4.5	1.6
svm.poly	0.72	0.08	0.55	0.12	3.3	2.7
svm.radial	0.78	0.06	0.55	0.13	3.0	3.6
rF	0.82	0.05	0.77	0.07	7.7	1.8
gbm	0.82	0.06	0.79	0.07	6.0	2.0

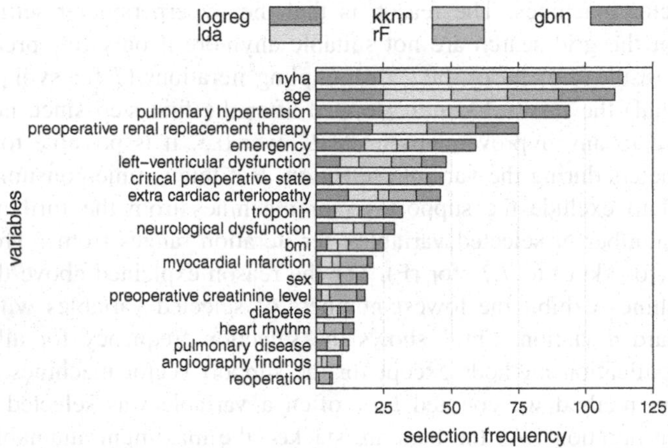

Fig. 1 Selection frequency of predictor variables for all classification methods except svm.poly and svm.radial

between results obtained with different classification methods and the differences between the variable ranking found for the clinic in Düsseldorf and in the KCH Score 3.0. On the other hand this means to further analyze the behavior of variable selection across the subsampling iterations and further investigate the relationships between the (selected) variables.

References

Bi, J., Bennett, K. P., Embrechts, M., Breneman, C. M., & Song, M. (2003). Dimensionality reduction via sparse support vector machines. *Journal of Machine Learning Research, 3*, 1229–1243.

Bischl, B. (2010). mlr: Machine learning in R. URL http://r-forge.r-project.org/projects/mlr/.

Fawcett, T. (2006). An introduction to ROC analysis. *Pattern Recognition Letters, 27*, 861–874.

Guyon, I., Gunn, S., Nikravesh, M., & Zadeh, L. (Eds.) (2006). *Feature extraction, foundations and applications* (Studies in fuzziness and soft computing). Berlin/Heidelberg: Springer.

Karatzoglou, A., Smola, A., Hornik, K., & Zeileis, A. (2004). kernlab – an S4 package for kernel methods in R. *Journal of Statistical Software, 11*(9), 1–20. URL http://www.jstatsoft.org/v11/i09/.

Liaw, A., & Wiener, M. (2002). Classification and regression by randomForest. *R News, 2*(3), 18–22. URL http://CRAN.R-project.org/doc/Rnews/.

R Development Core Team (2011). *R: A language and environment for statistical computing*. R Foundation for Statistical Computing, Vienna, Austria. URL http://www.R-project.org.

Ridgeway, G. (2010). gbm: Generalized boosted regression models. URL http://CRAN.R-project.org/package=gbm.

Schliep, K., & Hechenbichler, K., (2010). kknn: Weighted k-nearest neighbors. URL http://CRAN.R-project.org/package=kknn.

Sing, T., Sander, O., Beerenwinkel, N., & Lengauer, T. (2005). ROCR: Visualizing classifier performance in R. *Bioinformatics, 21*(20), 3940–3941.

Venables, W. N., & Ripley, B. D. (2002). *Modern applied statistics with S* (4th ed.). New York: Springer. URL http://www.stats.ox.ac.uk/pub/MASS4.

Part V
Archaeology and Geography, Psychology and Educational Sciences

Parallel Coordinate Plots in Archaeology

Irmela Herzog and Frank Siegmund

Abstract Parallel coordinate plots (PCPs) can be applied to explore multivariate data with more than three dimensions. This visualisation method is straightforward and easily intelligible for people without statistical background. However, to our knowledge, PCPs have not yet been applied to archaeological data. This paper presents some examples of archaeological classifications which are clearly visible in PCPs. For this purpose a program has been written offering some additional options which are not supported in standard software for PCP generation. Some of the functionality of Geographic Information Systems (GIS) was introduced for PCPs: this program is able to create a thematic display based on a user-selected variable, optionally multiple plots highlight each thematic colour. Another variable may control the width of the PCP lines. Moreover, an info-tool, zoom, and a find-function are supported. The resulting diagram can be saved in SVG format and in a GIS format.

1 Introduction

Most archaeologists have a limited knowledge of classification methods and statistics, and for this reason, only few of these methods are applied to archaeological data. A popular method is correspondence analysis, mainly used for identifying the relative chronological sequence of assemblages of objects like graves and the types of these objects (e.g. Nieveler and Siegmund 1999). Recently, German

I. Herzog (✉)
The Rhineland Commission for Archaeological Monuments and Sites, Endenicher Str. 133, 53115 Bonn, Germany
e-mail: i.herzog@lvr.de

F. Siegmund
Heinrich-Heine-Universität Düsseldorf, Historisches Seminar III: Lehrstuhl für Alte Geschichte, 23.31.05.37, Universitätsstr. 1, 40225 Düsseldorf, Germany

B. Lausen et al. (eds.), *Algorithms from and for Nature and Life*, Studies in Classification, Data Analysis, and Knowledge Organization, DOI 10.1007/978-3-319-00035-0_30,

archaeologists repeatedly applied canonical correspondence analysis for spatial analysis where the canonical variables are geographical coordinates or a linear combination of them (Kubach and Zimmermann 1997; Brather and Wotzka 2006; Müller-Scheeßel and Burmeister 2006; Hinz 2009). We felt that this approach is not appropriate, and discussion with a correspondence analysis expert at a previous GfKl conference confirmed our uneasiness. So we came to the conclusion that archaeologists need appropriate and easily intelligible methods for the explorative analysis of multivariate data and the visualisation of their classification results. In our view, parallel coordinate plots (PCPs) could serve this purpose. PCPs with brushing techniques are not new (e.g. Wegman 1990), but to our knowledge, this approach has not been applied to archaeological data before.

2 The Archaeological Data

Our test data consists of the grave goods of early Medieval cemeteries in Germany, the Netherlands, Belgium, France, and Switzerland (Siegmund 2000). The cemeteries are subdivided into three data sets according to the chronological time frame of the graves, i.e. data sets for the fifth, sixth and seventh century AD. For each of these data sets, Siegmund (2000) developed a classification assigning the cemeteries to different ethnic groups, mainly Frankish and Alemannic people. Our aim was to visualise these classifications by PCPs. In the present paper, we focus on the sixth century cemeteries. This data set consists of 82 cemeteries.

A frequency table lists for each cemetery five variables in the vessel category (total number of vessels, number of glass vessels, the amount of wheel-thrown, hand made, and Thuringian pottery) and six variables referring to weapons (swords, saxes, axes, shields, spearheads with slit and unslit sockets). Only cemeteries with at least four weapons are included in the data set. In this frequency table, the smallest grave group consists of three graves, while 169 graves from the sixth century were recorded for the largest cemetery.

It is quite obvious that proportions rather than frequency counts are required for the analysis. However, looting in ancient and in modern times severely affected nearly all the early Medieval cemeteries. Looters focus more on metal objects like fibulae or weapons than on pottery. For this reason, the proportion of a weapon category in relation to the total number of weapons is considered, whereas for vessels, the proportions are calculated with respect to the total number of graves. PCPs were used to visualise the ethnic classification (see below).

The sixth century data set is fairly complex, therefore we additionally selected a simpler tutorial data set. This data set is a frequency table listing 22 archaeological contexts with four amphorae types (Siegmund in press). These amphorae were used to trade liquids (e.g. wine, fish sauce) and fruits (e.g. olives) in Roman times. The time frame for the four amphorae types is in the range from 120 to 30 BC. Correspondence analysis was applied to the frequency table, the first axis provided a chronological sequence for the archaeological contexts. This sequence was subdivided into four phases: early, mid-a, mid-b, and late. A PCP based on the amphorae type percentages shows that the phases are well separated (see below).

3 Parallel Coordinate Plots

According to Friendly (2009), the first alignment diagrams using sets of parallel axes, rather than axes at right angles were already presented in 1884 by d'Ocagne. Independent of these early works, Inselberg (1985) proposed PCPs. The statistical theory and methods for PCPs were provided by Wegman (1990). The PCP allows to represent all dimensions of a multivariate data set in a 2D diagram by drawing parallel and equally spaced axes and connecting the coordinates on these axes, so that each point is represented by a sequence of continuous line segments. For a given case, the set of line segments could be thought of as a 'profile' (Brunsdon et al. 1997). Ladder plots or dot and line diagrams are variants of PCPs. They are mainly used to compare measurements or ranks before and after medical treatment, i.e. most applications show only two axes (e.g. Leuschner et al. 2008). If the treatment has no effect on the variables considered, the diagram looks like a ladder.

PCPs also bear some similarity to radar charts, also known as polar diagrams, spider charts, or star plots, for which the axes are drawn as spokes of a wheel. Such plots have been applied in archaeological research (e.g. Posluschny 2003). According to Friendly (2009), such diagrams were already used by Georg von Mayr in 1877. However, radar charts are mostly applied to present a low number of cases or group averages, and the scaling of the axes is generally different.

Another complete 2D representation of high-dimensional points is by Andrews curves (Andrews 1972). The impact of the variables placed on the low frequency terms is highest in this Fourier series representation (Scott 1992, p. 13). However, PCPs are more intelligible for archaeologists because they treat the variables in a symmetric fashion.

PCPs have been criticized for their limited applicability to large data sets, i.e. they create much ink (Scott 1992, p. 16). Scott discusses plotting random subsets as a method to address this issue but notes that this approach does not generate precisely reproducible results. Most archaeological data sets comprise of less than 1,000 cases which alleviates the 'much ink issue' somewhat.

4 First Experiments with Archaeological Data

At first, we investigated if PCPs can be applied successfully to visualise our test data. Therefore, we carried out some experiments with a short rudimentary program which generated simple PCPs and used the geographic information system MapInfo (http://www.pbinsight.eu/uk/products/location-intelligence/mapinfo-professional/mapinfo-professional/) for labelling, thematic displays, and data base queries. The visualisation results for the archaeological data created by this procedure and the exploration possibilities were promising.

However, we were aware that a geographic information system (GIS) like MapInfo is not quite the appropriate tool for this purpose. So we tried to find software which is (i) readily available for archaeologists, (ii) allows to reproduce the results created by our initial approach and (iii) supports some interactive querying and manipulation of the graphic results. Only very few archaeologists have access to expensive statistics packages, and the command line interface of R deters most colleagues.

At the time of our investigation, we tested only the open source software GGobi (www.ggobi.org), which is also available in R. Later on we found that the free software XDAT (http://www.xdat.org) as well as XmdvTool (http://davis.wpi.edu/xmdv) are alternatives. GGobi provides a large range of options and supports scatter plots linked to the PCPs. However, we found that it was quite complicated to reproduce the results we had created with our rudimentary program and subsequent manipulations in MapInfo. The GIS approach and the GGobi user interface are quite different, details are discussed below. Many archaeologists are used to working with a GIS, and therefore we thought of providing a software which creates PCPs and reproduces some of the functionality of a GIS.

5 The MultiPCP Program

We extended the initial rudimentary program so that it displays the PCPs and supports typical GIS features like zooming, an info tool, and thematic displays. The data input format is CSV, which can be generated easily with spreadsheet programs. The CSV format is configurable, i.e. field separators, field delimiters etc. can be entered by the user.

A configuration dialogue allows to assign roles like coordinate or thematic to each variable, to modify the axis sequence interactively, to select the colours for thematic display and to configure the legend, to select multiple plots and some other options. These parameters as well as the CSV configuration can be saved to a configuration file.

Similar to the functionality provided by a GIS, two options are supported for thematic display of a variable: (i) the option 'individual' lists all field entries and allows the user to assign a colour to each entry, (ii) the 'range' option is available for interval or ratio scale variables: the user can specify the number of ranges, the minimum and maximum value of each range as well as the corresponding colour.

In general, the drawing sequence of the profile lines in the thematic plot is not fixed so that a red profile line can overprint a green one or vice versa. With MultiPCP, if only one thematic colour except black was selected, the coloured profile lines are plotted in the foreground. When several thematic colours were chosen the multiple plot option generates a set of PCPs; each of these PCPs highlights one of the non-black colours in the foreground, and all other profile lines

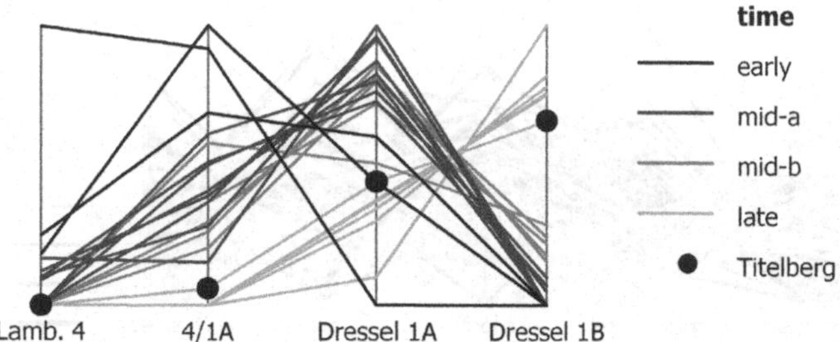

Fig. 1 Thematic PCP of the amphorae data set. The profile line corresponding to the Titelberg (Luxembourg) context is highlighted by black dots

are plotted in black (e.g. Fig. 2). This feature addresses the "much ink" issue. As far as we know, the multiple plot feature is unique and not available in other software creating PCPs.

The line width is set to a constant value for all profile lines, or alternatively, a numeric variable can be selected from the data table that controls this property. Either the variable stores the desired line width, or the first quantile of these field values is assigned the smallest line width, the second quantile is displayed with somewhat increased line width, and so on.

An alternative method to highlight a profile line or a group of these is by searching the table and marking the hits by dots. The colour of the dots can be selected by the user.

6 Application to Archaeological Data

Figure 1 presents the PCP created by MultiPCP for a frequency table listing 22 archaeological contexts with four amphorae types. In the plot, the axes for the amphorae types Lamboglia 4, Stöckli 4/1A, Dressel 1A, and Dressel 1B are shown sorted according to their chronological sequence from early to late. A thematic display including a legend was generated based on the time attribute. The Titelberg context, which is highlighted by black dots, is assigned to the late phase, but it is evident that all other late contexts are probably later because the Titelberg profile line is closer to the mid-b phase contexts.

In general, the late contexts are quite different from the earlier contexts due to a high proportion of Dressel 1B amphorae. The Dressel 1B axis also shows non-overlapping intervals for the two middle period context groups. The early contexts exhibit a higher percentage of both Lamboglia 4 and Stöckli 4/1A amphorae than all later contexts. So the PCP shows that the four phases form well separated groups, and the contexts belonging to the two mid phases are closer to each other than to the earlier or later contexts.

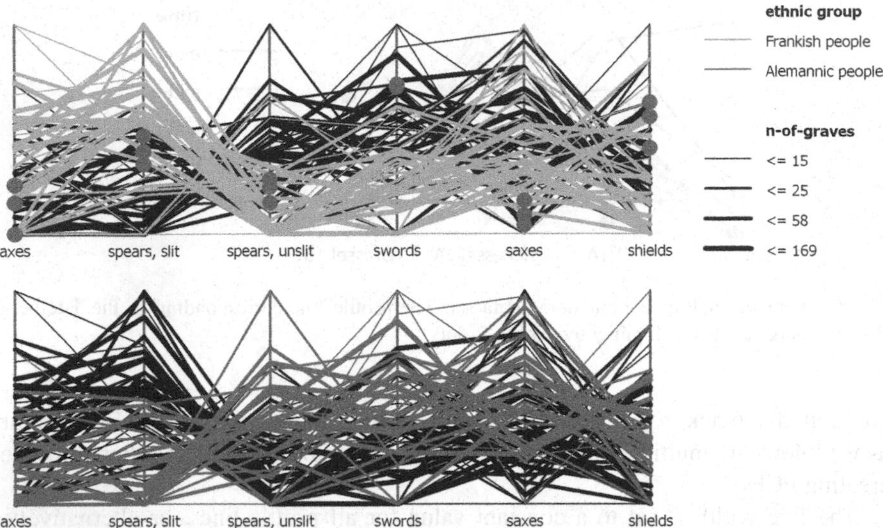

Fig. 2 Two linked PCPs for the weapons of the sixth century data set. In the first PCP, the Thuringian cemeteries are highlighted by grey dots

Figure 2 presents two thematic PCPs created with the multiple plot option. The plots show the ethnic group classification of sixth century cemeteries, focusing on the weapons. In the upper PCP, all records are depicted in black except for the Frankish cemeteries, which are represented by light grey lines in the foreground. Similarly, the profile lines corresponding to the Alemannic cemeteries are in the foreground of the lower plot. Moreover, the three Thuringian cemeteries are highlighted by dark grey dots in the upper PCP. We assume that the reliability of the data is higher for large cemeteries than for those including only a few graves. Therefore, a higher weight is assigned to the large cemeteries, i.e. the line width is controlled by the number of graves recorded for each sixth century cemetery. The four line width groups are calculated by MultiPCP on the basis of the quartiles.

The combination of both plots shows that Frankish cemeteries typically include high percentages of the axe weapon type, and the proportion of spearheads with a slit socket is in general higher than that of Alemannic cemeteries. A high proportion of spearheads with an unslit socket is typical for an Alemannic cemetery, along with a fairly large amount of swords. The proportion of saxes and shields varies within both ethnic groups. Two other ethnic groups are part of this data set: Thuringians and Saxon people. The grey dots marking the three Thuringian cemeteries indicate that these form a fairly homogeneous group with respect to weapon proportions. However, this does not apply to the eight Saxon cemeteries (not shown).

After zooming into the first plot, the info tool allows to identify the cut point on the "spears, slit" axis above which all cemeteries are Frankish. Figure 3 shows a thematic display based on this cut point value.

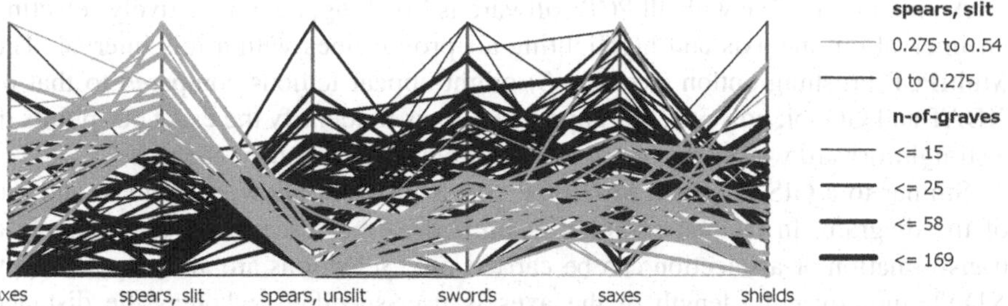

Fig. 3 Brushing the PCP by a thematic display based on a user-selected cut point

The result of brushing identifies a set of typical Frankish cemeteries with a low proportion of spearheads with unslit socket and also a low proportion of swords. Only one of these Frankish cemeteries has a fairly high proportion of swords, and the line width indicates that this cemetery is in the lowest quartile with respect to the number of graves. Using the info tool, this cemetery can be easily identified: Eberfingen in the German state Baden-Württemberg, with only seven graves containing a total of ten weapons. This example shows how MultiPCP supports the interactive exploration of multivariate data.

7 Comparison with Existing Programs

A detailed and comprehensive comparison is beyond the scope of this paper. The focus of this section is mainly on features where existing programs do not meet the requirements of archaeologists.

The input data to GGobi, XDAT, and MultiPCP is to be provided in CSV format. However, only XDAT and MultiPCP support some CSV configuration options. Therefore users working with a German spreadsheet program are faced with quite a few difficulties when trying to enter this data into GGobi. Moreover, the program crashes on reading a German Umlaut or a French accent. The program Excel2Xmdv converts Excel and CSV format files into the native format of XmdvTool. Neither the Readme-file nor the error messages of Excel2Xmdv provide details with respect to the required input format which must be "correct and valid". In our tests, conversion was successful only after deleting all columns with non-numeric data, though the test data provided with the program includes non-numeric data.

In general the multitude of options of the XmdvTool is impressive, yet difficult to comprehend. After spending hours on reading the help files and trying to understand the user interface, we gave up. It is well possible that XmdvTool is able to perform all the functionality we require. But the user interface is not quite suitable for archaeologists. For this reason, only the output possibilities of the tool are discussed in the paragraphs to come.

An important topic with all PCP software is brushing, i.e. interactively selecting an interval on one axis and highlighting all profile lines within this interval. The MultiPCP brushing option (see Fig. 3) might appear tedious compared to that of XDAT and GGobi; however, this approach allows to identify the exact cut-points in a straightforward way so that these are available for subsequent analysis.

Similar to a GIS program, MultiPCP and XmdvTool allow to zoom in and out of the diagram. In GGobi, zooming is limited to the screen size, alternatively a transformation or a selection can be carried out. Selections are also supported by XDAT, moreover the length of the axes is user-specified, whereas the distance between the axes is fix.

The info tool corresponds to the identification mode in GGobi that allows labelling individual records with one variable value (Cook and Swayne 2007, p. 41). However, only one attribute is shown. The info tool also shows information on the axis when its label is selected, the alternative programs provide this information by menu options. Identification options are poor in XDAT.

Axes for nominal variables are displayed in both GGobi and XDAT but not in MultiPCP. Suggesting an ordered sequence by depicting a nominal scale variable on an axis did not seem appropriate.

According to our tests, both GGobi and XDAT do not allow to save the resulting diagrams for publication. As mentioned above, using R for creating high resolution graphic output is beyond the scope of most archaeologists. So it seems that screen shots are the only possibility to save the image created, yet with a low resolution. In our view this is a serious drawback of both programs. XmdvTool is able to store the resulting raster image in the portable pixel map (ppm) format. In our tests, only the image visible on screen was exported, and the colours of the image created by XmdvTool did not agree with the display produced by our image viewer. So it seems, that screen shots provide better results. MultiPCP supports the BMP, the SVG and the MIF-format (MapInfo interchange format) for saving PCPs for publication or for processing in another program.

8 Future Work

Some of the options provided by GGobi or XDAT are also useful for archaeological data sets and can be implemented fairly easily: these include user configuration of axis parameters like colour and line width, axis tics and labelling of the tics, lower and upper limit of each axis, inverting an axis, and a vertical display.

Axes derived from standardizing the variables as described by Brunsdon et al. (1997) may provide a more accurate picture for variables on an interval or ratio scale than axes scaled on the basis of the minimum and maximum value for each variable.

As mentioned above, the radar chart bears some similarity to the PCP. A PCP can be converted to a radar chart by drawing the axes as spokes of a wheel. To emphasize differences in small values, axes could be inverted so that the maximum of each axis is at the centre of the chart. Moreover, radar charts are often used to depict average profile lines for groups rather than each individual profile line. Multiple PCPs could also show the average profile line for a group allowing the visual assessment of each variable's deviation within the group.

The impact of the ordering of the axes on detecting patterns in the PCP is crucial (Cook and Swayne 2007, p. 24). According to Brunsdon et al. (1997), high correlation between neighbouring axes is required to enhance outlier detection. In our view, the absolute value of rank correlation provides an appropriate dissimilarity measure for an optimal ordering of the axes. However, optimising the ordering in terms of some similarity measure between neighbouring axes is NP-complete (Ankerst et al. 1998). Due to the similarity with the travelling salesman problem, variants of the heuristics developed for this problem can be applied to find a near-optimal ordering of the axes. An alternative method of sorting the axes is proposed by Yang et al. (2003) based on a hierarchical ordering of the dimensions.

The MultiPCP program will be made available on the first author's web site www.stratify.org.

References

Andrews, D. F. (1972). Plots of high dimensional data. *Biometrics, 28*, 125–136.

Ankerst, M., Berchtold, S., & Keim, D. A. (1998). *Similarity clustering of dimensions for an enhanced visualization of multidimensional data.* In Proceedings of the IEEE symposium on information visualization, InfoVis'98, (pp. 52–60).

Brather, S., & Wotzka, H. P. (2006). Alemannen und Franken? Bestattungsmodi, ethnische Identitäten und wirtschaftliche Verhältnisse zur Merowingerzeit. In St. Burmeister & N. Müller-Scheeßel (Eds.), *Soziale Gruppen – kulturelle Grenzen. Die Interpretation sozialer Identitäten in der Prähistorischen Archäologie. Tübinger Archäologische Taschenbücher* (pp. 139–224). Münster: Waxmann.

Brunsdon, C., Fotheringham, A. S., & Charlton M. E. (1997). An investigation of methods for visualising highly multivariate datasets. http://www.agocg.ac.uk/reports/visual/casestud/brunsdon/brunsdon.pdf. Accessed 30 September 2011.

Cook, D., & Swayne, D. F. (2007). *Interactive and dynamic graphics for data analysis: With R and GGobi (Use R)* (pp. 24–34). New York: Springer.

Friendly, M. (2009). Milestones in the history of thematic cartography, statistical graphics, and data visualization. http://www.math.yorku.ca/SCS/Gallery/milestone/milestone.pdf. Accessed 17 September 2011.

Hinz, M. (2009). *Eine multivariate Analyse Aunjetitzer Fundgesellschaften.* Universitätsforschungen zur prähistorischen Archäologie, p. 173.

Inselberg, A. (1985). The plane with parallel coordinates. *The Visual Computer, 1*, 69–91.

Kubach, W., & Zimmermann, A. (1997). Eine kanonische Korrespondenzanalyse zur räumlichen Gliederung der hessischen Bronzezeit. In J. Müller & A. Zimmermann (Eds.), *Archäologie und Korrespondenzanalyse: Beispiele, Fragen, Perspektiven. Internationale Archäologie 23* (pp. 147–151). Espelkamp: Marie Leidorf.

Leuschner, F., Li, J., Göser, S., Reinhardt, L., Öttl, R., Bride, P., et al. (2008). Absence of auto-antibodies against cardiac troponin I predicts improvement of left ventricular function after acute myocardial infarction. *European Heart Journal, 29*, 1949–1955.

Müller-Scheessel, N., & Burmeister, St. (2006). Einführung: Die Identifizierung sozialer Gruppen. Die Erkenntnismöglichkeiten der Prähistorischen Archäologie auf dem Prüfstand. In St. Burmeister & N. Müller-Scheeßel (Eds.), *Soziale Gruppen – kulturelle Grenzen. Die Interpretation sozialer Identitäten in der Prähistorischen Archäologie. Tübinger Archäologische Taschenbücher* (pp. 9–38). Münster: Waxmann.

Nieveler, E., & Siegmund, F. (1999). The Merovingian chronology of the Lower Rhine area: Results and problems. In J. Hines, K. Høilund Nielsen, & F. Siegmund (Eds.), *The pace of change. Studies in early Merovingian chronology* (pp. 3–22). Oxford: Oxbow.

Posluschny, A. (2003). GIS in the vineyards: Settlement studies in Lower Franconia. In J. Kunow & J. Müller (Eds.), *Symposium »Landschaftsarchäologie und geographische Informationssysteme. Prognosekarten, Besiedlungsdynamik und prähistorische Raumordnung«, 15.–18. Oktober 2001, Wünsdorf. Forschungen zur Archäologie im Land Brandenburg 8. Archäoprognose Brandenburg 1* (pp. 251–258). Wünsdorf.

Scott, D. W. (1992). *Multivariate density estimation. Theory, practice and visualization*. New York: Wiley.

Siegmund, F. (2000). *Ergänzungsbände zum Reallexikon der Germanischen Altertumskunde, Band 23: Alemannen und Franken*. Berlin: De Gruyter.

Siegmund, F. (in press). *Basel-Gasfabrik und Basel-Münsterhügel: Amphorentypologie und Chronologie der Spätlatènezeit in Basel*. Germania, *89*(1), Jg. 2011.

Wegman, E. J. (1990). Hyperdimensional data analysis using parallel coordinates. *Journal of the American Statistical Association, 85*(411), 664–675.

Yang, J., Peng, W., Ward, M. O., & Rundensteiner, E. A. (2003). *Interactive hierarchical dimension ordering, spacing and filtering for exploration of high dimensional datasets*. INFOVIS 2003. Proceedings of the Ninth annual IEEE conference on Information visualization (pp. 105–112). http://davis.wpi.edu/~xmdv/docs/tr0313_osf.pdf. Accessed 30 September 2010.

Classification of Roman Tiles with Stamp PARDALIVS

Hans-Joachim Mucha, Jens Dolata, and Hans-Georg Bartel

Abstract Latterly, 14 Roman tiles were excavated in Nehren in the farthest eastern part of the former Roman province *Gallia Belgica*. Ten of them have the stamp PARDALIVS that was never seen before. First, this new set of tiles is compared with all currently determined provenances of Roman tile making in the northern part of *Germania Superior* based on their chemical composition. The discriminant analysis indicates that the set of tiles of Nehren is different. However, this class looks not homogeneous. Therefore, second, exploratory data analysis including data visualizations is performed. Additionally we investigated the tiles of a provenance with not yet identified location (NYI3) in detail because their statistical difference to the tiles of Nehren is the lowest among the considered provenances. A serious problem is the small sample size. In order to increase the latter one, we propose some combinations of bootstrapping and jittering to generate additional observations.

1 Introduction

Roofing with tiles and underfloor heating with brick-constructed *hypocausis* were most common in the north-western provinces of the Roman empire. Newsworthy with respect to Roman tiles research, the archaeometrical investigation of

---

H.-J. Mucha (✉)
Weierstrass Institute for Applied Analysis and Stochastics (WIAS), 10117, Berlin, Germany
e-mail: mucha@wias-berlin.de

J. Dolata
Head Office for Cultural Heritage Rhineland-Palatinate (GDKE), Große Langgasse 29, 55116, Mainz, Germany
e-mail: dolata@ziegelforschung.de

H.-G. Bartel
Department of Chemistry at Humboldt University, Brook-Taylor-Straße 2, 12489, Berlin, Germany
e-mail: hg.bartel@yahoo.de

B. Lausen et al. (eds.), *Algorithms from and for Nature and Life*, Studies in Classification, Data Analysis, and Knowledge Organization, DOI 10.1007/978-3-319-00035-0_31,

Fig. 1 A Roman tile with the stamp PARDALIVS

brickstamps of PARDALIVS (late Roman empire) can be reported. Figure 1 shows a fragment of a tile with this stamp. The location of the findings is Nehren on the River Mosel, nearby the former imperial residence Trier. The roof tiles belong to a Roman mausoleum equipped with a grave-chamber (Eiden 1982). Their chemical composition was measured by X-ray fluorescence analysis at the laboratory of Freie Universität Berlin (Gerwulf Schneider). Altogether nine main elements and ten trace elements were measured (for details, see Table 2 in the Appendix that contains all measurements).

First, a classification task was carried out: The new set of 14 tiles is compared with all currently known provenances of Roman tile making in Roman Southern Germany. Tiles of eight established provenances are the basis of the upcoming comparisons: see Mucha et al. (2002) and Mucha et al. (2008). For comparison purposes the linear discriminant analysis (LDA) is the first choice here. This method was described by Fisher (1936) for two classes and extended to more than two classes by Bryan (1951). Without any doubt, the class of tiles of Nehren looks very different. However, this class is not homogeneous. Therefore, second, exploratory data analysis including data visualizations is performed. Finally we investigated the tiles of the provenance "Not yet identified 3" (NYI3) (see, for instance, Mucha et al. 2002 and Bartel 2009) in more detail because their statistical difference to the tiles of Nehren is the lowest among all considered classes of tiles. Here, a serious problem is the small sample size, 14 and seven observations, respectively. The series of the analyzed tiles can nowadays not be increased by archaeologists, because already all known tiles have been made available. In order to increase the sample size, we propose some combinations of bootstrapping and jittering to generate additional observations.

2 Classification Results: The New Tiles as a Class

Here the new set of 14 tiles of Nehren is considered as a class that is defined from the archaeological point of view. It is compared with all currently known provenances of Roman tile making in the former province *Germania Superior*.

From the archaeological point of view, tiles with the same stamp could define a provenance (except the brickstamps of legions in the case that they operated globally). The stamp PARDALIVS suggests a private person in the background. Therefore, the main question arises: Is the class of tiles of Nehren significantly different from the known classes also from the statistical point of view? Or can they be assigned to a known provenance to a high degree?

For statistical data analysis, our starting point is the I x J data matrix $\mathbf{X} = (x_{ij})$ with I observations and J variables. Because here we have quantitative data of different scales, the following simple standardization of all variables to mean value equals 1 was figured out:

$$x_{ij} = \frac{u_{ij}}{\overline{u}_j} \quad i = 1, 2, \ldots, I, \quad j = 1, 2, \ldots, J, \tag{1}$$

where $\overline{u}_j$ is the average of the original measurements u_{ij} of the variable j. By the way, a transformation such as (1) is recommended for several multivariate methods (cluster analysis, principal components analysis based on the covariance matrix, K nearest neighbors method) but it does not affect the results of LDA.

First a LDA was carried out with the tiles of Nehren as a new class that has to be compared with the eight established provenances (concerning some basics on discriminant analysis, see Mucha 1992). The learned classifier yields no errors on the training set with respect to the class Nehren. However, cross-validation (concretely, the leave-one-out estimator) gives an error rate of about 2 % concerning the 14 tiles of Nehren. The result of pairwise statistical comparison of classes is that the provenance NYI3 is most similar to class Nehren among all known provenances. Usually, that was it: The new findings of Nehren seems to be a new, but inhomogeneous provenance also from the statistical point of view. But, what about the inhomogeneity? To answer this question, second, classification and clustering methods are performed that consider the tiles of Nehren as unseen observations.

3 Classification Results: Tiles of Nehren as Unseen Observations

We repeated the LDA with all eight provenances as given classes, but the 14 tiles of Nehren are considered as unseen observations. The latter do not affect the statistical analysis. Figure 2 shows the LDA plot of the eight classes C1,…, C8. Here additionally C9 represents the tiles of Nehren. These supplementary (i.e., non-active) observations were projected on the plane spanned by the first two canonical variables. The plot looks very similar to the principal components analysis plot (Bartel 2009). The LDA plot of the supplementary observations presents clear

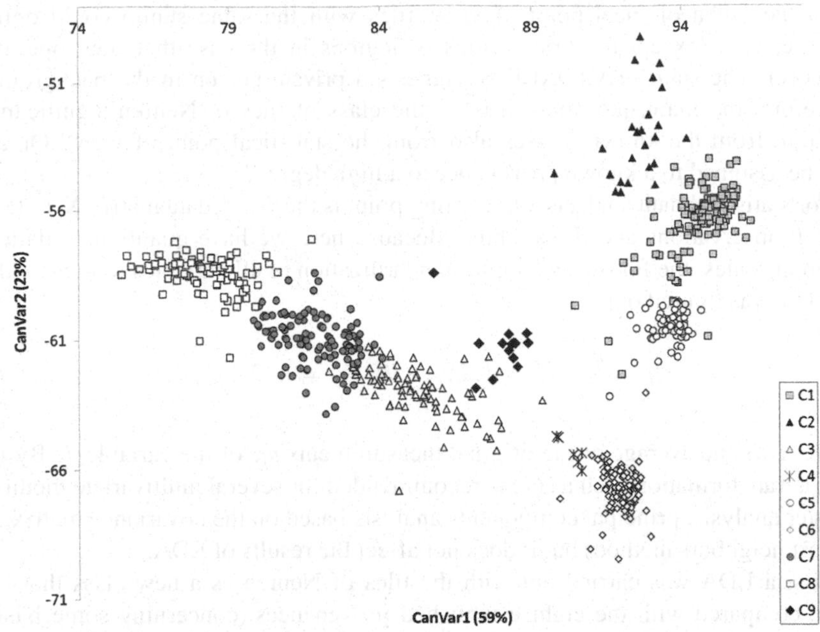

Fig. 2 LDA plot of all Roman tiles (eight provenances (classes C1–C8) and 14 tiles (C9) as supplementary observations)

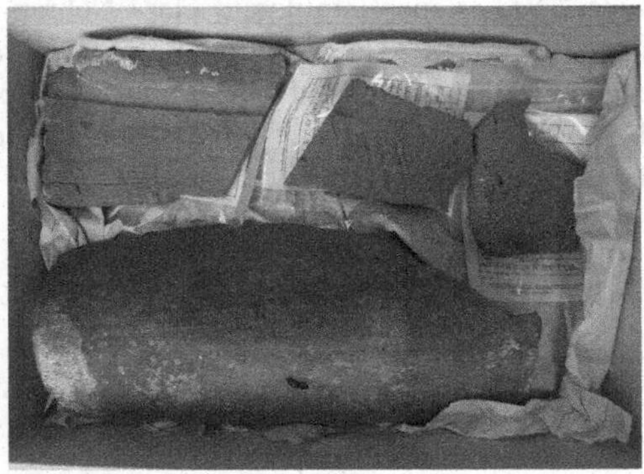

Fig. 3 The four fragments of tiles from Nehren without stamp under consideration

assignments of ten observations (N71–N78, N81, and N82) to class C4 (i.e., NYI3) and four observations (N79, N80, N83, and N84) to class C3 "Rheinzabern B".

The K nearest neighbors method presents the same result as LDA for different number of neighbors $K = 1$, 2, and 3: Four tiles belong to the provenance

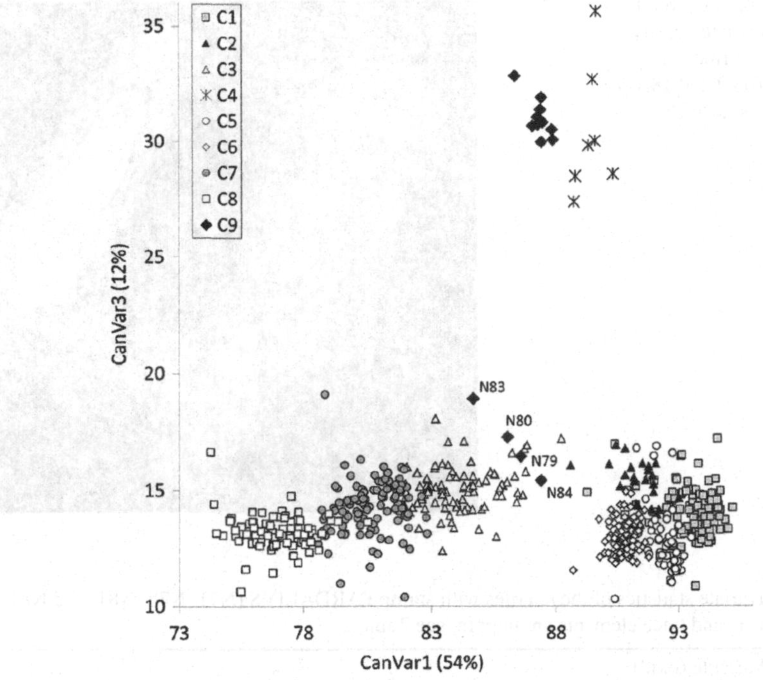

Fig. 4 LDA plot of all considered Roman tiles. Here the third canonical variable discriminates quite clearly between the two classes C9 ("Nehren") and C4 (NYI3) and all remaining classes

"Rheinzabern B" and ten to NYI3. (Concerning some basics on K nearest neighbors methods, see Mucha 1992.)

Also the same result is obtained by Ward's hierarchical clustering (Späth 1980): Four tiles belong to the provenance "Rheinzabern B" and ten to NYI3 if one cuts the dendrogram at $K = 8, 9, 10, 11$ clusters. When cutting the dendrogram at $K = 12$ clusters then ten tiles build an own cluster. This cluster is very stable because it does not disappear by cutting the dendrogram until $K = 45$ clusters. The question arises: Is there an archaeological reason for both the stability of the cluster of ten tiles and the inhomogeneity of the 14 tiles of Nehren? It comes out that exactly the four tiles that are grouped into "Rheinzabern B" are without any stamp. Figure 3 shows these four Roman tile fragments. Because of this knowledge and the outstanding stability another LDA was carried out with the ten stamped tiles PARDALIVS as a new class that has to be compared with the eight established provenances. The remaining four unstamped tiles are taken as unseen observations here. Figure 4 shows the LDA plane spanned by the first and third canonical variable. At the top, the subset of the class C9 consisting of the ten tiles PARDALIVS seems to be very homogeneous. Without any doubt, the four unstamped tiles N79, N80, N83, and N84 are assigned to the provenance "Rheinzabern B".

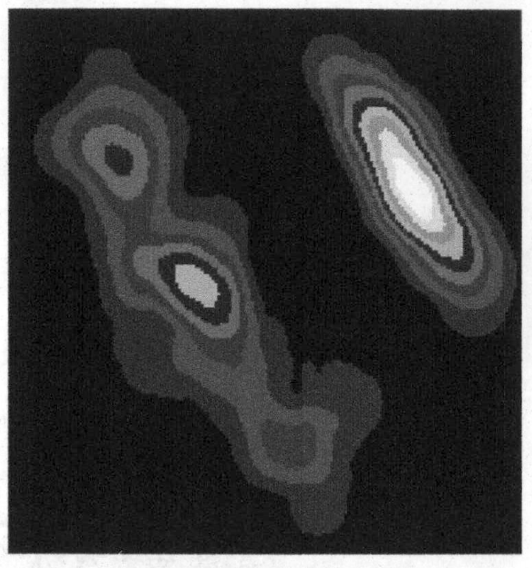

Fig. 5 Mapping of several cuts of the bivariate density of the first two principal components (63.2 and 18.1 % of variance, respectively)

Table 1 Univariate statistics of the ten tiles with stamp PARDALIVS (N71–N78, N81, and N82). Oxides are in %, and trace elements are in ppm, see Table 2

	Variable (oxid)								
Result	SiO_2	TiO_2	Al_2O_3	Fe_2O_3	MnO	MgO	CaO	Na_2O	K_2O
Min	67.08	0.856	16.27	6.010	0.0426	2.941	0.281	0.1914	4.119
Max	68.64	0.910	17.45	6.212	0.0527	3.866	0.477	0.2636	4.502
Median	67.44	0.876	17.05	6.146	0.0466	3.133	0.382	0.2369	4.406
Average	**67.68**	**0.877**	**17.00**	**6.134**	**0.0466**	**3.218**	**0.382**	**0.2321**	**4.371**
St.dev.	0.50	0.015	0.33	0.078	0.0032	0.258	0.070	0.0254	0.121

	Variable (trace element)									
Result	V	Cr	Ni	Zn	Rb	Sr	Y	Zr	Nb	Ba
Min	92.3	107.4	48.5	219.8	165.0	55.6	23.6	237.8	14.1	393.3
Max	113.2	119.7	59.5	261.8	189.2	61.2	30.0	259.4	19.3	441.0
Median	99.1	113.1	50.9	232.7	178.6	59.3	27.7	245.9	16.8	420.6
Average	**99.8**	**113.0**	**51.9**	**234.4**	**179.0**	**59.2**	**27.4**	**246.7**	**16.7**	**419.4**
St.dev.	5.6	4.2	3.4	12.2	7.2	1.9	1.8	7.6	1.6	12.6

4 Comparison Between the Classes PARDALIVS and "Not Yet Identified 3" (NYI3)

Finally we investigated the tiles of the provenance NYI3, because their statistical difference to the PARDALIVS-tiles is the lowest among all considered classes of tiles, see Fig. 4. A serious problem is the small sample size: ten (PARDALIVS) and seven (NYI3) observations, respectively. These two series of the analyzed tiles can nowadays not be increased by archaeologists because already all known tiles have been made available. In order to increase the sample size, we propose data jittering to generate additional observations by applying

$$x_{ijk} = x_{ijk} + (r_{ijk} - 0.5)s_{jk}z, \quad i = 1, 2, \ldots, I, \quad j = 1, 2, \ldots, J, \quad k = 1, 2, \tag{2}$$

where r_{ijk} are random generated numbers from the interval [0, 1) (uniform distribution), s_{jk} is the estimate of the standard deviation of variable j in class k, and the parameter z defines the degree of disturbance. Figure 5 shows several cuts of the bivariate density of the first two principal components. On the left hand side NYI3 and on the right PARDALIVS are located. They look well separated of each other. Here a tenfold increase by jittering is performed, and the degree of disturbance is chosen $z = 3$. Altogether the number of observations is 170 ($= 7 * 10 + 10 * 10$, $17 = 7 + 10$ of which are original observations). The class PARDALIVS looks very homogeneous, see also Table 1. Obviously, the roof tiles with stamp PARDALIVS define a new provenance. The tiles without a stamp of the mausoleum are quite different to PARDALIVS. They look similar to the provenance "Rheinzabern B", see Fig. 4.

5 Archaeological Valuation of Results

All known brickstamps of the brickmaker PARDALIVS belong to the same brickyard. The architectonic context of the tombs at Nehren hands down some other bricks and tiles without any stamps. These unstamped ones belong mainly to the roofing of the monuments. Some of them are used for plastering the walls and ceilings. These fit to the products of late bricksite of Rheinzabern in Palatinate. Perhaps they belong to a repairing set, and they are not part of the primary building of Nehren tombs-architecture. In any case these bricks and tiles give evidence to well functioning and prospering economic system in the environs of Emperors residence of *Augusta Treverorum* (Trier), protected by late river Rhine-limes. The statistical investigation has given decisive arguments for archaeological interpretation of the Roman tiles, excavated at Nehren.

Appendix

Table 2 contains the measurements of the tiles of Nehren. It completes the description of Roman tiles that was already published by Mucha et al. (2009).

Table 2 Measurements of 14 tiles of the Roman mausoleum. The oxides are in percent and the trace elements are in ppm

	Variable (oxid)								
Name	SiO_2	TiO_2	Al_2O_3	Fe_2O_3	MnO	MgO	CaO	Na_2O	K_2O
N71	67.08	0.867	17.45	6.212	0.0435	3.086	0.455	0.2529	4.502
N72	67.34	0.863	16.93	6.078	0.0426	3.866	0.477	0.2232	4.119
N73	67.90	0.885	17.11	6.088	0.0442	3.044	0.281	0.2310	4.370
N74	68.02	0.881	16.88	6.104	0.0494	2.941	0.463	0.1914	4.412
N75	67.43	0.910	17.14	6.188	0.0465	3.264	0.320	0.2583	4.401
N76	67.46	0.873	17.19	6.199	0.0485	3.084	0.415	0.2427	4.439
N77	67.33	0.877	17.29	6.208	0.0434	3.140	0.374	0.1918	4.492
N78	68.26	0.856	16.76	6.038	0.0467	3.126	0.315	0.2436	4.303
N79	61.03	0.969	21.55	7.493	0.0926	2.158	2.708	0.7330	3.075
N80	63.25	0.898	20.52	6.860	0.0849	2.516	1.899	0.7577	3.051
N81	67.39	0.883	17.00	6.211	0.0527	3.321	0.390	0.2636	4.439
N82	68.64	0.874	16.27	6.010	0.0487	3.312	0.328	0.2222	4.233
N83	63.87	0.802	17.73	7.525	0.1115	3.677	0.864	0.2612	5.053
N84	61.88	0.995	22.16	6.999	0.0847	1.805	2.118	0.6354	3.157
	Variable (trace element)								
Name	V	Cr	Ni	Zn	Rb	Sr	Y	Zr	Nb	Ba
N71	113.2	119.7	53.7	261.8	189.2	61.1	30.0	244.7	14.6	426.7
N72	96.8	116.9	49.5	241.1	170.8	57.4	23.6	237.8	15.6	419.3
N73	99.3	108.5	49.6	227.2	181.0	60.7	29.2	247.0	17.3	410.0
N74	98.9	112.0	49.2	230.6	178.9	61.2	27.0	259.4	18.2	420.7
N75	103.3	115.1	48.5	234.7	178.3	59.4	26.3	255.1	19.3	420.5
N76	99.8	114.1	59.5	243.5	187.3	61.0	27.6	241.5	16.2	428.5
N77	97.9	116.6	54.2	236.1	183.9	59.1	27.8	238.6	17.9	441.0
N78	95.4	111.1	53.2	224.4	178.0	58.2	28.3	239.7	17.0	413.0
N79	124.4	120.2	72.5	141.7	126.8	163.1	31.6	178.5	14.0	482.2
N80	108.4	109.1	70.0	127.6	125.6	147.9	24.6	161.6	15.1	431.8
N81	100.8	107.4	52.1	225.2	177.9	58.4	28.1	248.5	16.5	421.2
N82	92.3	108.2	49.5	219.8	165.0	55.6	25.9	254.5	14.1	393.3
N83	87.8	107.1	68.6	72.8	183.4	84.5	37.6	193.7	15.8	464.1
N84	116.7	130.9	73.9	132.2	132.0	141.4	27.6	168.3	17.3	430.6

References

Bartel, H. -G. (2009). Archäometrische Daten römischer Ziegel aus Germania Superior. In H. -J. Mucha & G. Ritter (Eds.), *Classification and clustering: Models, software and applications* (Rep. No. 26, pp. 50–72). Berlin: WIAS

Bryan, J. G. (1951). The generalized discriminant function: Mathematical foundations and computational routine. *Harvard Educational Review, 21*, 90–95.

Eiden, H. (1982). Die beiden spätantiken Grabbauten am Heidenkeller bei Nehren. *Trierer Zeitschrift Beiheft, 6*, 197–214.

Fisher, R. A. (1936). The use of multiple measurements in taxonomic problems. *Annals of Eugenics, 7*, 179–188.

Mucha, H. -J. (1992) *Clusteranalyse mit Microcomputern*. Berlin: Akademie.

Mucha, H. -J., Bartel, H. -G., & DOLATA, J. (2002). Exploring Roman brick and tile by cluster analysis with validation of results. In W. Gaul & G. Ritter (Eds.), *Classification, automation, and new media* (pp. 471–478). Berlin: Springer.

Mucha, H. -J., Bartel, H. -G., & Dolata, J. (2008). Finding Roman brickyards in Germania superior by model-based cluster analysis of archaeometric data. In A. Posluschny, K. Lambers, & I. Herzog (Eds.), *Proceedings of the 35th international conference on computer applications and quantitative methods in archaeology (CAA)*, Berlin, (pp. 360–365).

Mucha, H. -J., Bartel, H. -G., & Dolata, J. (2009). Vergleich der Klassifikationsergebnisse zum "Roman Tiles Data Set". In H. -J. Mucha & G. Ritter (Eds.), *Classification and clustering: Models, software and applications* (Rep. No. 26, pp. 108–126). Berlin: WIAS.

Späth, H. (1980). *Cluster analysis algorithms for data reduction and classification of objects*. Chichester: Ellis Horwood.

Applying Location Planning Algorithms to Schools: The Case of Special Education in Hesse (Germany)

Alexandra Schwarz

Abstract In Germany children with special educational needs are still predominantly taught in special schools. Although segregative schooling stands in conflict with the legal situation, only little development towards inclusive concepts has been made. One substantial argument are the costs involved by the required, extensive reorganization of the educational system. By using location planning procedures, this paper describes how organizational effects and the financial impact of implementing inclusive settings can be analyzed systematically. Schooling of children with learning disabilities in the German federal state of Hesse is used to exemplify and discuss the approach. The results indicate that especially in rural regions, where school enrollment is decreasing, preference should be given to inclusive concepts in order to provide a demand-oriented school system where schools are close to home for all students.

1 Introduction

In contrast to many other European countries like Denmark, Norway, Portugal and Spain, German children with special educational needs are still predominantly taught in special schools (European Agency for Development in Special Needs Education 2011; Sekretariat der Staendigen Konferenz der Kultusminister der Laender in der Bundesrepublik Deutschland 2011). In school year 2008/2009, 6.0 % of all German students are identified to have special educational needs. About 81 % of them attend special schools. The segregation of students with special needs stands in conflict with the legal situation in Germany as education acts give clear precedence to inclusive schooling. The United Nations convention on the rights of

A. Schwarz (✉)
German Institute for International Educational Research, Schloßstraße 29, D-60486, Frankfurt am Main, Germany
e-mail: a.schwarz@dipf.de

B. Lausen et al. (eds.), *Algorithms from and for Nature and Life*, Studies in Classification, Data Analysis, and Knowledge Organization, DOI 10.1007/978-3-319-00035-0_32,

persons with disabilities – which Germany ratified in 2009 – has reinforced the political and educational debate on implementing inclusive concepts.

Different special needs, like learning disabilities or sensory impairments, induce different instructional strategies and organizational requirements (e.g. handicapped accessible classrooms). Therefore, the costs incurred by the required reorganization of the school system are of great importance for policy makers. But until recently, the organizational effects and the financial impact of bringing inclusive education in practice have not been studied systematically on the micro level of students and schools. Is it really feasible – from an organizational perspective – to pass on special schools and send all students to regular schools? What would be the financial consequences, e.g. in terms of personnel costs, investment in equipment and devices and ways to school?

Hence, this paper does not focus on pedagogical motives, but on quantitative aspects of inclusive concepts. The empirical analysis described here accounts for varying preferences by comparing potential supply models for students with special needs. It directly incorporates a recipient-oriented model as it allows for a co-existence of special and regular schools. Schooling of children with learning disabilities in the German federal state of Hesse is investigated as an example.[1] Based on location planning concepts and a simulated allocation of students with special needs to regular schools the organizational consequences and financial effects of inclusive schooling are estimated.

2 Special Needs Education in Hesse

The analysis is based on administrative data of the Hessian school statistics for school year 2008/2009, which has been provided by the Hessian Ministry of Culture. In total 629,828 students attending any of the 1,802 public schools in Hesse are analyzed.[2] Using administrative data turns out to be a particular challenge: Firstly, schools are not labeled with respect to special needs education in a standardized manner. There is especially lack of information on which schools provide preventive or integrative measures (e.g. special classes for speech impaired students or joint classes). So in case of doubt, schools are classified by means of their students, i.e. regular schools are attended exclusively by students without special needs, whereas schools with integrative settings are attended by students with and without special needs (referred to as 'integrative schools'). The supported types of special needs are determined according to special lessons given in school (cf. Table 1).

[1]In Germany, education is controlled by the federal states (the "Laender"), i.e. each federal government is responsible for its educational policy and school system.

[2]Public schools are fully publicly funded and do not charge tuition. Privately funded schools are excluded as they do not report relevant information, e.g. on students per grade.

Table 1 Special educational needs and support by type of school

Special educational need	Estimated no. of students: Special need rate[a] (in %)	Primary level	Secondary level	Schools: Special	Integrative	Regular
Learning disability	46.18	4,179	7,113	105[b]	18	–
Visual impairment, blindness	1.27	60	135	6	0	–
Auditory impairment, deafness	3.33	244	459	6	0	–
Speech impairment	9.38	805	1, 379	14	7	–
Physical impairment	5.76	465	858	8	2	–
Mental disability	17.20	1,483	2, 575	41	1	–
Emotional/social development	9.19	779	1, 366	23	8	–
(Long-term) illness	7.69	660	1, 157	18	0	–
Schools total				185	655[c]	962
Students total	100.00	8,675	15,060	21,154	285,756	322,918

[a] Students with a specific need related to all students with special educational needs
[b] Including 97 'special schools for the learning-disabled'
[c] Joint lessons and preventive measures are most frequently not reported differentiated with respect to concrete special needs

Secondly, we do not have access to individual data on student level due to data protection. The total number of students with special needs is available for every school, the concrete need they have is not specified though. In addition, residential information is only reported per school on the municipality level; students' postal addresses are not provided. Therefore, the number of students with a concrete special need is estimated for each municipality by multiplying the number of students with special needs by the special need rate which is given for Hesse in total (Sekretariat der Staendigen Konferenz der Kultusminister der Laender in der Bundesrepublik Deutschland 2011). The results for students at primary level (grades one to four) and secondary level (grades five to nine) are given in Table 1.

3 Evaluation of Alternative Supply Models

The analysis aims at comparing alternative supply models for students with special educational needs, especially at evaluating quantitative effects of an ubiquitous implementation of inclusive schooling concepts. In this short paper, the evaluation is restricted to two scenarios which are defined for allocating students with special needs to schools:

- *Segregation (SEG):* Attendance of a special school supporting the special need at hand (reference model, currently prevailing concept).

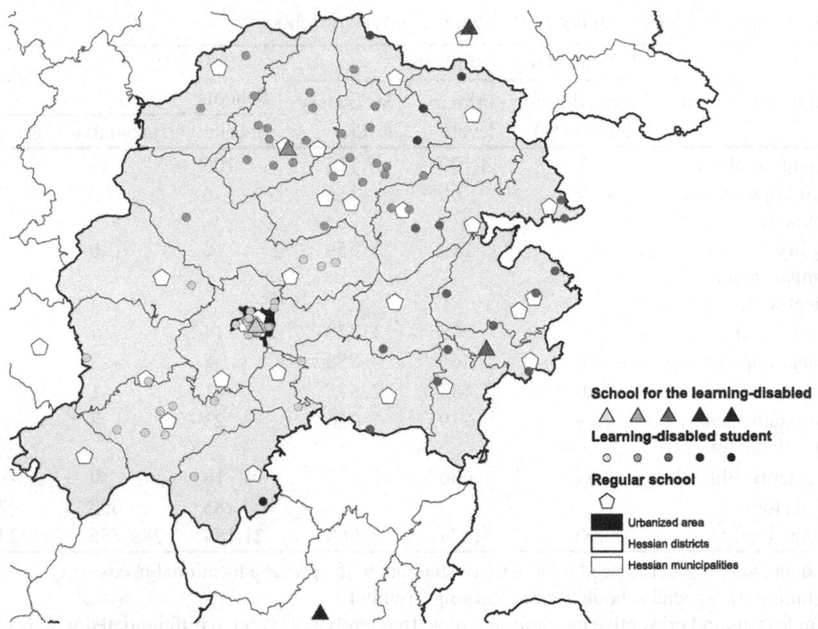

Fig. 1 Location of students with learning disabilities, regular schools and schools for the learning-disabled in the rural district of *Hersfeld-Rotenburg*. Students are shaded according to the special school they are assigned to

- *Inclusion (INC):* Attendance of any regular school (even if it is not yet supporting special needs) with joint lessons of students with and without special needs (reorganization).

By simulating the allocation of students with special needs to schools according to the defined scenarios it is possible to determine the effects of the two competing concepts. In a first approach, a finite set of schools is assumed, i.e. no schools are closed or established, and all schools accommodate students. Each school $s, s = 1, \ldots, S$ is of a defined type (special, integrative, regular). This set of school locations is restricted to a specific subset with regard to each scenario (*SEG*: Special schools supporting the observed need, *INC:* Regular and integrative schools). The demand for special needs education is measured on the municipality level. For each of $j = 1, \ldots, 426$ Hessian municipalities the number of students with $(n_{j,1})$ and without special needs $(n_{j,0})$ is observed. The number of students with special need k in municipality j $(n_{j,k})$ is estimated as explained above. As places of residence within municipalities are not available, we spread $n_{j,k}$ students randomly over the urbanized area of municipality j, using methods described in Beyer (2004). Figure 1 gives an example of the resulting random spatial distribution of students at the primary level.

By each school location s, fixed and variable costs are incurred. Fixed location costs f_s consist of all expenses for maintaining school s (e.g. rentals, insurances, personnel) which occur independently of the number of students. Hence, when optimizing allocation with respect to a finite set of schools, these costs are not part of the target function. In contrast to this, variable location costs $v_s(n_s)$ depend on the number and the composition of students assigned to a school. This especially applies to expenses for teachers and educators. In addition, we have to account for costs for transporting students to schools. Again, the fixed parts of these costs (personnel, maintenance of cars or busses) are not taken into consideration. Variable transport costs depend on the distance between a student's place of residence and the school location. If student i is located in an area by coordinates (x_i, y_i) and school s is located at (a_s, b_s), this distance is given by d_{is}; usually the Euclidean distance measure is used.

In finding the optimal allocation of students to schools in terms of the costs described we would be facing a typical transportation-allocation problem (Nahmias 2009; Domschke et al. 2008) which requires a simultaneous minimization of distances of students to schools *and* variable location costs. These costs are unknown so far, instead we are interested in estimating them. Hence, the costs of the different supply models are evaluated in a second step, based on a solution to the simplified allocation problem:

$$\text{Minimize} \quad F(a,b) = c \sum_{i=1}^{n} \sum_{s=1}^{S} (d_{is} | n_s < M_s) \qquad (1)$$

with $n_s = \sum_{i=1}^{n} I_i(s)$ where $I_i(s)$ is the indicator function for allocating students to schools:

$$I_i(s) = \begin{cases} 1 \text{ if student } i \text{ is allocated to school } s \\ 0 \text{ else} \end{cases} \quad \forall\, s = 1, \ldots, S \qquad (2)$$

In Eq. (1), c denotes (equal) transport costs per kilometer, and M_s denotes the maximum capacity of school s. It is important to note that students assigned to a regular or integrative school in scenario *INC* are composed of $n_{s,0}$ students without special needs who already attend the school and $n_{s,k}$ students with special need k who are assigned there ($n_s = n_{s,0} + n_{s,k}$).

4 Results for Students with Learning Disabilities at the Primary Level

Rather than analyzing all types of special needs at different levels of education, the focus of the analysis is on the most frequent special need – learning disabilities – of students at the primary level (grades one to four) in Hesse and on the more detailed description of the results for students and schools. To compare the two defined scenarios, 4,179 students (cf. Table 1) have to be allocated to a special school and to a regular school. The subset of special schools consists of 97 'special schools for the learning-disabled' (scenario *SEG*); in scenario *INC*, 1,213 regular schools supplying lessons at the primary level form the relevant subset of schools.

Concerning the maximum capacity M_s in Eq. 1 we assume that there is room enough to establish four additional classes at each regular school (one per grade). In joint lessons, the number of students per class should not exceed 20, where up to three students may have special educational needs (Hessisches Kultusministerium 2006). This leads to three possible models for joint lessons: 17/18/19 students without special needs plus 3/2/1 student(s) with special needs. As the number of actual classes at the primary level is not reported in our data, we estimate it by assuming that each class is currently attended by 25 students, which is the reference value given in Hessisches Kultusministerium (1992). The costs of the different models are evaluated in a second step after the specific allocation problem has been solved. These solutions are determined by an algorithm in which each student is assigned to the nearest regular school if this assignment still goes in line with the specific model for joint lessons (17 + 3, 18 + 2, 19 + 1) *and* the maximum capacity of this school. Otherwise, her/his allocation to the second-nearest school is tested in the same manner, and so on. The capacity of special schools is assumed to be unrestricted.

Table 2 summarizes the results of the simulated allocations. In the inclusive settings, the share of students with learning disabilities which is allocated to their nearest regular school depends more on the area where students live than on the model assumed for joint lessons. Using the 17 + 3-model, about 96 % of these students living in rural regions are assigned to the nearest school. Although schools in metropolitan areas are more likely to be fully booked, there are still about 85 % of the students with learning disabilities, who can be allocated to their nearest school. The higher share of students which is assigned to another than the nearest school influences the distribution of ways to school. The maximum distance between assigned school and place of residence is 7.7 km in the large Hessian cities and nearly 9 km in the urbanized areas surrounding these cities. In rural regions the maximum distance is 5.6 km only. But the mean distances to the assigned schools do not differ significantly, either between the regional types or between the different models of joint lessons.

The allocations to the nearest school for the learning-disabled indicate that distances to these schools are shorter in large cities, but – with respect to the total population of students – they are much larger than to regular schools. With respect to the way to school, special schools for the learning disabled may be an alternative for students in the city of Frankfurt am Main, for example, but for most students in rural districts they would not be. Figure 1 illustrates this by means of the Hessian district of Hersfeld-Rotenburg.

The results also suggest that not all schools would be affected by the implementation of inclusive settings. For example, for the 19 + 1-model of joint lessons we find that 206 out of 1,213 schools (about 17 %) are not assigned any students with learning disabilities. Another 65 schools do not need to establish additional classes, although they would have students with special needs (cf. Table 3, amounts to 271

Table 2 Allocations and distances to assigned regular and special school by type of region[a]

Municipalities	Students with learning disabilities	Scenario	Joint lessons	% Students assigned to ... nearest school 1st	2nd	3rd	Distance to assigned school (in km) Min.	Mean	Max.	Sum
Metropolitan areas										
186	2,548	*INC*	17 + 3	84.22	12.05	2.32	0.01	1.19	7.70	3,021
		INC	18 + 2	83.52	12.21	2.75	0.01	1.20	7.70	3,047
		INC	19 + 1	82.69	12.32	3.22	0.01	1.21	7.70	3,073
		SEG		100.00			0.01	3.92	23.86	29,987
Urbanized areas										
195	1,404	*INC*	17 + 3	94.02	4.91	0.78	0.03	1.73	7.36	2,434
		INC	18 + 2	93.30	5.41	0.85	0.03	1.75	8.46	2,457
		INC	19 + 1	92.59	5.98	1.00	0.03	1.76	8.77	2,473
		SEG		100.00			0.05	6.06	21.60	25,535
Rural areas										
43	227	*INC*	17 + 3	96.04	3.96		0.08	1.87	5.61	424
		INC	18 + 2	95.59	4.41		0.08	1.87	5.61	425
		INC	19 + 1	95.15	4.85		0.08	1.88	5.61	427
		SEG		100.00			0.34	5.26	14.54	3,584

[a] Type of region where student's place of residence has been located

Table 3 Additional classes and additional personnel requirements by model for joint lessons

Joint lessons	Additional classes 0	1	2	3	4	Total	Additional pedagogical hours: Per student and week Min.	Max.	All students, per month Min.	Max.	Sum of saved distances (km) per month
17 + 3	274	307	322	204	106	1,987	4	8	71,879	143,758	594,346
18 + 2	272	308	322	206	105	1,990	4	8	71,879	143,758	592,196
19 + 1	271	307	322	209	104	1,994	5	10	89,849	179,697	590,347

Note: All students: 4,179 students with learning disabilities; a month equals 4.3 weeks; a week equals 5 days at school

schools), because they have small classes at the primary level, which may be due to pedagogical concepts or (even more likely) to the demographic progress.

With respect to organizational and financial effects we further find for the 18 + 2-model of joint lessons, for example, that 1,990 additional classes have to be established. In case every class at the primary level needs a full-time teacher, this equals the number of additionally required teachers. If we assume salary costs of 61,000 Euros per teacher and year, this would lead to additional expenses of 10.149 Million Euros per month. Joint lessons require additional pedagogical personnel, regulated between a minimum and maximum of hours per week (Hessisches Kultusministerium 2006). For all students in the 18 + 2-model between about 72,000 and 144,000 additional hours a month could be assumed.

If these lessons would be given teachers or psychologists, the expenses for these additional hours would be between 2.947 and 5.894 Million Euros a month.

Due to the differing number of students with learning disabilities between urban and rural areas, about 60 % of the overall additional costs would incur in the metropolitan areas and only about 7 % in the rural areas of Hesse. If we divide the estimated salary costs by the total number of students with and without learning disabilities – and obviously we should do so if we are concerned with inclusion – the additional expenses are about 60–80 Euros per student and month. Added to the average expenses per student at primary level in Hesse (about 370 Euros, Baumann and Eichstaedt 2011), this would be an increase of between 16 and 21 %. And the expenses still have to be discounted, especially by saved transport costs. Based on the Hessian school law we can assume a reimbursement of 0.35 Euros per kilometer and these savings would amount to about 207,000 Euros per month (cf. Table 3). Hence, transport costs would play a minor role in discounting additional salary costs, though this result will differ markedly if supply models for children with physical or sensory impairment are considered. In addition, the simulations assume suspension of all special schools. Hence, teachers may be relocated as well (which significantly lowers extra salary costs) and unused school buildings may be for rent and may generate additional (public) revenues then. Such a more comprehensive estimation of variable and fixed costs should then serve as the input to solve the transportation-allocation problem in a next step of the analysis.

5 Discussion

Using students with learning disabilities at the primary level in Hesse (Germany) as an example, the paper describes a simulation-based approach to a systematic evaluation of the organizational consequences and financial impact of inclusive schooling. The results indicate that inclusive concepts should be preferred to provide a demand-oriented school system where schools are close to home for all students. This applies to rural regions in particular, where schools are being closed due to the demographic progress and where school enrollment will further decrease. Special schools may still be an option in large cities where schools get fully booked much faster than in thinly populated areas. Therefore, it is important to abandon the assumption of finite sets of schools in the next step of the analysis, i.e. to allow schools to be opened or closed. By now, we have to simulate the spatial distribution of students and we can only give a rough estimation of the costs incurred by inclusive concepts, especially because access to individual administrative data, e.g. on students' residence and teachers' qualification, is restricted. Hence, an ex-ante evaluation of alternative schooling concepts is rather a question of data quality than of statistical techniques for solving allocation problems.

References

Baumann, T., & Eichstaedt, H. (2011). Ausgaben je Schueler/-in in 2008. Wiesbaden: Statistisches Bundesamt (ed).

Beyer, H. L. (2004). Hawth's analysis tools for ArcGIS. Available at http://www.spatialecology.com/htools. Cited 03 Aug 2011.

Domschke, W., Drexl, A., & Mayer, G. (2008). Betriebliche Standortplanung. In A. Arnold et al. (Eds.), *Handbuch logistik* (3rd ed.). Berlin: Springer

European Agency for Development in Special Needs Education. (2011). Complete national overview – Germany. http://www.european-agency.org/country-information/germany/national-overview/complete-national-overview. Cited 31 July 2011.

Hessisches Kultusministerium. (1992). Verordnung ueber die Festlegung der Anzahl und der Groesse der Klassen, Gruppen und Kurse in allen Schulformen. Ordinance of 03 Dec 1992.

Hessisches Kultusministerium. (2006). Verordnung ueber die sonderpaedagogische Foerderung. Ordinance of 17 May 2006.

Nahmias, S. (2009). *Production and operations analysis* (6th ed.). New York: McGraw-Hill

Sekretariat der Staendigen Konferenz der Kultusminister der Laender in der Bundesrepublik Deutschland. (2011). Sonderpaedagogische Foerderung in Schulen. http://www.kmk.org/statistik/schule/statistische-veroeffentlichungen/sonderpaedagogische-foerderung-in-schulen.html. Cited 31 July 2011.

Detecting Person Heterogeneity in a Large-Scale Orthographic Test Using Item Response Models

Christine Hohensinn, Klaus D. Kubinger, and Manuel Reif

Abstract Achievement tests for students are constructed with the aim of measuring a specific competency uniformly for all examinees. This requires students to work on the items in a homogenous way. The dichotomous logistic Rasch model is the model of choice for assessing these assumptions during test construction. However, it is also possible that various subgroups of the population either apply different strategies for solving the items or make specific types of mistakes, or that different items measure different latent traits. These assumptions can be evaluated with extensions of the Rasch model or other Item Response models. In this paper, the test construction of a new large-scale German orthographic test for eighth grade students is presented. In the process of test construction and calibration, a pilot version was administered to 3,227 students in Austria. In the first step of analysis, items yielded a poor model fit to the dichotomous logistic Rasch model. Further analyses found homogenous subgroups in the sample which are characterized by different orthographic error patterns.

1 Introduction

Achievement tests in psychology and education are typically constructed with the aim of measuring a specific competency uniformly for all examinees. This requires

The test was constructed for the Austrian National Educational Standards which is a governmental project of the Austrian Federal Ministry for Education, Arts and Culture. By order of the Ministry the construction for the pilot version of the orthographic test was conducted by the Center of Testing and Consulting at the Faculty of Psychology, University of Vienna.

C. Hohensinn (✉) · K.D. Kubinger · M. Reif
Faculty of Psychology, Department of Psychological Assessment and Applied Psychometrics, University of Vienna, Vienna, Austria
e-mail: christine.hohensinn@univie.ac.at; klaus.kubinger@univie.ac.at; manuel.reif@univie.ac.at

B. Lausen et al. (eds.), *Algorithms from and for Nature and Life*, Studies in Classification, Data Analysis, and Knowledge Organization, DOI 10.1007/978-3-319-00035-0_33,

students to work on the test items in a homogenous way. In psychology these requirements are often modeled by the dichotomous logistic Rasch model (Rasch 1980) (also referred to as 1-PL model) which assumes a particular function between the observed response of an examinee and the latent trait:

$$P(X_{ij} = 1|\beta_i, \theta_j) = \frac{\exp(\theta_j - \beta_i)}{1 + \exp(\theta_j - \beta_i)} \tag{1}$$

With the Rasch model it is assumed that for examinee j the probability of solving item i only depends on the person parameter θ (which is interpreted as the ability of the examinee in this context) and the item parameter β (which is interpreted as the difficulty of the item). Thus applying the model postulates person and item homogeneity. The Rasch model has the advantageous characteristic that the test score is a sufficient statistic for the person parameter (as well as the item score is a sufficient statistic for the item parameter). Therefore conditional maximum likelihood (CML) estimation can be applied (Fischer 1973).

Various model tests and fit indices are available to test the model fit of the Rasch model on a data set. In the present study, Andersen's Likelihood Ratio test was applied for testing the global model fit (Andersen 1973):

$$Z = -2 \log \frac{cL}{\prod_{s=1}^{S} cL_s} \tag{2}$$

The conditional Likelihood (cL) of the sample is compared to the product of the conditional Likelihoods of S subsamples. Because of the sample homogeneity in the case of model fit of the Rasch model the product of the likelihoods of different subsamples must be (approximately) the same as for the whole sample.

The process of test construction begins with the development of an item pool to measure a specific latent trait. Subsequently these items are empiricially evaluated by administering them to a sample of typical test takers. Based on the results of pilot testing individual items with a poor fit can be excluded from the item pool and the resulting item pool can be tested again in a new sample as sort of cross validation (Kubinger and Draxler 2007). Of course, it is possible that not only a few individual items but the whole data set does not conform to the Rasch model. In this case conditions for the misfit of individual items can be identified by applying extended Rasch models or other Item Response models. The characteristic property of all models of Item Response Theory (IRT) is the assumption of a specific relationship between the observed item response and the latent trait(s) which the item measures (for an overview see for instance Embretson and Reise 2000).

The present article deals with the development of a new orthographic test for Austrian students. The goal of the pilot testing was to evaluate whether the items conform to the Rasch model. If items do not conform to the Rasch model, it is an important aspect for further test development to find reasons for the misfit.

1.1 Large-Scale Test for Orthography

The National Educational Standards in Austria are a governmental project for assessing students competencies in Math, German and English. In the process of developing these Educational Standards tests, a test was constructed to evaluate Austrian students' general knowledge of German orthography. This test was developed as a large-scale test, not as test for individual assessment. The orthographic test was developed by a team of teachers of German, experts of German teaching, a linguist and psychologists. The final pilot version of the test consisted of 35 items with each item representing a German word. The examinee's task was to find the spelling error in the word or to identify the word as correct if there was no error. The orthographic test included five categories of orthographic errors: "uppercase for nouns", "sharpening", "stretching", "words descending from foreign language" and "root word" as well as five correctly written words.

2 Data Analysis

The pilot version of the new orthographic test was administered in schools of all districts in Austria in 2008. The sample consisted of 3,227 8th grade students.

2.1 Rasch Model Analysis

Firstly, the model validness of the dichotomous logistic Rasch model was assessed by applying Andersen's Likelihood Ratio test (see Eq. 2) with the partition criteria: test score (score $>$ median versus score $\leq$ median), gender, native language (German versus other) and regional district (West Austria versus East Austria). To ensure some sort of cross validation after deleting non-fitting items, the sample was randomly split into a calibration sample ($n_c = 1{,}614$) and a testing sample($n_t = 1{,}613$). Rasch model analysis were conducted with the R-package eRm (Mair et al. 2011).

With the exception of the split by regional district, all Likelihood Ratio tests turned out to be significant. Thus, the ten poorest-fitting items were excluded stepwise from the data set, though this still left three of the four Likelihood Ratio tests significant (results are shown in Table 1).

Due to these significant results after excluding almost $\frac{1}{3}$ of the 35 items, it must be concluded that the Rasch model does not fit the data set. Therefore, further Item Response models were applied to explore reasons for the misfit.

Table 1 Andersen's Likelihood Ratio test for all items and after excluding the ten poorest-fitting items

Calibration sample all items			After excluding ten items	
Partition criterion	χ^2_{LRT}	$\chi^2_{34}(0.99)$	χ^2_{LRT}	$\chi^2_{24}(0.99)$
Score	535.58	56.06	111.12	42.98
Gender	135.78	56.06	58.85	42.98
Native language	133.13	56.06	74.79	42.98
Regional district	47.81	56.06	29.64	42.98

Table 2 Model fit for the mixed Rasch model with different number of latent classes

Number of classes	logL	$n_{parameters}$	BIC
1	−58703.89	69	117965.25
2	−56537.92	137	114182.70
3	−56099.04	205	113854.33
4	−55799.81	273	113805.27
5	−55709.71	341	114174.46

2.2 *Mixed Rasch Model Analysis*

As pointed out in Sect. 1, the Rasch model assumes person and item homogeneity. Thus a misfit could occur because of different underlying subgroups in the population. This happens, if there are groups of students who have difficulties with specific types of orthographic errors. This would imply different relative item difficulties between spelling error types for different students.

In general, different underlying subpopulations are modeled by finite mixture models:

$$P(X_{ij}) = \sum_{g=1}^{K} \pi_g \mathrm{p}(\mathrm{X_{ij}}|\pi_\mathrm{g}) \qquad \text{with} \sum_{g=1}^{K} \pi_g = 1 \tag{3}$$

with π_g denoting the proportional size of class g. Plugging in Eq. 1 for the probability function leads to the mixed Rasch model proposed by Rost (1990). As for the dichotomous logistic Rasch model CML estimation is also possible for the mixed Rasch model. The mixed Rasch model was calculated with the software Winmira (von Davier 2001) using CML parameter estimation. The BIC was used to compare models since it is more reliable than the AIC for this kind of model (Preinerstorfer and Formann 2011). The results displayed in Table 2 show, that a mixed Rasch model with four classes has the lowest BIC.

It already was hypothesized that there are groups (latent classes) of students who have difficulties with specific error types. If this is true, it is important to determine whether the model offers a reasonable and consistent interpretation, i.e. to evaluate whether specific "error profiles" can be found. These would occur if particular orthographic error types are consistently more (proportionally) difficult

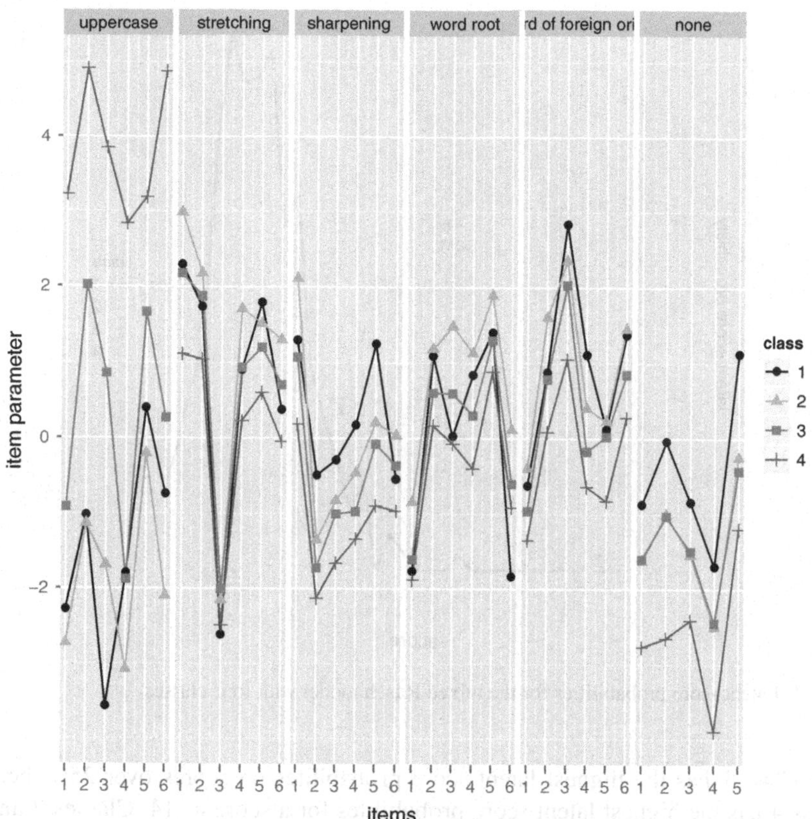

Fig. 1 Item difficulties sorted according to error type for each latent class of the four class mixed Rasch model

for a specific latent class. Figure 1 shows the item parameters for each latent group $\beta_{i|g}$. Because the item parameters in each latent group are constrained to $\sum_{i=1}^{I} \beta_{i|g} = 0$, the comparison of item difficulty between the latent groups is only of relative size. The diagram reveals error-specific profiles: Class 4 has (proportionally to the other items) the most difficulties solving the uppercase errors; compared to the other classes, Class 1 has problems identifying the correct words as correct and has some difficulties with the sharpening words. For Class 2, items with the error type "word root" are proportionally more difficult (in comparison to the other groups) whereas Class 3 has also some problems with the uppercase words (though not as distinct as Class 4). The mixed Rasch model with four classes has the lowest BIC; moreover, it allows for a reasonable interpretation. Each latent class seems to have different (relative) difficulties with various error-types. Thus, the four-class solution seems an appropriate model.

To find out whether there are differences in the overall skill level of the four latent classes, the latent score probabilites for each class were plotted (Fig. 2). This shows

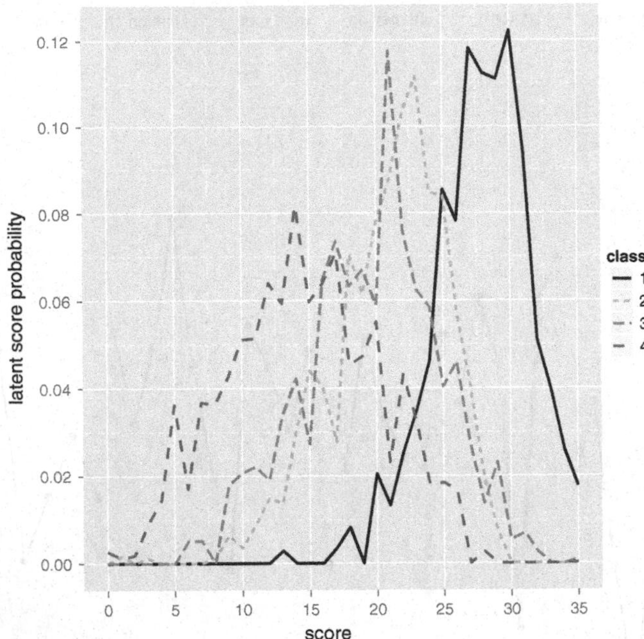

Fig. 2 Latent score probabilities for the mixed Rasch model with four classes

that Class 1 has the highest latent score probabilites on scores over 25, whereas Class 4 has the highest latent score probabilites for a score of 14. Classes 2 and 3 have very similar latent score probabilities, with score probability distributions lying between Classes 1 and 4.

2.3 Multidimensional Item Response Model

The four-classes mixed Rasch model offers a very reasonable interpretation of the test results. Nevertheless model fit was evaluated comparing only models that assumed person heterogeneity. Instead, it is also possible that items with different error types measure different orthographic skills uniformly for all examinees. This hypothesis can be tested using a multidimensional Item Response model (Adams et al. 1997). For dichotomous data this model estimates a person parameter θ_{js} for each assumed skill s. The assignment of items to the different skill dimensions s must be fixed a priori by matrix A. d_i denotes the difficulty of item i:

$$P(X_{ij} = 1|\mathbf{a_i}, d_i, \boldsymbol{\theta}_\mathbf{j}) = \frac{\exp(\mathbf{a_i}\boldsymbol{\theta}'_\mathbf{j} + d_i)}{1 + \exp(\mathbf{a_i}\boldsymbol{\theta}'_\mathbf{j} + d_i)} \tag{4}$$

Table 3 Multidimensional IRT models compared to the four-class mixed Rasch model

Model specification	logL	$n_{parameters}$	BIC
Four class mixed Rasch model	−55856.00	156	112,972
Two skills: uppercase, other	−58028.54	41	116,388
Three skills: uppercase, none, other	−56308.80	46	112,989
Four skills: uppercase, none, sharpen, other	−56315.09	51	113,042
Five skills: uppercase, none, sharpen, foreign, other	−56409.42	57	113,279

In order to reduce the number of estimated models, first those items that showed the largest differences between latent groups in the mixed Rasch model were modeled as measuring separate dimensions (uppercase error words, then correct words, etc.). The multidimensional models were estimated by using the software mdltm (von Davier, 2005) with a marginal maximum likelihood (MML) estimation. For the purpose of comparing the multidimensional IRT models to the mixture models, the four-class mixed Rasch model was re-estimated by means of MML parameter estimation (this is why the number of parameters and the logLikelihood of the four-class mixed Rasch model are now different to the results of the CML estimation in Table 2). Results in Table 3 reveal that the four-class mixed Rasch model has a lower BIC than the multidimensional IRT models. The multidimensional model with three skill dimensions (uppercase error, correct words, remaining words) has a BIC close to that of the mixed Rasch model; nevertheless, modeling person heterogeneity with four latent classes turns out to fit the data better.

3 Summary and Discussion

A newly developed large-scale test for measuring the orthographic competency of Austrian students was evaluated using Item Response models in a first pilot testing. Because the Rasch model was not valid for the given data, the ability of person or item heterogeneity to explain this lack of model fit was examined. Results show that a four-class mixed Rasch model had the best model fit. Among the multidimensional Item Response models, the three-skill-dimension model has a BIC relatively close to that of the mixed Rasch model. The difference in BIC of these two models is small and therefore raises the question of whether the mixed Rasch model is really the more appropriate one. Because information criteria do not offer a "critical" level of difference, the question of whether the difference in BIC is significant cannot be conclusively answered. However, for the present data the four-class mixed model has the highest goodness-of-fit value and therefore seems the most appropiate choice. Of course, if the goal of the study were to establish a generally valid model of the structure of orthographic skills, the result would have to be replicated in a different sample. For the present purpose of improving the test during its development, the mixed Rasch model seems helpful and provides a reasonable result with various groups of students having difficulties with different orthographic error types. Inspecting the different item parameters in each latent group of the four-class mixed Rasch model reveals that the most distinct differences between groups concern the uppercase words and the correct words. Furthermore, the three-dimensional model in which solving uppercase error and correct word items constituted separate skill dimensions, was the best fitting model among the multidimensional models. However, it seems that these two kinds of items induced item responses that contradict the assumptions of the Rasch model. According to the results of the mixed Rasch model the group of students with higher latent score probabilities and therefore higher overall orthographic competencies seems to have more difficulties identifying the correct words. A possible reason for this somewhat unexpected result could be that these students did not thoroughly read the test instruction which mentioned the possibility that words might be correct. Alternatively, these students might not have "trusted" the instructions and instead been confident that every test item would have an error. In contrast, another latent class of students had more problems than other examinees finding uppercase errors. As a whole, the IRT analyses reveal issues which must be considered in order to construct a test that is fair for all students: the inclusion of correct words and words with an uppercase error must be discussed and/or the test instruction needs to be improved.

References

Adams, R. J., Wilson, M., & Wang, W. C. (1997). The multidimensional random coefficients multinomial logit model. *Applied Psychological Measurement, 21*, 1–23.

Andersen, E. B. (1973). A goodness of fit test for the Rasch model. *Psychometrika, 38*, 123–140.

Embretson, S., & Reise, S. P. (2000). *Item response theory for psychologists*. Mahwah: Erlbaum.

Fischer, G. H. (1973). *Einführung in die Theorie psychologischer Tests* [Introduction to psychological test theory]. Bern: Huber.

Kubinger, K. D., & Draxler, C. (2007). Probleme bei der Testkonstruktion nach dem Rasch-modell [Difficulties in test development applying the Rasch model]. *Diagnostica, 53*, 131–143.

Mair, P., Hatzinger, R., Maier, M., & Gilbey, J. (2011). eRm: extended Rasch models. R package version 0.13–4. http://cran.r-project.org/web/packages/eRm.

Preinerstorfer, D., & Formann, A. K. (2011). Parameter recovery and model selection in mixed Rasch models. *British Journal of Mathematical and Statistical Psychology, 65*, 251–262.

Rasch, G. (1980). Probabilistic models for some intelligence and attainment test (Reprint from 1960). Chicago: University of Chicago Press.

Rost, J. (1990). Rasch models in latent classes: An integration of two approaches to item analysis. *Applied Psychological Measurement, 14*, 271–282.

von Davier, M. (2001). WINMIRA 2001 – A windows-program for analyses with the Rasch model, with the latent class analysis and with the mixed Rasch model [Computer software], ASC-Assessment Systems Corporation USA and Science Plus Group, Groningen.

von Davier, M. (2005). mdltm: Software for the general diagnostic model and for estimating mixtures of multidimensional discrete latent traits models [Computer software]. Princeton: ETS.

Linear Logistic Models with Relaxed Assumptions in R

Thomas Rusch, Marco J. Maier, and Reinhold Hatzinger

Abstract Linear logistic models with relaxed assumptions (LLRA) are a flexible tool for item-based measurement of change or multidimensional Rasch models. Their key features are to allow for multidimensional items and mutual dependencies of items as well as imposing no assumptions on the distribution of the latent trait in the population. Inference for such models becomes possible within a framework of conditional maximum likelihood estimation. In this paper we introduce and illustrate new functionality from the R package **eRm** for fitting, comparing and plotting of LLRA models for dichotomous and polytomous responses with any number of time points, treatment groups and categorical covariates.

1 Introduction

Linear logistic models with relaxed assumptions (LLRA; see Fischer 1993; Fischer and Ponocny 1993) can be thought of as generalised Rasch models with multidimensional latent trait parameters where change is modelled as a function of treatment (main) effects, treatment interactions and trend effects. Relaxed assumptions mean that neither unidimensionality of the items needs to be assumed nor are there any distributional assumptions made about the population of subjects. Conditional maximum likelihood estimation (CML) allows for the separation of treatment effect parameters and nuisance trait parameters. Consequently, given the prerequisites for LLRA hold, results about the effect parameters are completely independent of the trait parameters in the sample of subjects.

The LLRA has some very useful properties for the measurement of change, such as the ratio scale properties of the estimated parameters, $\hat{\eta}$ (Fischer 1993). It is

T. Rusch (✉) · M.J. Maier · R. Hatzinger
Institute for Statistics and Mathematics, WU Vienna University of Economics and Business, Augasse 2–6, 1090 Vienna, Austria
e-mail: thomas.rusch@wu.ac.at; marco.maier@wu.ac.at

B. Lausen et al. (eds.), *Algorithms from and for Nature and Life*, Studies in Classification, Data Analysis, and Knowledge Organization, DOI 10.1007/978-3-319-00035-0_34,

therefore possible to assess the relative effectiveness of treatments (e.g., a treatment might be twice as effective). Furthermore, independence of effect parameters from trait parameters allows for generalisability beyond the current experimental situation, which is desirable if treatments are to be compared in a competitive way.

2 Linear Logistic Models with Relaxed Assumptions

Whereas the LLRA originally referred to measurement of change with dichotomous items, our definition also includes models for measuring change with polytomous items with possibly different numbers of categories per item.

More specifically, let θ_{vit} denote the location of subject v for item i at time point T_t, let h refer to the h-th response category ($h = 0, \ldots, m_i$) and let ω_{ih} stand for the parameter for category h for item i. The logistic model at the baseline T_1 is

$$P(X_{vih1} = 1|T_1) = \frac{\exp(h\theta_{vi1} + \omega_{ih})}{\sum_{l=0}^{m_i} \exp(l\theta_{vi1} + \omega_{il})} \tag{1}$$

At any subsequent measurement point T_t ($t \in \{2, 3, \ldots\}$) the logistic model is

$$P(X_{viht} = 1|T_t) = \frac{\exp(h\theta_{vit} + \omega_{ih})}{\sum_{l=0}^{m_i} \exp(l\theta_{vit} + \omega_{il})} = \frac{\exp(h(\theta_{vi1} + \delta_{vit}) + \omega_{ih})}{\sum_{l=0}^{m_i} \exp(l(\theta_{vi1} + \delta_{vit}) + \omega_{il})} \tag{2}$$

with $\delta_{vit} = \theta_{vit} - \theta_{vi1}$ denoting the amount of change of person v for trait i between time T_1 and T_t. This model is the most general one in terms of dimensions as each item is seen as measuring a single latent trait. In the following, we will assume all items to measure mutually exclusive traits, which can be simplified by specifying groups of items to measure the same trait if desired.

The flexiblity of LLRA models arises from a (linear) reparameterisation of δ_{vit} to include different effects:

$$\delta_{vit} = \mathbf{w}_{it}^T \eta \tag{3}$$

Here, $\mathbf{w}_{it}^T$ denotes a row of design matrix W for item/trait i up to T_t. The parameterisation of η can be written as

$$\delta_{vit} = \sum_j q_{vjit}\lambda_{jit} + \tau_{it} + \sum_{j<l} q_{vjit} q_{vlit} \rho_{jlit} \tag{4}$$

where q_{vjit} stands for a dosage or indicator of a covariate or a factor level j for trait i between T_1 and T_t, λ_{jit} denoting a main effect of the covariate/factor level j on trait i at T_t, τ_{it} is the parameter for the trend effect on trait i between T_1 and T_t, and ρ_{jlit} are the parameters for interaction effects of treatments j and l on trait i at T_t. This multidimensional formulation allows for any restriction concerning effects such as generalisations of effects over different traits or groups.

3 Implementation

The LLRA functionality as described here is available in version 0.14–4 or higher[1] of the **eRm** package (Mair and Hatzinger 2007) for R (R Core Development Team 2011) version 2.12.0 and higher. The functions described in this paper automatise the approach laid out in Hatzinger and Rusch (2009) as well as allow plotting of effects with **lattice** (Sarkar 2008). The user interface for LLRA is modelled after other **eRm** functions. The following functions have been devised for LLRA models:

`LLRA` This is the main function. It automatically fits a quasi-saturated LLRA model. Data structure, design matrix and group assignment are set up by the function and model fitting with `LPCM` is then carried out. It returns an object of class `llra`.

`print` Standard S3 print method for objects of class `llra`.

`collapse_W` A convenience function for collapsing columns of a design matrix to simplify a LLRA model or to generalise effects over time or items or groups or any combination of them.

`summary` Standard S3 summary method for objects of class `llra`. It displays more details of the results of the model fit. This function can also be used to extract confidence intervals.

`anova` S3 anova function for class `llra`. Conducts a likelihood ratio test for nested LLRA models.

`plotGR` **lattice** plots of group or covariate effects over time for all items.

`plotTR` **lattice** plot of trend effects over time for all items.

Some of these functions call other functions (e.g. for building the design matrix) internally, but for most users these should not be of interest. Please note that currently only categorical covariates can be passed to `LLRA` via the `groups` argument.[2] Continuous covariates currently either need to be discretised first or have to be set up manually with a specific design matrix.

4 Illustration

We use an artifical data set from (Hatzinger and Rusch 2009) to illustrate the usage of LLRA functionality in **eRm**. After the package is installed and made available with `library("eRm")`, the data can be accessed in R via `data(llraDat2)`. The data consist of responses of 70 subjects to 5 items. Item 1 is dichotomous, all others are polytomous with 3, 4, 5, and 6 categories respectively. The subjects belong to 3 groups, a control group (CG) of size 40, and 2 treatment groups (TG1 of size 20 and TG2 of size 10). Each item was presented to each subject at 4 different

[1]The most recent version can be obtained from http://r-forge.r-project.org/projects/erm/.

[2]We plan to support continuous covariates in a future version.

times. To fit LLRA, the data need to be in "wide" format for repeated measurements as displayed in Table 1.

In our example, the first 20 columns are the responses and the last column encodes which group a subject belongs to. We saved them in new objects `dats` and `groups` for simplicity.

Fitting the LLRA model is straightforward. The function `LLRA` takes as its first argument the data frame of responses, followed by the number of measurement points and the group membership as well as the reference group (if no baseline group is supplied, by default the group with the lowest alpha-numerical score for the group name will be used):

```
R> llra1 <- LLRA(dats, mpoints = 4, groups = groups,
+                baseline = "CG")
```

By default, the `LLRA` function always fits a quasi-saturated LLRA unless a design matrix is passed as an argument. With polytomous items the function will print a warning message that the first two category parameters are equated for each item. Doing this is motivated pragmatically to save parameters and is standard in the usage of LLRA. However, this decision can also be justified with a theoretical argument (Hatzinger and Rusch 2009). The function `LLRA` returns an object of class `llra` for which some standard S3 functions have been implemented. The `summary` function displays estimated parameters and fit information (only output for the first two items at time point 2 is displayed below):

```
R> summary(llra1)

Results of LLRA via LPCM estimation:

Call: LLRA(X=dats, mpoints=4, groups=groups, baseline="CG")

Conditional log-likelihood: -1143.422
Number of iterations: 69
Number of parameters: 55

Estimated parameters with 0.95 CI:
             Estimate Std.Error lower.CI upper.CI
TG2.I1.t2       0.467     0.832   -1.163    2.097
TG1.I1.t2      -0.658     0.613   -1.859    0.544
TG2.I2.t2      -0.024     0.450   -0.905    0.857
TG1.I2.t2      -0.321     0.358   -1.022    0.380
...
trend.I1.t2     1.262     0.355    0.565    1.959
trend.I2.t2     0.621     0.234    0.162    1.079
...
c2.I2           0.669     0.244    0.191    1.148
...
Reference Group:  CG
```

Table 1 Wide data format for repeated measurements

Real persons	T_1				T_2			
S_1	x_{111}	x_{121}	...	x_{1k1}	x_{112}	x_{122}	...	x_{1k2}
S_2	x_{211}	x_{221}	...	x_{2k1}	x_{212}	x_{222}	...	x_{1k2}
$\vdots$	$\vdots$				$\vdots$			
S_n	x_{n11}	x_{n21}	...	x_{nk1}	x_{n12}	x_{n22}	...	x_{nk2}

For our example, we see that compared to the reference group CG, item 1 gets easier for TG2 at time point 2 whereas it becomes more difficult for TG1. However both changes are not significantly different from zero at a 5 % significance level, as the 95 %-confidence interval (CI) indicates. For item 2 both groups display a negative non-significant change. The trend effects for both items between time points 1 and 2, however, are significant ($\alpha = 0.05$). This means we reject the hypothesis that there is no general change for all groups, i.e., in this case the items get easier over time. We see that the trend effect for item 1 is twice the effect for item 2. Hence, both items get easier over time but item 1 changes twice as much. Additionally the estimated category parameter for the third category of item 2 is listed.

The `summary` function can be used to extract point estimates, standard errors or CI for the parameters. For example, the 99 %-CI can be extracted like this

```
R> cis <- summary(llra1, gamma = 0.99)$ci
R> cis[1:3, ]

                0.5 %    99.5 %
TG2.I1.t2 -1.675638 2.6092456
TG1.I1.t2 -2.236630 0.9212073
TG2.I2.t2 -1.181980 1.1342965
```

The relative trend and group or covariate effects for each item from the quasi-saturated model can be displayed with the **lattice** plot functions `plotGR` for covariate or group effects and `plotTR` for trend effects. This is convenient since often for LLRA models, a large number of parameters is estimated which produces long output when using the `summary` or `print` functions. The plot functions help to identify positive or negative changes over time and may provide hints for possible model simplifications. For our example, the group effect plots can be found in Fig. 1 and the trend effects plot in Fig. 2.

For the quasi-saturated LLRA, we need to estimate 55 parameters, which is a lot. However, we can try to simplify the model and test hypotheses by generalising effects. To do this, the function `collapse_W` allows to collapse specific columns of the design matrix. For example, the results as displayed in Fig. 2 indicate a linear trend effect for item 2. We might therefore substitute separate estimates for each time point by a single linear trend which will save us three parameters. To that end, the according columns of the design matrix of the quasi-saturated LLRA that need

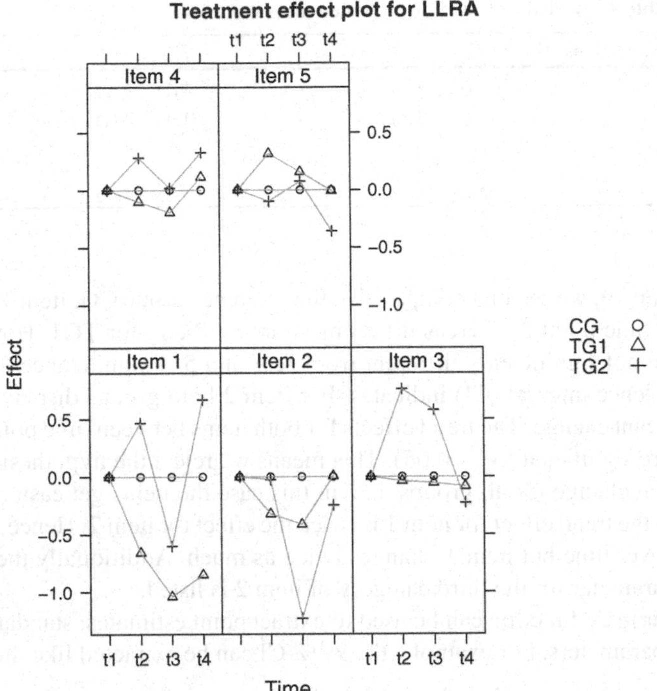

Fig. 1 Lattice plot of the group effects for all items. At the reference measurement at T_1 all effects are zero. For subsequent time points the effects are displayed relative to T_1 and the baseline group (constant change of zero)

to be collapsed are columns 32, 37, and 42. The function `collapse_W` allows to do that conveniently. It requires the design matrix and a list of columns to be collapsed. The quasi-saturated LLRA design matrix can be extracted via `$W` from the object returned by `LLRA`, here `llra1`. The list of columns to be collapsed, e.g. for time points 2, 3, and 4 for item 2, can be specified such

```
R> collItems1 <- list(c(32, 37, 42))
```

Then the collapsed design matrix `Wstar1` can be obtained by

```
R> Wstar1 <- collapse_W(llra1$W, collItems1)
```

and can be passed as an argument to `LLRA` to fit the LLRA with a linear trend for item 2:

```
R> llra2 <- LLRA(dats, W = Wstar1, mpoints = 4,
+                groups = groups)
```

Since collapsed models are all nested within the quasi-saturated model, we can use a likelihood ratio test with `anova` to find out if the simplification is admissable.

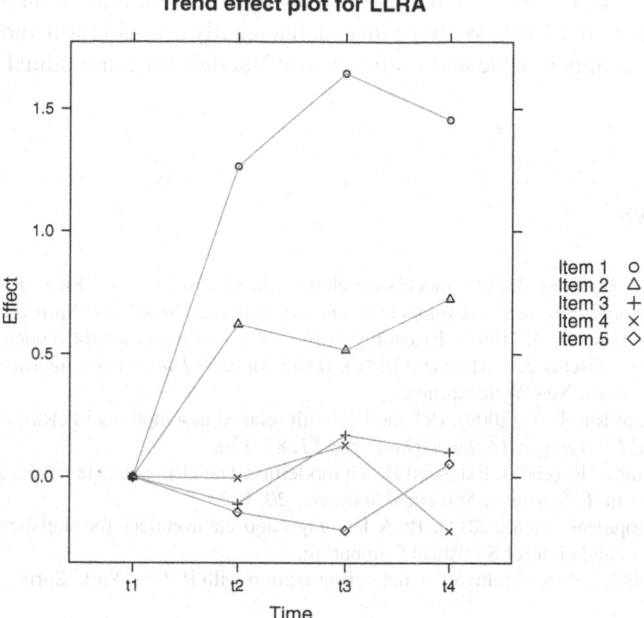

Fig. 2 Lattice plot of the trend effects for all items. Trend effects are assumed to be the same for all groups. At the reference measurement at T_1 all effects are zero. For subsequent time points the effects are displayed relative to T_1

```
R> anova(llra1, llra2)

Analysis of Deviance Table

  Npar  logLik df     -2LR Asymp.p-Value
1    53 -1143.7
2    55 -1143.4  2 0.62478        0.7317
```

It turns out that item 2 can be seen as displaying a linear trend without a significant loss of information.

5 Conclusion

We presented and illustrated functionality to fit linear logistic models with relaxed assumptions with the **eRm** package for R. To the best of our knowledge, it is the only ready-made implementation to fit LLRA. In principle, all software packages that can fit Linear Partial Credit Models with CML can be used to fit LLRA models as well, but that usually comes along with tedious restructuring of the data and setting up of complicated design matrices. The presented functionality tries to alleviate that.

We aimed at user-friendliness to provide a low threshold for practitioners with basic R knowledge to fit LLRA. We hope that such a readily available software will spark new interest in this flexible and useful class of models for longitudinal categorical data.

References

Fischer, G. (1993). Linear logistic models for change. In G. Fischer & I. Molenaar (Eds.), *Rasch models: Foundations, recent developments and applications*. New York: Springer.

Fischer, G., & Ponocny, I. (1993). Extending rating scale and partial credit model for assessing change. In G. Fischer & I. Molenaar (Eds.), *Rasch models: Foundations, recent developments and applications*. New York: Springer.

Hatzinger, R., & Rusch, T. (2009). IRT models with relaxed assumptions in eRm: A manual-like instruction. *Psychological Science Quarterly, 51*, 87–120.

Mair, P., Hatzinger, R. (2007). Extended Rasch modeling: The eRm package for the application of IRT models in R. *Journal of Statistical Software, 20*, 1–20

R Core Development Team. (2011). R: A language and environment for statistical computing. Vienna: R Foundation for Statistical Computing.

Sarkar, D. (2008). Lattice: Multivariate data visualization with R. New York: Springer.

Part VI
Text Mining, Social Networks and Clustering

An Approach for Topic Trend Detection

Wolfgang Gaul and Dominique Vincent

Abstract The detection of topic trends is an important issue in textual data mining. For this task textual documents collected over a certain time period are analysed by grouping them into homogeneous time window dependent clusters. We use a vector space model and a straight-forward vector cosine measure to evaluate document-document similarities in a time window and discuss how cluster-cluster similarities between subsequent windows can help to detect alterations of topic trends over time. Our method is demonstrated by using an empirical data set of about 250 pre-classified time-stamped documents. Results allow to assess which method specific parameters are valuable for further research.

1 Introduction

In order to detect emerging topics (see, e.g., Allan et al. 1998, Kontostathis et al. 2004, and Kumaran et al. 2004) or to monitor existing topics it is of interest to analyse document streams (see, e.g., Wang et al. 2007 and Wang et al. 2009) which are emitted, e.g., by news sites like spiegel.de, zeit.de, or nytimes.com. For topic detection in text mining it is common to use document clustering (see, e.g., Allan et al. 1998, Larsen and Aone 1999, and Manning et al. 2009). Let us remind SMART, the System for the Mechanical Analysis and Retrieval of Text (see, e.g., Salton 1989) as an early example for the analysis and retrieval of information by computers.

In the next Sect. 2 notation and some background information will be provided. Section 3 describes the suggested approach while in Sect. 4 an example is presented

W. Gaul (✉) · D. Vincent
Institute of Decision Theory and Management Science, Karlsruhe Institute of Technology (KIT), Kaiserstr. 12, 76128 Karlsruhe, Germany
e-mail: wolfgang.gaul@kit.edu; dominique.vincent@kit.edu

B. Lausen et al. (eds.), *Algorithms from and for Nature and Life*, Studies in Classification, Data Analysis, and Knowledge Organization, DOI 10.1007/978-3-319-00035-0_35,

to demonstrate what can be expected from our topic trend detection technique. Section 5 contains concluding remarks.

2 Notation and Background Information

We need a dictionary which is created using a corpus (see, e.g., Allan et al. 2002) composed of a set of documents $d \in D$. For every term w in the dictionary this corpus is used to compute the term frequencies tf_w as well as inverse document frequencies idf_w given by

$$idf_w = \log \frac{|D|}{|\{d \,:\, w \in d\,\}|}$$

where $|M|$ denotes the cardinality of a set M.

The well-known vector space model (see, e.g., Salton et al. 1975) is applied for representing the text documents that we want to analyse. Vector components in the vector space are used to reflect the importance of corresponding terms from the dictionary. The dimension z of the vector space is crucial (the smaller the dimension of z can be chosen the faster the computation).

One of the best known weighting schemes is tf-idf weighting (see, e.g., Salton and Buckley 1988 and Allan et al. 2000) which we also examined – among others – for the underlying situation.

Finally, we applied the cosine measure as (dis)similarity between documents (document – document similarities) as well as between clusters of documents (cluster-cluster similarities). If C_k^t respectively C_l^{t+1} denote clusters of documents at subsequent time windows t and $t+1$, c_k^t the centroid of C_k^t with $c_{k_w}^t$ as vector component for term w of centroid c_k^t (where a centroid is just the average vector of the documents associated with the corresponding cluster) we have

$$cos(c_k^t, c_l^{t+1}) = \frac{\sum_{w=1}^{z} c_{k_w}^t * c_{l_w}^{t+1}}{\sqrt{\sum_{w=1}^{z} (c_{k_w}^t)^2} * \sqrt{\sum_{w=1}^{z} (c_{l_w}^{t+1})^2}}$$

for the cluster-cluster measure.

With M_t as set of documents in time window t we use the cosine similarity of documents to compute a $|M_t| \times |M_t|$ matrix of dissimilarities $dis^t(i, j)$, $i, j \in M_t$, between all documents of time window t from which we get a clustering $\mathcal{K}^t = \{C_1^t, \ldots, C_k^t, \ldots, C_{|\mathcal{K}^t|}^t\}$ by application of a hierarchical cluster analysis procedure together with the number of classes $|\mathcal{K}_t|$ (which is one of the reasons to use hierachical clustering).

With the clusterings $\mathcal{K}^t$ and $\mathcal{K}^{t+1}$ from two subsequent time windows we are able to compute the dissimilarities between the corresponding sets of clusters.

Table 1 Dissimilarity matrix w.r.t. $\mathscr{K}^t$ and $\mathscr{K}^{t+1}$

	C_1^{t+1}	...	C_l^{t+1}	...	$C_{K_{t,t+1}}^{t+1}$
C_1^t					
$\vdots$					
C_k^t			$dis^{t,t+1}(C_k^t, C_l^{t+1})$		
$\vdots$					
$C_{K_{t,t+1}}^t$					

Table 2 Dissimilarity matrix with missing values, a vanishing cluster C_k^t, and a newly arising cluster C_l^{t+1}

	C_1^{t+1}	...	C_3^{t+1}	...	...	C_l^{t+1}	...	C_5^{t+1}	...	$C_{K_{t,t+1}}^{t+1}$
C_1^t						$\min\limits_{C \in \mathscr{K}^t} \{dis^{t,t+1}(C, C_l^{t+1})\} >$ threshold				
C_2^t										
$\vdots$										
C_k^t			$\min\limits_{C \in \mathscr{K}^{t+1}} \{dis^{t,t+1}(C_k^t, C)\}$					$>$ threshold		
$\vdots$										
$C_{\lvert\mathscr{K}^t\rvert}$										
$\vdots$					Missing values					
$C_{K_{t,t+1}}^t$										

The matrix of dissimilarities $dis^{t,t+1}(C_k^t, C_l^{t+1})$ (determined with the help of $cos(c_k^t, c_l^{t+1})$) has size $K_{t,t+1}$ which just is the maximum of $|\mathscr{K}^t|$ and $|\mathscr{K}^{t+1}|$ (cf. Table 1).

Assume that $|\mathscr{K}^t|$ is less than $|\mathscr{K}^{t+1}|$. In this case the rows of the dissimilarity matrix from $C_{|\mathscr{K}^t|+1}^t$ to $C_{K_{t,t+1}}^t$ have missing values (Likewise, if $|\mathscr{K}^t|$ is greater than $|\mathscr{K}^{t+1}|$ the columns from $C_{|\mathscr{K}^{t+1}|+1}^{t+1}$ to $C_{K_{t,t+1}}^{t+1}$ have missing values.) which indicates that the number of clusters from different time windows don't need to coincide.

In case a value $dis^{t,t+1}(C_k^t, C_l^{t+1})$ of a pair of clusters C_k^t and C_l^{t+1} in the dissimilarity matrix is "small" cluster C_k^t corresponds to cluster C_l^{t+1}, i.e., we assume that cluster C_k^t at time window t can be assigned to cluster C_l^{t+1}.

Additionally, it can happen that the minimum of the dissimilarities of cluster C_k^t to all clusters of $\mathscr{K}^{t+1}$ is greater than a predefined threshold from which one can conclude that cluster C_k^t is not similar to any of the clusters of time window

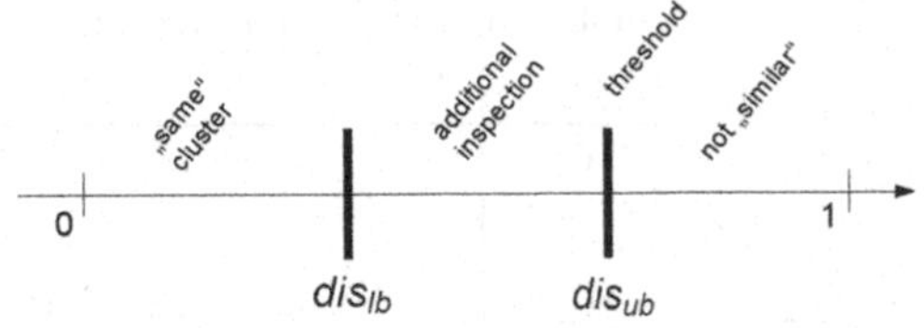

Fig. 1 *Lower* and *upper* bounds for dissimilarity checks

$t + 1$, i.e., topic C_k^t has vanished (death in t) and is no longer in clustering $\mathscr{K}^{t+1}$. Another case appears if all dissimilarities in the column of C_l^{t+1} are greater than the threshold, i.e., C_l^{t+1} is a newly arising topic which was not in clustering $\mathscr{K}^t$ (birth or reappearance in $t + 1$). These possibilities are depicted in Table 2.

Figure 1 tries to describe the underlying situation. When cluster-cluster dissimilarities between two clustes C_k^t and C_l^{t+1} of subsequent time windows are smaller than a problem-specific lower bound dis_{lb} it is assumed that the documents of C_k^t and C_l^{t+1} belong to the "same" cluster. However, if a problem-specific threshold dis_{ub} as an upper bound is exceeded by all cluster-cluster dissimilarities in a row (or column) of the matrix a trend has vanished (a new trend is born). In the area between dis_{lb} and dis_{ub} an additional inspection is necessary.

3 Approach

Given the explanations of the last section the following approach to support topic trend detection is suggested:

- Collect the set M_t of documents in time window t.
- Compute the $|M_t| \times |M_t|$ matrix of document-document dissimilarities $dis^t(i, j)$, $i, j \in M_t$.
- Perform hierarchical clustering to get $\mathscr{K}^t = \{C_1^t, \ldots, C_k^t, \ldots, C_{|\mathscr{K}^t|}^t\}$.
- With $K_{t,t+1} = max\{|\mathscr{K}^t|, |\mathscr{K}^{t+1}|\}$ compute the $K_{t,t+1} \times K_{t,t+1}$ matrix of cluster-cluster dissimilarities $dis^{t,t+1}(C_k^t, C_l^{t+1})$.
- Choose problem-specific dissimilarity-bounds and check for the birth (or reappearance) of topics, the death of no longer interesting topics, or the continuation of trends. In case that different lower and upper dissimilarity-bounds have to be considered additional inspection is needed to classify critical cases for which dissimilarities are situated within the bounds.

4 Example

Our test data set is a sample drawn from a set of time-stamped documents (see, e.g., Kupietz and Keibel 2009 and Kupietz et al. 2010) of the Institut für Deutsche Sprache IDS, located in Mannheim. The test documents are from newspapers

Table 3 Test configuration

$\|M_t\|$	52	65	50	35	52
time window t	$t = 1$	$t = 2$	$t = 3$	$t = 4$	$t = 5$
Cluster 1	C_1^1	C_1^2	C_1^3	C_1^4	C_1^5
Cluster 2	C_2^1	C_2^2	C_2^3	C_2^4	C_2^5
Cluster 3	C_3^1	C_3^2	C_3^3	–	C_3^5
Cluster 4	–	C_4^2		–	–

categorized by IDS into the four topics politics (P), sport (S), technique, industry, and transportation (TIT), and economy and finance (EF) which could be assigned to five time windows.

We used a dictionary with about 2 million terms and restricted our test runs to the 200, 2000, respectively 20,000 most frequent terms of that dictionary as dimension z of the vector space.

The test configuration of 254 documents is shown in Table 3. At time window $t = 1$ we had a subsample of 52 documents which could be assigned to three of the IDS topics. At time windows $t = 2$ and $t = 3$ the subsamples of 65 and 50 documents were from four respectively three topics. At time windows $t = 4$ and $t = 5$ two topics respectively three topics could be assigned. The next section will reveal which topics are hidden behind the general C_k^t-notation of Table 3.

5 Results

As writing restrictions do not allow to describe all results of the example we just explain the activities in the time windows $t = 1$ and $t = 2$ as well as the transitions between the time windows $1 \rightarrow 2$ and $3 \rightarrow 4$. We conclude with an overall view on topic trend detection situations.

5.1 Transition Between Time Windows $1 \rightarrow 2$

The Fig. 2(a), (b) show the dendrograms at time windows $t = 1$ and $t = 2$.

Three clusters at $t = 1$ and four clusters at $t = 2$ are marked by circles as interesting topics. In Table 4 the dissimilarity matrix between the clusterings $\mathcal{K}^1$ and $\mathcal{K}^2$ is shown. The marked cells with lowest dissimilarity values in the matrix indicate which document clusters are most similar to each other ($C_2^1 \leftrightarrow C_4^2$, $C_3^1 \leftrightarrow C_1^2$, $C_1^1 \leftrightarrow C_2^2$ although C_4^2 and C_1^1 have also a low dissimilarity). All values in the column of C_3^2 are "large", i.e., C_3^2 is a newly arising cluster, and row 4 has missing values.

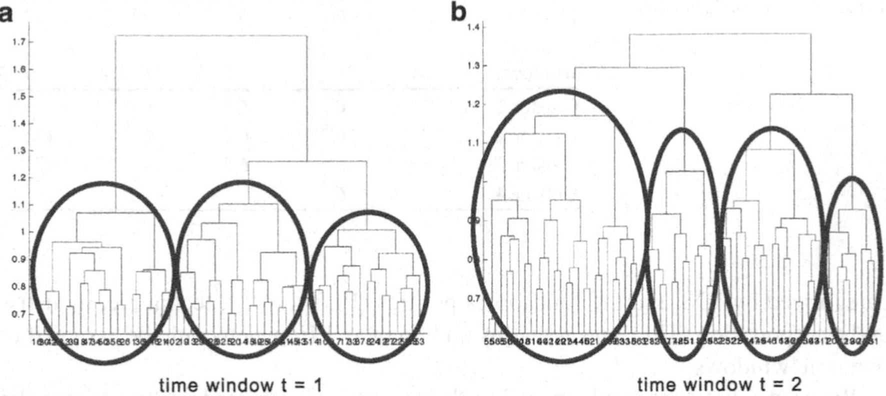

Fig. 2 Dendrograms. (**a**) Time window $t = 1$. (**b**) Time window $t = 2$

Table 4 Dissimilarity matrix w.r.t. $\mathscr{K}^1$ and $\mathscr{K}^2$

	C_1^2	C_2^2	C_3^2	C_4^2
C_1^1	0.4713	0.2807	0.3961	0.2866
C_2^1	0.5696	0.4144	0.4275	0.2210
C_3^1	0.2587	0.4170	0.5375	0.4909
C_4^1	Missing values			

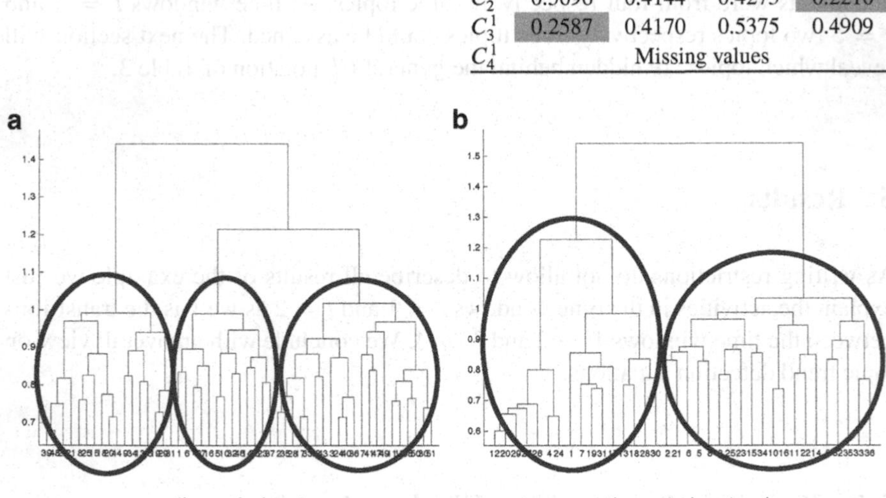

Fig. 3 Dendrograms. (**a**) Time window $t = 3$. (**b**) Time window $t = 4$

5.2 Transition Between Time Windows 3 → 4

Again, Fig. 3(a), (b) depict the dendrograms at time windows $t = 3$ and $t = 4$ together with the circles which show that a three-cluster-solution $\mathscr{K}^3$ and a two-

Table 5 Dissimilarity matrix w.r.t. $\mathscr{K}^3$ and $\mathscr{K}^4$

	C_1^4	C_2^4	C_3^4
C_1^3	0.3597	0.5690	Missing values
C_2^3	0.2634	0.5062	
C_3^3	0.3994	0.5574	

Table 6 General result

Time window t	$t = 1$	$t = 2$	$t = 3$	$t = 4$	$t = 5$
Cluster 1	P	S	EF	P	EF
Cluster 2	EF	P	P	S	P
Cluster 3	S	TIT	TIT		S
Cluster 4		EF			

cluster-solution $\mathscr{K}^4$ were chosen. This time we have $C_2^3 \leftrightarrow C_1^4$, all values in the column of C_2^4 are "large", i.e., C_2^4 is a newly arising cluster, and column 3 has missing values (because of $|\mathscr{K}^4| = 2$). Additionally, one can see that C_1^3 and C_3^3 will vanish (cf. Table 5).

5.3 *Overall View*

All in all we get the results depicted in Table 6 (see also Table 3).

Topic P (politics) exists in all time window dependent clusterings $\mathscr{K}^t$.

Topic S (sport) has vanished in time window $t = 3$, but reappeared in $t = 4$. To check whether a topic is newly arising in time window t we have to compare the centroid of that topic to the centroids of all clusters in the preceeding time windows $\tau \leq t - 2$. If we find a cluster in an earlier time window the dissimilarity of which to the actual cluster is smaller than the lower bound dis_{lb} we assume that the actual cluster is not new, if all dissimilarities are greater than the upper bound dis_{ub} we assume that a newly arising topic has been found.

The topic TIT (technique, industry, and transportation) is newly arising at time window $t = 2$ in our sample of documents but vanishes again in the time windows $t = 4$ and $t = 5$.

The chosen example was small on purpose to be able to demonstrate how the topic trend detection approach works where reappearance checks in earlier time windows are of importance in case that a topic is newly arising in a certain time window.

6 Conclusion

We described an approach for Topic Trend Detection and mentioned the problem to find an accurate threshold respectively lower and upper bounds between which an additional inspection should be performed.

The size of the vector space has an impact on the parameters mentioned. The greater the dimension of the vector space the more less frequent terms from the dictionary might have to be considered and the larger the threshold must be chosen.

References

Allan, J., Carbonell, J., Doddington, G., Yamron, J., & Yang, Y. (1998). Topic detection and tracking pilot study final report. In *Proceedings of the DARPA broadcast news transcription and understanding workshop*, Lansdowne (pp. 194–218).

Allan, J., Lavrenko, V., Frey, D. & Vikas, K. (2000). UMass at TDT 2000. In *Topic detection and tracking workshop notebook* (pp. 109–115).

Allan, J., Lavrenko, V. & Swan, R. (2002). Explorations within topic tracking and detection. In J. Allan (Ed.), *Topic detection and tracking: event-based information organization* (pp. 197–222). Norwell: Kluwer Academic.

Kontostathis, A., Galitsky, L., Pottenger, W. M., Roy, S., & Phelps, D. J. (2004). A survey of emerging trend detection in textual data mining. In M. W. Berry (Ed.), *A comprehensive survey of text mining - Clustering, classification, and retrieval*. New York: Springer.

Kumaran, G., Allan, J., & McCallum, A. (2004). Classification models for new event detection. In *International conference on information and knowledge management (CIKM2004)*. ACM.

Kupietz, M., Belica, C., Keibel, H., & Witt, A. (2010). The German reference corpus DEREKO: A primordial sample for linguistic research. In N. Calzolari, et al. (Eds.), *Proceedings of the 7th conference on international language resources and evaluation (LREC 2010)*, Valletta (pp. 1848–1854). Valletta, Malta: European language resources association (ELRA).

Kupietz, M., & Keibel, H. (2009). The Mannheim German reference corpus (DeReKo) as a basis for empirical linguistic research. In M. Minegishi, Y. Kawaguchi, (Eds.), *Working papers in corpus-based linguistics and language education*, No. 3 (pp. 53–59). Tokyo: Tokyo University of Foreign Studies (TUFS).

Larsen, B., & Aone, C. (1999). Fast and effective text mining using linear-time document clustering. In *Proceedings of the fifth ACM SIGKDD international conference on knowledge discovery and data mining KDD '99*, New York (pp. 16–22). ACM.

Manning, C.D., Raghavan, P., & Schuetze, H. (2009). *An introduction to information retrieval*. Cambridge: Cambridge University Press.

Salton, G. (1989). *Automatic text processing: the transformation, analysis, and retrieval of information by computer*. Boston: Addison-Wesley.

Salton, G., & Buckley, C. (1988). Term-weighting approaches in automatic text retrieval. *Information Processing and Management, 24*(5), 513–523.

Salton, G., Wong, A., & Yang, C.S. (1975). A vector space model for automatic indexing. *Communications of the ACM, 18*(11), 613–620.

Wang, X., Jin, X., Zhang, K., & Shen, D. (2009). Mining common topics from multiple asynchronous text streams. In *International conference on web search and data mining*, New York (pp. 192–201). ACM.

Wang, X., Zhai, C., Hu, X., & Sproat, R. (2007). Mining correlated bursty topic patterns from coordinated text streams. In *International conference on knowledge discovery and data mining*, New York (pp. 784–793). ACM.

Modified Randomized Modularity Clustering: Adapting the Resolution Limit

Andreas Geyer-Schulz, Michael Ovelgönne, and Martin Stein

Abstract Fortunato and Barthélemy (Proc Nat Acad Sci USA 104(1):36–41, 2007) investigated the resolution limit of modularity clustering algorithms. They showed that the goal function of the standard modularity clustering algorithm of Newman and Girvan (Phys Rev E 69(2):026113, 2004) implies that the number of clusters chosen by the algorithm is approximately the square root of the number of edges. The existence of the resolution limit shows that the discovery of the number of clusters is not automatic. In this paper we report on two contributions to solve the problem of automatic cluster detection in graph clustering: We parametrize the goal function of modularity clustering by considering the number of edges as free parameter and we introduce permutation invariance as a general formal diagnostic to recognize good partitions of a graph. The second contribution results from the study of scaling bounds combined with the stability of graph partitions on various types of regular graphs. In this study the connection of the stability of graph partitions with the automorphism group of the graph was discovered.

1 Introduction

Detecting cohesive subgroups of vertices in graphs at different scales is an important problem in graph clustering. Many natural networks have a hierarchical community structure (Ravasz et al. 2002). As inherent to their functional principle, detecting communities by optimizing a (fixed) objective function can only reveal a single hierarchical level.

A. Geyer-Schulz (✉) · M. Ovelgönne · M. Stein
Information Services and Electronic Markets, IISM, Karlsruhe Institute of Technology, Kaiserstrasse 12, D-76128 Karlsruhe, Germany
e-mail: andreas.geyer-schulz@kit.edu; michael.ovelgoenne@kit.edu; martin.stein@kit.edu

B. Lausen et al. (eds.), *Algorithms from and for Nature and Life*, Studies in Classification, Data Analysis, and Knowledge Organization, DOI 10.1007/978-3-319-00035-0_36,

There are two fundamental problems of multi-resolution community detection: How to parametrize an objective function? Partitions at all scale levels should be identifiable when using appropriate parameters. And how to validate that a clustering result for a specific parameter is a 'good' partition? I.e. how to decide whether a partition corresponds to a natural level of hierarchy or whether a partition describes some arbitrary grouping between two natural levels.

Previous work on parametrized objective functions for graph clustering addresses these problems insufficiently. In Sect. 2, we present a parametrized variant of modularity that is able to cover the full scale range. Furthermore, we discuss a new diagnostic approach to decide whether an identified graph partition is (partially) arbitrary or not in Sect. 3. In Sect. 4 we review existing work on multi-resolution community detection.

2 The Link Parametrized Modularity Function

In the following we consider the undirected, loop-free graph $G = (V, E)$ and a partition $C = \{C_1, \ldots, C_p\}$ of V. The adjacency matrix M of G is defined by $m_{xy} = m_{yx} = 1$ if $\{v_x, v_y\} \in E$ and 0 otherwise. Newman and Girvan (2004) defined the modularity $Q(G, C)$ originally as

$$Q(G,C) = \sum_{i=1}^{p} (e_{ii} - a_i^2) \tag{1}$$

with $e_{ij} = \frac{\sum_{v_x \in C_i} \sum_{v_y \in C_j} m_{xy}}{\sum_{v_x \in V} \sum_{v_y \in V} m_{xy}}$ and $a_i = \sum_j e_{ij}$.

e_{ii} is the observed fraction of edge endpoints that connect vertices in the cluster C_i. a_i^2 is the expected fraction of edge endpoints belonging to edges that connect vertices in C_i of a randomly generated graph with the same vertex degree distribution as G. Modularity measures the non-randomness of a graph partition.

The number of edges in the adjacency matrix is $L = 1/2 \sum_{v_x \in V} \sum_{v_y \in V} m_{xy}$ and the number of edges in cluster C_i is $l_i = 1/2 \sum_{v_x \in C_i} \sum_{v_y \in C_i} m_{xy}$. Let l_i^{out} denote the number of edges connnecting vertices in C_i with vertices in the rest of the graph. We rewrite e_{ii} in terms of the number of edges L in G and the number of edges l_i in C_i (intra-cluster edges) as $e_{ii} = l_i / L$. And we rewrite a_i as the fraction of $d_i = 2l_i + l_i^{out}$ (the total degree of vertices in C_i) and twice the number of edges L in G as $a_i = (d_i / 2L)^2$. Substitution in $Q(G, C)$ (Eq. 1) gives

$$Q(G,C) = \sum_{i=1}^{p} \left(\frac{l_i}{L} - \left(\frac{d_i}{2L} \right)^2 \right) \tag{2}$$

Cluster C_i makes a positive contribution to $Q(G, C)$ (Eq. 2) if $l_i / L - (d_i / 2L)^2 > 0$. According to Fortunato and Barthélemy (2007), expressing l_i^{out} as a proportion b of l_i we get $l_i^{out} = bl_i$ and $d_i = (b + 2)l_i$. After substitution of d_i into the above inequality and rearranging of terms we get

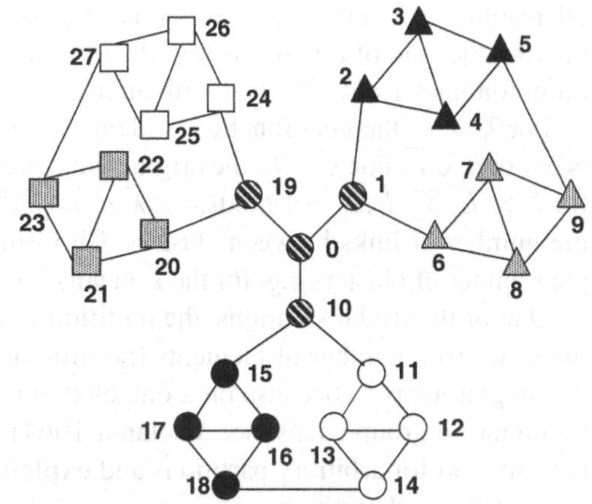

Fig. 1 A cubic graph with $k = 2$ at high resolution ($\lambda = 48$ in Eq. (4))

$$l_i < \frac{4L}{(b+2)^2} \tag{3}$$

The resolution limit (inequality (3)) shows that $Q(G,C)$ depends on the whole network and that there is an upper limit on l_i for $b > 0$ (Fortunato and Barthélemy 2007). When we analyze inequality (3), we see that it crucially depends on L, the total number of edges. To eliminate the resolution limit from modularity clustering for graphs with a homogeneous scale we parametrize $Q(G,C)$ with a parameter λ which substitutes the number of edges L in the graph:

$$Q(G,C,\lambda) = \sum_{i=1}^{p} \left(\frac{l_i}{\lambda} - \left(\frac{d_i}{2\lambda} \right)^2 \right) = \frac{1}{\lambda} \sum_{i=1}^{p} \left(l_i - \frac{d_i^2}{4\lambda} \right) \tag{4}$$

We see immediately that by maximizing $Q(G,C,\lambda)$ for an appropriate $\lambda \in N$ we can find partitions maximizing $Q(G,C,\lambda)$ at the resolution level of λ. Since $l_i < \frac{L}{4}$ and $0 \leq b < 2$ are sufficient conditions that C_i can be considered as a cluster (Fortunato and Barthélemy 2007), we see that by setting $\lambda \leq 4$ we force the size of C_i to 1 (the trivial partition of the graph in its singletons) and by setting $\lambda >> 4L$ we get a single cluster which contains the whole graph.

An example for a graph with an almost (except for the connecting tree) homogeneous scale parameter is the cubic graph with $k = 2$ shown in Fig. 1. Equation (4) works well for families of graphs with a single scale parameter (e.g. the parameter k of the family of cubic graphs with $4(3k + 1)$ vertices, where k controls the number of small clusters in each of the three rings). However, by introducing local scale parameters (e.g. in a cubic graph: k_1, k_2, k_3 controlling each ring separately) we can immediately construct counter-examples.

An interpretation for setting λ which can be justified e.g. for very large networks like the Internet is that the total number of edges of the graph is unobservable. However, when we allow $\lambda \in R^+$, we may study modularity maximization over all resolution levels as a continuous one parameter scale transformation with λ playing the role of a non-linear scale parameter. The number of clusters decreases monotonously from $|V|$ to 1 with an increase in λ.

For $\lambda = L$, the goal function is identical to $Q(G, C)$ and the number of clusters is of order $\sqrt{L}$. For $\lambda = L$, the original interpretation of modularity holds. However, for $\lambda > L$, $\sum_i l_i/\lambda < 1$ holds, for $\lambda < L - l^{between}$, $\sum_i l_i/\lambda > 1$ holds. $l^{between}$ is the number of links between clusters. Choosing λ almost corresponds to selecting the number of clusters e.g. for the k-means algorithm.

One of the trivial solutions, the partition of single vertices, has the property that the order (the number of elements (permutations)) of the automorphism group on the singletons is 1, because on a one element set only the identity mapping of the permutation group exists (see Wielandt 1964). We will extend this property in the next section for arbitrary partitions and exploit it to define an information measure to identify good partitions.

We have implemented parametrized modularity clustering by parameterizing the randomized greedy algorithm of Ovelgönne and Geyer-Schulz (2010) and applied this algorithm to the family of cubic graphs with $4(3k + 1)$ vertices (see e.g. Cvetkovic et al. 1997, p. 165). By choosing k we can change the scaling properties of the graph and thus demonstrate scaling effects in a controlled way.

The family of cubic graphs consists of vertices with degree 3 and two levels of structure. E.g. for $k = 2$, the graph has 28 vertices and 42 edges. The top level of the graph is a tree joining three ring-shaped clusters. Each ring is made up of 2 groups of vertices and the tree leaf linking the ring to the center vertex. On the low resolution level, we should find two variants of 4 clusters, namely first three chains of two groups of four vertices linked by a tree, and second, three rings with two groups of four vertices and a single vertex linked by a single center vertex. On the high resolution level, 7 clusters exist (Fig. 1). All three solutions are permutation invariant with respect to the permutation group generated by the permutation g of the vertices of the the cubic graph shown in Fig. 1:

0	1	2	3	4	5	6	7	8	9	10	11	12	13	14	15	16	17	18	19	20	21	22	23	24	25	26	27
0	10	11	12	13	14	15	16	17	18	19	20	21	22	23	24	25	26	27	1	2	3	4	5	6	7	8	9

g is the generator of an automorphism group of order 3 of the cubic graph. Its elements are the identity map, g, and g^2 – all of which are isomorphisms of the graph. Permutation invariance of a graph-partition under the operations of the automorphism group of a graph means that the image of the graph-partition is isomorphic to the graph-partition under the automorphism group of the graph.

Figure 1 shows the result of the best of 100 runs of the parametrized modularity maximization algorithm with $\lambda = L = 48$ which is the optimal partition at the high resolution level. However, for this example the parametrized modularity algorithm reduces to modularity maximization, because $\sqrt{48} = 6.928$ is slightly below 7 the expected number of clusters in the graph. Figure 2a shows a near optimal partition of the same cubic graph with $\lambda = 500$. The center vertex should form the 4th

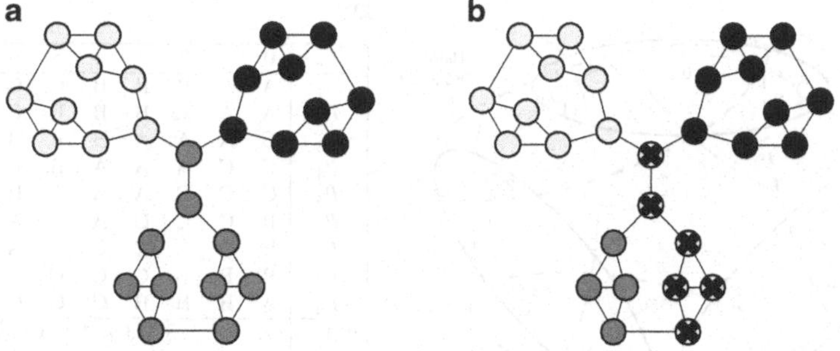

Fig. 2 A cubic graph with $k = 2$. (**a**) At low resolution ($\lambda = 500$ in Eq. (4)). (**b**) At intermediate resolution ($\lambda = 91$ in Eq. (4)).

cluster or the tree found as center cluster at the high resolution level. Finally, Fig. 2b shows a partition at an intermediate resolution level. We see that this solution is not permutation invariant, because of possible rotations around the center vertex. Note, that the lower cycle of the graph is split in two clusters and also the center tree is not identified.

3 Recognizing Good Partitions

Let us attack the question "What is a good partition of a graph?" by considering the opposite question, namely "What are the properties of a graph which should not be partitioned?". Examples of this last type of structure are cycles of arbitrary length (rings), complete graphs (cliques), and Petersen graphs. The reason is that these graphs are completely regular in the sense that there is no information in the structure that can be revealed by a partition. For a proof of this we consider the case of a ring with 9 vertices and 9 edges. Applying modularity clustering by maximizing $Q(G, C)$ leads to a partition with three clusters, each consisting of a chain with 3 vertices with $Q = 0.33$. However, as Fig. 3b shows, there exist nine labelled partitions with the same Q generated by shifting the partition on the circle. P_1, P_4, P_7 refer to the unlabelled partition (801; 234; 567) shown in Fig. 3a.

The table in Fig. 3b shows that we can generate all possible partitions (the three partitions shown in Fig. 3a) by shifting P_1 over the 9-element ring. The permutation (912345678) is the generator of the automorphism group of the graph of the 9-element ring. We define for each vertex of G the frequency distribution over the clusters of the set of partitions in the automorphism group of the graph and we measure the information content of each vertex v by Shannon's entropy $H(v) = -\sum_{i=1}^{p} P(i, v) \log_2 P(i, v)$ with p denoting the number of clusters and $P(i, v)$ the probability that vertex v is in cluster C_i for all $Aut(P)$. $Aut(P)$ is the set of all partitions generated from the partition P by applying all permutations in the automorphism group of G to P. We have shown this for our example

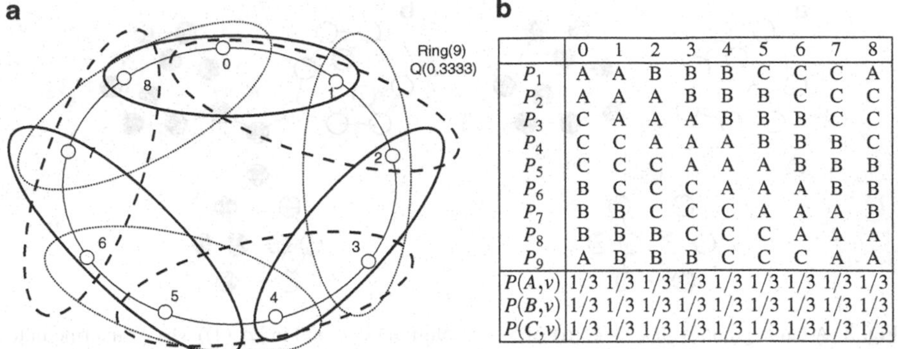

	0	1	2	3	4	5	6	7	8
P_1	A	A	B	B	B	C	C	C	A
P_2	A	A	A	B	B	B	C	C	C
P_3	C	A	A	A	B	B	B	C	C
P_4	C	C	A	A	A	B	B	B	C
P_5	C	C	C	A	A	A	B	B	B
P_6	B	C	C	C	A	A	A	B	B
P_7	B	B	C	C	C	A	A	A	B
P_8	B	B	B	C	C	C	A	A	A
P_9	A	B	B	B	C	C	C	A	A
$P(A,v)$	1/3	1/3	1/3	1/3	1/3	1/3	1/3	1/3	1/3
$P(B,v)$	1/3	1/3	1/3	1/3	1/3	1/3	1/3	1/3	1/3
$P(C,v)$	1/3	1/3	1/3	1/3	1/3	1/3	1/3	1/3	1/3

Fig. 3 Three partitions with the same modularity

in Fig. 3b. The total entropy over the set of vertices for $Aut(P)$ is then simply $H(Aut(P)) = \sum_{v \in V} H(v)$. For the table in Fig. 3 we get $H(v) = 1.585$ for all v, and $H(Aut(P)) = 14.265$ which is the maximal entropy possible. It reveals that the partitions found reveal no information at all. As the reader may convince himself, only the partition with single vertices and the partition with a single cluster containing all vertices have the minimal entropy of $H(Aut(P)) = H(v) = 0$, but also a lower Q.

For the identification of a good partition of a graph, we propose the following approach which is not yet implemented:

1. For all λ, compute a partition of the graph maximizing $Q(G, C, \lambda)$.
2. Define an automorphism group of the graph $Aut(G)$. Compute $Aut(P)$. If $|Aut(P)| = 1$, we have found a unique partition at $Q(G, C, \lambda)$ with an entropy of 0 (with maximal information).
3. Compute $H(Aut(P))$ and $H(v) \quad \forall v$. The information on $H(v)$ can be used to detect instable regions in the graph. E.g. for the cubic tree depicted in Fig. 2a $Aut(G)$ is defined by the automorphism group generated by g. Clearly, the center vertex has an entropy of $H(center) = 1.585$ (will be assigned to all three partitions) and the entropy of all other vertices is 0.
4. Identify the partitions with minimal $H(Aut(P))$ with a locally maximal $Q(G, C, \lambda)$.

4 Related Work

Several objective functions have been proposed to address the resolution limit problem of the standard modularity definition. Muff et al. (2005) proposed the localized modularity measure

$$LQ(G,C) = \sum_{i=1}^{p} \left(\frac{l_i}{L_{i_N}} - \left(\frac{d_i}{2L_{i_N}} \right)^2 \right),$$

where L_{i_N} denotes the number of edges in the subgraph that is induced by vertex i and its direct neighbors. They argued, that modularity assumes that connections between all vertices are equally probable, but in many complex real-world networks this is not true. This approach is equivalent to Eq. (4), except for the fact that Muff et al. (2005) do not consider L_{i_N} as a free parameter, but as a fixed local parameter. As a consequence, not all scale levels can be resolved, e.g. coarser partitions than those identified by standard modularity cannot be discovered: E.g. the partition depicted in Fig. 2a cannot be discovered by the approach of Muff et al. (2005).

Reichardt and Bornholdt (2006) introduced a resolution parameter to the modularity function by weighting the term that represents the expected value

$$Q_a(G,C,\gamma) = \sum_{i=1}^{p} \left(\frac{l_i}{L} - \gamma \left(\frac{d_i}{2L} \right)^2 \right).$$

For $\gamma = 1$, Q_a reduced to standard modularity. Setting γ to values < 1 results in a coarser resolution and setting γ to values > 1 to a finer resolution. A different approach to dealing with the resolution limit of the modularity function has been chosen by Li et al. (2008). Their modularity density function is local:

$$Q_d(G,C,\lambda) = \sum_{i=1}^{p} \frac{2\lambda l_i - 2(1-\lambda)\bar{l}_i}{|C_i|}$$

where l_i denotes as before the number of edges inside of cluster C_i and $\bar{l}_i$ denotes the number of edges that connect a vertex in cluster C_i with an vertex outside the cluster. The difference between the number of intra- and inter-cluster edges adjacent to a vertex of a cluster is set in relation to the clusters size. For $\lambda = 1/2$ Q_d reduces to the unweighted difference of intra- and inter-cluster edges. The unweighted difference as the standard modularity density definition is the arithmetic mean of the ratio cut ($\lambda = 0$) and the ratio association ($\lambda = 1$) (see Shi and Malik 2000).

The problem of identifying resolution parameters with 'good' partitions has been discussed by Ronhovde and Nussinov (2009). They proposed to quantitatively estimate the best resolution(s) for multiscale community detection algorithms by creating several (typically 8 to 12) partitions for every candidate resolution parameter. If the partitions identified for a candidate resolution parameter are especially stable (measured by information based metrics), the resolution parameter is regarded as identified. From the set of partitions for an identified resolution parameter, the partition with the best objective function value is selected as the solution.

5 Conclusion and Outlook

The main innovations of this article are the link-parametrization of the modularity measure and the definition of two information measures on the application of the automorphism group of a graph to the partition at a locally maximal modularity. Both of these measures are invariant with respect to the group action of the automorphism group.

For the definition and computation of the automorphism group of the graph at a locally maximal modularity measure (which is the most difficult task suggested in this paper) recent advances in the theory of finite (permutation) groups (e.g. Cameron 1999) and algorithms of computational group theory will be used (e.g. Seress 2003). However, a (problematic) shortcut consists in sampling randomly generated locally maximal partitions and to compute the information measures on the sample. The problems with this are twofold: First, sampling combinatorial configurations usually is highly nonlinear and severely biased. Second, information measures which are not taking the complete automorphism group of the graph into account may be biased to a considerable degree and they cannot be properly interpreted. E.g. the mutual information of two partitions on a cycle will vary depending on the number of shifts and the length of the cycle. This effect is at work with all measures for comparing clusterings, e.g. the measures investigated by Meila (2007).

References

Cameron, P. J. (1999). *Permutation groups* (London mathematical society student texts, Vol. 45). Cambridge: Cambridge University Press.

Cvetkovic, D., Rowlinson, P., & Simic, S. (1997). Eigenspaces of graphs (Encyclopedia for mathematics and its applications, Vol. 66). Cambridge: Cambridge University Press.

Fortunato, S., & Barthélemy, M. (2007). Resolution limit in community detection. In *Proceedings of the National Academy of Sciences of the United States of America, 104*(1), 36–41.

Li, Z., Zhang, S., Wang, R. S., Zhang, X. S., & Chen, L. (2008). Quantitative function for community detection. *Physical Review E, 77*, 036109

Meila, M. (2007). Comparing clusterings – An information based distance. *Journal of Multivariate Analysis, 98*(5), 873–895.

Muff. S., Rao, F., & Caflisch, A. (2005). Local modularity measure for network clusterizations. *Physical Review E, 72*, 056107.

Newman, M. E. J., & Girvan, M. (2004). Finding and evaluating community structure in networks. *Physical Review E, 69*(2), 026113.

Ovelgönne, M., & Geyer-Schulz, A. (2010). Cluster cores and modularity maximization. In W. Fan, W. Hsu, G. I. Webb, B. Liu, C. Zhang, D. Gunopulos, X. Wu (Eds.), *ICDMW '10. 10th IEEE international conference on data mining workshops* (Sydney, Australia) (pp. 1204–1213). Los Alamitos: IEEE Computer Society.

Ravasz, E., Somera, A. L., Mongru, D. A., Oltvai, Z. N., & Barabasi, A. L. (2002). Hierarchical organization of modularity in metabolic networks. *Science, 297*(5586), 1551–1555.

Reichardt, J., & Bornholdt, S. (2006). Statistical mechanics of community detection. *Physical Review E, 74*, 016110.

Ronhovde, P., & Nussinov, Z., (2009). Multiresolution community detection for megascale networks by information-based replica correlations. *Physical Review E, 80*, 016109.
Seress, A. (2003). Permutation group algorithms (Cambridge Tracts in Mathematics, Vol. 152). Cambridge: Cambridge University Press.
Shi, J., & Malik, J. (2000). Normalized cuts and image segmentation. *IEEE Transactions on Pattern Analysis and Machine Intelligence, 22*(8), 888–905.
Wielandt, H. (1964). *Finite permutation groups*. New York: Academic.

Cluster It! Semiautomatic Splitting and Naming of Classification Concepts

Dominik Stork, Kai Eckert, and Heiner Stuckenschmidt

Abstract In this paper, we present a semiautomatic approach to split overpopulated classification concepts (i.e. classes) into subconcepts and propose suitable names for the new concepts. Our approach consists of three steps: In a first step, meaningful term clusters are created and presented to the user for further curation and selection of possible new subconcepts. A graph representation and simple *tf-idf* weighting is used to create the cluster suggestions. The term clusters are used as seeds for the subsequent content-based clustering of the documents using k-Means. At last, the resulting clusters are evaluated based on their correlation with the preselected term clusters and proper terms for the naming of the clusters are proposed. We show that this approach efficiently supports the maintainer while avoiding the usual quality problems of fully automatic clustering approaches, especially with respect to the handling of outliers and determination of the number of target clusters. The documents of the parent concept are directly assigned to the new subconcepts favoring high precision.

1 Introduction

The maintenance of classification hierarchies is still an expensive and time-consuming process. There has been a lot of research towards (semi-) automatic classification construction and enhancement, but maintenance by and large is still a manual task as many decisions and changes still require human interaction. The problem of splitting a concept – i.e. a class – into useful subconcepts is akin to the problem of clustering a set of documents into useful clusters and finding a suitable name for each cluster to define the new subconcept. We rely on the

D. Stork (✉) · K. Eckert · H. Stuckenschmidt
KR & KM Research Group, University of Mannheim, Mannheim, Germany
e-mail: dominik.stork@gmx.de; Kai@informatik.uni-mannheim.de; Heiner@informatik.uni-mannheim.de

B. Lausen et al. (eds.), *Algorithms from and for Nature and Life*, Studies in Classification, Data Analysis, and Knowledge Organization, DOI 10.1007/978-3-319-00035-0_37,

human maintainer of the classification and include two interaction steps in the process that allow the maintainer to influence the result and subsequently get better recommendations for new subconcepts.

In short, we use a straight-forward clustering algorithm to cluster the documents based on their content. The following naming step employs simple *tf-idf* weighting. The contribution of our approach is the combination of these steps with a first preparatory step that extracts meaningful terms from the documents to be clustered which are used to "push" the clustering in the desired direction. This can be seen as a variant of the so-called *description-comes-first* (DCF) paradigm (Osinski et al. 2004) that states that it might be preferable to find descriptive cluster labels before the documents are clustered, i.e. assigned to these labels. We believe that this approach in its pure form misses the opportunity to use the strength of clustering approaches to find similarities between documents even if synonymous terms are used for the same concepts. Thus, we try to use the best of both worlds. This paper is organized as follows: we first explain in detail our proposed approach in Sect. 2, followed by an illustrating example (Sect. 3). Afterwards, we briefly summarize related publications in Sect. 4.

2 Concept Splitting and Naming

Our approach comprises three steps (Fig. 1): First, we let the maintainer determine the desired number of new subconcepts by means of term based suggestions (Step I). Next, the documents are clustered based on their contents, but biased by the predetermined subconcepts (Step II). At last, the clusters together with name suggestions based on the predetermined term clusters are presented to the maintainer as suggestions for new subconcepts (Step III).

2.1 Term-Based Cluster Preselection (Step I)

The motivation for Step I is twofold: On the one hand, prior experiments showed that generally cluster algorithms that use an a-priori defined number of target clusters perform better for our purpose (Stork 2011). So we need this step to determine the desired number of clusters. On the other hand, we that way incorporate the DCF paradigm which has an additional advantage: For the maintainer, it is easier to evaluate and select possible clusters based on a limited set of terms than on the actual content of the documents in the clusters.

To identify meaningful terms within the documents to be split, we use a weighting scheme based on *tf-idf*. The modification solely lies in the definition of the document sets employed. First of all, we concatenate all documents assigned to concept c into an artificial document, denoted as C. We further define S as the set of all documents that belong to sibling concepts, plus our artificial document C (Fig. 2). The weight for each occurring term t in C is then calculated

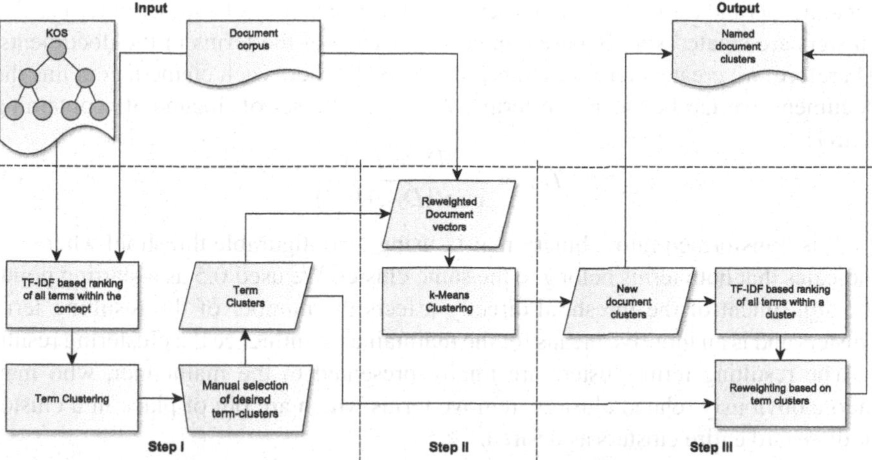

Fig. 1 Concept splitting process

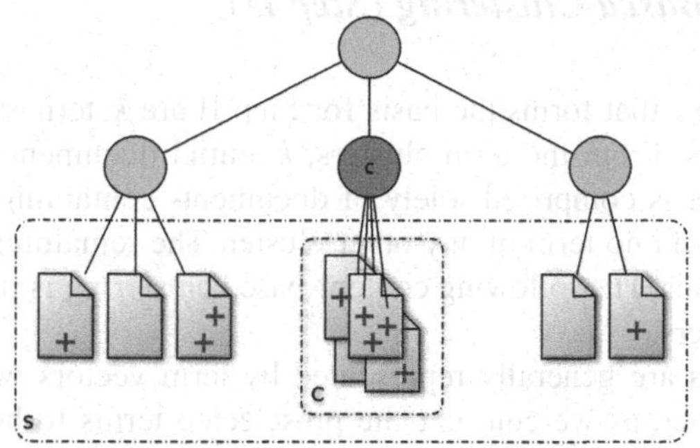

Fig. 2 Weight calculation: $|S| = 6, df_{t,S} = 4, tf_{t,C} = 5$

straight-forward:

$$w_{t,C} = tf_{t,C} \cdot \left(1 + \log \frac{|S|}{df_{t,S}}\right) \tag{1}$$

with term frequency $tf_{t,C}$ denoting the number of occurrences of term t in C and document frequency $df_{t,S}$ denoting the number of documents inside S that contain term t.

With this approach, we identify meaningful terms describing the broad, overall topic of the concept, as well as terms that are meaningful, but not representative for the whole concept. The latter are the interesting terms that allow to discriminate between clusters and possibly account for new subconcepts.

To support the maintainer in a proper term cluster selection, the top n highest weighted terms have to be pre-clustered. n is configurable and depends on the setting; we used $n = 50$, considering the fact that we expect approximately 5 new subconcepts and estimate 10 terms per cluster as a meaningful number of terms. The clusters are created simply based on co-occurrence of the terms in the documents. Therefore, we create a term-relationship matrix T where each element contains the document overlap between two terms, D_t being the set of documents containing term t:

$$T_{i,j} = \frac{|D_i \cap D_j|}{\min(|D_i|, |D_j|)} \tag{2}$$

T is transformed into a binary matrix using a configurable threshold where a 1 indicates that both terms belong to the same cluster. We used 0.5 as a starting point, the adjustment of the threshold directly affects the number of the resulting term clusters and is an intuitive means for the maintainer to influence the clustering result.

The resulting term clusters are finally presented to the maintainer, who may merge obviously related clusters, remove terms which are out of place in a cluster or disregard entire clusters as desired.

2.2 Content-Based Clustering (Step II)

The result of Step I that forms the basis for Step II are k term clusters containing a total of m terms. From the term clusters, k initial document clusters are built, where each cluster is comprised solely of documents containing at least one term of a term cluster and no term of any other cluster. The remaining documents form an additional cluster. The following content-based clustering is performed on these $k + 1$ initial clusters.

The documents are generally represented by term vectors with standard *tf-idf* weighting. However, as we consider the preselected terms to be more important, we increase their value by setting $df = 2$ – this is the lowest occurring document frequency, as terms that occur only in a single document do not affect the clustering and are therefore removed.

The actual clustering is performed with k-Means, despite two requirements of this algorithms that often cast its application into doubt: the number of target clusters has to be defined beforehand and the result depends on the choice of initial seed points. Both requirements are met in our case using the results from Step I: we use the number of selected term clusters – plus one outlier cluster – as the specified number of output clusters; instead of single seed points we use pre-initialized clusters, based on documents solely containing terms from one term cluster. After the maintainer's curation of term clusters, these initial clusters can reasonably be expected to be thematically homogeneous.

With these provisions together with the increased weight of the m preselected terms, we ensure that the clustering result is in line with the input of the maintainer, while we still harness the benefits of a content-based clustering approach.

2.3 *Cluster Naming (Step III)*

In the last step, the term-based clusters and the content-based clusters have to be combined to generate the final suggestions for the maintainer. First, we calculate the most meaningful terms in each *cluster*, using the same approach as described in Step I (Eq. 1). The *tf-idf* values calculated in this manner are sorted in descending order and a new term list is built for each cluster. The number of considered terms depends on the number of terms in the corresponding term cluster which was used to initialize the cluster in question. This list of new terms is presented to the maintainer in combination with the list of original terms in form of a diff visualisation, i.e. both new and dropped terms are highlighted and displacements are marked.

With this visualization, it is very easy to evaluate the final clusters and the maintainer is able to judge, whether or not the content-based clusters are created as expected. From these clusters, the resulting subconcepts can directly be created, using one or more labels from the proposed term list, or provided with a better-suited, possibly superordinate term selected by the maintainer. The documents of the cluster are assigned to the new subconcept directly.

3 Hands On: A Bench Test

Due to the lack of publicly available classifications with full texts, we used the 20 *newsgroups* collection which is a popular dataset for the evaluation and testing of clustering algorithms. Our experiments are based on the version by Jason Rennie[1] with duplicates and most headers removed. The newsgroups are organized in a hierarchy, creating a classification where each newsgroup forms a concept in. For this test, we create an artificially broad concept by merging the newsgroup messages (i.e., our documents) of all groups below *science* (sci.*): *space*, *medical*, *cryptography* and *electronics*, amounting to 2,373 documents. The remaining 16 groups (8,941 documents) in the collection found above are viewed as sibling concepts. The task is to use our method to cluster the documents belonging to our artificial *science* concept. The original classification based on the four subgroups is used as gold standard to evaluate the results.

The first result that is presented to the maintainer are the term clusters of the 50 highest weighted terms from the artificial *science* concept, together with the number of associated documents (Table 1). The remaining 12 terms (*db*, *don*, *health*, *medical*, *orbit*, *patients*, *program*, *research*, *sci*, *science*, *technology* and *time*) are not related to any other term, they are presented to the maintainer as additional terms. In this test, we expect the maintainer to conduct the following refinements to

[1]http://people.csail.mit.edu/jrennie/20Newsgroups/

Table 1 Term clusters and number of associated documents, as presented to the maintainer

Cluster 1	Cluster 2	Cluster 3	Cluster 4
clipper, encryption, key, chip, crypto, privacy, data, security, information, keys, des, algorithm, system, cryptography, escrow, public, ripem, government, available, secret, pgp, nsa, rsa, people, announcement, wiretap, secure, encrypted	tapped, code	space, nasa, launch, shuttle, spacecraft	moon, lunar
858	34	187	43

Table 2 Term clusters and number of associated documents, as presented to the maintainer after the naming step

Cluster 1	Cluster 3
clipper(±0), **encryption**(±0), **key**(±0), **chip**(±0), **security**(+2), **keys**(+2), **privacy**(−1), **des**(+1), **crypto**(−4), **cryptography**(+1), **algorithm**(−1), **nsa**(+5), ~~**rsa**~~, ~~**wiretap**~~, **secret**(+2), **escrow**(−2), **government**(+3), **ripem**(−3), tapped, **secure**(+2), code, **pgp**(−4), announcement, ~~encrypted~~	**space**(±0), **launch**(+3), **moon**(+1), **lunar**(+2), **nasa**(−3), **orbit**(−3), **shuttle**(±0), **spacecraft**(±0)

define the result of Step I:

- Remove *announcement, public, people, system, available, information* and *data* from Cluster 1, which is concerned with cryptography;
- Merge Cluster 4 into Cluster 3, as both are dealing with space;
- Discard Cluster 2; and
- Add term *orbit* (amongst the remaining 12 terms) to Cluster 3.

The remaining term clusters 1 and 3 are used to create the initial clusters as input for the content-based clustering (Step II). Based on the k-Means clustering, the next result is presented to the maintainer: the document clusters, together with meaningful terms that can be used for the final naming of the desired clusters (Table 2). As the term clusters already contained terms selected by the maintainer, this final terms are presented for an efficient review: terms that already belonged to the first term cluster are marked in bold, the number in brackets indicates the difference in position in the sorted terms; new terms are in plain text; and terms that appear in the first term cluster but not in the new list are striked-through.

Both term lists exhibit a great degree of overlap to the original term clusters, indicating that the content-based clustering was performed according to the curated term clusters. In Table 3, we list a general evaluation for these clusters according to the gold-standard, without further curation by the maintainer. While we failed to identify the two other topics contained in the *science* concept, namely *electronics* and *medical*, the two subconcepts *cryptography* and *space* were correctly identified by the term clusters created in Step I. For the latter topics, the created subconcepts exhibit a very high precision, which is in line with our goal.

Table 3 The produced clusters with some central measurements

Cluster	# of documents	Precision	Recall	F1-measure
1	563	**0.961**	0.909	0.934
3	452	**0.969**	0.739	0.838

It is worth noting that executing our approach after new subconcepts have been introduced will generate new term clusters and possibly aid in detecting topics that were hidden by dominating topics during earlier executions. Based on the cluster presentation, the maintainer can easily select an appropriate name for the new subconcepts. With these two simple steps, the maintainer created two new subconcepts containing about 1,000 documents, at an average precision of 96.5 %.

4 Related Work

Our work was mainly motivated by Brank et al. (2008) who use machine learning to predict the additions of new concepts in a classification. They list the assignments of documents by means of clustering and especially the naming of the new concepts based on these clusters as possible extensions.

Clustering of documents is a common task and many other approaches have been developed, e.g. Suffix Tree Clustering (Zamir and Etzioni 1998) (STC), an incremental algorithm which creates clusters on the basis of common phrases between documents. That way, descriptions for each cluster can directly be taken from these common phrases. However, with STC, only documents containing common phrases are grouped together, neglecting thematical overlap using varying terms. STC focuses on isolated document sets (or text snippets) and is suitable to extract key phrases to be used for further exploration of the documents. As STC allows overlapping clusters, it can not be used in a classification context. Moreover, as it favors longer cluster labels, STC tends to produce a high number of rather small clusters.

Some of these drawbacks are addressed by *SHOC* (Semantic, Hierarchical Online Clustering) (Zhang and Dong 2004), an extension to STC designed to cluster a set of web search results with meaningful cluster labels. It is based on *Latent Semantic Indexing* (Deerwester et al. 1990), an indexing and retrieval method that employs *Singular Value Decomposition* to discover hidden concepts contained in a body of text. By identifying these semantic concepts in a document corpus, the shortcomings of a clustering algorithm solely depending on lexical matches can be mitigated. SHOC introduces the notion of *complete phrases* to identify possible cluster label candidates with the help of suffix arrays. In our scenario, SHOC has the disadvantage that it behaves like a black-box. As the discovery of cluster labels is performed after the clusters have been created, it is not possible to let the maintainer support the process in an intuitive way.

The *Description-comes-first* paradigm is employed by Lingo (Osinski et al. 2004), an algorithm inspired by SHOC. In contrast to SHOC, the clustering step is executed after the discovery of the cluster labels which are used to assign documents to clusters. Similar to STC, Lingo's preference for longer cluster labels leads to a large number of clusters representing topics at a higher granularity than desired for our purpose.

Another supposedly DCF-based approach is Descriptive k-Means (Stefanowski and Weiss 2007) (DKM) that at first sight looks similar to our approach. The authors extract cluster labels with two different approaches, frequent phrase extraction (implemented with suffix trees), and simple linguistic processing (noun phrase extraction with a trained statistical chunker). However, the k-Means clustering is performed independently, subsequently the cluster labels are assigned to clusters based on their similarity to the cluster centroids and the contents of the documents in the cluster. Clusters without an assigned label are discarded. The number of target clusters has to be selected beforehand, the initial seeding points are created from randomly selected documents in a way that the most diverse documents of this subset are used. It can be questioned, if DKM follows the DCF paradigm, as the descriptions are not used to influence the clustering, rather they are used to filter the clusters. Nevertheless, the approach follows the same motivation as ours: the identification of labeled clusters favoring a high precision.

5 Conclusion

In this paper, we presented a workflow in three steps to create recommendations for new subconcepts (i.e. subclasses) in a hierarchical classification system. The creation is mainly performed by clustering documents associated to the concepts to be split (Step II). We improved the result by incorporating the human maintainer of the classification: once before the clustering takes place, when the maintainer selects term clusters in order to influence the clustering; once afterwards, when the actual subconcepts are created based on the recommendations. We have shown that our approach works with promising results under laboratory conditions and are confident that it can be used in a productive setting. The strength of our approach lies in the transparency for the user who can influence the result easily based on comprehensible term clusters, while the actual recommendations are still created on the document contents and not just on a term basis.

References

Brank, J., Grobelnik, M., & Mladenic, D. (2008). Predicting category additions in a topic hierarchy. In J. Domingue & C. Anutariya (Eds.), *ASWC*, Bangkok, Thailand (Lecture notes in computer science, Vol. 5367, pp. 315–329). Berlin, Germany: Springer.

Deerwester, S.C., Dumais, S.T., Landauer, T.K., Furnas, G.W., & Harshman, R.A. (1990). Indexing by latent semantic analysis. *Journal of the American Society for Information Science, 41*(6), 391–407.

Osinski, S., Stefanowski, J., & Weiss, D. (2004). *Lingo: Search results clustering algorithm based on singular value decomposition* (pp. 359–368). Berlin Heidelberg, Germany: Springer.

Stefanowski, J., & Weiss, D. (2007) Comprehensible and accurate cluster labels in text clustering. In *Large scale semantic access to content (text, image, video, and sound)* (pp. 198–209). RIAO '07, Le centre de hautes etudes internationales d'informatique documentaire, Paris, France.

Stork, D. (2010). Automatic concept splitting and naming for thesaurus maintenance. Master's thesis, University of Mannheim.

Zamir, O., & Etzioni, O. (1998). Web document clustering: A feasibility demonstration. In *SIGIR*, Melbourne, Australia (pp. 46–54). New York: ACM.

Zhang, D., & Dong, Y. (2004). Semantic, hierarchical, online clustering of web search results. In J.X. Yu, X. Lin, H. Lu, & Y. Zhang (Eds.), *APWeb*, Hangzhou, China (Lecture notes in computer science, Vol. 3007, pp. 69–78) New York/Berlin, Germany: Springer.

Part VII
Banking and Finance

A Theoretical and Empirical Analysis of the Black-Litterman Model

Wolfgang Bessler and Dominik Wolff

Abstract The Black-Litterman (BL) model aims to enhance asset allocation decisions by overcoming the weaknesses of standard mean-variance (MV) portfolio optimization. In this study we propose a method that enables the implementation of the BL model on a multi-asset portfolio allocation decision. Further, we empirically test the out-of-sample portfolio performance of BL optimized portfolios in comparison to mean-variance (MV), minimum-variance, and adequate benchmark portfolios. Using an investment universe of global stock markets, bonds, and commodities, we find that for the period from January 2000 to August 2011 out-of-sample BL optimized portfolios provide superior Sharpe ratios, even after controlling for different levels of risk aversion, realistic investment constraints, and transaction costs. Further the BL approach is suitable to alleviate most of the shortcomings of MV optimization, in that the resulting portfolios are more diversified, less extreme, and hence, economically more intuitive.

1 Introduction and Literature Review

The traditional mean-variance (MV) optimization based on Markowitz (1952) is critically viewed by most portfolio managers (Drobetz 2003), mainly due to three severe shortcomings. First, the estimation of the required input data such as expected returns and the variance-covariance-matrix is problematic. Inevitably, all estimates are subject to estimation errors that distort optimal portfolio allocations. Estimation errors of returns, however, are much more critical than those of the variance-covariance-matrix (Chopra and Ziemba 1993). In the MV optimization framework, assets with the largest estimation errors tend to obtain the highest

W. Bessler (✉) · D. Wolff
Center for Finance and Banking, University of Giessen, Licher Strasse 74, 35394 Giessen, Germany
e-mail: wolfgang.bessler@wirtschaft.uni-giessen.de; dominik.wolff@wirtschaft.uni-giessen.de

B. Lausen et al. (eds.), *Algorithms from and for Nature and Life*, Studies in Classification, Data Analysis, and Knowledge Organization, DOI 10.1007/978-3-319-00035-0_38,

portfolio weights, resulting in 'estimation error maximization' (Michaud 1989). Second, the MV approach tends to generate extreme portfolio allocations and, hence, a low level of diversification across asset classes (Broadie 1993), i.e. the optimized portfolios involve corner solutions. Third, the optimized portfolio weights are very sensitive to changes in the input parameters which results in radical portfolio reallocations even for small variations in expected return estimates (Best and Grauer 1991). High transaction costs resulting from substantial portfolio reallocations might contribute to the low acceptance of MV optimization among practitioners. Black and Litterman (1992) extend the MV approach to alleviate these problems. By combining subjective and neutral return estimates, the high sensitivity of portfolio weights is reduced. In contrast to MV, however, the investors may provide return estimates for each asset and also incorporate the reliability of these estimates. Hence, the investor is able to distinguish between qualified estimates and pure guesses. So far there is hardly any empirical evidence for the superiority of the BL model. Although several studies analyze the economic rationale of the BL model and use it to derive efficient frontiers, there is no evidence that the BL model generates superior portfolio allocations relative to the MV, minimum-variance or other benchmark portfolios in out-of-sample optimizations. In addition, the literature on the BL model does not provide a satisfying answer on how to derive 'subjective' return estimates and how to quantify the quality of these estimates. Most studies assume exogenously given estimates (Drobetz 2003) and suggest confidence intervals of the return estimates as a measure of uncertainty (Black and Litterman 1992). The challenging task for investors is to provide return estimates and confidence intervals which hinder a successful implementation.

We contribute to the literature by testing the BL model empirically. We conduct an out-of-sample multi-asset portfolio optimization for the period from 2000 to 2011 by using the BL approach, MV, and minimum variance optimization and evaluate the portfolio allocations by using several performance measures and calculating adequate benchmark portfolios. We analyze whether the BL model is able to alleviate the problems of MV optimization and whether it leads to a superior portfolio performance. Further, the literature is extended by implementing the BL model on multi-asset instead of stock-only portfolios.

2 Methodology

The BL model combines two sources of information to obtain return estimates: neutral return estimates implied in market weights also referred to as 'implied returns' and subjective return estimates also referred to as 'views'. The simple assumption behind implied returns is that the observed market weights of the assets are the result of a risk-return optimization. In fact it is assumed that all market participants maximize the utility function U:

$$\max U(\omega) = \omega^T \Pi_e - \frac{\delta}{2}\omega^T \Sigma \omega, \tag{1}$$

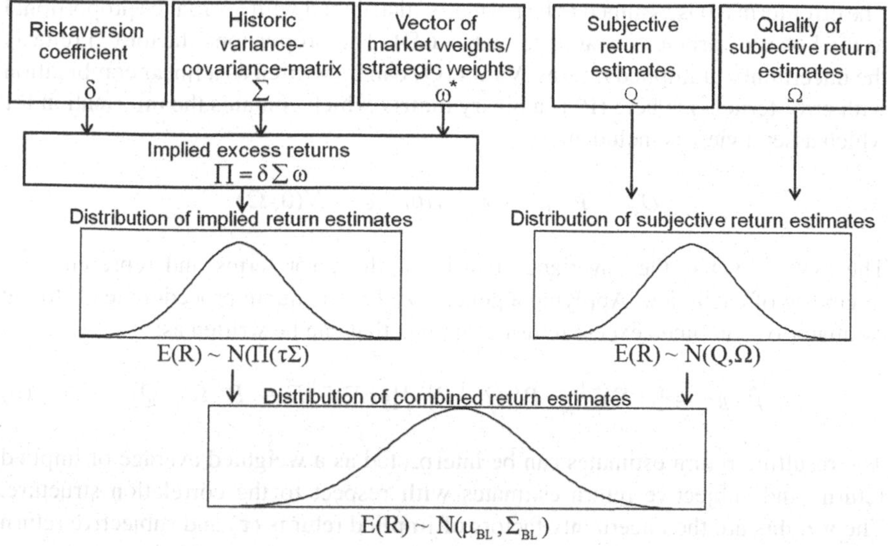

Fig. 1 The procedure of the Black-Litterman approach

where (ω) is the vector of portfolio weights, (Π_e) the vector of implied asset excess returns, (Σ) the variance-covariance-matrix, and (δ) is the investor's risk aversion coefficient. Maximizing the unrestricted utility function results in the optimal portfolio weights:

$$\omega^* = (\delta\Sigma)^{-1}\Pi_e. \tag{2}$$

Assuming that the observable market weights (ω) are the average optimized portfolio weights of investors, the average excess return estimates of the market can be calculated as:

$$\Pi_e = \delta\Sigma\omega. \tag{3}$$

In the BL framework the vector of implied returns (Π) is combined with the vector of views (Q), incorporating the reliability of each view. To derive the combined return estimates, Black and Litterman (1992) apply the Theil's mixed estimation model, while several authors also suggest a Bayesian estimation model (e.g. Drobetz 2003). Figure 1 illustrates the procedure of the BL approach.

We briefly describe the intuition of combining the return estimates following Theil's mixed estimation approach. It is assumed that implied returns (Π) and views (Q) are estimators for the mathematically correct return estimates (μ). Hence the correct excess return estimates (μ_e) can be written as implied excess return estimates (Π_e) plus an error term (η), where (I) is the identity matrix:

$$\Pi_e = I \cdot \mu_e + \eta \quad with \quad \eta \sim N(0, \tau\Sigma). \tag{4}$$

The error term (η) is assumed to be normally distributed with a variance proportional to the historic variance-covariance-matrix (Σ). The proportional factor (τ) reflects the uncertainty of implied returns. Views (Q) can be written as a linear combination with error term (ϵ), where (P) is a binary matrix which contains the information for which asset a view is included:

$$Q_e = P \cdot \mu_e + \epsilon \quad with \quad \epsilon \sim N(0, \Omega). \tag{5}$$

The matrix (Ω) is the covariance matrix of the error terms and represents the reliability of each view. Applying a generalized least square procedure leads to the estimator of combined excess return estimates that can be written as:

$$\hat{\mu}_{e,BL} = [(\tau\Sigma)^{-1} + P^T\Omega^{-1}P]^{-1}[(\tau\Sigma)^{-1}\Pi + P^T\Omega^{-1}Q]. \tag{6}$$

The resulting return estimates can be interpreted as a weighted average of implied returns and subjective return estimates with respect to the correlation structure. The weights are the uncertainty factors of implied returns (τ) and subjective return estimates (Ω). The posterior variance-covariance-matrix is:

$$\Sigma_{BL} = \Sigma + [(\tau\Sigma)^{-1} + P^T\Omega^{-1}P]^{-1}. \tag{7}$$

After computing combined return estimates and the posterior variance-covariance-matrix a traditional risk-return optimization is conducted, maximizing the investor's utility function as defined in Eq. (1). As constraints we implement a budget restriction, an exclusion of short-selling, and an upper bound on the portfolio volatility. The latter allows us to differentiate between different investor types in terms of their desired portfolio risk rather than risk aversion coefficients, which are intuitively difficult to quantify. We keep the risk aversion coefficient constant at a level of 2. For MV optimization, we implement the same optimization procedure. The only difference is that for the MV approach the vector of mean historic excess returns and the historic variance-covariance-matrix are used while in the BL framework combined excess return estimates and the posterior variance-covariance-matrix are employed. For the time period January 2000 to August 2011 we calculate monthly out-of-sample optimized portfolios at every first trading day of the month, using the BL, MV and minimum-variance approaches.

Performance measures. We calculate several performance measures to evaluate the optimized portfolios. We estimate the portfolio's net Sharpe ratio (after transaction costs) as the fraction of the out-of-sample mean net excess-return divided by the standard deviation of out-of-sample returns. We use the two-sample statistic for comparing Sharpe ratios as proposed by Opdyke (2007) to test if the difference in Sharpe ratios of two portfolios is significant. This test allows for correlation between the portfolio returns and non-normal distributions. As a risk measure we calculate the maximum drawdown, which does not require any assumption on the return distribution and reflects the maximum accumulated loss that an investor may suffer in the worst case during the whole investment period. Further, we calculate the portfolio turnover, which quantifies the amount of trading required to implement a particular strategy. In line with DeMiguel et al. (2009), we account for trading costs by assuming proportional transaction costs of 50 basis points of the transaction volume.

3 Data

To construct multi-asset portfolios we use six geographical Datastream stock indices covering both developed and emerging markets: North America, Europe, Pacific, Latin America, Asia and Emerging Europe. Further, we include the Datastream US-Government Bond Index (all maturities) as a low risk investment and the S&P GSCI index representing investments in commodities. We obtain monthly total return index data for the time period from January 1995 to August 2011 and calculate monthly index returns. To implement an out-of-sample portfolio optimization with the BL and MV approaches, we estimate the variance-covariance matrix by using historic monthly return data and a moving estimation window of 36 months in the base case. Different window sizes are used as a robustness check. In the MV approach, historic mean returns are used as return estimates. We use a 12-month moving estimation window in the base case and analyze different estimation windows in the robustness check. Since the focus is on comparing the portfolio performance of MV and BL optimized portfolios, we use the same historic mean returns as in the MV for the BL approach as views (Q). However, the BL approach requires additional input parameters: The reliability of views (Ω), implied returns (Π) and the reliability of implied returns (τ). We measure the reliability (Ω) of each view (i) by computing the historic variance of the error terms (ϵ), where ($\epsilon_{i,t}$) is the difference of the subjective return estimate ($q_{i,t}$) for an asset (i) in month (t) and the realized return ($r_{i,t}$) of asset (i) in month (t), using a 12-months moving estimation window.

The idea is that in stable market conditions when the last month's views are close to the realized returns, the view for the next month is more reliable. For the parameter (τ) we employ a value of 0.1 which is in the range of 0.025–0.3 that we found to be commonly used in the literature. To calculate implied returns the market weights of stocks at the rebalancing date are employed. However, we cannot use market weights for bonds and commodities. For commodities, these weights would be difficult to measure, while for bonds market weights would be problematic due to their relatively heavy weight in comparison to stocks. This would imply that investors allocate a high proportion of their assets to bonds, if they do not have 'subjective' return estimates or if the reliability of these estimates is rather small. Since this might not be an adequate assumption for all investors, we use strategic weights for bonds and commodities. We account for various investor types – 'conservative', 'moderate' and offensive' – and set different strategic weights for bonds, commodities, and stocks as presented in Table 1. These strategic weights are used to construct benchmark portfolios for each investor type as well. Next,

Table 1 Strategic weights and benchmark portfolios

Investment type	Benchmark portfolio weights			Historic volatility of benchmark portfolios	Optimization constraint: max. portfolio volatility
	Bonds	Commodities	Stocks		
Conservative	80 %	5 %	15 %	3.98 % p.a.	5.00 % p.a.
Moderate	40 %	15 %	45 %	7.45 % p.a.	10.00 % p.a.
Offensive	0 %	25 %	75 %	12.15 % p.a.	15.00 % p.a.

we calculate implicit return estimates for all assets according to Eq. (3) for each investor type. To derive the maximum allowed portfolio volatility as an optimization constraint for each investor type we rely on historic benchmark volatilities (1995–2000) and add roughly 25 % to allow for a reasonable deviation from the benchmark. Thus, we assume maximum desired portfolio volatilities for the 'conservative', 'moderate', and 'offensive' investor clienteles of 5, 10, and 15 % p.a., respectively (see Table 1).

4 Empirical Results

Table 2 summarizes the empirical results for the evaluation period from January 2000 to August 2011 for BL, MV, and minimum-variance optimized portfolios as well as for two benchmark portfolios. Benchmarks I and II are computed according to the asset weights in Table 1. For benchmark I, regional stock indices are equally weighted within the asset class stocks, while for benchmark II regional stock indices are market weighted. All portfolios are rebalanced at the first trading day of every month. Panel I of Table 2 shows the net Sharpe ratio (after transaction costs) for each portfolio. The results reveal that BL optimized portfolios exhibit higher net Sharpe ratios and hence better performance than MV, minimum-variance and both benchmark portfolios for all investor types. For the moderate and offensive investors these results are significant. The insignificant result for the conservative investor is not surprising since both BL and MV optimization converge to the minimum-variance portfolio for high risk aversions. The risk measure 'maximum drawdown' in panel II of Table 2 indicates a consistently lower risk of the BL optimized portfolio in comparison to all other approaches and independently of the investor type. The average portfolio turnover in section III is an indicator for the amount of trading and, hence, transaction costs generated by implementing a certain optimization strategy. However, transaction costs are already priced in the net Sharpe ratio measure. The results show that for less risk-averse investors the BL approach tends to exhibit a lower portfolio turnover and, hence, lower transaction costs and less extreme reallocations of the optimized portfolios in comparison to the MV approach. The fourth part of Table 2 reveals that BL portfolios, on average, include a higher number of assets than MV and minimum variance portfolios. Consequently, BL portfolios tend to be better diversified across asset classes.

Table 2 Empirical results

(I) Net Sharpe ratio (after transaction costs)					
Investor type	Black-Litterman	Mean-Variance	Minimum-Variance	Benchmark I	Benchmark II
conservative	0.96	0.80	0.70	0.75	0.67
moderate	$0.87^{*\#\#\dagger\dagger}$	$0.60^{\dagger}$	/	0.31	0.18
offensive	$0.88^{*\#\#\#\dagger\dagger\dagger}$	$0.56^{\#\dagger\dagger}$	/	0.18	0.04
(II) Maximum drawdown (after transaction costs)					
Investor type	Black-Litterman	Mean-Variance	Minimum-Variance	Benchmark I	Benchmark II
conservative	4.44 %	11.94 %	6.50 %	11.09 %	10.03 %
moderate	8.67 %	21.03 %	/	40.27 %	38.09 %
offensive	14.14 %	27.61 %	/	59.78 %	57.31 %
(III) Average portfolio turnover					
Investor type	Black-Litterman	Mean-Variance	Minimum-Variance	Benchmark I	Benchmark II
conservative	1.18	1.19	0.44	0.20	0.18
moderate	2.30	2.80	/	0.40	0.33
offensive	2.89	4.17	/	0.40	0.27
(IV) Average number of assets in portfolio					
Investor type	Black-Litterman	Mean-Variance	Minimum-Variance	Benchmark I	Benchmark II
conservative	4.15	3.59	4.08	8.00	8.00
moderate	5.24	3.07	/	8.00	8.00
offensive	4.80	2.71	/	8.00	8.00

*/**/***, ($^{\#}$/$^{\#\#}$/$^{\#\#\#}$), [†/††/†††] represents a significantly higher Sharpe Ratio compared to Mean-Variance, (Benchmark I), [Benchmark II] at the 10 %-/5 %-/1 %-level, respectively

Table 3 Robustness check

(I) Variation of maximum allowed portfolio volatility						
Maximum Volatility p.a.	5.00 %	7.50 %	10.00 %	12.50 %	15.00 %	20.00 %
Net Sharpe Ratio Black-Litterman	$0.87^{\#\#\dagger\dagger}$	$0.81^{\#\dagger\dagger}$	$0.87^{*\#\#\dagger\dagger}$	$0.89^{*\#\#\dagger\dagger}$	$0.92^{*\#\#\dagger\dagger}$	$0.85^{**\#\dagger\dagger}$
Net Sharpe Ratio Mean-Variance	$0.80^{\#\dagger\dagger}$	$0.65^{\dagger}$	0.60†	$0.60^{\dagger}$	0.57	0.47
(II) Variation of estimation window for variance-covariance matrix						
Estimation window # months.		12	24	36	48	
Net Sharpe Ratio Black-Litterman		$0.77^{\#\dagger\dagger}$	$0.89^{**\#\#\dagger\dagger}$	$0.87^{*\#\#\dagger\dagger}$	$0.85^{*\#\dagger\dagger}$	
Net Sharpe Ratio Mean-Variance		0.52	0.53	$0.60^{\dagger}$	0.54	
(III) Variation of estimation window for historic return estimates						
Estimation window # months.		3	6	12	18	
Net Sharpe Ratio Black-Litterman		0.57	$0.84^{\#\dagger\dagger}$	$0.87^{*\#\#\dagger\dagger}$	0.57	
Net Sharpe Ratio Mean-Variance		0.42	$0.71^{\dagger}$	$0.60^{\dagger}$	0.33	
Net Sharpe Ratio Benchmark I: 40 %, 15 %, 45 % stocks equally weighted						0.31
Net Sharpe Ratio Benchmark II: 40 %, 15 %, 45 % stocks market weighted						0.18

Base Case: Moderate investor. Max. portfolio volatility 10 % p.a. Strategic weights: Bonds: 40 %, Commodities: 15 %; Stocks: 45 %. */**/***, ($^{\#}$/$^{\#\#}$/$^{\#\#\#}$), [†/††/†††] represents a significantly higher Sharpe Ratio compared to Mean-Variance, (Benchmark I), [Benchmark II] at the 10 %-/5 %-/1 %-level, respectively

Next, we perform a sensitivity analysis to check if our results are robust to changes in the input data. To test if the outperformance of the BL approach is driven by the strategic weights of asset classes (see Table 1), we vary the maximum allowed portfolio volatility while keeping the strategic weights of bonds, commodities, and stocks for the calculation of implied returns constant at the level of 40, 15 and 45 %, respectively. Panel I of Table 3 shows that the BL approach generates consistently higher net Sharpe ratios than MV optimization and benchmark portfolios for all considered volatility constraints. The insignificant results for small portfolio volatilities (5 and 7.5 % p.a.) are due to the convergence of BL and MV to the minimum-variance portfolio for high risk aversion. In panels II and III of Table 3, we present the results for different estimation windows. Again, for all scenarios we find consistently higher net Sharpe ratios for the BL approach than for MV optimization and both benchmark portfolios.

5 Conclusion

Using strategic weights, the BL model can successfully be applied to asset allocation decisions for multi-asset portfolios. For the period between January 2000 and August 2011, we find consistently higher out-of-sample Sharpe ratios for the BL approach than for MV optimized portfolios. For risk-averse investors, the difference between Sharpe ratios is insignificant as both approaches converge to the minimum-variance portfolio for high risk aversion. Compared to adequate benchmark portfolios, the BL optimized portfolios reveal significantly higher Sharpe ratios in all cases. Further, out-of-sample BL portfolios exhibit lower risk in terms of smaller maximum drawdowns than MV and benchmark portfolios. In addition, BL optimized portfolios include, on average, more assets than MV optimized portfolios and, hence, tend to be better diversified across asset classes.

References

Best, M. J., & Grauer, R. R. (1991). On the sensitivity of mean-variance-efficient portfolios to changes in asset means: Some analytical and computational results. *Review of Financial Studies, 4*(2), 315.

Black, F., & Litterman, R. (1992). Global portfolio optimization. *Financial Analysts Journal, 48*(5), 28–43.

Broadie, M. (1993). Computing efficient frontiers using estimated parameters. *Annals of Operations Research, 45*, 21–58.

Chopra, V. K., & Ziemba, W. T. (1993). The effect of errors in means, variances, and covariances on optimal portfolio choice. *Journal of Portfolio Management, 19*(2), 6–11.

DeMiguel, V., Garlappi, L., & Uppal, R. (2009). Optimal versus naive diversification: How inefficient is the 1/N portfolio strategy? *Review of Financial Studies, 22*, 1915–1953.

Drobetz, W. (2003). Einsatz des Black-Litterman-Verfahrens in der Asset Allocation. In H. Dichtl, J. M. Kleeberg, & C. Schlenger (Eds.), *Handbuch Asset Allocation: Innovative Konzepte zur systematischen Portfolioplanung*, Bad Soden, Germany: Uhlenbruch, 211–239.

Markowitz, H. (1952). Portfolio selection. *Journal of Finance, 7*(1), 77–91.

Michaud, R. O. (1989). The Markowitz optimization enigma: Is the optimized optimal? *Financial Analysts Journal, 45*(1), 31–42.

Opdyke, J. D. (2007). Comparing Sharpe ratios: So where are the p-values? *Journal of Asset Management, 8*(5), 308–336.

Vulnerability of Copula-VaR to Misspecification of Margins and Dependence Structure

Katarzyna Kuziak

Abstract Copula functions as tools for modeling multivariate distributions are well known in theory of statistics and over the last decade have been gathering more and more popularity also in the field of finance. A Copula-based model of multivariate distribution includes both dependence structure and marginal distributions in such a way that the first may be analyzed separately from the later. Its main advantage is an elasticity allowing to merge margins of one type with a copula function of another one, or even bound margins of various types by a common copula into a single multivariate distribution. In this article copula functions are used to estimate Value at Risk (VaR). The goal is to investigate how misspecification of marginal distributions and dependence structure affects VaR. As dependence structure normal and student-t copula are considered. The analysis is based on simulation studies.

1 Risk Measurement: VaR

There are many different concepts of measuring risk, like Value at Risk (VaR), Expected Shortfall (ES) or coherent measures of risk, to name just a few (e.g. Artzner et al. 1999; Crouhy 2001; Gregoriou 2010; McNeil et al. 2005). We will use the VaR as basic risk measure in simulations. The definiton of VaR is as follows:

$$P\left(R_t \leq F_{R,t}^{-1}(q)\right) = q \tag{1}$$

$$VaR_{R,t}(q) = -F_{R,t}^{-1}(q) \tag{2}$$

K. Kuziak (✉)
Department of Financial Investments and Risk Management, Wroclaw University of Economics, ul. Komandorska 118/120, Wroclaw, Poland
e-mail: katarzyna.kuziak@ue.wroc.pl

B. Lausen et al. (eds.), *Algorithms from and for Nature and Life*, Studies in Classification, Data Analysis, and Knowledge Organization, DOI 10.1007/978-3-319-00035-0_39,

where: R_t is a rate of return (periodic or logarithmic) and $F^{-1}()$ is a quantile of return distribution related to the probability of $1 - q$ (where q has to be assumed).

VaR is quantile of return distribution related to the probability $1 - q$ (where q is prior defined tolerance level). VaR measures the maximum loss associated with a certain statistical level of confidence (e.g. 95 % or 99 %) assuming 1 day horizon and normal market conditions.

The main problem in portfolio VaR estimation is measurement of dependence. One may consider two approaches (Crouhy 2001; Embrechts et al. 2002; Franke et al. 2011; Gregoriou 2010):

1. Correlation – in finance and in risk management (e.g. traditional portfolio theory, the Capital Asset Pricing Model, a set of VaR methods);
2. Copula function – modern approach in risk management (in some VaR methods).

Unfortunately, in some applications correlation is an unsatisfactory measure of dependence, because it takes into account only linear dependence. Correlation also does not measure tail-dependence, which is what risk management should focus on. From risk management point of view, failure to model correctly tail-dependence may cause serious problems (under- or over-estimation of risk). One of modern approaches to measuring dependence is to use copula functions. The essential idea of the copula approach is that a joint distribution can be factored into the marginals and a dependence function called a copula (Gregoriou 2010).

2 Dependence Measurement: Copula Function

The use of copula function allows to overcome the issue of estimating the multivariate distribution function, thanks to the operation of splitting it into two parts:

- Determine the margins $F_1, \ldots, F_n$ which represent the distribution of each risk factor, and estimate their parameters;
- Determine the dependence structure of the random variables $X_1, \ldots, X_n$, specifying a copula function.

The dependence relationship is entirely determined by the copula, while scaling and shape (mean, standard deviation, skewness, and kurtosis) are entirely determined by the marginals (Nelsen 2006).

Copulas can be useful for combining risks when the marginal distributions are estimated individually. This is sometimes referred to as obtaining a joint density with "fixed marginals." Using a copula, marginals that are initially estimated separately can then be combined in a joint density that preserves the characteristics of the marginals.

An n-dimensional copula is a multivariate d.f. C with uniform distributed margins in $[0, 1]$ ($U(0, 1)$) and the following properties (Cherubini et al. 2004; Nelsen 2006):

1. $C : [0, 1]^n \to [0, 1]$.
2. C is grounded and n-increasing.
3. C has margins C_i which satisfy:
   $C_i(u) = C(1, \ldots, u, 1, \ldots, 1) = u$ for all $0 \leq u \leq 1$ the copula is equal to u if one argument is u and all others 1.
4. $C(u_1, u_2, \ldots u_{k-1}, 0, u_{k+1}, \ldots, u_n) = 0$, the copula is zero if one of the arguments is zero.

Theorem 1 (Sklar Theorem). *The dependence between the real-valued random variables is completely described by their joint cumulative probability distribution function:*

$$\mathbf{F}(x_1, \ldots, x_n) = \mathbf{P}(X_1 \leq x_1, \ldots, X_n \leq x_n) \tag{3}$$

$$\mathbf{F}(x_1, \ldots, x_n) = \mathbf{C}(\mathbf{F}_1(x_1), \ldots, \mathbf{F}_n(x_n)) \tag{4}$$

The function $C(.)$ is called a Copula: it connects marginal distributions in a way giving as a results a joint probability distribution. Sklar's theorem says, that under quite general assumptions there always exists a copula function.

There are two main groups of methods of copula function estimation:

1. Parametric:
   Maximum likelihood ML (e.g. Cherubini et al. 2004)
   Inference functions for margins IFM (e.g. Joe and Xu 1996)
   Based on dependence measure (e.g. Lehmann and Casella 1998)
2. Non parametric:
   Empirical copula (e.g. Deheuvels 1978, 1981; Nelsen 2006)

The problem with the ML method is that it could be computational intensive in the case of high dimension, because it requires to estimate jointly the parameters of the margins and the parameters of dependence structure. IFM method could be viewed as a CML (canonical maximum likelihood) method with some assumptions/conditions.

2.1 Multivariate Copulas Family

The two popular parametric families of n-dimensional copulas are elliptical (e.g. the multivariate Gaussian, the multivariate Student-t) and Archimedean. The Gaussian copula is the traditional candidate for modeling dependence, but the Student-t copula can capture dependence in the tails. The Archimedean copulas find application because they are easy to construct. Secondly there is a very wide range of copulas which belong to this class and therefore a great variety of properties which can be modeled using them. There is however also a drawback. The number of free parameters is very low, usually one or two parameters can be adjusted. This makes a more detailed modeling of dependence structure impossible.

The Multivariate Gaussian Copula (MGC) model works well because it is easy to take samples from, and because it can model a fully defined multivariate distribution for any number of marginal distributions. MGC is defined as follows (Cherubini et al. 2004):

$$C\left(u_1, u_2, \ldots, u_n; \Sigma\right) = \Phi_{\Sigma}\left(\Phi_1^{-1}\left(u_1\right), \Phi_2^{-1}\left(u_2\right), \ldots, \Phi_n^{-1}\left(u_n\right)\right) \tag{5}$$

where Φ^{-1} is the inverse of the standard univariate normal distribution function, Φ_{Σ} is the joint cumulative distribution function of a multivariate normal distribution with mean vector zero and covariance matrix equal to the correlation matrix Σ. The limitation of Gaussian copula is that it allows to model only classical Gaussian dependence structure, described with covariance matrix. Its advantage over multivariate normal distribution consists, however, in the fact that marginal distributions may be modeled separately and may be of any type. Each marginal distribution may belong to another type and even utterly different family of univariate distributions.

The second example of elliptical copulas is the t-copula (MTC), which uses multivariate Student t distribution. MTC is defined as follows:

$$C_{k,\Sigma}\left(u_1, u_2, \ldots u_n\right) = t_{k,\Sigma}{}^{n}\left(t_k^{-1}\left(u_1\right), t_k^{-1}\left(u_2\right), \ldots, t_n^{-1}\left(u_n\right)\right) \tag{6}$$

where $t_{k,\Sigma}{}^{n}\left(u_1, \ldots u_n\right)$ is standardized multivariate Student t distribution function with correlation matrix and k degrees of freedom.

A useful property of t-copula is a "tail dependence", when even for most extreme events assets stay dependent, while for normal copula they become asymptotically independent. This leads to higher probability of joint defaults, which is important for large portfolios of assets (Cherubini et al. 2004). The dependence in elliptical distributions is determined by covariances. Covariances are rather poor tools for non-Gaussian distributions in particular for their extremal dependence. Unfortunately, comparing copulas to other high-dimensional models, copulas usually increase number of parameters to be fitted (e.g. for t-distribution one has to estimate correlation matrix and the degrees of freedom, for t-copula in addition parameters of margins). The other problem is the choice of the functional form for the copula (Mikosch 2006).

The problem of measuring the risk of a portfolio may be divided into two stages:

1. The modeling of the joint evolution of risk factor returns affecting portfolio profit and loss distribution during a specified holding period,
2. The modeling of the impact of changes in risk factor returns on the value of portfolio instruments using pricing models.

In this work the focus is on first stage of the problem.

The copula methodology can be applied both to compute Value-at-Risk (VaR) and to perform stress testing (Embrechts et al. 2002; Gregoriou 2010).

2.2 *What May Go Wrong?*

Many approaches to risk measurement start from the assumption of a given model, which is basically equivalent to ignoring model risk. Model risk is the risk of error in our estimated risk measures due to inadequacies in our risk models. The issue of model risk is still underestimated, but all market participants are exposed to mark-to-model risk. The degree of exposure depends upon the type and concentration of less-readily priced and non-readily priced instruments in individual and aggregate portfolios and on the type of valuation, risk and strategy models used (Crouhy 2001; Derman 1996; Kato and Yoshiba 2000). There is a set of sources of model risk. Here we focus on two of them, namely:

1. Erroneous model and model misspecification – errors in analytical solution; misspecifying the underlying stochastic process; missing of risk factor; missing of consideration such as transaction costs and liquidity.
2. Incorrect model calibration – competing statistical estimation techniques; estimation errors; outliers; estimation intervals; calibration and revision of estimated parameters (Crouhy 2001).

3 Simulation Experiment

To verify how VaR estimation is sensitive to misspecification of margins and dependence structure for copula function, we construct two scenarios in our simulation experiment[1] with following steps:

1. Simulating values from given copula (dependence structure and marginals distributions) for three risk factors.
2. Calculating Value at Risk_true (at given confidence level) assuming equally weighted risk factors.
3. Fitting possible copula (four scenarios) to simulated values.
4. Calculating of Value at Risk_sim (at given confidence level) assuming equally weighted risk factors.
5. Calculating of differences between true and simulated values.

In this experiment we analyzed two sources of model risk: uncertainty about the structure and uncertainty about estimation of parameters. For data simulated from known dependence structure, with known marginals we fit incorrect dependence structure or wrong marginals.

[1] Tool: Matlab ver. 7.9.0.529 (2009b).

Assumptions:

- Three types of marginal distributions (normal, beta, and gamma),
- dependence structure:

correlation matrix $\rho = \begin{bmatrix} 1 & 0.25 & -0.4 \\ 0.25 & 1 & 0.7 \\ -0.4 & 0.7 & 1 \end{bmatrix}$,

First scenario: Gaussian copula, second scenario: Student-t copula $v = 6$,
- VaR is quantile of joint distribution, confidence level: 0.95,
- Estimation method: ML method,
- 500,000 observations.

We consider two scenarios: the first starting with Gaussian copula as a correct model, and the second with assumption of t-copula. We compare results by calculating absolute value of difference between "true" VaR (result of given copula, three different marginals) and simulated VaR (result of possible copula and marginals). The Gaussian and t-copula capture dependency structure by correlation matrix, and this makes simulation experiment based on these copulas very easy, with different marginal distributions.

For the first scenario (data generated from Gaussian copula with three different marginals: gamma, beta, normal) we use four models in VaR estimation:

1. The correct model;
2. Incorrect model (with t-copula, normal marginals);
3. Incorrect model (with t-copula, three marginals – gamma, beta, normal);
4. Incorrect model (with Gaussian copula, normal marginals).

Results of VaR are collected in Table 1.

For the second scenario (data generated from t-copula with three different marginals: gamma, beta, normal) we use four models in VaR estimation:

1. The correct model;
2. Incorrect model (with t-copula, normal marginals);
3. Incorrect model (with Gaussian copula, three marginals – gamma, beta, normal);
4. Incorrect model (with Gaussian copula, normal marginals).

Results of VaR are collected in Table 2.

It is observed that differences of VaR for the same scenario (Tables 1 and 2) are growing with the confidence level for the true as well as incorrect models (here results for 0.95 confidence level are only included). As expected, results of VaR in both scenarios are comparable (values for MGC are lower). The smallest differences are in incorrect model III (with wrong copula and right marginal distribution types) in both scenarios. In the first scenario the largest differences are in model IV (with correct copula and wrong marginal distribution types), the same is observed in the second scanario (for model II with correct copula and wrong marginal distribution types). In both scenarios differences in model II and IV have similar level regardless of VaR value. Results of VaR are very sensitive to parameters of

Table 1 Results of VaR 0.05 for MGC (Source: own calculations)

Parameters of marginal distributions				VaR_true – VaR_sym			
beta	normal	gamma	VaR0.05_true	I	II	III	IV
2, 2	2, 7	8, 1	−4.2315	0.0418	0.3772	0.0398	0.4291
2, 2	2, 7	6, 1	−5.7480	0.0152	0.3696	0.0293	0.2978
2, 2	2, 7	5, 1	−6.4510	0.0091	0.3807	0.0442	0.3728
2, 2	2, 6	8, 1	−2.6265	0.0556	0.4099	0.0478	0.4531
2, 2	2, 6	6, 1	−4.0993	0.0093	0.3995	0.0193	0.3914
2, 2	2, 6	5, 1	−4.8318	0.0226	0.3881	0.0036	0.3276
2, 2	2, 5	8, 1	−0.9964	0.0034	0.4389	0.0479	0.4296
2, 2	2, 5	6, 1	−2.4884	0.0160	0.4150	0.0185	0.4129
2, 2	2, 5	5, 1	−3.1967	0.0199	0.4280	0.0068	0.3932
2, 1	2, 7	8, 1	−4.0691	0.0074	0.3355	0.0007	0.4398
2, 1	2, 7	6, 1	−5.5693	0.0349	0.4074	0.0347	0.4162
2, 1	2, 7	5, 1	−6.2889	0.0292	0.4071	0.0289	0.4311
2, 1	2, 6	8, 1	−2.4442	0.0084	0.3904	0.0030	0.3607
2, 1	2, 6	6, 1	−3.9396	0.0179	0.3745	0.0087	0.4238
2, 1	2, 6	5, 1	−4.6447	0.0483	0.4067	0.0675	0.3643
2, 1	2, 5	8, 1	−0.8309	0.0148	0.4437	0.0175	0.4360
2, 1	2, 5	6, 1	−2.3080	0.0257	0.4295	0.0001	0.4198
2, 1	2, 5	5, 1	−3.0347	0.0233	0.3587	0.0272	0.4132
2, 3	2, 7	8, 1	−4.3141	0.0323	0.3727	0.0320	0.3780
2, 3	2, 7	6, 1	−5.8377	0.0129	0.3777	0.0111	0.3863
2, 3	2, 7	5, 1	−6.5483	0.0115	0.3739	0.0051	0.3708
2, 3	2, 6	8, 1	−2.6945	0.0252	0.4494	0.0255	0.3835
2, 3	2, 6	6, 1	−4.1822	0.0255	0.3958	0.0046	0.3898
2, 3	2, 6	5, 1	−4.9161	0.0082	0.3858	0.0019	0.3908
2, 3	2, 5	8, 1	−1.0900	0.0034	0.4243	0.0079	0.4716
2, 3	2, 5	6, 1	−2.5412	0.0097	0.4656	0.0431	0.4255
2, 3	2, 5	5, 1	−3.2787	0.0182	0.4150	0.0084	0.4185

marginal distributions (e.g. Table 1, for (2,2 2,7 5,1) VaR = −6.4510; (2,2 2,5 5,1) VaR = −3.1967).

Generally, even in correct model we make an error (model I in both scenarios). Assumption of copula type for analyzed copulas is less important comparing to assumption of marginal distribution types. There are differences between copulas (larger errors are for Gaussian copula comparing to t-copula). Assumed parameters of dependence structure were better captured by t-copula (smaller errors). Conclusions are the same analyzing relative differences (absolute value between VaR_true and VaR_sim divided by VaR_true).

Research indicates that a misspecified Gaussian copula performs similar with respect to estimation risk compared to the true Student-t copula (Hamerle and Rösch 2005), but still a lot of work has to be done on estimation risk.

There is no a recommended way of checking whether the dependency structure of a data set is appropriately modeled by the chosen copula. Sensitivity studies of

Table 2 Results of VaR 0.05 for MTC (Source: own calculations)

Parameters of marginal distributions				VaR_true – VaR_sym			
beta	normal	gamma	VaR0.05_true	I	II	III	IV
2, 2	2, 7	8, 1	−4.2362	0.0148	0.4243	0.0210	0.3813
2, 2	2, 7	6, 1	−5.7121	0.0128	0.4189	0.0347	0.3571
2, 2	2, 7	5, 1	−6.4461	0.0260	0.3998	0.0047	0.3529
2, 2	2, 6	8, 1	−2.6140	0.0351	0.3631	0.0403	0.4198
2, 2	2, 6	6, 1	−4.0771	0.0016	0.4160	0.0224	0.4189
2, 2	2, 6	5, 1	−4.8122	0.0183	0.3988	0.0126	0.3966
2, 2	2, 5	8, 1	−1.1039	0.1305	0.3612	0.1149	0.1075
2, 2	2, 5	6, 1	−2.4684	0.0039	0.4106	0.0061	0.4377
2, 2	2, 5	5, 1	−3.1978	0.0202	0.4059	0.0026	0.4068
2, 1	2, 7	8, 1	−4.0156	0.0048	0.4263	0.0582	0.3846
2, 1	2, 7	6, 1	−5.5669	0.0091	0.3847	0.0383	0.4240
2, 1	2, 7	5, 1	−6.2840	0.0671	0.3344	0.0357	0.3980
2, 1	2, 6	8, 1	−2.4477	0.0286	0.3587	0.0163	0.4033
2, 1	2, 6	6, 1	−3.9312	0.0665	0.4119	0.0187	0.4811
2, 1	2, 6	5, 1	−4.6550	0.0239	0.3461	0.0622	0.3511
2, 1	2, 5	8, 1	−0.8217	0.0175	0.4244	0.0193	0.3795
2, 1	2, 5	6, 1	−2.2958	0.0015	0.3959	0.0419	0.4876
2, 1	2, 5	5, 1	−3.0373	0.0018	0.4031	0.0226	0.4133
2, 3	2, 7	8, 1	−4.3093	0.0057	0.4023	0.0212	0.3725
2, 3	2, 7	6, 1	−5.8063	0.0788	0.4043	0.0289	0.3588
2, 3	2, 7	5, 1	−6.5366	0.0381	0.3779	0.0107	0.3725
2, 3	2, 6	8, 1	−2.7121	0.0210	0.4149	0.0806	0.4298
2, 3	2, 6	6, 1	−4.1795	0.0019	0.4319	0.0039	0.4314
2, 3	2, 6	5, 1	−4.9006	0.0107	0.4517	0.0598	0.3247
2, 3	2, 5	8, 1	−1.0855	0.0168	0.4891	0.0086	0.4594
2, 3	2, 5	6, 1	−2.5695	0.0614	0.4193	0.0046	0.4463
2, 3	2, 5	5, 1	−3.2793	0.0329	0.3829	0.0310	0.3902

estimation procedures and works on goodness-of-fit tests for copulas have been intensively investigated during the last years. Several goodness-of-fit approaches for copulas have been proposed in literature (e.g. Genest et al. 2006; Panchenko 2005; Weiss 2011).

References

Artzner, P., Delbaen, F., Eber, J. M., & Heath, D. (1999). Coherent measures of risk. *Mathematical Finance, 9*, 203–228.

Cherubini, U., Luciano, E., & Vecchiato, W. (2004). *Copula methods in finance*. New York: Wiley.

Crouhy, M., Galai, D., & Mark, R. (2001). Risk Management. New York: McGraw-Hill.

Deheuvels, P. (1978). Caractérisation complète des Luis Extrêmes Multivariées et de la Covergence des Types Extrêmes. *Publications de l'Institut de Statistique de l'Université de Paris, 23*(3), 1–36.

Deheuvels, P. (1981). A nonparametric test for independence. *Publications de l'Institut de Statistique de l'Université de Paris, 26*(2), 29–50.

Derman, E. (1996). Model risk. *Risk, 9*, 34–37.

Embrechts, P., Hoing, A., & Juri, A. (2002). *Using copulae to bound the Value-at-Risk for functions of dependent risks* (Report), ETH Zurich.

Franke, J., Härdle, W. K., Hafner, Ch. M. (2011). Copulae and value at risk. In J. Franke, W. Härdle, & C. Hafner (Eds.), *Statistics of financial markets. An introduction* (pp. 405–446). Berlin/Heidelberg, Germany: Springer.

Genest, Ch., Quessy, J. F., & Remillard, B. (2006). Goodness-of-fit procedures for copula models based on the probability integral transformation. *Scandinavian Journal of Statistics, 33*(2), 337–366.

Gregoriou, G., Hoppe, C., & Wehn, C. (2010). *The risk modeling evaluation handbook: Rethinking financial risk management methodologies in the global capital markets*. New York: McGraw-Hill.

Hamerle, A., & Rösch, D. (2005). Misspecified copulas in credit risk models: How good is Gaussian? *Journal of Risk, 8*(1), 41–58.

Joe, H., & Xu, J. J. (1996). *The estimation method of inference functions for margins for multivariate models* (Tech. Rep. No. 166). Department of Statistics, University of British Columbia.

Kato, T., & Yoshiba, T. (2000). *Model risk and its control* (pp. 129–156). Tokyo: Monetary and Economic Studies.

Lehmann, E. L., & Casella, G. (1998). *Theory of point estimation*. New York: Springer.

McNeil, A., Frey, R., & Embrechts, P. (2005). *Quantitative risk management: Concepts, techniques, and tools*. Princeton, NJ: Princeton University Press.

Mikosch, T. (2006). Copulas: Tales and facts. *Extremes, 9*, 3–20.

Nelsen, R. (2006). *An introduction to copulas* (2nd ed.). New York: Springer.

Panchenko V (2005) Goodness-of-t test for copulas. *Physica A 355*(1), 176–182.

Weiss, G. N. F. (2011) Are Copula-GoF-tests of any practical use? Empirical evidence for stocks, commodities and FX futures. *The Quarterly Review of Economics and Finance 51*(2), 173–188.

Dynamic Principal Component Analysis: A Banking Customer Satisfaction Evaluation

Caterina Liberati and Paolo Mariani

Abstract An empirical study, based on a sample of 27.000 retail customers, has been curried out: the management of a national bank with a spread network across Italian regions wanted to analyze the loss in competition of its retail services, probably due to a loss in customer satisfaction. The survey has the aim to analyze weaknesses of retail services, individuate possible recovery actions and evaluate their effectiveness across different waves (3 time lags). Such issues head our study towards a definition of a new path measure which exploits a dimension reduction obtained with Dynamic Principal Component Analysis (DPCA). Results which shown customer satisfaction configurations are discussed in the light of the possible marketing actions.

1 Scenario

Since the 1990s, the increasing level of information and competence of the average retail customer has lead to a more complex needs architecture and a demand of diverse financial services. Therefore, today, the bank cannot act regardless of its customer feelings. Customer concerns and wishes change continuously, and necessitate therefore, non-stop improvement, quality enhancement, and their effectiveness by customer satisfaction testing. The bank's ability in focusing on how `to serve the client` and not on what the client receives clearly creates

C. Liberati (✉)
DEMS, University of Milano-Bicocca, P.zza Ateneo Nuovo n.1, 20126 Milan, Italy
e-mail: caterina.liberati@unimib.it

P. Mariani
DEMS, University of Milano-Bicocca, via Bicocca degli Arcimboldi, n.8, 20126 Milan, Italy
e-mail: paolo.mariani@unimib.it

B. Lausen et al. (eds.), *Algorithms from and for Nature and Life*, Studies in Classification, Data Analysis, and Knowledge Organization, DOI 10.1007/978-3-319-00035-0_40,

customer satisfaction dissatisfaction and consequently his loyalty (ABI: Dimensione Cliente 2009). Today every financial institution measures customer satisfaction with a high level of precision, with attention to client segmentation and his changing needs. Evidently satisfied customers tend to diffuse a positive image of the bank which reinforces competitive strength. A good competitive positioning influences the banks profitability by increasing margins and recovering satisfaction costs. According to the most recognized concept, the SERVQUAL (Berry et al. 1985, 1988), satisfaction is the customer answer to the perceived discrepancies between pre-consumption expectations and product/service effective performance. In particular satisfaction derives from the positive confirmation of expectations on a certain product/services and it is as high as the distance expectation-performance (Oliver 1977). Satisfaction is a valuation output that influences and is influenced by perceived quality. The value perceived by the customer is a summary of perceived quality and other personal elements of evaluation (Berry et al. 1991, 1993). For these reasons, in this paper we explored the main aspects of the banking sector in terms of customer expectation and satisfaction, the identification of improvement areas. In particular, we focus our attention on the dynamic aspect of the customer satisfaction which represents a key asset for addressing management recovery actions. The main idea pursued consisted in identifying customer trajectories and deriving synthetic measures of customer evolutions. We performed the analysis via a three way factor technique and we also introduced a new system of weights in order to adjust the factor plan according with the goodness of representation of every point. The solution found is coherent with the aim of the analysis and retain some of the properties of the original factor plan. Results of the analysis employed on a real case of study shows the powerful of such approach which is very promising.

2 Methodological Framework

Customer satisfaction data sets can be multidimensional and have a complex structure: especially when they are collected as sets (tables) of objects and variables obtained under different sampling periods (as in our case of study). Dynamic multivariate techniques allow the analysis of complex data structures in order to study a given instance phenomenon in both a structural (fixing base relationships among interesting objects (variables)) and a dynamic way (to identify change and development of in accordance to the occasions referred to). When a sufficiently long term series is not available a Multiway Factor Analysis (MFA) turns to be a suitable tool for the study of variable dynamics over various time periods (Tucker 1996; Escofier and Pagès 1982; Kroonenberg 1993; Kroonenberg and De Leeuw 1980; Kiers 1989, 1991). MFA main idea is to compare different data tables (matrices) obtained under various experimental conditions, but containing the same number of rows and/or columns. By analogy to N-way methods, the three-way data set is denoted by X with dimensions n, p and k, corresponding to the number of rows (individuals), columns (variables) and tables (occasions), respectively (Rizzi and

Vichi 1995). Thus, an element of X is x_{ijh}, where $i = 1, \ldots, n$, $j = 1, \ldots, p$ and $h = 1, \ldots, k$. Following (Escofier and Pagès 1994) we built a common factorial space, the "compromise space", in which the elements are represented according to their average configuration relative to data volume as a whole. This space is obtained by means of a Principal Component Analysis of the unfolded matrix X, solving the following the minimum equation:

$$\min ||X - A\Lambda B'||^2 \tag{1}$$

where Λ is a matrix $(m \times m)$ of eigenvalues, A is a matrix $(m \times m)$ of eigenvectors of the quadratic form $X'X$ and B is a matrix $(n \times n)$ of eigenvectors of the quadratic form XX'. The compromise plan (composed by the first ad the second factor axes) is the space spanned by a linear combination of three factor analysis. Such plan is the mean of the covariance matrices as in two-way PCA of unfolded matrix. Distribution of the subjects belonging to different occasions can be visualized in the space spanned by the principal components where the center is located according with the volume of the n objects observed in $k = 3$ occasions. Such distribution could be also explored via graphical analysis of subjects trajectories which consists in drawing instances route paths composing adjacent vectors in order to highlight the evolution, across the three waves, of the subjects position respect to the compromise plane (Carlier 1986; D'Urso 2000; D'Urso and Vichi 1998; Coppi and D'Urso 2001).

3 Weighted Factor Analysis and Trajectories Study

As we underlined in Sect. 2 Multiway Factor Analysis is a suitable technique for summarizing the variability of a complex phenomenon by highlighting both similarities/dissimilarities among the occasions considered and the main components of the average behavior in the time interval chosen. Visual inspection of the objects plotted onto the principal space derived via PCA on the unfolded X matrix (compromise plan) allows to drawn subjects routes covered in the three waves. Such solution has to be reviewed in the light of some limitations and geometrical properties of the orthogonal projection. As it is well known mapping data from a high dimensional space to a lower dimensional space (compromise plane) might cause new scatter point configuration. Thus, in such new plot, we obtain a representation where points do not have the same distance from the factor plan itself. According with this remark, in this work we propose, as new system of weights, the usage of the quality of representation of each point (Fig. 1) defined as follows:

$$QR(i) = \cos^2_\theta(x_i, u_\alpha) = \frac{c^2_\alpha(i)}{\sum_{\alpha=1}^{p} c^2_\alpha} \tag{2}$$

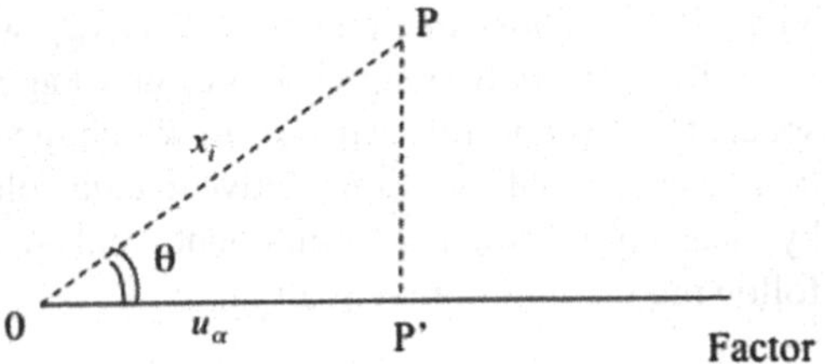

Fig. 1 Quality of representation

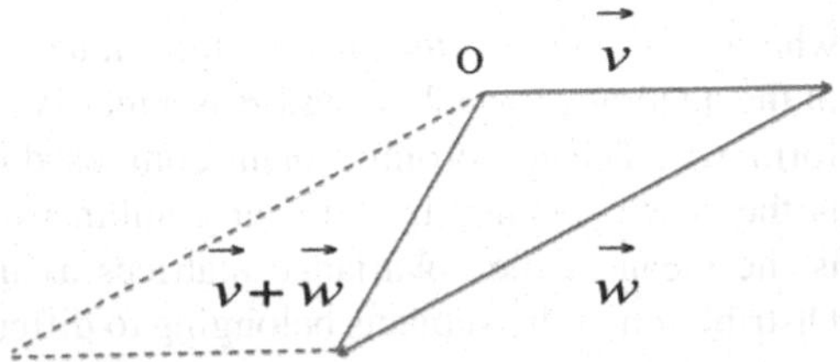

Fig. 2 Sum of two vectors

which is a measure of closeness of a point to the axe itself. Thus, in order to adjust the multiway solution according with relevance of the point projection we re-weight each coordinates of the compromise plane with a linear combination in a fashion as follows:

$$c_{QR\alpha}(i) = QR_\alpha(i) \cdot c_\alpha(i) \tag{3}$$

Such transformation produces a rescaling in terms of value but not in terms of sign. Then subjects trajectories study has been performed based on the new system of axes: we derived a synthetic measure of distance which takes into account route and direction of every single line segment. We employed the parallelogram law which provides a straightforward means to perform vector addition with two vectors in two-dimensional space (Fig. 2). Simply it states that the sum of two two-dimensional vectors $\vec{v}$ and $\vec{w}$ can be determined as follows:

Definition 1. If two vectors are represented by two adjacent sides of a parallelogram, then the diagonal of parallelogram through the common point represents the sum of the two vectors in both magnitude and direction.

4 Case of Study

The managing board of an Italian bank, with a distribution network throughout the country, wanted to analyze the loss in the market share in some clusters. The first recovery action was to improve product features: the board revised the product catalogues and enriched the product attributes in an attempt to give the customer with the maximum flexibility. This lead to a highly complex product catalogue which disoriented bank customers. Hence, it became clear to the board that the loss of competitiveness was due to service level. After a progressive loss of some customer segments, the management decided to conduct a survey, choosing a sample of 27.000 retail customers who effected at least 5 retail requests, conjoint with other contact points of the bank (call center, e-banking,...) within a year. The same sample was monitored for three questionnaire waves to observe both customer preference evolution and management action impact on satisfaction. The questionnaire was framed according to SERVQUAL model, therefore, with five dimensions to analyze perceived quality and expectation of the banking service. All the scores are on a Likert scale 1–10. There are 16 questions (type A), measuring expectations/importance of items, and 16 questions (type B), catching evaluations on perceived quality of a particular item. One final question aims summarizing the entire banking service satisfaction. The same questionnaire has been applied to the same sample for three waves(occasions). A descriptive analysis of the sample shows a homogeneous distribution across different ages, sex, instruction levels and profession segments. This reflects the Italian banking population: more than 60 % is between 26 and 55 years old; the sample is equally distributed between the two sexes and shows a medium low level of instruction. The sample has been analyzed across nine different professional clusters: entrepreneurs, managers, employees, workers, farmers, pensioners, housewives, student, others. The sample is well distributed across the different professional segments employees 24 %, pensioners 22 %, housewives 14 %. The customer satisfaction was analyzed according to three different indicators to avoid dependency on the metrics used. For the three indicators, satisfaction scores are high (above 7/10 in the three cases) with the same trend across the waves. There is an increase in satisfaction from the first to the second wave and a decrease in the third wave. A gap analysis between questions A and B shows that expectations/item importance are always higher than the perception of that item. A dynamic analysis will show evidence of a gap decrease between the first and third wave.

5 Results: Professional Clusters Trajectories and Routes

In this section, we evaluate the effectiveness of our Weighted Factor Analysis and we present the results on professional clusters trajectories and distances.

The first step of the study consists in estimating, the compromise matrix in order to represent the average position of the professional clusters with respect to the selected variables (regardless of the different occasions). The PCA results are very sturdy: the first two components explain about 84 % of the total variance and in particular the first one explains 54 % alone. The compromise matrix based on the first eigenvalue is robust and can provide a realistic view of the evolution of variables and individual positioning in the time horizon considered. Also KMO index (0,8) and Bartlett test (1.312,7; sig. .000) show the quality of the factorial model created. Representing the compromise matrix and the occasion-points of the k matrices on the factorial plan created by the first two components, a clear polarization of the variables on the two axis showed. The first component is characterized by the

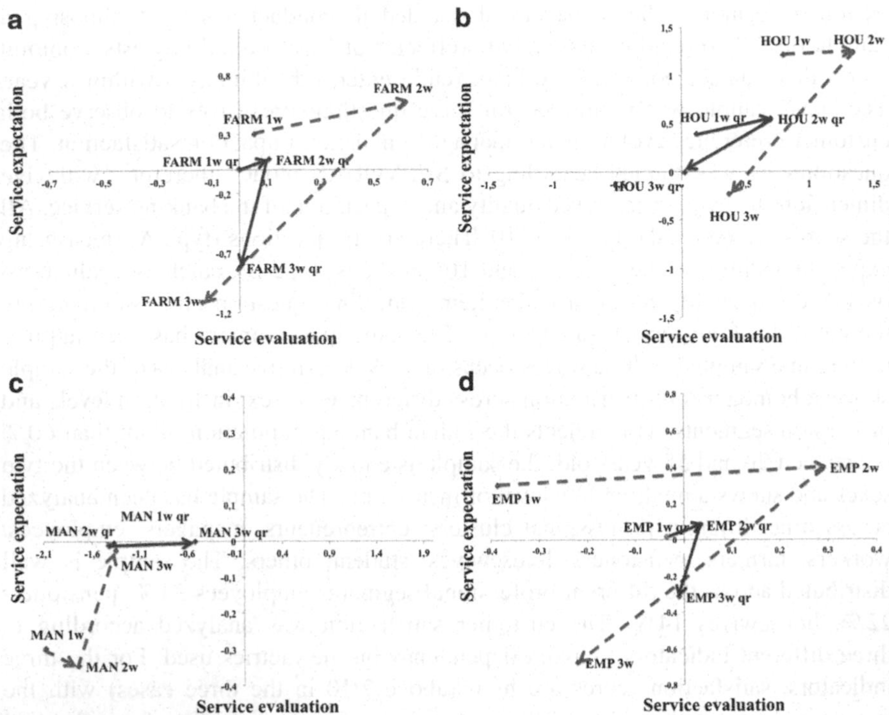

Fig. 3 Professional cluster trajectories: panel (**a**) Employees, panel (**b**) Housewives, panel (**c**) Managers, panel (**d**) Farmers

positive pole B variables (valuations on the different aspects of the service) and from the negative pole by A variables; the second component shows the contrary. In accordance to these results, the first axis is called service evaluation and the second service expectations. By projecting the unities on the factorial axis of the compromise phase, movements with respect to the different waves can be analyzed (Fig. 3).

Figure 3 illustrates evolutions of four professional categories: Employees Housewives Farmers and Managers computed according with the simple multiway coordinates (dashed red trajectory) and the weighted multiway coordinates (blue trajectory). Visual inspection of the plots showed non continuos trends respect to both factor coordinates: in particular the first line-segment (passage from first to second wave) is hardly compressed in case of Employees Housewives and Managers, but this is due to the fact that the compromise plane of the MFA does not take into account the points inertia which are better represented by the third and/or the fourth factor axe (Table 1). Thus, it highlights the effectiveness of such transformation which retains the original value of the factor coordinates only if the quality of representation of a point reaches its maximum value ($QR(i) = 1$) otherwise it reduces it according with its closeness to the factor itself. Obviously the

Table 1 Professional clusters quality of representation and multiway coordinates

Category per wave	QR_1	QR_2	QR_3	QR_4	MFA f1	MFA f2	WMFA f1	WMFA f2
Emp 1w	0.0926	0.0562	0.0001	0.8510	−0.3856	0.3005	−0.0357	0.0169
Emp 2w	0.1129	0.2095	0.6714	0.0062	0.2907	0.3960	0.0328	0.0830
Emp 3w	0.0498	0.4648	0.0157	0.4697	−0.2209	−0.6753	−0.0110	−0.3139
Farm 1w	0.0009	0.0265	0.0696	0.9031	0.0550	0.3013	0.0000	0.0080
Farm 2w	0.1785	0.1564	0.0296	0.6356	0.6165	0.5772	0.1100	0.0903
Farm 3w	0.0114	0.7377	0.1845	0.0664	−0.1381	−1.1096	−0.0016	−0.8186
Hou 1w	0.1420	0.3236	0.2445	0.2899	0.8004	1.2085	0.1136	0.3911
Hou 2w	0.5233	0.4448	0.0281	0.0037	1.3412	1.2364	0.7019	0.5500
Hou 3w	0.0643	0.0251	0.6114	0.2992	0.3682	−0.2298	0.0237	−0.0058
Man 1w	0.6581	0.0133	0.3049	0.0237	−2.0813	−0.2961	−1.3697	−0.0039
Man 2w	0.8149	0.0318	0.1480	0.0053	−1.7449	−0.3447	−1.4218	−0.0110
Man 3w	0.3827	0.0001	0.3328	0.2843	−1.3501	−0.0250	−0.5168	0.0000

weights system might cause a rotation of the factor axes that could not be orthogonal any more, but if the percentage of the inertia explained by the compromise plane is high we are confident that such rescaling does not effect the orthogonality.[1]

The study of the trajectories shows that satisfaction increased, in almost all the professional clusters, from the first to the second wave. This is coherent with the management actions focused on queuing time optimization and service rapidity by retail space rationalization avoiding high capital expenditure. The drop of the satisfaction due to evaluation service occurs during the third wave in almost every professional cluster (Fig. 3): it is negative feedback caused by a stop in banking service monitoring. This trend could be summarizes in a synthetic index (computed according to Definition 1) which measures the paths covered by each category. Such index provides a raw estimate of the total route covered by each professional cluster without any indication about shift positioning. Values obtained can be interpreted as a proxi of customer instability: therefore more distance is covered more satisfaction volatility is recorded. Again index has been computed employing the two system of coordinates and results are collected in Table 2. According with the original factors the categories which show high values of distance covered are the Farmers (1.424) then the Housewives (1.072), Employees (0.990) and finally the Managers (0.780). If we use our coordinates we obtain a new ranking of the values which highlights a situation completely different from the one earlier illustrated. This time the most "mobile" category is Housewives, therefore it has to be carefully monitored.

[1] The significance of the correlation value between Weighted Multiway Factor Axes WMFA f1 and WMFAf2 has been computed via Pearson test ($\rho = -0.0163$ $p - value = 0.9356$) and Spearman test ($\rho = 0.0537$ $p - value = 0.7898$). It turned to be not zero but its value was not significative.

Table 2 Professional clusters routes

Professional categories	Route on MFA	Route on WMFA
Employees	0.990	0.332
Farmers	1.424	0.827
Housewives	1.072	1.273
Managers	0.780	0.853

6 Conclusions

In this paper we presented a novel approach to perform a dynamic customer satisfaction analysis based on a three way factors analysis. We also introduced a clever system of weights which exploits the quality of representation of each point to adjust the original factor solution and to obtain more reliable trajectories. Additionally we have proposed a synthesis of the subject trajectories as a route covered which measures the effects of satisfaction/dissatisfaction in terms of customer instability. A valuable extension of this work would be to derive a new index which takes into account both paths and shifting direction for operative usage.

References

ABI: Dimensione Cliente (2009). Bancaria Editrice, Roma.

Berry, L. L., Parasuraman, A., & Zeithaml, V. A. (1985). A conceptual model of service qualit and its implications for future research. *The Journal of Marketing, 49*, 41–50.

Berry, L. L., Parasuraman, A., & Zeithaml, V. A. (1988). SERVQUAL: A multiple-item scale for measuring consumer perceptions of service quality. *Journal of Retailing, 64*, 12–37.

Berry, L. L., Parasuraman, A., & Zeithaml, V. A. (1991). Refinement and reassessment of the SERVQUAL scale. *Journal of Retailing, 67*, 420–450.

Berry, L. L., Parasuraman, A., & Zeithaml, V.A. (1993). More on improving service quality measurement. *Journal of Retailing, 69*, 140–147.

Escofier, B., & Pagès, J. (1982). *Comparaison de groupes des variables dèfiniers sue le mème ensenble d'individus* (Rapport IRISA n.166). Rennes.

Escofier, B., & Pag ès, J. (1994). Multiple factor analysis (AFMULT package). *Computational Statistics & Data Analysis, 18*, 121–140.

Carlier, A. (1986). Factor analysis of evolution and cluster methods on trajectories. In F. de Antoni, N. Lauro, & A. Rizzi (Eds.), *COMPSTAT86*, Heidelberg, Germany; Wien: Physica-Verlag.

Coppi, R., & D'Urso, P. (2001). The geometric approach to the comparison of multivariate time trajectories. In S. Borra, R. Rocci, M. Vichi, & M. Schader (Eds.), *Advances in data science and classification*. Heidelberg, Germany: Springer.

D'Urso, P. (2000). Dissimilarity measures for time trajectories. *Journal of the Italian Statistical Society, 9*, 53–83.

D'Urso, P., & Vichi, M. (1998). Dissimilarities between trajectories of a three-way longitudinal data set. In A. Rizzi, M. Vichi, & H. H. Bock (Eds.), *Advances in data science and classification*. Heidelberg, Germany: Springer.

Kiers, H. (1989). *Three way method for the analysis of qualitative and quantitive two way data.* Leiden: DSWO Press

Kiers, H. (1991). Hierarchical relations among three way methods. *Psychometrika, 56*, 449–470.

Kroonenberg, P.M. (1993). *Principal component analysis of three-mode data*. Leiden: DSWO Press.

Kroonenberg, P.M., & De Leeuw, J. (1980). Principal component analysis of three-mode data by means of alternating least squares algorithm. *Psychometrika, 65*, 69–97.

Oliver, R.L. (1977). Effect of expectation and disconfirmation on post-exposure product evaluations: An alternative interpretation. *Journal of Applied Psychology, 4*, 480–486.

Rizzi, A., & Vichi, M. (1995). Representation, synthesis, variability and data preprocessing of three-way data set. *Computational Statistics & Data Analysis, 19*, 203–222.

Tucker, R.L. (1966). Some mathematical notes on three-mode factor analysis. *Psychometrika, 31*, 279–311.

Comparison of Some Chosen Tests of Independence of Value-at-Risk Violations

Krzysztof Piontek

Abstract Backtesting is the necessary statistical procedure to evaluate performance of Value-at-Risk models. A satisfactory test should allow to detect both deviations from the correct probability of violations, as well as their clustering. Many researchers and practitioners underline the importance of the lack of any dependence in the hit series over time. If the independence condition is not met, it may be a signal that the Value at Risk model reacts too slowly to changes in the market. If the violation sequence exhibits a dependence other than first order Markov dependence, the classical test of Christoffersen would fail to detect it. This article presents a test based on analysis of duration, having power against more general forms of dependence, based on the same set of information as the Christoffersen test, i.e. hit series.

The aim of this article is to analyze presented backtests, focusing on the aspect of limited data sets and the power of tests. Simulated data representing asset returns are used here. This paper is a continuation of earlier research done by the author.

1 Introduction

> **Value at Risk** (*VaR*) is such a loss in market value of a portfolio that the probability that it occurs or is exceeded over a given time horizon is equal to a prior defined tolerance level q.

It is still one of the most popular risk measures used by financial institutions (Jorion 2001; Campbell 2005). The sense of this definition is presented below:

$$P\left(W \leq W_0 - VaR\left(q\right)\right) = q, \tag{1}$$

K. Piontek (✉)
Department of Financial Investments and Risk Management, Wroclaw University of Economics, ul. Komandorska 118/120, 53-345 Wroclaw, Poland
e-mail: krzysztof.piontek@ue.wroc.pl

B. Lausen et al. (eds.), *Algorithms from and for Nature and Life*, Studies in Classification, Data Analysis, and Knowledge Organization, DOI 10.1007/978-3-319-00035-0_41,

where: W_0 is a value of a financial instrument (portfolio) at present, W is a value at the end of the investment horizon and q is a prior defined tolerance level.

However, this definition is quite general and does not inform in what way a *VaR* measure should actually be estimated (Jorion 2001). Because of it, there are many approaches which can give different values (Jorion 2001; Campbell 2005). The risk managers never know a priori, which approach will be the best or even correct, and should use several models and then backtest them, what should be the critical issue in the acceptance of internal models.

> **Backtesting** is an ex-post comparison of a risk measure generated by a risk model against actual changes in portfolio value over a given period of time, both to evaluate a new model and to reassess the accuracy of existing models (Jorion 2001; Piontek 2010).

A lot of different backtesting procedures have been proposed. They have different level of usage difficulty and they need different sets of information (Campbell 2005; Piontek 2010; Christoffersen and Pelletier 2004; Haas 2001).

The most popular tool for validation of *VaR* models (for the length of the backtesting time period equal to T units) is the hit (or failure) process $[I_t(q)]_{t=1}^{t=T}$. The hit function is defined as follows (Jorion 2001; Piontek 2010):

$$I_t(q) = \begin{cases} 1; & W - W_0 \leq -VaR(q) \quad \text{if a violation occurs} \\ 0; & W - W_0 > -VaR(q) \quad \text{if no violation occurs} \end{cases} \tag{2}$$

and tallies the history of whether or not the exceptions have been realized.

A satisfactory backtest should be able to detect both deviations from the correct probability of violations, as well as their clustering. Usually we test the number of violations, what is often not so easy for small samples (Piontek 2010). However, many researchers and practitioners underline the importance of the lack of any dependence in the violation series over time. If the independence condition is not met, it may be a signal that the *VaR* model reacts too slowly to changes in the market. It is important to recognize that the frequency and independence properties of the hit sequence are separate and distinct and must both be satisfied by a correct *VaR* model.

The main interest of this paper is a comparison of properties of two tests of independence based on the hit sequence – the most known Christoffersen test and the approach based on the duration between violations proposed by Haas (2001, 2005).

2 Backtesting the independence of violations

2.1 *Christoffersen test of independence*

The most popular method for examining the independence of exceptions (*ind*) is the Christoffersen test (Jorion 2001; Piontek 2010), which examines whether or not the likelihood of *VaR* violations depends on whether or not a *VaR* violation occurred on the previous day:

$$H_0 : q_{01} = q_{11}. \tag{3}$$

$$LR_{ind}^{Ch} = -2\ln\left(\frac{(1-\bar{q})^{T_{00}+T_{10}}\bar{q}^{T_{01}+T_{11}}}{(1-\hat{q}_{01})^{T_{00}}\hat{q}_{01}^{T_{01}}(1-\hat{q}_{11})^{T_{10}}\hat{q}_{11}^{T_{11}}}\right) \sim \chi_1^2, \tag{4}$$

where:

$$\hat{q}_{ij} = \frac{T_{ij}}{T_{i0}+T_{i1}},\ \bar{q} = \frac{T_{01}+T_{11}}{T_{00}+T_{01}+T_{10}+T_{11}}, \tag{5}$$

and T_{ij} is a number of i values followed by a j value in the hit series.

Christoffersen and many others criticize first order Markovian process as a limited alternative compared to other forms of clustering (Piontek 2010; Haas 2001) but this approach is easy to implement and, therefore, it is still the most popular one.

2.2 *Duration based backtesting procedure*

Tests that have also power against more general forms of dependence, but still rely on simple information in the hit sequence and on estimating only a few parameters were needed. One of the answer was a new tool for backtesting based on the duration of days between the violations (Christoffersen and Pelletier 2004; Haas 2001, 2005).

The no-hit duration can be simply defined as follows:

$$D_i = t_{I(i)} - t_{I(i-1)} \tag{6}$$

where $t(i)$ is the number of days on which the violation number i occurred.

It is known, that if the model is correct, the distribution of the number of days between violations is geometric and the probability density function is given by:

$$P(D=d) = f_G(d;q) = (1-q)^{d-1}q, \qquad d \in N. \tag{7}$$

If violations are clustered, the number of relatively short and relatively long durations is such that it would be unlikely to obtain it under a proper duration distribution. This property might be tested (Christoffersen and Pelletier 2004).

The concept of the test can be described by a hazard function, which, in the discrete framework, has the simple interpretation of the conditional probability of a violation on day D after we have gone $D-1$ days without a violation (Christoffersen and Pelletier 2004; Haas 2001):

$$\lambda(d) = \frac{P(D = d)}{1 - P(D < d)} = \frac{f(d)}{S(d)}, \tag{8}$$

where $f(d)$ – probability density function, $S(d)$ – survival function.

The geometric distribution has no-memory property, which means that the probability of observing a hit today does not depend on the number of days that have elapsed since the last violation. If the model is correct, one should expect the geometric distribution of durations and a flat hazard function, i.e. $\lambda_G(d) = q$ (Christoffersen and Pelletier 2004).

Tests based on the shape of the estimated hazard function have been proposed.

Christoffersen and Pelletier use the exponential distribution as the continuous analogue of geometric distribution. The exponential distribution is memory-free and has (by definition) a flat hazard function (Christoffersen and Pelletier 2004):

$$f_E(d; q) = q \exp(-qd). \tag{9}$$

To be able to test the independence, they use continuous Weibull distribution as an alternative, because it allows for different shapes of the hazard function and allows for description of duration dependence or independence by parameter choice.

$$f_{CW}(d; a, b) = a^b b d^{b-1} \exp\left[-(ad)^b\right]. \tag{10}$$

The Weibull distribution has the advantage that the hazard function has a closed form representation:

$$\lambda_{CW}(d; a, b) = a^b b d^{b-1}. \tag{11}$$

The exponential distribution appears as a special case when the b parameter is equal to 1. The hazard function is flat, i.e. $\lambda_{CW}(d; a, 1) = a$. We test the value of the shape parameter b to test the independence of violations.

The discrete counterpart (closer to the nature of the problem) was proposed by Haas (2001, 2005). He used a special parameterization of the discrete Weibull (DW) distribution proposed by Nakagawa and Osaki with the density function given by:

$$f_{DW}(d; a, b) = \exp\left[-a^b (d-1)^b\right] - \exp\left[-a^b d^b\right]. \tag{12}$$

The survivor and hazard function are given also by simple equations:

$$S_{DW}(d; a, b) = \exp\left[-a^b (d-1)^b\right], \tag{13}$$

$$\lambda_{DW}(d; a, b) = 1 - \exp\left\{-a^b \left[d^b - (d-1)^b\right]\right\}. \tag{14}$$

The geometric distribution is nested in the Discrete Weibull distribution for $b = 1$, when the hazard function does not depend on d, i.e. $\lambda_{DW}(d;a,1) = 1 - \exp(-a)$. We can test the independence property checking the value of the parameter b.

Because of the nature of the problem this is more popular approach than the one based on the continuous density function and it will be used in the empirical part of this paper.

Parameters of the hazard function are usually estimated with the maximum likelihood method (Haas 2005). In order to implement the test we first need to transform the hit sequence into a duration series. Log-likelihood function which considers censored and uncensored durations can be written as follows (Haas 2005):

$$\ln L(d;a,b) = C_1 \ln S(d_1) + (1 - C_1) \ln f(d_1) + \dots \tag{15}$$
$$\dots + \sum_{i=2}^{N-1} \ln f(d_i) + C_N \ln S(d_N) + (1 - C_N) \ln f(d_N).$$

Now one can build tests of violation properties. The interest of this paper is focused only on the test of the independence of violations. The null hypothesis of the independence claims that:

$$H_{0,IND}: \quad b = 1. \tag{16}$$

We test the hypothesis with the log likelihood ratio test for unrestricted and restricted models:

$$LR_{IND}^{DW} = 2\left[\ln L\left(d;\hat{a},\hat{b}\right) - \ln L(d;\hat{a},1)\right] \sim \chi_1^2. \tag{17}$$

3 Backtesting Errors

Testing the internal *VaR* model one should remember that two types of errors can occur: a correct model can be rejected or a wrong one may be not rejected (Jorion 2001; Piontek 2010).

All presented approaches control the probability of rejecting the *VaR* model when the model is correct. It means that the error of the first kind is known (i.e. α error, significance level of the test). This type of wrong decisions leads to the necessity of searching for another model, which is just wasting time and money. But the error of the second kind (acceptance of the incorrect model, i.e. *beta* error) is a severe misjudgement because it can result in the use of an inadequate *VaR* model that can lead to substantial negative consequences (Piontek 2010).

Because of this, one needs to analyze performance of the selected tests, concerning to the error of the second kind, in order to select the best one for different (but usually small) numbers of observations and model misspecifications. We compare tests using the power of a statistical test idea. It is the probability that the test will reject the null hypothesis when the null hypothesis is actually false (i.e. the probability of not committing a *beta* error).

$$\text{Power} = P\,(\text{reject } H_0 | H_0 \text{ is false}) = 1 - P\,(\beta\text{-error}) = 1 - \beta \qquad (18)$$

The higher values of the power (for different misspecifications) the better test.

4 Approximation of the Power of Tests: Simulation Study

To evaluate the power of the tests, it is necessary to know exactly the properties of the process the asset returns come from. This data generating process can differ from the assumed *VaR* model.

To obtain different volatility clustering effects and, thus, different strength of violation clustering, the GARCH process with different values of parameters was used. The generated returns follow a simple Generalized Autoregressive Conditional Heteroscedastic (*GARCH*) process with Gaussian innovations (Tsay 2001):

$$r_t = \sqrt{h_t} z_t, \qquad h_t = \omega + \alpha r_{t-1}^2 + \beta h_{t-1}, \qquad z_t \sim N\,(0,1)\,. \qquad (19)$$

The data series of the length of 100, 250, 500, 750, 1,000 and 2,500 observations were simulated.

To obtain the proper number of violations, *VaR* was estimated as a unconditional percentile of generated return series (GARCH series) with rolling window approach:

$$VaR_{t|t-1} = \text{percentile}\left(\{r_i\}_{i=t-M}^{t-1}, 100q\right), \qquad (20)$$

where the tolerance level q was equal to 5 % and the rolling window parameter M was 500. This procedure leads to the correct unconditional coverage, which means the correct number of violations for long time series. For each series a hit process was determined.

Depending on GARCH parameters the volatility clustering effect can be found or not. So the violation dependence property may be met or not.

To illustrate the idea, the example of series without volatility clustering (independence of violations) is presented in the Fig. 1. The durations are geometrically distributed. One needs to remember, that it is just one and quite short generated series, but it would generally look like this. The Fig. 2 presents the GARCH effect and violation clusters (violations are not independent). The longer the no-hit period, the higher probability that we are outside the high volatility cluster, and the lower conditional probability of violation on the next day. The hazard function is decreasing in this example, what can be tested.

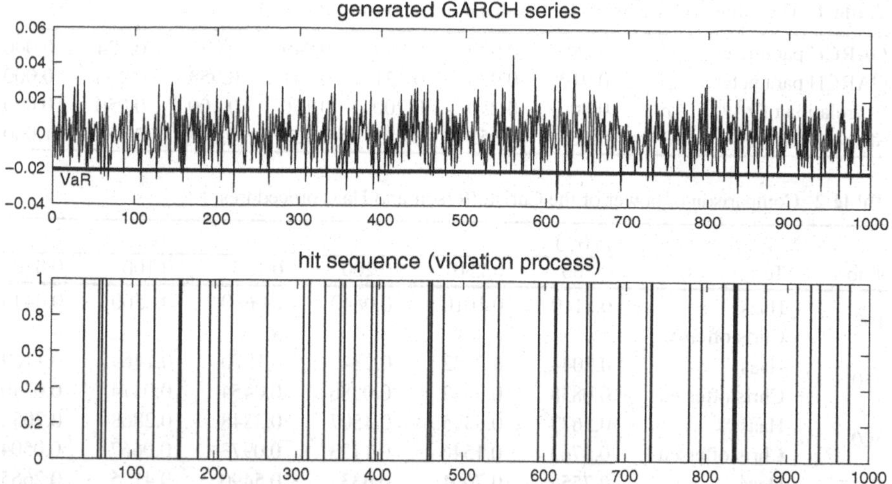

Fig. 1 Lack of clustering effect

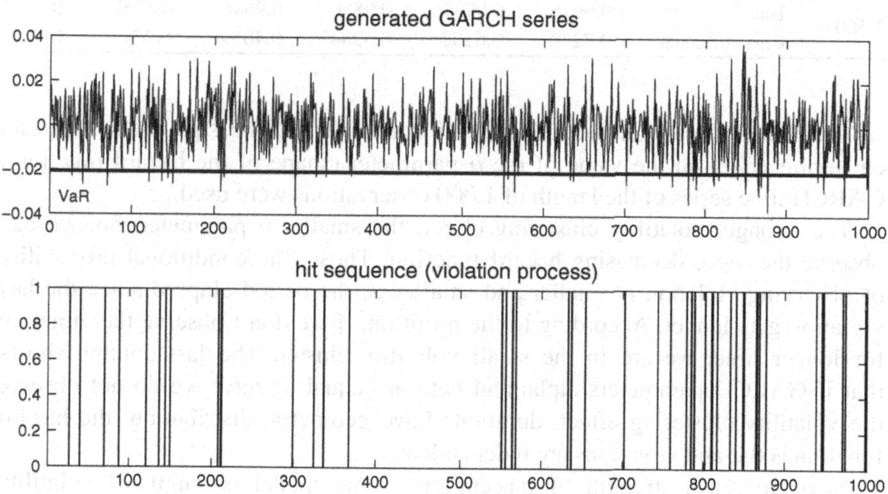

Fig. 2 Clustering effect

The correlation of the squared GARCH observations has been chosen as a measure of the volatility clustering. By changing GARCH parameters, we obtain different values of theoretical correlation (first and higher orders):

$$\rho_1(r_t^2) = \alpha_1 + \frac{\alpha_1^2 \beta_1}{1 - 2\alpha_1\beta_1 - \beta_1^2}, \qquad \rho_k(r_t^2) = \rho_1(r_t^2)(\alpha_1 + \beta_1)^{k-1} \tag{21}$$

Table 1 The influence of clusters on the shape of the hazard function

GARCH parameter	α_1	0.080	0.070	0.059	0.049	0.037	0.024	0.000
GARCH parameter	β_1	0.910	0.920	0.931	0.941	0.953	0.966	0.000
Correlation of r_t^2	ρ_1	0.300	0.250	0.200	0.150	0.100	0.050	0.000
Shape parameter	b	0.624	0.652	0.685	0.731	0.792	0.858	1.000

Table 2 Comparison of power of the Christoffersen and Haas procedures

		$\rho_1(r_t^2)$					
# obs.	Test	0.300	0.250	0.200	0.150	0.100	0.050
100	Haas	0.1160	0.1010	0.0840	0.0680	0.0560	0.0440
	Christoffersen	–	–	–	–	–	–
250	Haas	0.3044	0.2622	0.2342	0.1870	0.1469	0.1169
	Christoffersen	0.0838	0.0632	0.0586	0.0454	0.0414	0.0340
500	Haas	0.5677	0.5315	0.4507	0.3848	0.2868	0.2052
	Christoffersen	0.1763	0.1546	0.1193	0.0973	0.0667	0.0504
750	Haas	0.7554	0.7001	0.6335	0.5490	0.4125	0.2665
	Christoffersen	0.2719	0.2261	0.1812	0.1407	0.0907	0.0686
1,000	Haas	0.8594	0.8246	0.7616	0.6729	0.5245	0.3228
	Christoffersen	0.3523	0.2987	0.2333	0.1804	0.1247	0.0799
2,500	Haas	0.9978	0.9937	0.9886	0.9646	0.8732	0.5822
	Christoffersen	0.7232	0.6108	0.5284	0.4059	0.2517	0.1271

The Table 1 presents the influence of correlation of the squared returns and clustering effect on the value of the b parameter (shape of the hazard function). GARCH time series of the length of 1,000 observations were used.

The stronger volatility clustering effect, the smaller b parameter. So, we can observe the more decreasing hazard function. Those, the conditional probability of observing violation is smaller and smaller as the period elapsed since the last violation gets longer. According to the intuition, if we don't observe the violation for longer time, we are in the small volatility cluster. The last column shows that if GARCH parameters alpha and beta are equal to zero, we do not observe the volatility clustering effect, durations have geometric distribution, the hazard function is flat and violations are independent.

Next, for each strength of inaccuracy of the model (strength of volatility clustering effect measured with first order autocorrelation of squared observations) and for each of the specified lengths of the data series Monte Carlo simulations with 100.000 draws were done. It enabled us to calculate test statistics and to estimate the frequency at which the null hypothesis was rejected for wrong models. The last may be treated as an approximation of the power of test. The Dufour approach was used to improve the Monte Carlo procedure of power of test calculating (Haas 2005).

The Table 2 compares the power of the Christoffersen and Haas procedure, what was the main aim of this paper.

The power of the Haas procedure based on the hazard function of the discrete Weibull distribution is considerably higher than for Christoffersen approach.

One should remember that both approaches are based only on information contained in the hit sequence. The superiority of the Haas procedure is observed even for short lengths of time series (which is typical for risk management) and for small values of correlation of squared observations. In case of short time series, testing for independence using the Christoffersen approach is ineffective.

The best choice is using the Haas procedure for all lengths of series, but we should be aware of the error of the first and second kind.

5 Some final conclusions

Many approaches to backtesting has been proposed so far and this area of research is still developing. New procedures are suggested and the old ones are assessed and reviewed. Based on the current research (including research done by the author Piontek (2010)) some general suggestions about testing the number and the independence of violations can be formulated.

(1) There is no perfect procedure. Don't relay on one backtest. (2) Don't use the mixed tests. Choose and use separately the tests you know to have the highest power. Find the set of acceptable models using at least the Berkowitz approach (to test the number of violations) and the Haas procedure (to test the independence). (3) If there are two *VaR* estimation procedures which have passed the tests, use the loss function approach to choose the best model out the of correct models. (4) Be especially careful when using short time series to test models and drawing conclusions about them. (5) Try to test models for different *VaR* tolerance levels, positions and data sets. (6) Check how the test hypothesis is formulated (what is meant under the term 'independence', for instance).

But first of all, in order to use backtesting, it comes out that it is necessary to determine how low the power of *VaR* backtests may be, so that they can still remain applicable in typical cases. It seems extremely important to discuss the minimum value of the test power and to focus on the type II error. It is still an open problem.

References

Campbell, S. (2005). A review of backtesting and backtesting procedures. Federal Reserve Board, Washington, D.C. www.federalreserve.gov/Pubs/Feds/2005/.

Christoffersen, P., & Pelletier, D. (2004). Backtesting value-at-risk: a duration-based approach. *Journal of Financial Econometrics, 2*(1), 84–108.

Haas, M. (2001). *New methods in backtesting*. Research Discussion Paper. Financial Engineering Research Center Caesar, Bonn.

Haas, M. (2005). Improved duration-based backtesting of value-at-risk. *Journal of Risk, 8*(2), 17–38.
Jorion, P. (2007). *Value at risk* (3rd ed.). New York: McGraw-Hill
Piontek, K. (2010). Analysis of power for some chosen VaR backtesting procedures – simulation approach. In A. Fink, B. Lausen, W. Seidel, & A. Ultsch (Eds.), *Advances in data analysis, data handling and business intelligence. Studies in classification, data analysis, and knowledge organization*. Part 7, pp. 481–490.
Tsay, R. (2001). *Analysis of financial time series*. Chicago: Wiley

Fundamental Portfolio Construction Based on Mahalanobis Distance

Anna Rutkowska-Ziarko

Abstract In the classic Markowitz model, at an assumed profitability level, the portfolio risk is minimized. The fundamental portfolio introduces an additional condition aimed at ensuring that the portfolio is only composed of companies in good economic condition. A synthetic indicator is constructed for each company, describing its economic and financial situation. There are many methods for constructing synthetic measures. This article applies the standard method of linear order. In models of fundamental portfolio construction, companies are most often organised in order on the basis of Euclidean distance. Due to possible correlation between economic variables, the most appropriate measure of distance between enterprises is the Mahalanobis distance.

The aim of the article is to compare the composition of fundamental portfolios constructed on the basis of Euclidean distance with portfolios determined using the Mahalanobis distance.

1 Introduction

Although the classic Markowitz model enables effective risk diversification, it does not enable taking into account the fundamental factors which characterise a listed company. Studies carried out on developed and emerging capital markets indicate that there is a possibility of gaining additional benefits based on information about the financial and economic situation of listed companies (Tarczyński and Łuniewska 2005; Łapińska and Rutkowska-Ziarko 2003; Tarczyński 2002; Ritchie 1996). Therefore, fundamental analysis plays an important role in methods of research applied on capital markets. Fundamental analysis is restricted by the absence of risk

A. Rutkowska-Ziarko (✉)
Faculty of Economic Sciences, University of Warmia and Mazury, Oczapowskiego 4, 10-719 Olsztyn, Poland
e-mail: aniarek@uwm.edu.pl

B. Lausen et al. (eds.), *Algorithms from and for Nature and Life*, Studies in Classification, Data Analysis, and Knowledge Organization, DOI 10.1007/978-3-319-00035-0_42,

diversification which is possible in portfolio analysis. Models of the choice of an investment variant which combine elements of portfolio analysis and fundamental analysis would be very welcomed by investors. An attempt at combining the two types of analysis was made by Tarczyński (1995), who presented a model of a fundamental portfolio. When the portfolio is developed based on fundamental analysis, it is necessary to quantitatively determine the economic and financial situation of the company. In practical terms, financial ratios are used which describe different aspects of the company's operation. The problem is to present the economic and financial situation of a company with just one index. A synthetic measure can be determined by one of linear ranking methods. A synthetic development measure was first used for creating a portfolio by Tarczyński (1995). He referred to the measure as *the taxonomic measure of attractiveness of investments—TMAI* (Tarczyński 1995; Łuniewska 2005). In models for creating a fundamental portfolio, companies are usually arranged according to the Euclidean distance (Tarczyński 1995, 2002; Tarczyński and Łuniewska 2005; Rutkowska-Ziarko 2011). Due to a possible correlation among diagnostic variables, the Mahalanobis distance (Everitt 1993) is a more appropriate distance measure between companies.

The aim of the article is to compare the composition of fundamental portfolios constructed on the basis of Euclidean distance with portfolios determined using the Mahalanobis distance.

2 Fundamental Portfolio

Portfolio analysis enables effective risk diversification by using the appropriate optimisation methods and information contained in historical quotations of listed companies. The portfolios constructed are effective in the sense of minimisation of a selected risk measure with set limitations, the most important of which is the required rate of return. The models used in portfolio analysis take into account the links between profitability of individual portfolio components, which allows for effective reduction of the total investment risk. The analysis skips information about the economic and financial situation of the quoted companies under analysis.

Fundamental analysis based on financial statements enables one to choose companies with good economic and financial standing. However, this approach leaves out the investment risk, which—along with profitability—is the main parameter which describes an investment variant. According to the portfolio theory, investors should choose effective portfolios, i.e. ones associated with the lowest risk at the set level of profitability. Fundamental analysis leaves out information contained in the historic quotations of the companies under analysis. The information relates to average past profitability of the companies as well as links between the profitability of the portfolio components.

A classic model of construction of a share portfolio is Markowitz's model. The issue of determination of the contribution of shares in Markowitz's model (Markowitz 1952) is reduced to solving the following optimisation problem:

minimising the variance of portfolio rate of return:

$$S_p^2 = \sum_{i=1}^{k} \sum_{j=1}^{k} x_i x_j \operatorname{cov}_{ij} , \tag{1}$$

with the limitations:

$$\sum_{i=1}^{k} x_i = 1 , \tag{2}$$

$$\sum_{i=1}^{k} x_i \bar{z}_i \geq \gamma , \tag{3}$$

$$x_i \geq 0, \quad i = 1, \ldots, k , \tag{4}$$

where: S_p^2 is variance of rate of return; cov_{ij} is the covarince between security i and security j, γ—target rate of return, assuming that $\gamma \leq \max \bar{z}_i$; $\bar{z}_i$—mean rate of return on security i; x_i—contribution by value of the i-th share in the portfolio.

By combining elements of the portfolio and fundamental analysis, an additional condition can be introduced to the classic Markowitz model which ensures that the portfolio will contain only companies with good economic and financial standing. The model of construction of a fundamental portfolio, which has been used in the study, is a modification of the classic Markowitz model. A limiting condition has been introduced to the portfolio construction model, according to which the *TMAI* average, weighted by contribution of shares of a specific company in the portfolio, must achieve at least the level set by the investor. It means that one more limitation should be added to the classic Markowitz model (1)–(4) (Rutkowska-Ziarko 2011):

$$TMAI_p = \sum_{i=1}^{k} TMAI_i \, x_i \geq TMAI_\gamma , \tag{5}$$

where: $TMAI_\gamma$—the sum of *TMAI*, required by the investor, weighted by the contribution of shares in the portfolio.

The model minimises the risk at limitations which concern achieving the assumed values by the portfolio rate of return and the average *TMAI*. Similar to Tarczyński's model (Tarczyński 1995), it ensures that only companies with good financial standing will be added to the portfolio. It also identifies a portfolio with the minimum variance at the set limitations (Rutkowska-Ziarko 2011).The fundamental portfolio is an efficient portfolio if, for a specific average rate of return and average *TMAI*, the variance calculated for it is the lowest.

3 Taxonomic Measure of Attractiveness of Investments

Tarczyński employs the term coined by himself—*taxonomic measure of attractiveness of investments* (*TMAI*)—for a synthetic measure based on the Euclidean distance (Tarczyński 1995). This phrase will be used to denote the measure based on Mahalanobis distance. In order to distinguish between the two measures, that constructed based on the Euclidean distance will be referred to as *ETMAI*, and that based on the Mahalanobis distance—as *MTMAI*. The *TMAI* abbreviation will be used when both synthetic development measures discussed in the paper will be considered. Other synthetic measures have been used in examination of capital markets lately, such as: *generalized distance measure* (*GDM*) and *absolute development level ratios* (*BZW*) (Łuniewska 2005). The advantage of the *GDM* measure lies in its ability to be used for variables measured in weaker scales (Walesiak and Dudek 2010).

In order to determine *TMAI*, a set of diagnostic variables should be chosen as well as a method of object arrangement. Diagnostic variables are financial indexes determined from financial statements; in this case, quoted companies are objects. After selecting the diagnostic features, they should be divided into stimulants, destimulants and nominants. Subsequently, transformations are made which allow the expression of the diagnostic variables with stimulants.

Four financial ratios were taken as diagnostic variables in the paper. Three of them described the examined companies' financial standing: current ratio (*CR*), return of assets ratio (*ROA*) and debt ratio (*DR*). The study also took into account the market price-earning ratio (P/E). Studies of capital markets have revealed a negative correlation between the value and future share price increases (Basu 1977). Therefore, P/E was regarded as a destimulant and replaced with E/P:

$$E/P = \frac{1}{P/E}\,. \tag{6}$$

CR and *ROA* ratios were regarded as stimulants, whereas *DR* as a destimulant. *DR* was replaced with a corresponding stimulant (DR').

$$DR' = \frac{1}{DR}\,. \tag{7}$$

Subsequently, diagnostic variables were standardised and a reference standard was created with which the analysed companies were compared. Let w_{il} denote values of standardised variables, where $l = 1, \ldots, m$ is the number of considered diagnostic variables. For each standardised diagnostic variables, the highest observed value of (w_{0l}) is sought (Tarczyński 2002, p. 97):

$$w_{0l} = \max_i \{w_{il}\}\,. \tag{8}$$

An abstract point $P_0(w_{0l})$ was taken as the reference standard; its coordinates assume the highest values of the diagnostic variables after standardisation and transformation of variables.

Subsequently, the Euclidean distance (EQ_i) from the reference standard was calculated for each analysed company:

$$EQ_i = \sqrt{\frac{1}{m}\sum_{l=1}^{m}(w_{il} - w_{0l})^2}\,. \tag{9}$$

The Euclidean distance was used to determine the *taxonomic measure of attractiveness of investments* ($ETMAI_i$) for each company:

$$ETMAI_i = 1 - \frac{EQ_i}{\max_i\{EQ_i\}}\,. \tag{10}$$

MTMAI values are determined in a similar manner, but the distance between objects is calculated differently. Calculation of Mahalanobis distance takes into account the links between diagnostic measures. The Mahalanobis distance is a generalisation of the Euclidean distance.

When a correlation matrix of the diagnostic variables is an identity matrix, Mahalanobis distance equals Euclidean distance based on standardized variables (McGarigal et al. 2000). The Mahalanobis distance could be calculated as follows (Mahalanobis 1936):

$$MQ_i = \sqrt{(W_i - W_l)\cdot C^{-1}\cdot (W_i - W_l)^T}\,, \tag{11}$$

where W_i is a row vector, $W_i = [w_{i1}, \ldots, w_{i4}]$, W_l is a row vector, representing "the ideal quoted company" $W_l = [w_{01}, \ldots, w_{04}]$, C is covariance matrix for diagnostic variables.

It is not necessary to standardise the variables earlier when calculating the Mahalanobis distance (Balicki 2009).

The Mahalanobis distance can be used similarly to the Euclidean distance to calculate *taxonomic measure of attractiveness of investments* (*MTMAI*) for each company:

$$MTMAI_i = 1 - \frac{MQ_i}{\max_i\{MQ_i\}}\,. \tag{12}$$

4 Empirical Results

The study covered 10 of the largest and most liquid companies listed on The Warsaw Stock Exchange (included in the WIG20 index), excluding financial institutions.[1] The study was based on quarterly rates of return calculated based on daily closing

[1]The three-letter abbreviations used at the Warsaw Stock Exchange are used in the paper instead of the full names of stock issuers.

Table 1 Correlation matrix for diagnostic variables

	CR	*ROA*	$1/DR$	E/P
CR	1.0000	0.4492	0.2875	0.0544
ROA	0.4492	1.0000	0.2914	**0.6894**
$1/DR$	0.2875	0.2914	1.0000	0.2943
E/P	0.0544	**0.6894**	0.2943	1.0000

Source: the author's own calculation with *Statistica*

prices during the period from January 1, 2010 until March 22, 2011. Rates of return were computed as relative increases in prices of stocks according to the formula:

$$R_{it} = \frac{N_{i,t+s} - N_{it}}{N_{it}} \cdot 100\,\% \,, \tag{13}$$

where R_{it} is the rate of return on security i at time t, s is the length of the investment process expressed in days, N_{it} is the listed value of security i at time t, $N_{i,t+s}$ is the listed value of security i after s days of investing started at time t.

The share closing price on March 22, 2011, was taken as the market share price of a company. Financial ratios were calculated for each company based on annual financial reports for 2010. The correlation coefficients for diagnostic variables were calculated and the correlation matrix is presented in Table 1.

There is significant linear correlation (significance level $\alpha = 0.05$) between *ROA* and E/P. In such a case, Euclidean distance should not be applied to calculate the *taxonomic measure of attractiveness of investments*, because the assumption of absence of linear correlation between variables is not fulfilled. The Mahalanobis distance is a better distance measure.

For each of the analysed companies, *taxonomic measure of attractiveness of investments* was determined in two variants, using the Euclidean distance (*ETMAI*) or the Mahalanobis distance (*MTMAI*). Based on time series of rates of return, mean rate of return and variance were calculated. Profitability, risk and *taxonomic measures of attractiveness of investments* are presented in Table 2.

During the analysed period, the most profitable companies were at the same time the most risky ones. The two methods of calculation of *taxonomic measure of attractiveness of investments* indicate a similar arrangement of companies. Correlation coefficient between *MTMAI* and *ETMAI* for the analysed companies is 0.921. The Spearman's rank correlation coefficient is slightly lower and is equal to 0.903. KGHM is the best company with regard to both the *taxonomic measures of attractiveness of investments*, whereas PKN is the worst one. For seven companies, the absolute difference between ranks calculated from *ETMAI* and *MTMAI* was equal to one. The largest difference is for PGN, which is ranked fourth with respect to *ETMAI*, and seventh with respect to *MTMAI*.

The values of *ETMAI* for all the analysed companies were higher than those of *MTMAI*. The existence of correlations between diagnostic variables caused *ETMAI* to increase the fundamental strength of the companies under examination. The calculated values of *ETMAI* and *MTMAI* were used to create the fundamental

Table 2 Profitability, risk and *taxonomic measures of attractiveness of investments*

Company	Mean (%)	Variance	*ETMAI*	Rank	*MTMAI*	Rank
ACP	−2.25	18.25	0.7777	2	0.5766	3
KGH	16.26	299.22	0.8672	1	0.7195	1
LTS	9.22	137.19	0.3593	9	0.1519	8
LWB	12.41	208.32	0.4533	6	0.4287	5
PBG	0.03	36.78	0.7668	3	0.6674	2
PGE	0.39	27.81	0.4705	5	0.4526	4
PGN	8.37	53.24	0.4710	4	0.1949	7
PKN	1.61	86.57	0	10	0	10
TPS	−0.15	72.87	0.4455	7	0.2819	6
TVN	3.11	132.35	0.3682	8	0.1166	9

Source: the author's own calculation

effective portfolios in accordance with model (1)–(5), for selected levels of target rate of return ($\gamma = 4, 8, 12\,\%$). The model takes into account, alternatively, two levels of average *TMAI* required by the investor for the companies in the portfolio ($TMAI_\gamma = 0.5, 0.6$). The limitation for the mean rate of return for all the portfolios was an active limitation, therefore each time the mean rate of return was equal to the target rate of return; the values are listed in one line in Table 3. Efficient fundamental portfolios calculated based on Euclidean distance will be referred to as *EFP* and those calculated from the Mahalanobis distance—as *MFP*. The compositions of *EFP* and *MPF* portfolios were compared to determine how much they are composed of the same shares (similarity). The composition of the determined portfolios (by value) and their selected characteristics are specified in Tables 3 and 4.

For two portfolios, *FTP* for $TMAI_\gamma = 0.5$ ($\gamma = 4, 12\,\%$). The limitation for $TMAI_p$ (6), was not active and the constructed portfolios had higher values of $ETMAI_p$ than $TMAI_\gamma = 0.5$.

In none of the portfolios were shares of TPS or PGE present. The compositions of the *EFP* and *MFP* portfolios were different.

The most similar were the compositions of the *EFP* and *MFP* portfolios for $TMAI_\gamma = 0.5$ and $\gamma = 12\,\%$, their composition was similar in nearly 70 %, considering the portfolio value. The lowest similarity was between *EFP* and *MFP* for $TMAI_\gamma = 0.6$ and $\gamma = 12\,\%$, which were similar only in 56 %. The values of $MTMAI_p$ for all the *EFP* portfolios were lower than 0.5. Optimisation based on *ETMAI* did not guarantee sufficient fundamental strength of selected portfolios, taking into account *MTMAI*, which is a more appropriate synthetic development measure for linear correlation between diagnostic variables. The portfolios created for $TMAI_\gamma = 0.6$ had higher values of variance than those of the corresponding portfolios for $TMAI_\gamma = 0.5$. In the classic Markowitz model, an increase in profitability of an efficient portfolio is accompanied by an increase in its risk. Similarly, increasing the portfolio's *TMAI* for an efficient fundamental portfolio, with an unchanged condition for the portfolio's profitability, results in an increase in the risk for the entire portfolio.

Table 3 Efficient fundamental portfolio for $TMAI_{\gamma} = 0.5$

	Portfolio					
Company	*EFP*	*MFP*	*EFP*	*MFP*	*EFP*	*MFP*
ACP	0.1620	0.3519	0.0423	0.1470	–	–
KGH	–	0.1218	0.0980	0.2959	0.2808	0.4689
LTS	0.1287	0.0670	0.1234	0.0614	0.1958	0.0465
LWB	0.2169	0.1686	0.1974	0.1893	0.3092	0.2181
PGE	0.2541	0.1325	–	0.0829	–	–
PGN	–	–	–	–	–	–
PKN	–	–	0.3484	0.0746	0.2142	0.1505
TPS	–	–	–	–	–	–
PBG	0.0807	0.1216	0.1818	0.1489	–	0.1160
TVN	0.1574	0.0365	0.0087	–	–	–
Similarity of portfolios composition						
Similarity	0.6474		0.6145		0.6959	
Risk, profitability and taxonomic measure of attractiveness of investment						
Variance	14.969	20.2794	32.0534	56.1232	96.8991	115.0571
Mean, γ (%)	4	4	8	8	12	12
$ETMAI_p$	**0.5594**	0.6491	0.5000	0.6439	**0.5549**	0.6448
$MTMAI_p$	0.4167	0.5000	0.3185	0.5000	0.4061	0.5000

Source: the author's own calculation with *WinQSB*

Table 4 Efficient fundamental portfolio for $TMAI_{\gamma} = 0.6$

	Portfolio					
Company	*EFP*	*MFP*	*EFP*	*MFP*	*EFP*	*MFP*
ACP	0.3516	0.3869	0.1550	0.2225	–	0.0304
KGH	0.0627	0.2126	0.2304	0.4099	0.3516	0.5897
LTS	0.0774	–	0.0284	–	0.0604	–
LWB	0.1563	0.1138	0.1340	0.1487	0.1998	0.1777
PGE	0.0551	0.2448	–	0.1332	–	0.0726
PGN	–	–	–	–	–	–
PKN	0.1169	–	0.3181	–	0.3882	0.0344
TPS	–	–	–	–	–	–
PBG	0.1292	0.0420	0.1228	0.0857	–	0.0951
TVN	0.0510	–	0.0113	–	–	–
Similarity of portfolios composition						
Similarity	0.6251		0.6051		0.56	
Risk, profitability and taxonomic measure of attractiveness of investment						
Variance	14.8215	33.4458	44.9388	76.7250	99.7682	143.4651
Mean, γ (%)	4	4	8	8	12	12
$ETMAI_p$	0.6000	0.7432	0.6000	0.7362	0.6000	0.7299
$MTMAI_p$	0.4285	0.6000	0.4149	0.6000	0.4235	0.6000

Source: the author's own calculation with *WinQSB*

Table 5 Markowitz portfolios—risk, profitability and taxonomic measure of attractiveness of investment

Variance	10.8606	26.4998	96.8991
Mean, γ (%)	4	8	12
$ETMAI_p$	0.3936	0.3007	0.3743
$MTMAI_p$	0.2687	0.2172	0.3284

Source: the author's own calculation with *WinQSB*

Table 6 Rates of return for all constructed portfolios

	M	*EFP*(0.5)	*MFP*(0.5)	*EFP*(0.6)	*MFP*(0.6)
4 %	−6.3286	−1.0064	−2.2630	−3.5206	0.9445
8 %	−3.9762	−4.0033	−1.8355	−2.6016	0.2394
12 %	0.2667	0.2667	−1.0.881	0.4973	0.6957

Source: the author's own calculation

The variance, mean and *TMAI* for classic Markowitz portfolios has been calculated and are presented in Table 5.

Markowitz portfolios had lower values of variance and *TMAI* than fundamental portfolios.

In order to compare the profitability of all constructed portfolios, the assumtion was made that portfolio was bought on March 23, 2011 and sold on June 23, 2011.

In the considered period the fundamental portfolios constructed using Mahlanobis distance for $TMAI_\gamma = 0.6$ were the most profitable (Table 6).

5 Conclusion

There was a linear correlation among the diagnostic variables for the analysed companies. The correlation was not strong; it was significant only between *ROA* and E/P and it was not significant in the other cases. The ranking of the companies based on the Euclidean distance was similar to that based on the Mahalanobis distance; for one company, there was a difference of three places between the rankings. The values of *ETMAI* were highly correlated with those of *MTMAI*, which might suggest that similar results are achieved with both distance measures. However, the values of *ETMAI* were clearly higher than *MTMAI*. If a linear correlation exists, using the Euclidean distance may indicate a higher fundamental strength of the analysed companies than the actual strength. This may result in irrational investment decisions. The portfolios constructed based on Mahalanobis distance had different composition than those constructed based on the Euclidean distance. Studies indicate that an increase in the average value of *TMAI* is accompanied by an increase in the portfolio risk. In given example the fundamental portfolios for Mahlanobis and $TMAI_\gamma = 0.6$ had the highest profitability.

When carrying out analysis of capital markets, investors should use the Mahalanobis distance or a different measure which takes into account the existence of correlations among diagnostic variables, rather than the Euclidean distance.

References

Balicki, A. (2009). Statystyczna Analiza Wielowymiarowa i jej Zastosowania Społeczno-Ekonomiczne. Gdask: Wydawnictwo Uniwersytetu Gdańskiego.

Basu, S. (1977). Investment performance of common stocks in relation to their price-earnings ratios: a test of the efficient market hypothesis. *The Journal of Finance, 32*(3), 663–682.

Everitt, B. S. (1993). *Applied multivariate data analysis* (3rd ed.). London: Edward Arnold

Łapińska, A., & Rutkowska-Ziarko, A. (2003). Wykorzystanie wybranych wskaźników analizy finansowej do budowy portfela akcji na przykładzie spółek notowanych na Giełdzie Papierów wartościowych w Warszawie. *Biul. Naukowy UWM, 23*, 57–70.

Łuniewska, M. (2005). Evaluation of selected methods of classification for the Warsaw stock exchange. *International Advances in Economic Research, 11*, 469–481.

Mahalanobis, P. C. (1936). On the generalized distance in statistics. *Proceedings of the National Institute of Sciences of India, 12*, 49–55.

Markowitz, H. (1952). Portfolio selection. *The Journal of Finance, 7*, 77–91.

McGarigal, K., Cushman, S., & Stafford, S. (2000). *Multivariate statistics for wildlife and ecology research*. New York: Springer

Ritchie, J. (1996). *Fundamental analysis. A back-to the basics investment guide to selecting quality stocks*. Chicago: Irwin Professional

Rutkowska-Ziarko, A. (2011). The alternative method of building of the fundamental portfolio of securities. In K. Jajuga & M. Walesiak (Eds.) *Taksonomia 18. Klasyfikacja i analiza danych – teoria i zastosowania*. Research Papers of Wrocław University of Economics No. 176 (pp. 551–559). Wydawnictwo Uniwersytetu Ekonomicznego we Wrocławiu, Wrocław.

Tarczyński, W. (1995). O pewnym sposobie wyznaczania składu portfela papierów wartościowych. *Stat. Rev., 1*, 91–106.

Tarczyński, W. (2002). *Fundamentalny portfel papierów wartościowych*. Warszawa: PWE.

Tarczyński, W., & Łuniewska, M. (2005). Analiza portfelowa na podstawie wskaźników rynkowych i wskaźników ekonomiczno-finansowych na Giełdzie Papierów Wartościowych w Warszawie. *Pr. Kated. Ekonom. Stat., 16*, 257–271.

Walesiak, M., & Dudek, A. (2010). Finding groups in ordinal data: an examination of some clustering procedures. In H. Locarek-Junge & C. Weihs (Eds.) *Classification as a tool for research*. Studies in classification, data analysis, and knowledge organization, Part 2 (pp. 185–192). Berlin/Heidelberg: Springer.

Multivariate Modelling of Cross-Commodity Price Relations Along the Petrochemical Value Chain

Myriam Thömmes and Peter Winker

Abstract We aim to shed light on the relationship between the prices of crude oil and oil-based products along the petrochemical value chain. The analyzed commodities are tied in an integrated production process. This characteristic motivates the existence of long-run equilibrium price relationships. An understanding of the complex price relations between input and output products is important for petrochemical companies, which are exposed to price risk on both sides of their business. Their profitability is linked to the spread between input and output prices. Therefore, information about price relations along the value chain is valuable for risk management decisions. Using vector error correction models (VECM), we explore cross-commodity price relationships. We find that all prices downstream the value chain are cointegrated with the crude oil price, which is the driving price in the system. Furthermore, we assess whether the information about long-run cross-commodity relations, which is incorporated in the VECMs, can be utilized for forecasting prices of oil-based products. Rolling out-of-sample forecasts are computed and the forecasting performance of the VECMs is compared to the performance of naive forecasting models. Our study offers new insights into how economic relations between commodities linked in a production process can be used for price forecasts and offers implications for risk management in the petrochemical industry.

M. Thömmes (✉)
Humboldt University of Berlin, Spandauer Str. 1, 10099 Berlin, Germany
e-mail: thoemmem@hu-berlin.de

P. Winker
Department of Statistics and Econometrics, Justus-Liebig-University Giessen, Licher Str. 64, 35394 Giessen, Germany
e-mail: peter.winker@wirtschaft.uni-giessen.de

B. Lausen et al. (eds.), *Algorithms from and for Nature and Life*, Studies in Classification, Data Analysis, and Knowledge Organization, DOI 10.1007/978-3-319-00035-0_43,

1 Introduction

We aim to shed light on the relationship between the prices of the raw material crude oil and derived products along the petrochemical value chain. Using vector error correction models (VECM), we explore cross-commodity price relationships in the petrochemical industry in Europe.

Petrochemical products are derived from two major raw materials, crude oils and natural gas. Before being processed into final products, petrochemicals undergo several transformations. European polymer production is almost solely based on crude oil as primary raw material and its derivative naphtha as cracker feedstock. Due to the predominance of naphtha as feedstock in the European petrochemical industry,[1] the following analysis focuses on the naphtha based steam cracking process. Naphtha is one of the products that emerge from refining crude oil. In the first step of the petrochemical value chain, naphtha is fed into a steam cracker and cracked into different fractions. In terms of volume, the major olefins made from naphtha are ethylene (about 30 %) and propylene (about 16 %). These olefins are the key building blocks of the petrochemical industry. By-products that emerge from the cracking of naphtha are, among others, butadiene and butenes (C4 fraction) as well as aromatic hydrocarbons, such as benzene (C6 fraction). In a second step of the value chain, the olefin products are polymerized to polyolefins, such as polyethylene (major derivative from ethylene) (Behr et al. 2010). The secondary petrochemical products are in turn used as feed for the production of chemical products at higher value stages.

Economic theory suggests that cross-commodity equilibrium relations exist, since the analyzed commodities are tied in an integrated production process. An understanding of the complex price relations between input and output products is important for petrochemical companies, which are exposed to price risk on both sides of their business. Their profitability is linked to the spread between input and output prices. The information about price relations and margin development is valuable for risk management decisions, operational optimization and for strategic decisions, such as investment in new cracker capacities.

The majority of the research focuses on the price relationship between crude oil and retail gasoline, see e.g. Kilian (2010). Some studies also consider other oil-derived commodities that emerge from refining crude oil, such as kerosene and naphtha. For instance, Chng (2010) studies economic linkages among commodity futures for crude oil, gasoline and kerosene in a VEC framework. However, commodities at higher stages of the value chain and especially petrochemical products are rarely considered in the literature.

This paper is organized as follows. The dataset is described in Sect. 2, followed by the empirical analysis of the cross-commodity price relations in Sect. 3, where

[1] According to the *Association of Petrochemicals Producers in Europe* about 75 % of the European ethylene production is naphtha-based. Retrieved from http://www.petrochemistry.net/ethylene-production-consumption-and-trade-balance.html.

we present the estimation results and rolling impulse response functions. Out-of-sample ethylene price forecasts are evaluated in Sect. 4. Section 5 concludes.

2 Data

The sample includes monthly commodity prices for the time period from June 1991 to June 2011. Data on petrochemical prices is provided by ICIS Pricing and obtained from Thomson Reuters Datastream. The sample for the Northwest European market covers spot prices[2] of eight commodities over several steps of the value chain, beginning with the raw material crude oil. Two crude oils commonly serve as benchmark, West Texas Intermediate and Brent. Since our analysis focuses on the European market, we use Brent, a crude which originates from the North Sea, as marker. There is no uniform measurement unit for the analyzed commodities. All petrochemical product prices are converted to US Dollar per metric ton (USD/MT), whereas the crude oil price is given in USD/barrel. For the further analysis, we used log price series. Advantages are (i) reduced skewness and (ii) that one does not need to account explicitly for the different measurement units (volume measure barrel vs. mass measure metric ton).

3 Long-Run Cross-Commodity Price Relations and Error Correction Mechanisms Along the Value Chain

In this section, we explore the cross-commodity price relations between petrochemical products along the value chain, from the raw material crude oil over major basic petrochemicals to higher-value products. The economic equilibrium mechanism can be described by error correction models. On one hand, a VECM captures the long-run relationship between the prices and the adjustment mechanisms that help to restore an equilibrium. On the other hand, a VECM also describes the short-run dynamics.

First, we examine the unit root properties of the time series employed. For the standard ADF test, the null hypothesis of non-stationarity cannot be rejected at any conventional level of significance for any of the price series. Graphical analysis suggests that the series may contain structural breaks within the sample period. Therefore, we also apply a unit root test that accounts explicitly for structural breaks

[2]Contract prices are also available but the use of contract price data is problematic because price assessment of contracts changed during the sample period. Up to 2008, contract prices in Europe have been commonly fixed on a quarterly basis. In 2009, a monthly contract price system has been implemented and for most petrochemical commodities, the quarterly contract quotes were discontinued.

(Saikkonen and Lütkepohl 2002). Since the exact break date for the price series is unknown, it was searched for in the inferred range from January 2007 to June 2009, based on a procedure proposed by Lanne et al. (2003). The null of a unit root cannot be rejected at any conventional level of significance for any of the eight time series. For the first difference of the log price series, the null of a unit root is rejected. We thus conclude that all variables are I(1).

Second, we address the issues variable selection and model specification. Consider the following error correction representation of a VAR(p) where the variables of interest, in our case commodity prices, are contained in a $k \times 1$ vector $\mathbf{P}_t = [P_{1t}, P_{2t}, \ldots, P_{kt}]'$ with $t = 1, \ldots, T$:

$$\Delta \mathbf{P}_t = \boldsymbol{\mu} + \boldsymbol{\Pi} \mathbf{P}_{t-1} + \boldsymbol{\Gamma}_1 \Delta \mathbf{P}_{t-1} + \ldots + \boldsymbol{\Gamma}_{p-1} \Delta \mathbf{P}_{t-p+1} + \boldsymbol{\varepsilon}_t \,. \tag{1}$$

The vector $\boldsymbol{\varepsilon}_t$ is a k-dimensional white noise process, i.e. $E(\boldsymbol{\varepsilon}_t) = 0$, $E(\boldsymbol{\varepsilon}_t, \boldsymbol{\varepsilon}_t') = \boldsymbol{\Sigma}$ and $E(\boldsymbol{\varepsilon}_t, \boldsymbol{\varepsilon}_s') = 0$ for $s \neq t$. The deterministic components are contained in $\boldsymbol{\mu}$. The variables collected in $\mathbf{P}_t$ are considered to be I(1) based on the unit root test results. The Johansen procedure (Johansen and Juselius 1990) is applied to determine the number of cointegrating vectors, i.e. the rank of the $\boldsymbol{\Pi}$ matrix. The optimal lag length p is determined based on the Hannan Quinn information criterion (Hannan and Quinn 1979) and we include all lags up to lag p.

Due to the multitude of products along the value chain several types of models are employed: first, we estimate seven bivariate models that include two variables each, the primary raw material crude oil and one of the seven product prices. Second, a larger model is estimated, focusing on the ethylene production process. The model is referred to as model *step012* because it incorporates the prices of the raw material crude oil, the cracker feed naphtha, ethylene as primary petrochemical product, and polyethylene as secondary petrochemical product. A Johansen trace test with structural break (Johansen et al. 2000) is performed to determine the number of cointegrating relations. As deterministic component, we include intercepts in the cointegrating relations, which implies that some equilibrium means may be nonzero. For all seven pairs of product and crude oil log prices, the trace test indicates one cointegrating relation at the 5 % level of significance. Hence the variables share one common stochastic trend. Two cointegrating relations are indicated for model *step012*, and thus two stochastic trends. In sum, the Johansen trace tests give evidence that all prices downstream the value chain are cointegrated with the crude oil price.

The estimation results for the bivariate VECMs (Table 1) suggest that the speed of adjustment differs across products at different steps of the value chain. The estimated loading coefficients indicate that the adjustment speed declines as one moves downstream the value chain. While the price correction mechanism is relatively fast for naphtha, prices of products at higher stages of the value chain adjust more slowly to deviations from the long-run equilibrium. The relatively slow price transmission to higher-value products can be explained by the fact that feedstocks are predominantly purchased through long-term contracts and thus price changes are smoothed. The crude oil price is found to be the driving price in the

Table 1 Estimated loading coefficients and long-run elasticities for the bivariate VECMs. All coefficients are significant at the 5 % level. Detailed estimation results are available from the authors upon request

	Step 0	Step 1				Step 2	
Product	Naphtha	Ethylene	Propylene	Benzene	Butadiene	Polyethylene	Polypropylene
LR elasticity	0.9185	0.6692	0.7966	0.7613	1.0724	0.5107	0.4109
Adjustment speed	−0.3632	−0.0951	−0.1302	−0.1292	−0.1197	−0.0899	−0.0700

system: A likelihood-ratio test is employed to test for long-run weak exogeneity of the crude oil price. The results indicate that only the prices of refined products move in order to compensate for deviations from the long-run equilibrium. Explanations are the market structure in the oil industry and the fact that only a minor share of the crude oil production goes into the petrochemical industry. Furthermore, we infer that (i) long-run elasticities for products at the same value step appear to lie close together and (ii) long-run elasticities for higher-value products are lower than for basic products.[3] A possible explanation is that the vast majority of cost for higher-value products is not feedstock-related, but can be attributed to other cost components, e.g. capital cost.

To be able to interpret and understand the dynamic patterns of model *step012*, we compute rolling impulse response functions (IRF). IRFs allow to visualize the transmission of a shock through the system. In particular, rolling impulse responses are useful to get a better notion of the structural (in)constancy in the system and to capture gradual changes of price relations. We do so by using a fixed window of 80 months, with the first window starting June 1991 and being moved forward by one month until the sample end is reached. We use orthogonalized IRFs and order the variables according to their occurrence downstream the value chain. Hence, crude oil is first, followed by naphtha, ethylene and polyethylene. Figure 1 shows the rolling IRFs in a three-dimensional space. The notation is *impulse variable → response variable*.

The IRFs for *brent → ethylene* exhibit similar patterns for most IRFs in the sample period. An exception is the IRF for the subsample running from Feb 2002 to Oct 2008. The end of this subsample coincides with the beginning of the sharp price decline at the end of 2008. As expected, a shock in the oil price has a positive effect on the ethylene price. Strikingly, the effect of a shock in the brent price does not die out to zero. Such a long-term effect of a one-time shock reflects the nonstationarity of the system. The IRFs *polyethylene → ethylene* exhibit a substantial variation in their patterns. This is an indication for structural inconstancy. The feedback effect from the higher-value product polyethylene to the lower value product ethylene changes substantially over the sample period. For

[3]Due to potential structural breaks, we also estimate bivariate VECM for a subsample from 1991 to 2007. The findings remain qualitatively unchanged.

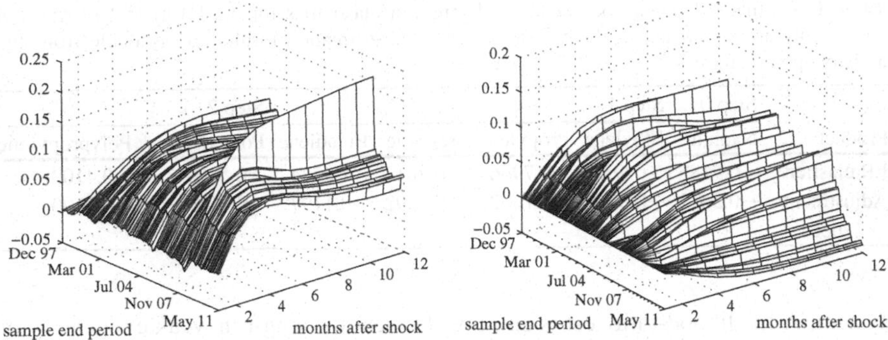

Fig. 1 Rolling impulse response functions for model *step012*. The IRFs *brent* → *ethylene* are depicted on the *left hand side*, IRFs *polyethylene* → *ethylene* are displayed on the *right hand side*

the subsamples up to the third quarter 2008, a polyethylene price shock leads to a positive ethylene price response. This effect becomes smaller as the estimation period is moved forward. The response turns negative for subsamples including data after Q3 2008. Note that for the subsamples with negative responses, the zero line is included in bootstrapped confidence bands (not depicted), thus the response is not significantly different from zero. To further investigate the causes of the changing price relations it would be useful to incorporate information on quantities, such as cracker capacities, production and trading volume. However, limited data availability so far restricts a further empirical analysis.

4 Evaluating Ethylene Price Forecasts

We analyze whether information on current deviations from the equilibrium, which is incorporated in our VECMs, can be used for price forecasts. The focus is on ethylene price forecasts because ethylene is the most important feedstock for chemical products at higher steps of the value chain. Furthermore, we focus explicitly on short-horizon forecasts since these are most relevant for practical application. We employ two of the models described in Sect. 3: a bivariate VECM (log brent price and log ethylene price as endogenous variables) and a model that accounts for interrelations of the prices of products on different steps in the value chain. The second model incorporates log prices of four commodities: brent crude oil, naphtha, ethylene and polyethylene. A series of dynamic out-of-sample forecasts are carried out for time horizons 1–3 months for the time period January 2000 to June 2011, resulting in 136 (3 months) to 138 (1 month) forecasts. We use a fixed-length rolling window approach (window size: 103 months) which is advantageous if the process is subject to structural breaks.

Table 2 Quantitative and direction-of-change forecast evaluation for VECMs. For mDM test and χ^2 test *(**) indicates significance at 10 % (5 %) level

				Theil's U		m. Diebold Mariano		χ^2 test of indep.	
				Benchmark:		Benchmark:			
Modell	h	RMSE	MAE	(a) RW	(b) VAR	(a) RW	(b) VAR	Success rate (%)	χ^2 stat
VEC1	1	86.58	64.30	0.82	0.97	−1.8894*	−1.4386	66	11.7363**
	2	152.17	112.58	0.86	0.96	−1.6587*	−1.6086	62	5.9150**
	3	203.69	148.01	0.91	0.96	−1.5695	−1.5434	56	0.9523
VEC2	1	92.45	69.10	0.87	1.04	−1.4007	0.9649	68	15.8429**
	2	165.11	121.68	0.94	1.04	−0.8462	0.8725	67	13.7102**
	3	221.86	157.29	0.99	1.05	−0.1516	0.9059	55	0.8801

First, the forecasts are evaluated with regard to quantitative forecast performance by computing root mean squared error (RMSE), as well as mean absolute error (MAE). Furthermore, we consider a bivariate VAR(2) model in first differences of the log prices and a random walk without drift (RW) as benchmark models. Theil's U is applied to evaluate the usefulness of our forecasting models relative to the benchmark models. It is computed as the ratio of the RMSE of our forecasting model to the RMSE of the benchmark model.

The results of the performance evaluation of the ethylene price forecasts are presented in Table 2. More accurate forecasts produce smaller RMSE/MAE. Both VECMs exhibit smaller RMSE than the RW, yet the performance records of VECMs and the VAR(2) are very similar. Given the very small differences in forecast performances, the question arises as to whether the differences are statistically significant. On account of this, we perform a modified Diebold Mariano (mDM) test (Diebold and Mariano 1995). The null hypothesis of equal forecast accuracy of RW and bivariate VECM is rejected at the 10 % level for 1-month and 2-months ahead forecasts. Thus, the bivariate VECM performs significantly better than the RW benchmark for short forecasting horizons. Comparing VECMs and the VAR(2), we do not find significant differences in forecast accuracy. The similarity in forecast performance of VECMs and VAR(2) is an indication that the VECMs, which are restricted VARs, are reasonably specified.

Second, we evaluate how well the models predict the direction of price changes, with the variable of interest being the change in the ethylene price. If our forecasting models are able to predict the direction of the price change, the models would be valuable for risk management and decision making processes. A Chi-squared test of independence is employed. It assesses whether paired outcomes of two binary variables are independent. For this application, the binary variables actual and predicted price change are set to one if $P_t > P_{t-h}$ and to zero if $P_t < P_{t-h}$, with forecast horizon $h = 1,\dots,3$ months and $t = 1,\dots,136$. Under the null, independence, the test statistic is approximately χ_1^2 distributed.

For both VECMs, the null of statistically independent outcomes is rejected at the 5 % level for 1- and 2-months-ahead forecasts (Table 2). In contrast, the null hypothesis cannot be rejected for a forecast horizon of 3 months. Based on these test results, we infer that the VECMs are useful in predicting the direction of the price change for short horizons. The models predict about 66–68 % of the signs of the price changes from 1 month to the next month correctly. Interestingly, the second forecasting model VEC2 achieves higher success rates for directional forecasts whereas the quantitative performance does not stand out. A possible explanation is that the incremental information about price relations along the value chain enhances directional forecast performance, but at the price of increased variance, which becomes apparent in slightly higher RMSE and MAE compared to VEC1 or VAR(2).

5 Conclusion

In this study, the attention is drawn on the relationship between prices of crude oil and products in the petrochemical industry in Europe. Using a multivariate and seven bivariate VECMs, cross-commodity price relations along the chemical value chain are examined. We find that the crude oil price is the driving price in the system and only the prices of derived products move in order to compensate for deviations from the long-run equilibrium. Furthermore, while the price correction mechanism is relatively fast for naphtha, prices of products at higher steps of the value chain revert more slowly towards the long-run equilibrium. Also, long-run elasticities for higher-value products are lower than for basic products. These findings are of value for petrochemical producers' risk management: the information about deviations from the long-run equilibrium and likely price movements back towards the equilibrium can be incorporated into the analysis of price risk. The evaluation of the VECM's forecast performance shows that the bivariate VECM is significantly better than a random walk for short horizon forecasts, and both forecasting models produce valuable forecasts of the direction of the ethylene price change.

An avenue for future research are nonlinearities in the models. For instance, the adjustment process towards the equilibrium may be nonlinear and the long-run price relations may show threshold effects. Such effects could be related to capacity changes or to changing margins. The latter reflects the idea that the overall profit situation in the industry may play a role in the adjustments of downstream prices to crude oil price changes.

References

Behr, A., Agar, D.W., & Jörissen, J. (2010). *Einführung in die Technische Chemie*. Heidelberg: Spektrum.

Chng, M. (2010). Comparing different economic linkages among commodity futures. *Journal of Business Finance and Accounting, 37*, 1348–1389.

Diebold, F., & Mariano, R. S. (1995). Comparing predictive accuracy. *Journal of Business & Economic Statistics, 13*, 253–263.

Hannan, E., & Quinn, B. (1979). The determination of the order of an autoregression. *Journal of the Royal Statistical Society. Series B, 41*, 190–195.

Johansen, S., & Juselius, K. (1990). Maximum likelihood estimation and inference on cointegration. *Oxford Bulletin of Economics and Statistics, 52*, 169–210.

Johansen, S., Mosconi, R., & Nielsen, B. (2000). Cointegration analysis in the presence of structural breaks in the deterministic trend. *Econometrics Journal, 3*, 216–249.

Kilian, L. (2010). Explaining Fluctuations in Gasoline Prices: A Joint Model of the Global Crude Oil Market and the U.S. Retail Gasoline Market. *The Energy Journal, 31*(2), 87–112.

Lanne, M., Lütkepohl, H., & Saikkonen, P. (2003). Test procedures for unit roots in time series with level shifts at unknown time. *Oxford Bulletin of Economics and Statistics, 65*, 91–115.

Saikkonen, P., & Lütkepohl, H. (2002). Testing for a unit root in a time series with a level shift at unknown time. *Econometric Theory, 18*, 313–348.

Part VIII
Marketing and Management

Lifestyle Segmentation Based on Contents of Uploaded Images Versus Ratings of Items

Ines Daniel and Daniel Baier

Abstract Clustering algorithms are standard tools for marketing purposes. So, e.g., in market segmentation, they are applied to derive homogeneous customer groups. However, recently, the available resources for this purpose have extended. So, e.g., in social networks potential customers provide images which reflect their activities, interests, and opinions. To compare whether contents of uploaded images lead to similar lifestyle segmentations as ratings of items, a comparison study was conducted among 478 people. In this paper we discuss the results of this study that suggests that similar lifestyle segmentations can be found. We discuss advantages and disadvantages of the new approach to lifestyle segmentation.

1 Introduction

Since Smith's introductory article (Smith 1956), market segmentation is associated with the division of a heterogeneous market into a number of smaller homogeneous markets, in order to provide a possibility to satisfy the customers' desires more precisely. Lifestyle segmentation is one possibility for this purpose when markets with similar activities, interests and opinions are desired. It has received a wide attention among consumer researchers in the last 50 years (Wedel and Kamakura 2000; Goller et al. 2002; Boejgaard and Ellegaard 2010). They have proposed and applied a huge variety of approaches, most of them using questionnaires with lifestyle items whose ratings formed the basis for segmentation. However, in recent studies, the authors have demonstrated that – instead – it is also possible to group customers based on the contents of uploaded digital images that describe the consumers' preferred activities, opinions, and interests (see Baier and Daniel 2012).

I. Daniel (✉) · D. Baier
Institute of Business Administration and Economics, Brandenburg University of Technology Cottbus, 03013 Cottbus, Germany
e-mail: ines.daniel@tu-cottbus.de; daniel.baier@tu-cottbus.de

B. Lausen et al. (eds.), *Algorithms from and for Nature and Life*, Studies in Classification, Data Analysis, and Knowledge Organization, DOI 10.1007/978-3-319-00035-0_44,

For this purpose, a huge variety of image feature selection and clustering algorithms exist in Content Based Image Retrieval (Rubner and Tomasi 2001). They have been compared with respect to their ability to derive homogeneous lifestyle segments (see Daniel and Baier (2012)). However, in these publications it was not analyzed whether the derived segmentations are similar to the ones that could be obtained in the traditional way using ratings of items in questionnaires. Therefore, the research question for this paper is the following: Is it possible to replace or reduce item batteries in lifestyle segmentation by ratings of uploaded or predefined images? To which extent does this replacement lead to a similar lifestyle segmentation?

The paper is organized as follows: After this introduction, Sect. 2 is dedicated to a short description of lifestyle segmentation and its popular traditional approaches. Then, in Sect. 3, we present our empirical investigation where lifestyle segmentation based on ratings of items is compared to lifestyle segmentation based on ratings of images and clustering of preferred images. Finally, in Sect. 4 we discuss additional advantages and disadvantages of using image contents for lifestyle approaches.

2 Lifestyle Segmentation

Lifestyle is a field of interest in different areas of science, such as sociology, psychology, geography, economics and marketing with various definitions. At first the sociologist Georg Simmel used the term lifestyle in 1900 in a case of social differentiation (Simmel 1990). Another important concept of lifestyle was introduced by Max Weber in 1922. He distinguished classes and status groups. Classes are defined by economical factors, status groups by conduct of life (Weber 1922).

In marketing, the lifestyle concept was introduced by William Lazer in 1964. He defined lifestyle as "… a systems concept. It refers to a distinctive or characteristic mode of living, in its aggregate and broadest sense, of a whole society or segment thereof … The aggregate of consumer purchases, and the manner in which they are consumed, reflect a society's lifestyle." (Lazer 1964, p. 130). Since this time many methods for measuring lifestyle were developed. One of the most famous and often used methods consists of measuring activities, interests and opinions (AIO). Activities describe how to spend one's time, e.g., work, hobbies, shopping, doing sports. Interests describe how to set value on the surroundings, e.g., on family, home, community, opinions how to view oneself, politics, education and so on (Plummer 1974). A representative sample of consumers is grouped on the basis of their ratings w.r.t. a battery of such AIO items. The results are homogeneous groups of consumers, so-called lifestyle segments.

The AIO approach was introduced by Wells and Tiggert (see Wells and Tigert 1971). An overview about applications of AIO studies is given in Gonzalez and Bello (2002) or Ahmad et al. (2010). Especially the early studies included more then

200 AIO items. For example the study of Wells and Tigert (1971) used 300 items. These items were analyzed by factor analysis and reduced to 22 dimensions (Wedel and Kamakura 2000). In later studies the number of items was reduced even more severely. For example Lee et al. (2009) only used 18 lifestyle items in a study of technology products. Some problems are associated with typical AIO studies. One problem is the large number of items and that every item has to be rated by a scoring scale. So the respondent must think about every item, which produces failures. The questionnaires are also self administered. If a respondent does not understand a question, it is not possible to clarify it. These facts cause a long time to complete the questionnaires. Because of the large amount of items multiple statistical analyses are used to reduce the number of variables. Causing that a part of information is not handled (Gonzalez and Bello 2002).

Another way to analyze lifestyles is to use ratings of values as the basis for the psychographic segmentation. In this area researchers argue that values are closer to behavior than personality traits and refer more to motivation than attitudes. They also have fewer items (Wedel and Kamakura 2000). One of the important surveys is the Rokeach Value Survey (RVS). Rokeach defines a value as: "an enduring belief that a specific mode of conduct or end-state of existence is personally or socially preferable to an opposite or converse mode of conduct or state of existence" (Rokeach 1973). He identified two types of values: terminal values and instrumental values. Terminal values are the goals which a person would achieve during the lifetime like a world at peace, pleasure, or family security. Instrumental values are modes of behavior like loving, polite, independent or capable. At each value type consumers have to rank 18 values by their importance. In many studies Rockeach's Value Scale was used to describe the value structure of populations. Applications of the model were also used to explain differences in value systems among market segments (Wedel and Kamakura 2000).

3 Empirical Investigation

In Sect. 2 we presented some approaches for lifestyle segmentation in marketing. All these approaches do not use image contents to analyze the consumer behavior. In our research we detected only one research project which uses images to get a better classification of consumers. The Sinus Sociovision accomplished a survey about the living style of German people. During this survey they taken some pictures of the apartments of the people. With the help of these pictures it was possible to get better descriptions about the consumers living styles, product needs, and habits (Sinus-Institut Heidelberg 2011). In this section we present our results to answer the open research question: Is it possible to replace or reduce item batteries in lifestyle segmentation by uploading or rating predefined images (as a proxy for uploading this image)? For this purpose a traditional lifestyle questionnaire with a section with items concerning activities, interests and opinions was extended by a section where similar activities, interests and opinions were presented via images and the consumers had to rate them instead. So, in the analysis, the groupings based on the ratings of the items could be compared to the groupings based on the ratings of the images. Moreover, the image contents could be used – via image feature extraction and clustering w.r.t. these features – to group the images (to be concrete: the most preferred images) and to compare these results to the already derived groupings.

3.1 Research Design

For our investigation a questionnaire with typical lifestyle statements and additional images was created. The survey was constrained to the field holiday – activities, interests and opinions in this field – and concentrates on the three different alternatives mountains (climbing and hiking), sunset (swimming and sun bathing) as well as city-lights (entertainment in the city). For each alternative (mountains, sun on the beach, and city light), possible activities, interests and opinions were described using five items as well as five images. Additional items and images were used to disguise this concentration.

Overall 519 participants filled out the questionnaire. In a first step we analyzed the quality of the survey. With the help of standard selection criteria the number of usable responses was reduced to 478 participants. In this process respondents with too many missing answers and with too high inner an item variance (over four in our case) were eliminated. In a second step we analyzed the reliability and validity of the items using first generation assessment according to Homburg and Giering (1996). We also checked the Pearson correlation coefficient. In a third step on the one hand we analyzed the answers about the images and at the other hand we analyzed the answers about the holiday statements. All items used a seven point scale ("definitely not applicable for me (=1)" to "definitely applicable for me (=7)"). In both cases we executed a cluster analysis using k-means with respect to these ratings. After that we compared the matching of both analyses. In a fourth step we analyzed the used images by feature extraction and clustering w.r.t. these features. In this case we implemented Lab-cubes of the images (see details in Chap. 3.2) and compare them – according to the propositions in Baier and Daniel (2012) – by hierarchical cluster analysis using ward algorithm and quadratic euclidean distance. These results were compared with the results of the hierarchical cluster analysis of the transposed data matrix with the ratings of the 15 images.

3.2 Results

After adjusting the responses we analyzed the data by exploratory factor analysis, Cronbach's Alpha and item-to-total correlation. In Table 1 the results of the analysis

Table 1 Results of exploratory factor analysis, Cronbach's Alpha and item-to-total correlation w.r.t. to the ratings of images

	Rated image	Loading	Variance accounted for (%)	Corrected item to total correlation	Cronbach's alpha
Mountains	Image 1	0.879	31,397	0.806	0.920
	Image 2	0.922		0.868	
	Image 3	0.914		0.850	
	Image 4	0.670		0.581	
	Image 5	0.919		0.869	
Sunset	Image 1	0.834	23,845	0.755	0.923
	Image 2	0.876		0.805	
	Image 3	0.885		0.815	
	Image 4	0.853		0.795	
	Image 5	0.881		0.829	
City-lights	Image 1	0,701	17,404	0,550	0.844
	Image 2	0.836		0.713	
	Image 3	0.828		0.703	
	Image 4	0.670		0.554	
	Image 5	0.851		0.735	

Table 2 Results of exploratory factor analysis, Cronbach's Alpha and item-to-total correlation w.r.t. to the ratings of items

	Rated item	Loading	Variance accounted for (%)	Corrected item to total correlation	Cronbach's alpha
Mountains	Item 1	0,814	31,792	0,732	0.910
	Item 2	0.863		0.802	
	Item 3	0.866		0.783	
	Item 4	0.841		0.745	
	Item 5	0.891		0.824	
City-lights	Item 1	0,826	24,900	0,710	0.858
	Item 2	0.849		0.742	
	Item 3	0.754		0.619	
	Item 4	0.712		0.583	
	Item 5	0.829		0.722	
Sunset	Item 1	0.907	15,644	0.805	0.880
	Item 2	0.897		0.791	
	Item 3	0.868		0.709	

of the image ratings are shown. In Table 2 the results of the lifestyle items are shown. In both cases the exploratory factor analysis led to three factors, which can be described as mountains, sunset and city-lights. The case of the images the three factors explained 72.646 % of the variance, in the case of statements the three factors explained 72.336 % of the variance. After analyzing Cronbach's Alpha and the item-to-total correlation only two items of the sunset statements had to be deleted.

Table 3 Comparison of consumer groupings on the basis of the ratings of items and of images (numbers and percentages of consumers)

	Numbers of consumers		Percentage of customers		Identical
Grouping	Item based	Image based	Item based (%)	Image based	consumers
A	114	98	23.8	20.5 %	64
B	156	137	32.6	28.7 %	99
C	208	243	43.5	50.8	192
					74.26 %

They do not fulfill the requirements of Cronbach's Alpha larger than 0.7 (see Nunnally (1978)) and an item-to-total correlation larger than 0.5 (see Bearden et al. 1989).

After that we calculated the Pearson correlation between the images and the items. W.r.t. mountains we got a correlation of 0.773, w.r.t. sunset 0.681 and w.r.t. city-lights 0.651. All coefficients are significantly different from 0.

Our main question in this investigation was to find out, if it is possible to replace lifestyle statements by images. Therefore, we compared the grouping based on the rating of the items with the grouping based on the rating of the images. In both cases we derived a three-segment-solution ("A", "B", "C") which was labeled according to a maximum number of consumers which are both groupings in the "A","B", and "C" group. Table 3 shows the sizes and the consensus of these two groupings. 74.26 % of the consumers are in groups with the same label w.r.t. to both groupings, suggesting that – in spite of the different approaches – the groupings are close to each other. So it is possible to say, that in the case of holiday surveys it is possible to use images instead of questions to get a reliable classification of tourists.

Moreover, we compared the image contents by using feature extraction methods with the ratings of the images. For feature extraction, three-dimensional color histograms (cubes) were used: Color values are typically coded by intensities w.r.t. three underlying color dimensions, e.g. red, green and blue. The color cube is a partitioned fixed sized representation of the color distribution. Every partition reflects the percentage of pixels within the predefined intensity interval. According to Wyszecki and Stiles (2000) as well as Daniel and Baier (2012) the CIE-L*a*b color model was used for describing the colors of each pixel. This model was used because Euclidean distances better reflect the human's perception of color differences than the usual coding via the RGB (red-green-blue) color model. The color cubes and the ratings of the images were analyzed by hierarchical cluster analysis (Ward algorithm and quadratic euclidean distance). Figure 1 displays the results. In both cases we received three clusters with the grouping we had expected. So it is possible to say, clustering of images by the feature extraction methods is an additional way to classify favorite pictures of consumers.

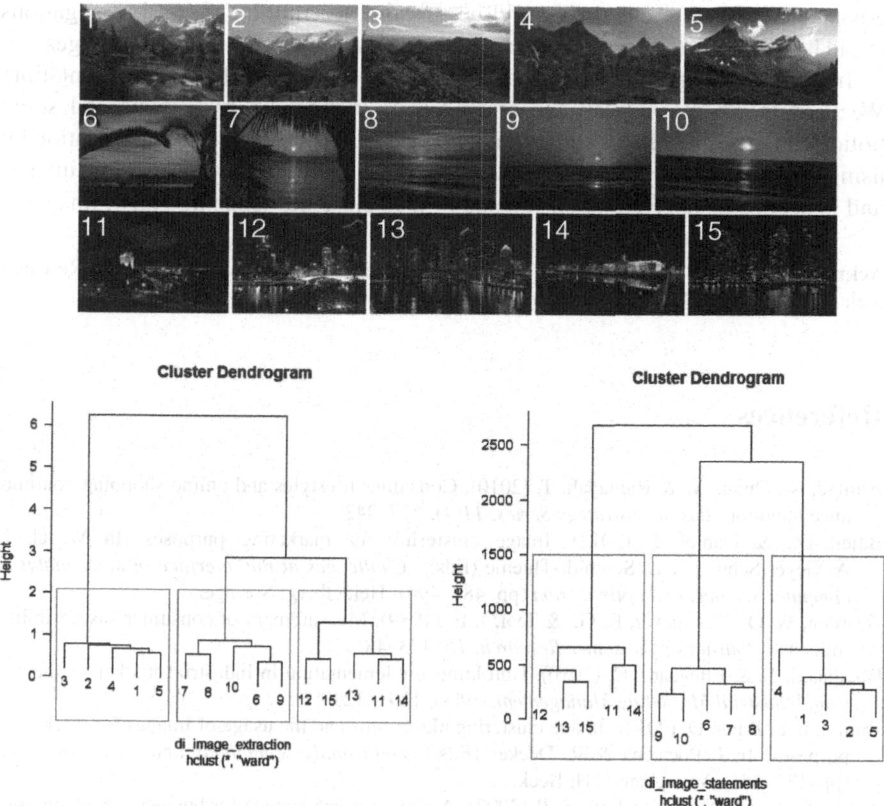

Fig. 1 Images (1–5: mountains, 6–10: sunset, 11–15: city lights) and results of clustering (Ward, quadratic Euclidean) based on image ratings (*left*) and image features (*right*)

4 Conclusion and Outlook

In our investigation we tested if it is possible to analyze lifestyles by using images. The results show, that this is possible in the case of holidays. You can derive the same classifications. The advantage by using images is that the respondent does not read a high number of questions about a theme, the respondent only has to look at some images and has to say, this is a typical situation of my life or not. Another advantage is that images are easy to understand. Often respondents have problems by answering items, because they do not know, what is meant by this question. But using images also has a disadvantage. In preparation of the investigation it is very important to use meaningful images. For example if you want to know some preferences about city trips it is a difference if you use an image of old monuments or a busy city. Some people like big cities but do not like old monuments, so it is important to differentiate between this images. It is also very important to make some pretests about the meaning of images before investigation. To get the advantages of both methods it could be possible to combine them. Investigations could be shorter and mixed up by using a combination of questions and images.

In our investigation we analyzed the usage of images in lifestyle segmentation. We used the theme holiday and created some typical lifestyle items and some holiday images. Our results show, that it is possible to get a lifestyle classification by using images. In following researches it has to be analyzed how to combine images and items in a balanced relationship and how to close the semantic gap of images.

Acknowledgements This research is funded by Federal Ministry for Education and Research under grants 03FO3072. The author is responsible for the content of this paper.

References

Ahmad, N., Omar, A., & Ramayah, T. (2010). Consumer lifestyles and online shopping continuance intention. *Business Strategy Series, 11*(4), 227–243.

Baier, D., & Daniel, I. (2012). Image clustering for marketing purposes. In W. Gaul, A. Geyer-Schulz, & L. Schmidt-Thieme (Eds.), *Challenges at the interface of data analysis, computer science, and optimization* (pp. 487–494). Heidelberg: Springer.

Bearden, W. O., Netemeyer, R. G., & Teel, J. E. (1989). Measurement of consumer susceptibility influence. *Journal of Consumer Research, 15*, 473–481.

Boejgaard, J., & Ellegaard, C. (2010). Unfolding implementation in industrial market segmentation. *Industrial Marketing Management, 39*(8), 1291–1299.

Daniel, I., & Baier, D. (2012). Image clustering algorithms and the usage of images for marketing purposes. In J. Pociecha & R. Decker (Eds.), *Data analysis methods and its applications* (pp. 171–181). Warschau: C.H. Beck.

Goller, S., Hogg, A., & Kalafatis, S. P. (2002). A new research agenda for business segmentation. *European Journal of Marketing, 36*(1/2), 252–271.

Gonzalez, A. M., & Bello, L. (2002). The construct lifestyle in market segmentation – the behavior of tourist consumers. *European Journal of Marketing, 36*, 51–85.

Homburg, Ch., & Giering, A. (1996). Konzeptualisierung und Operationalisierung komplexer Konstrukte – Ein Leitfaden fuer die Marketingforschung. *Marketing Zeitschrift fuer Forschung und Praxis, 18*(1), 5–24.

Lazer, W. (1964). Life style concepts and marketing. In A. Greyser (Ed.) *Toward scientific marketing* (pp. 130–139). Chicago: American Marketing Association.

Lee, H. -J., Lim, H., Jolly, L. D., & Lee, J. (2009). Consumer lifestyles and adoption of high-technology products: a case of South Korea. *Journal of International Consumer Marketing, 21*(2), 153–167.

Nunnally, J. C. (1978). *Psychometric theory* (2nd ed.). New York: McGraw-Hill.

Plummer, J. T. (1974). The concept and application of life style segmentation. *The Journal of Marketing, 38*(1), 33–37.

Rokeach, M. (1973). *The nature of human values* (1st ed.). New York: Free Press.

Rubner, Y., & Tomasi, C. (2001). *Perceptual metrics for image database navigation*. Boston: Kluwer.

Simmel, G. (1900). *Philosophie des Geldes* (1st ed.). Berlin: Duncker und Humblot.

Sinus-Institut Heidelberg (2011). *Informationen zu den Sinus-Milieus 2011*, http://www.sinus-institut.de/uploads/tx_mpdownloadcenter/Informationen_Sinus-Milieus_042011.pdf (checked on March 11, 2013).

Smith, W. (1956). Product differentiation and market segmentation as alternative marketing strategies. *The Journal of Marketing, 21*, 3–8.

Weber, M. (1922). *Wirtschaft und Gesellschaft* (1st ed.). Tübingen: J.C.B. Mohr (P. Siebeck).

Wedel, M., & Kamakura, W. A. (2000). *Market segmentation: conceptual and methodological foundations* [International series in Quantitative marketing] (2nd ed.). Boston: Kluwer.

Wells, W. D., & Tigert, D. (1971). Activities, interests and opinions. *Journal of Advertising Research, 11*, 27–35.

Wyszecki, G., & Stiles, W. (2000). *Color science. Concepts and methods, quantitative data and formulae* (2nd ed.). New York: Wiley.

Optimal Network Revenue Management Decisions Including Flexible Demand Data and Overbooking

Wolfgang Gaul and Christoph Winkler

Abstract In aviation network revenue management it is helpful to address consumers who are flexible w.r.t. certain flight characteristics, e.g., departure times, number of intermediate stops, and booking class assignments.

While overbooking has some tradition and the offering of so-called flexible products, in which some of the mentioned characteristics are not fixed in advance, is gaining increasing importance, the simultaneous handling of both aspects is new. We develop a DLP (deterministic linear programming) model that considers flexible products and overbooking and use an empirical example for the explanation of our findings.

1 Introduction

Revenue management deals with the sale of products in a certain time period with the objective of maximizing profit margins. A restrictive aspect is that after the end of the period the sale of the offered products is often not possible, e.g., in the airline industry an untaken seat in a plane after takeoff is useless.

We will concentrate on aviation network revenue management, for which quite a lot of various data have to be considered. Important data are demand and prices w.r.t. the offered products (see Talluri and Van Ryzin (2004)). In case of overbooking (the sale of seats exceeds the physical capacity) further data are the amounts of no-shows (passengers that don't appear for take-off) and costs for rejected customers (see Bertsimas and Popescu (2003), Karaesmen and Van Ryzin (2004a), Karaesmen and Van Ryzin (2004b)).

W. Gaul (✉) · C. Winkler
Karlsruhe Institute of Technology (KIT), 76128 Karlsruhe, Germany
e-mail: wolfgang.gaul@kit.edu; christoph.winkler@kit.edu

B. Lausen et al. (eds.), *Algorithms from and for Nature and Life*, Studies in Classification, Data Analysis, and Knowledge Organization, DOI 10.1007/978-3-319-00035-0_45,

In our contribution we use the distinction in specific and flexible offerings. The sale of a specific product means that the consumer buys an exactly predefined product which implies that utilization time and other restrictions (e.g., route of the flight, booking-class) are fixed. By contrast, if the consumer buys a flexible product the product consists of a list of several alternatives that the consumer is willing to accept. Then, the provider can make an assignment to one of the specific products that fulfils the constraints by which the flexible product is described. For a detailed characterization of this concept see Gallego et al. (2004), Gallego and Phillips (2004), Petrick et al. (2010), and Petrick et al. (2009).

We state a DLP (Deterministic Linear Programming) model in which the concepts of flexible products and overbooking are handled simultaneously. In addition we extend this model to an incorporation of different booking-classes.

The remainder of the paper is organized as follows: In Sect. 2 we state our DLP-model and give explanations of the most important model-characteristics. We test our model in Sect. 3 with data that describe an airline network example and close in Sect. 4 with a summary and conclusion.

2 Model Description

DLP-models were introduced in network revenue management by Simpson (1989) and Williamson (1992) and further investigated by Liu and Van Ryzin (2008), Talluri and Van Ryzin (1998), and Talluri and Van Ryzin (1999). After notations in Sect. 2.1 we formulate and explain our model in Sect. 2.2 and state important model-characteristics in Sect. 2.3.

2.1 Notation

The following notations are needed:

With $|M|$ as cardinality of a set M we have a given set of flights $H = \{1,\ldots,|H|\}$ offered by an airline and a set of available classes $K = \{1,\ldots,|K|\}$. Each class k on flight h has a given capacity c_{hk}. We consider a set of specific products $I = \{1,\ldots,|I|\}$ with revenue r_i for a specific product $i \in I$ and a set of flexible products $J = \{1,\ldots,|J|\}$ with revenue f_j for $j \in J$. The set of execution-modes $M_j \subseteq I$ stands for possible allocations of the flexible product j to a subset of the given set I of specific products.

A matrix A indicates which flights are needed by a specific product i respectively execution-mode m ($a_{hi} = 1$, if product i needs flight h, $= 0$, otherwise, respectively $a_{hm} = 1$, if execution-mode m needs flight h, $= 0$, otherwise, for all $h \in H, i \in I$, and $m \in M_j$ for $j \in J$).

A class-matrix B indicates which class is available for which specific or flexible product ($b_{ik} = 1$, if class k is available for product i, $= 0$, otherwise, respectively

$b_{mk} = 1$, if class k is available for execution-mode m, $= 0$, otherwise, for all $h \in H, i \in I$, and $m \in M_j$ for $j \in J$).

As data for demand we do not have exact values but an approximation of expected demand. The expected aggregated demand is expressed by $\overline{D}^s_{it}$ for the specific product i and by $\overline{D}^f_{jt}$ for the flexible product j up to time t.

Decision variables x_i respectively y_{jm} describe the number of bookings for the specific product $i \in I$ respectively for the flexible product $j \in J$ with execution-mode $m \in M_j$ that are possible w.r.t. the given data.

Values p^s_{hk} and p^f_{hk} denote the probability of usage (divided in specific and flexible bookings) for a seat in class k on flight h. d_{hk} gives the costs for a denied service on flight h and class k. The overbooking-limit for flight h and class k is denoted by z_{hk} while u_{hk} describes the denied boardings on flight h and class k. Finally, we define a parameter ρ_k which stands for the fraction of flexible customers w.r.t. the overall demand in class k.

2.2 Model Formulation

Our class-specific deterministic linear programming model for flexible products and overbooking (DLP$_t$-flex-over) can, now, be formulated as follows:

$$max \sum_{i \in I} r_i \cdot x_i + \sum_{j \in J} f_j \cdot \sum_{m \in M_j} y_{jm} - \sum_{h \in H} \sum_{k \in K} d_{hk} \cdot u_{hk} \tag{1}$$

$$s.t.: \sum_{i \in I} a_{hi} \cdot b_{ik} \cdot x_i + \sum_{j \in J} \sum_{m \in M_j} a_{hm} \cdot b_{mk} \cdot y_{jm} \leq z_{hk} \quad \forall h \in H, k \in K \tag{2}$$

$$x_i \leq \overline{D}^s_{it} \quad \forall i \in I \tag{3}$$

$$(DLP_t - flex - over) \qquad x_i \geq 0 \quad \forall i \in I \tag{4}$$

$$\sum_{m \in M_j} y_{jm} \leq \overline{D}^f_{jt} \quad \forall j \in J \tag{5}$$

$$y_{jm} \geq 0 \quad \forall j \in J, m \in M_j \tag{6}$$

$$u_{hk} \geq z_{hk} \cdot \left(\rho_k \cdot p^f_{hk} + (1 - \rho_k)\, p^s_{hk} \right) - c_{hk} \quad \forall h \in H, k \in K \tag{7}$$

$$z_{hk} \geq c_{hk} \quad \forall h \in H, k \in K \tag{8}$$

$$u_{hk} \geq 0 \quad \forall h \in H, k \in K \tag{9}$$

The objective function (1) maximizes the expected revenue at time t consisting of the revenues caused by the acceptance of specific and flexible products from which the expected costs for denied services are subtracted. Constraints of type (2) secure that the sum of sold flexible and specific products does not exceed the overbooking-limits (Note that we do not have inserted the physical capacities c_{hk} but the overbooking-limits z_{hk} as right hand side of (2)). Constraints (3) and (5) consider the expected demand w.r.t. specific and flexible products while (4) and (6) are nonnegative constraints.

Conditions (7)–(9) model the overbooking situation: Constraints (7) restrict the denied services via the difference between weighted overbooking-limits minus physical capacities. Constraints (8) secure that the overbooking-limits exceed the physical capacities while (9) ensures that the number of denied customers is nonnegative.

2.3 Characteristics of the Model

The most important features of the DLP$_t$-flex-over model are:

- The joint treatment of the concepts of flexible and specific products as well as an extension to overbooking,
- the separation of different booking-classes, which is new in DLP-models for overbooking (e.g., it is possible to exclude the first class from assignments to flexible products),
- the more accurate specification of denied services in combination with overbooking-limits (The terms u_{hk} count the number of rejected consumers, the overbooking-limits z_{hk} are multiplied with the show-probability of customers interested in flexible or specific products.),
- the possibility to calculate optimal overbooking-limits directly from the DLP-program w.r.t. different demand situations.

3 Example

After the description of the data-input in Sect. 3.1 we present the results of an aviation network example obtained by application of the DLP$_t$-flex-over model in Sect. 3.2.

3.1 Data-Input

We use an easy to describe situation of flight connections operated by an airline within a certain day.

Figure 1 shows an example for a time-space network with three cities A, B, and C, and four flights (1), (2), (3), and (4). The continuous arrows describe flights, the

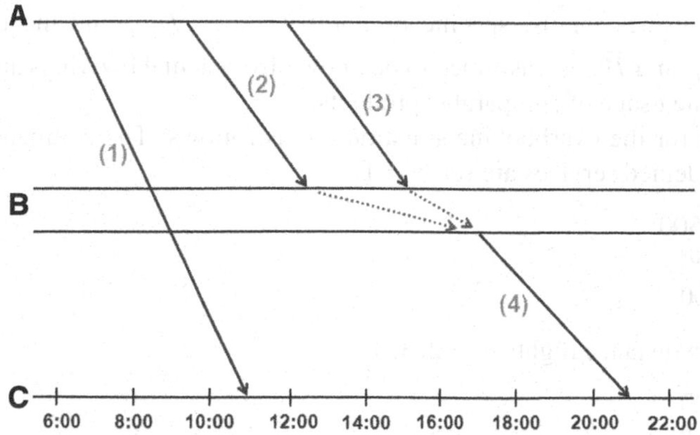

Fig. 1 Time-space network description

dotted ones show the possible transfers in the airport of B between flights. There are six transport possibilities:

- A to B: (2); (3),
- B to C: (4),
- A to C: (1) (without stopover in B),
- A to B to C: (2)→(4); (3)→(4).

Someone travelling from A to C can take the direct flight (1) or a connection via one transfer in B: (2)→(4) or (3)→(4). With 3 booking-classes per flight (first/business/economy) there are 18 specific products.
In addition we define two different flexible products:

- Flight from A to B with assignment to flight (2) or (3) in business- or economy-class (four execution-modes),
- flight from A to C with assignment to flight (1) or (2)→(4) or (3)→(4), again, in business- or economy-class (six execution-modes).

Note that we have excluded the possibility to assign flexible customers to first class seats, i.e., $\rho_1 = 0$.

Further data are as follows: The capacity for the long distance flight (1) is 400 seats, for the short distance flights (2), (3), and (4) we assume 200 seats in each plane. These capacities are partitioned in first/business/economy at the ratio of 5/15/80 %.

Direct flights from A to C are priced with 1,200, 600, and 300 monetary units (mu), depending on the three booking-classes. There is a discount if travellers from A to C accept a transfer (30 %). For short distance flights the prices are 720 mu, 360 mu, and 180 mu, again, depending on the three booking-classes. The revenues for the two flexible products are 250 mu for the flexible product from A to C and 150 mu for the flexible product from A to B.

Demand values for the specific products $\overline{D}^s_{1t}, \ldots, \overline{D}^s_{18t}$ and the two flexible products $\overline{D}^f_{1t}$ and $\overline{D}^f_{2t}$ are assumed to be known from actual bookings and historical data about the usage of comparable products.

The data for the overbooking situation are as follows: The compensation-costs (in mu) for denied services are for $h = 1$

- $d_{h1} = 1{,}500$
- $d_{h2} = 750$
- $d_{h3} = 400$

and for short distance flights $h = 2, 3, 4$

- $d_{h1} = 800$
- $d_{h2} = 400$
- $d_{h3} = 200$

And, finally, the probability of usage, divided in flexible and specific bookings, has to be specified: For the show-probability of flexible passengers p^f_{hk} we take a value of 1.0 (because we argue, that if someone can make a flexible booking s(he) is such flexible that s(he) can always (or nearly always) come to departure). For the show-rate of specific customers p^s_{hk} we deliberately take values smaller than 1.0 as explained in the discussion of the solution of this example.

3.2 Results

As writing restrictions do not allow to describe alterations of the results depending on different data-inputs we just explain one solution based on a reasonable parameter constellation.

Figure 2 illustrates the results of the application of our model to the just described example with different show-probabilities for specific products p^s_{hk} between 0.9 and 0.99 and a fraction of $\rho_1 = 0$ and $\rho_k = 0.2, k = 2, 3$, for flexible customers. The number of accepted specific products (with and without transfer) and of flexible products (from A to B and from A to C) are depicted and separated for the different connections (single flights, combined flights) by dashed lines in Fig. 2. With the given data-input no customer should be rejected. In case of low show-rates (e.g., 0.9 for flight (1)) the overbooking-limits z_{hk} exceed the capacities c_{hk} by nearly 10 % while in case of high show-rates (e.g., 0.99 for flight (2)) the z_{hk} values exceed the capacities c_{hk} only by 0.2 % in business and economy-class.

Of course, dependent on different data-inputs the solutions will vary but, although the example was kept simple on purpose, already the discussion of the results of Fig. 2 shows which kind of support can be provided for revenue management decisions.

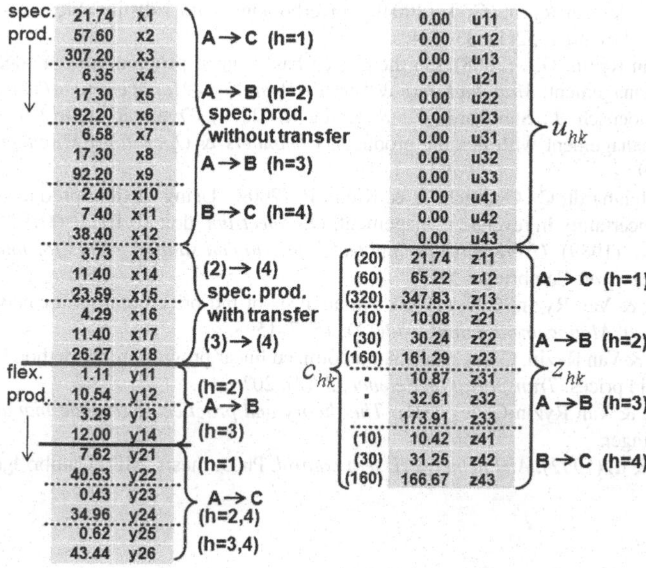

Fig. 2 Results for $\rho_1 = 0$, $\rho_k = 0.2, k = 2, 3$; $p^s_{hk} = 0.9/0.99/0.9/0.95$ for $h = 1, 2, 3, 4$

4 Summary and Conclusion

We have explained the concepts of flexible offerings and overbooking in aviation network revenue management and formulated a deterministic linear programming model (DLP$_t$-flex-over) in which these two aspects could be handled simultaneously. In addition, we extended this model to an incorporation of different booking-classes in order to exclude higher-valued classes from an assignment to flexible products.

We applied our model in a revenue management network example and calculated optimal overbooking-limits for different demand situations.

References

Bertsimas, D., & Popescu, I. (2003). Revenue management in a dynamic network environment. *Transportation Science, 37*(3), 257–277.

Gallego, G., Iyengar, G., Phillips, R., & Dubey, A. (2004). *Managing flexible products on a network* (CORC Technical Report TR-2004-01). IEOR Department, University of Columbia.

Gallego, G., & Phillips, R. (2004). Revenue management of flexible products. *Manufacturing & Service Operations Management, 6*(4), 321–337.

Karaesmen, I., & Van Ryzin, G. J. (2004a). *Coordinating overbooking and capacity control decisions on a network* (Technical Report), Columbia Business School.

Karaesmen, I., & Van Ryzin, G. J. (2004b). Overbooking with substitutable inventory classes. *Operations Research, 52*(1), 83–104.

Liu, Q., & Van Ryzin, G.J. (2008). On the choice-based linear programming model for network revenue management. *Manufacturing & Service Operations Management, 10*(2), 288–310.

Petrick, A., Goensch, J., Steinhardt, C., & Klein, R. (2010). Dynamic control mechanisms for revenue management with flexible products. *Computers & Operations Management, 37*(11), 2027–2039.

Petrick, A., Steinhardt, C., Goensch, J., & Klein, R. (2009). Using flexible products to cope with demand uncertainty in revenue management. *OR Spectrum*. doi: 10.1007/s00291-009-0188-1.

Simpson, R. W. (1989). *Using network flow techniques to find shadow prices for market and seat inventory control*. Cambridge: MIT.

Talluri, K. T., & Van Ryzin, G. J. (1998). An analysis of bid-price controls for network revenue management. *Management Science, 44*(11), 1577–1593.

Talluri, K. T., & Van Ryzin, G. J. (1999). A randomized linear programming method for computing network bid prices. *Transportation Science, 33*(2), 207–216.

Talluri, K. T., & Van Ryzin G. J. (2004). *The theory and practice of revenue management*. New York: Springer.

Williamson, E. L. (1992). *Airline network seat control*. Ph.D. thesis, MIT, Cambridge.

Non-symmetrical Correspondence Analysis of Abbreviated Hard Laddering Interviews

Eugene Kaciak and Adam Sagan

Abstract Hard laddering is a kind of a semi structured interview in a quantitative means-end chain (MEC) approach that yields a summary implication matrix (SIM). The SIM is based on pairwise associations between attributes (A), consequences (C), and personal values (V), and constitutes a base for developing hierarchical value maps (HVM). A new summary data presentation of the A-C-V triplets that form a summary ladder matrix (SLM) is presented. The structure of the SLM is examined with the use of non-symmetrical correspondence analysis. This approach permits us to identify the dependence structure in the SLM as well as the ladders that contribute most to the system's inertia.

1 Hard Laddering and Summary Ladder Matrix

1.1 Introduction

One of the main issues related to the analysis of consumer behavior is the identification of the cognitive and motivational structures that underlie the consumers' choices of the products on the B2C market. This results partly from personal values (push factors) and partly from the external stimuli associated with attributes and functional benefits of the product, advertising and other forms of promotional activities (pull factors). The most promising research approach, which allows for the identification of these dependencies between values, consequences, and

E. Kaciak (✉)
Brock University, St. Catharines, ON, Canada
e-mail: ekaciak@brocku.ca

A. Sagan
Cracow University of Economics, Krakow, Poland
e-mail: sagana@uek.krakow.pl

B. Lausen et al. (eds.), *Algorithms from and for Nature and Life*, Studies in Classification, Data Analysis, and Knowledge Organization, DOI 10.1007/978-3-319-00035-0_46,

attributes, is the means-end chain (MEC) theory and related laddering interviews. The laddering interviews can take the form of either soft, qualitative approaches or various versions of hard, quantitative, semistructured techniques. The identification of A-C-V's structures as an output of laddering is based on summary implication (SIM) or summary ladder (SLM) matrices that represent the frequencies of pairwise or triadic hierarchical relationships between attributes, consequences and values. In the analysis of SIM/SLM data, several methods are proposed. Hierarchical value maps as a graphical tool of presentation of the results are accompanied by other multidimensional methods: multidimensional scaling, correspondence analysis, factor analysis or nonlinear canonical correlation. However, most of them are used for the analysis of interdependence structures among sets of variables. The proposed application of non-symmetrical correspondence analysis (NSCA), as a dependence analysis, clearly distinguishes between dependent (attributes) and independent (consequences and values) variables in the analysis of SLM/SIM matrices.

1.2 Abbreviated Hard Laddering

Kaciak and Cullen (2009) proposed a method of abbreviating (shortening) a hard laddering interview without a substantial reduction in the number of ladders generated by the respondents during the procedure. They proposed dividing the hard laddering formats into two categories: (1) the $p \times q$ and (2) the $p \times (1 + k + k \times m)$ formats. The most popular $p \times q$ format consists of p sequences (in rows) of q boxes (in columns) connected by arrows, that take a respondent from the product's attribute to consequences and values, thus explaining why this attribute is important to him/her. In the $p \times (1 + k + k \times m)$ formats, a respondent is first requested to provide the most important attribute of the product and then asked to reveal up to k most important consequences of this attribute. In the last step, for each of the k consequences, the respondent is encouraged to give up to m personal values. The procedure is repeated for p attributes. Usually, the $3 \times (1 + 3 + 3 \times 3)$ questionnaire format is used and Kaciak and Cullen (2009) have developed their procedure specifically for such a format. For the first attribute, the respondent is asked for the first associated consequence and only up to two underlying values, thus producing triads (1,1,1), (1,1,2), where (r,s,t) = (rth attribute, sth consequence, tth value). Thus, the third triad (1,1,3) is not needed during the abbreviated procedure. Then, the second associated consequence is followed by only one underlying value – triad (1,2,1).The respondent is not asked for the third associated consequence nor for any underlying personal value. For the second attribute, the first associated consequence is elicited, followed by up to two underlying values, thus yielding triads (2,1,1) and (2,1,2). Then, the second consequence is followed by just one value – triad (2,2,1). Finally, for the third attribute, the first associated consequence is identified, followed by just one, the most important, value – triad (3,1,1).

1.3 Summary Ladder Matrix

The most popular presentation format of the laddering (hard or soft) interview results is the hierarchical value map (HVM) proposed by Reynolds and Gutman (1988). The HVM is a graphical representation of the most significant relationships (mostly those that exceed a certain percentage rate, such as 5–10 % of the sample size) between the attributes, consequences, and values. The basis for the map construction is the distribution of dual linkages between categories A-C, C-V and A-V, reflected in the so-called summary implication matrix (SIM), which is an analog of a contingency table. This type of a table (and the resulting HVM) shows only the dyadic linkages between the attributes, consequences and values. On the other hand, the identification of the A-C-V triadic relationships must be based on triplets (triads) of the categories rather than the dyads. Kaciak and Cullen (2006) introduced a method for assessing the presence of triadic (A-C-V) relationships among the categories of means-end chain structures and presented these relations in the so-called summary ladder matrix (SLM). The SLM is a three-way contingency table indicating the relationship frequencies between triplets of A-C-V.

2 Non-symmetrical Correspondence Analysis of SLM

2.1 Non-symmetrical Correspondence Analysis

The analysis of SIM or SLM tables is predominantly based on multidimensional interdependence methods like correspondence analysis, multidimensional scaling or nonlinear canonical correlation. All of these methods are applied for identification of interrelationships between the variables in a reduced space. However, laddering interviews produce dependence structures in the means-end chains, where the independent (ends) and dependent (means) variables are clearly distinguished. Therefore, we suggest that laddering data be analyzed with the non-symmetrical correspondence analysis (NSCA) rather than the standard correspondence analysis (CA). We illustrate our approach with hard laddering data obtained in a survey of Polish smokers conducted by Kaciak and Cullen (2006). The data was generated by hard laddering interviews, elicited from the respondents. The total number of ladders was 1,828, which were coded and analyzed using content analysis. The final set of laddering categories consisted of seven attributes (A1 – mild, A2 – taste, A3 – quality, A4 – cheap, A5 – strong, A6 – filter, A7 – aroma), eight consequences (C1 – pleasure, C2 – physically better, C3 – less health damage, C4 – project good image, C5 – socially acceptable, C6 – save money, C7 – smoke less, C8 – kill nicotine hunger) and seven personal values (V1 – health, V2 – self-direction, V3 – hedonism, V4 – achievement, V5 – social recognition, V6 – conformity, V7 – benevolence). Following our approach, we applied the non-symmetric correspondence analysis to contingency tables in which a dependency between the rows and the columns of the contingency table is the main focus in the analysis.

In such a table, the product attributes are affected by consequences and values. This asymmetry causes the problem of prediction and visualization of the predictive role of the consequences and values on attributes (Kroonenberg and Lombardo 1999; D'ambra and Lauro 1992). The Pearson chi-square statistic cannot be applied because of this asymmetric relationship between the rows and columns of the contingency table. Instead, the Goodman-Kruskal τ and C-statistic of Light and Margolin are commonly proposed (Beh and D'ambra 2009). The latter statistic (derived from the categorical analysis of variance for contingency tables – CATANOVA) is preferred in NSCA because the low value of τ does not always reflect the low association between the two variables and seems to be more stable when sparse tables are analyzed. The aim of the non-symmetrical correspondence analysis is predicting the values of the dependent variable from the independent variables, and reducing the uncertainty about the values of response categories given the predictor variable. The associations between the variables and the reduction of uncertainty can be measured using Goodman-Kruskal *tau*:

$$\tau = \frac{\sum\sum p_{.j}\left(\frac{p_{ij}}{p_{.j}} - p_{i.}\right)^2}{1 - \sum p_{i.}^2} = \sum \tau_j \tag{1}$$

where $\pi_{ij} = \frac{p_{ij}}{p_{.j}} - p_{i.}$ is the matrix of differences between the unconditional marginal prediction described by marginal profiles and the conditional prediction (conditional column/row profiles). If the conditional distribution of response variables is identical to the overall, unconditional, marginal distribution, then there is a relative increase in predictability and thus $\tau = 0$. The opposite situation has a perfect predictability of response categories given the independent categories, and thus gives $\tau = 1$. So, Goodman-Kruskal τ is the measure of proportional reduction in error for the prediction of the response (row) marginal probability related to the total error of prediction. The other measures of dependencies in contingency tables (also for three-way tables) are Gray-Williams index, Marcotorchino's τ_M index, Simonetti's Δ, Aggregate Prediction Index and Tallur's index. Margolin- Light C statistic is used to test the hypothesis that the marginal and conditional distributions are equal and, therefore, to test the significance of predictive relations indicated by τ.

$$C = (n-1)(I-1)\tau \tag{2}$$

The use of nonsymmetric correspondence analysis (instead of the well known version of CA) and the decomposition of the correspondence table (Table 1) is dictated by the asymmetry of the relationship between the rows and columns. In the structure of triads, the perception of attributes results from values and psycho-social characteristics of smokers. Therefore, the row categories (cigarette attributes) depend on column categories (consequences and values) in the correspondence table. In the non-symmetric correspondence analysis, a singular value

Table 1 Correspondence table of ladders

Items	Means	C1–V2	C1–V3	C1–V6	C2–V3	C2–V4	C3–V1	C3–V2	C3–V6	C4–V2	C4–V5
A1	0.21	−0.21	−0.11	−0.21	0.25	−0.21	0.41	0.79	0.79	−0.21	−0.21
A2	0.09	0.23	0.19	−0.09	0.13	−0.09	−0.09	−0.09	−0.09	−0.09	−0.09
A3	0.15	0.06	0.02	−0.15	−0.15	−0.15	0.04	−0.15	−0.15	0.85	0.59
A4	0.14	−0.14	−0.14	−0.14	−0.14	−0.14	−0.14	−0.14	−0.14	−0.14	−0.14
A5	0.26	−0.13	0.07	−0.26	0.05	0.74	−0.26	−0.26	−0.26	−0.26	0.01
A6	0.03	−0.03	−0.03	−0.03	−0.03	−0.03	0.16	−0.03	−0.03	−0.03	−0.03
A7	0.11	0.11	0.14	0.89	−0.11	−0.11	−0.11	−0.11	−0.11	−0.11	−0.11

Items	Means	C5–V6	C6–V1	C6–V2	C6–V7	C7–V1	C7–V2	C7–V7	C8–V2	C8–V3	C8–V4
A1	0.21	0.18	−0.21	−0.21	−0.21	−0.21	−0.21	−0.21	−0.21	−0.21	−0.21
A2	0.09	−0.09	−0.09	−0.09	−0.09	−0.09	−0.09	−0.09	−0.09	−0.09	−0.09
A3	0.15	−0.15	0.85	−0.15	−0.15	−0.15	−0.15	−0.15	−0.15	−0.15	−0.15
A4	0.14	−0.14	−0.14	0.86	0.86	−0.14	−0.14	−0.14	−0.14	−0.14	−0.14
A5	0.26	−0.26	−0.26	−0.26	−0.26	0.74	0.74	0.74	0.74	0.74	0.74
A6	0.03	−0.03	−0.03	−0.03	−0.03	−0.03	−0.03	−0.03	−0.03	−0.03	−0.03
A7	0.11	0.49	−0.11	−0.11	−0.11	−0.11	−0.11	−0.11	−0.11	−0.11	−0.11

decomposition (SVD) is applied to decompose the residual matrix of predictions (differences between the unconditional and conditional predictions).

$$\pi_{ij} = \frac{p_{ij}}{p_{.j}} - p_{i.} = \sum u_{im}\lambda_m v_{jm} \tag{3}$$

where u_{im} and v_{jm} are left and right singular vectors associated with ith row and jth column categories, and λ_m is the is the mth singular value of π_{ij}.

Centered column profiles are the main interest in NSCA. They reflect the information as to which cell has a higher or lower proportion in comparison to the marginal one, and also indicate the increase (or decrease) in the predictability given the column category. A result of the SVD is the presentation of response and predictor variables in a reduced number of dimensions in a way that maximizes the predictive power of the independent variables. The main difference between CA and NSCA is that the non-symmetrical CA is less sensitive to the small marginal proportions in the response variable (and more sparse tables) because the weight factor ($1/p_{i.}$) is not taken into account in the calculation of the inertia.

2.2 NSCA for Smokers Data

Table 1 presents the centered column profiles that are conditional distributions of attributes given consequences/values.[1]

[1] NSCA has been performed using Kroonenberg's Asymtab 2.0c and XLSTAT 2006 packages.

Table 2 Eigenvalues of correspondence table

Eigenvalues and inertia explained	F1	F2	F3	F4	F5	F6
Eigenvalues	0.16	0.14	0.07	0.05	0.01	0.003
Row dependence (%)	37.16	32.19	15.94	11.00	3.06	0.65
Cumulated %	37.16	69.34	85.29	96.29	99.35	100.00

Table 3 Proportional contribution to *tau* of each row and column

A1	A2	A3	A4	A5	A6	A7			
0.19	0.04	0.11	0.28	0.28	0.01	0.08			
C1–V2	C1–V3	C1–V6	C2–V3	C2–V4	C3–V1	C3–V2	C3–V6	C4–V2	C4–V5
0.02	0.04	0.02	0.02	0.02	0.11	0.03	0.04	0.02	0.07
C5–V6	C6–V1	C6–V2	C6–V7	C7–V1	C7–V2	C7–V7	C8–V2	C8–V3	C8–V4
0.05	0.02	0.16	0.13	0.01	0.04	0.02	0.04	0.09	0.03

It shows the deviates between marginal distributions of the dependent variables (Means) and conditional distribution of attributes given the independent variables. Clearly, some categories of consequences/values can improve the prediction of particular attributes (i.e., C3–V1, C3–V2 help sufficiently well to predict the attribute A1). The predictive coefficient of Goodman-Kruskal τ equals 0.527, which reflects the prediction of attributes' categories by values' categories. Also, the C statistic seems to be significant (Light-Margolin C statistic $= 4{,}246.57$, p-value $=$ 0.000). The rows dependent CATANOVA test decomposes the total mean square ($MS_T = 0.411$) into the "between" mean square ($MS_B = 0.217$) and "within" mean square ($MS_W = 0.195$). The total inertia of the deviance table is equal $2 \times MS_B$ ($= 2 \times 0.434$), while the total inertia in symmetric CA is 2.71. The C statistic is preferred over Chi-Square in the case of asymmetrical relations and is more suitable for sparse tables. Additionally, *tau* also equals MS_B divided by MS_T. These deviates are decomposed using the SVD. The structure of eigenvalues is shown in Table 2.

As shown in Table 2, the first two dimensions explain around 70% of the total inertia (which is equal to the numerator of τ). The eigenvalues are usually smaller in τ_{num} decomposition because only the predictive power of the table is decomposed. One of the important objectives of asymmetric correspondence analysis is to evaluate the predictive power of the categories of the independent variables. Table 3 shows the proportional contribution to τ of the columns and rows of the correspondence table.

Contributions to τ show to what extent the predictor categories improve prediction of the response categories. Table 3 shows that the four concrete attribute categories – "mild", "quality", "cheap" and "strong" – and two functional consequences – "less health damage" and "save money" – have relatively the greatest contributions to τ in the two-dimensional array (however no specific threshold is used to distinguish between low or high contribution).

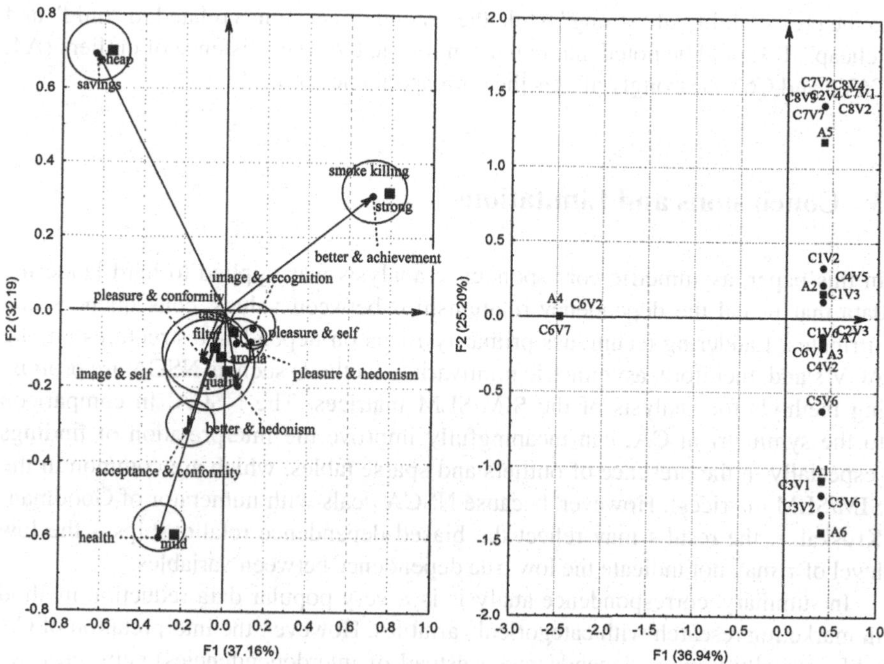

Fig. 1 NSCA vs. CA biplot
Key: *A1* – mild, *A2* – taste, *A3* – quality, *A4* – cheap, *A5* – strong, *A6* – filter, *A7* – aroma, *C1* – pleasure, *C2* – physically better, *C3* – less health damage, *C4* – project good image, *C5* – socially acceptable, *C6* – save money, *C7* – smoke less, *C8* – kill nicotine hunger, *V1* – health, *V2* – self-direction, *V3* – hedonism, *V4* – achievement, *V5* – social recognition, *V6* – conformity, *V7* – benevolence

Figure 1 presents the correspondence biplot with 95 % confidence circles and, for comparison purposes, the degenerated result of symmetric CA. Both analyses are based on the same data.

The origin of the plot represents the marginal distribution of columns (predictors) and the Euclidean distances between values with respect to the origin, thus showing their deviations from the marginal distribution. For the attributes, standard coordinates were calculated, and for the consequences/values – principal coordinates. Correspondence biplot shows that the first dominant dimension (with the largest contribution to the explained inertia) predicts the relationships between "low price", "mildness", "self-reliance", and "health". The second dimension has the strongest impact on "taste", "strength" and "hedonism". The bundle of points with confidence circles around the origin of the plot indicates that, on average, most of the non-functional consequences are insignificant predictors for cigarettes' attributes. The dominant predictors (with relatively smaller radii of the confidence circles) are "smoke killing", "savings" and "health". Additionally, the proportional contribution of the axes to the increase of predictability of rows shows that the two dimensional plots explain 70 % of the *tau* numerator. The first dimension is related to the predictability of "strong" while the second dimension is related to "mild" and "cheap". It should be noted that in the symmetric CA, the existence of outliers (A4, C6V2 and C6V7) strongly affects the correspondence plot.

3 Conclusions and Limitations

In this paper, asymmetric correspondence analysis was applied to hard laddering data that reveal the dependency relationships between values, consequences and attributes. Laddering techniques primarily focus on dependence structures among ACV's and, therefore, asymmetric multivariate methods such as NSCA are promising methods for analysis of the SIM/SLM matrices. The NSCA, in comparison to the symmetrical CA, can meaningfully improve the interpretation of findings (especially in the presence of outliers and sparse tables, which are common in the SIM/SLM matrices). However, because NSCA deals with numerator of Goodman-Kruskal τ, the results may reflect the biased dependence relationships – the low level of τ may not indicate the low true dependence between variables

In summary, correspondence analysis is a very popular data reduction method in marketing research with categorical variables. However, the interpretation of CA tables usually implies dependencies (instead of interdependencies) between rows and columns of contingency tables. Hence, an improper use of correspondence analysis in identifying dependencies in data tables may result in misleading conclusions.

References

Beh, E. J., & D'ambra, L. (2009). Some interpretative tools for non-symmetrical correspondence analysis. *Journal of Classification, 26*, 55–76.

D'ambra, L., & Lauro, N. C. (1992). Non symmetrical exploratory data analysis. *Statistica Applicata, 4*, 511–529.

Kaciak, E., & Cullen, C. W. (2006). Analysis of means-end chain data in marketing research. *Journal of Targeting, Measurement and Analysis for Marketing, 15*(1), 12–20.

Kaciak, E., & Cullen, C. (2009). A method of abbreviating a laddering survey. *Journal of Targeting, Measurement and Analysis for Marketing, 17*(2), 105–113.

Kroonenberg, P., & Lombardo, R. (1999). Nonsymmetric correspondence analysis: a tool for analysing contingency tables with a dependence structure. *Multivariate Behavioral Research, 34*, 367–396.

Reynolds, T. J., & Gutman, J. (1988). Laddering theory, method, analysis, and interpretation. *Journal of Advertising Research, February/March*, 11–31.

Antecedents and Outcomes of Participation in Social Networking Sites

Sandra Loureiro

Abstract This study seeks to understand factors that influence the participation in online social networks and outcomes. The proposed model integrates variables such as identification, satisfaction, degree of influence, usefulness and ease of use into a comprehensive framework. The empirical approach was based on an online survey of 336 young adults in Portugal, undertaken during November/December 2010. Research findings showed that identification, perceived usefulness, interaction preference, and extroversion are the most important factors in order to influence the members' participation. The degree of influence and the identification have an indirect effect on the participation through perceived usefulness. Participation in social networking sites, in turn, is linked to higher levels of loyalty, actual use, and word-of-mouth. The results of this study have implications for researchers and practitioners.

1 Introduction

The concept of community emerges from the Sociology perspective as a group of people linked by social ties, sharing common values and interests and having common meanings and expectations among them (Muniz and O'Guinn 2001). Social networking sites (SNS) can be regarded as communities. Boyd and Ellison (2007) allude that SNS allow individuals to build a public or semi-public profile within a bounded system, interact and share connection with other users in a list, and view and traverse their list of connections and those made by others within the system. However, what leads members to participate in SNS is not well know.

S. Loureiro (✉)
Marketing, Operations and General Management Department, ISCTE-IUL business school, Av. Forças Armadas 1649-026 Lisbon, Portugal
e-mail: sandra.loureiro@iscte.pt; sandramloureiro@netcabo.pt

B. Lausen et al. (eds.), *Algorithms from and for Nature and Life*, Studies in Classification, Data Analysis, and Knowledge Organization, DOI 10.1007/978-3-319-00035-0_47,

Therefore, the major goal of this study is to analyse antecedents and outcomes of participation in SNS. The proposed model integrates technology acceptance variables with relational and personality variables.

2 Literature Review and Hypothesis

The Technology Acceptance Model (TAM) is based on the Theory of Reasoned Action, which states that human beliefs influence attitudes and shape behavioural intentions (Davis 1989). In the core of the model there are two specific beliefs: perceived usefulness and perceived ease of use. Perceived usefulness is a belief that the information technology or system will help the user in performing his/her task (Davis 1989). Perceived ease of use is defined as the degree to which individuals believe that using a particular system would be free of effort (Chung and Tan 2004). The two variables are linked, since several studies demonstrate the significant influence of perceived ease of use on perceived usefulness (e.g., Lin 2009). So, it is argued that:

H1: Ease of use has a positive effect on usefulness.

Later, external variables of perceived usefulness and perceived ease of use (Lee et al. 2003) have been introduced in order to examine the usage-context factors that may influence the users' acceptance (Moon and Kim 2001). Nevertheless, in order to better understand the antecedents of participation in SNS, this study regards constructs such as identification, satisfaction, degree of influence, interaction preference, extroversion, usefulness and ease of use. The first three variables are based on the psychological sense of community and are regarded as being drivers for ensuring a brand community's customer participation (Algesheimer et al. 2005; Von Loewenfeld 2006), as well as virtual brand community's member participation (Woisetschlägera et al. 2008). The identification is a concept established in the field of psychology and has been adapted into the marketing context. Thus, brand identification can be defined as the consumer's perception of similarity between him/herself and the brand (Bagozzi and Dholakia 2006). The consumer's identification with an organisation is based on his/her perceptions of its core values, mission, and leadership (Bhattacharya and Sen 2003). Therefore, social identification is the perception of belonging to a group with the result that an individual identifies with that group (Bhattacharya et al. 1995). The individual specific characteristics and interests are similar to the group. Bagozzi and Dholakia (2006) found that social identity is the cognitive self-awareness of membership in brand community, affective commitment, and perceived importance of membership. In the case of SNS, the group is the social network community. In order to be identified with SNS, members should feel as they belong to the community, perceive the importance of membership, and see themselves in many of the aspects of the community. Satisfaction is recognised as one of the important and widely studied concepts in marketing. The concept is regarded as an emotional reaction to a specific product/service encounter, and this reaction comes from disconfirmation between the consumer's perceived performance and consumer's expectation (Mano and Oliver 1993). It is also viewed as the overall evaluation performance and based on prior experiences (McDougall and Levesque 2000). Although traditionally the studies of satisfaction/dissatisfaction assume a cognitive focus (Oliver 1980), nowadays researchers consider it a cognitive phenomenon with affective elements. The degree of influence is related to the concept of self-efficacy. The psychological theory of psychological sense of community, proposed by Chavis et al. (1986), discusses the concept of influence for members as the feeling that they have some control and influence within the community. This theory was first developed in a neighbourhood context, however later was applied to relational communities, such as Internet communities (Obst et al. 2002). As Bandura (1986) proposes, high degree of influence should be more willing to engage in the community. Interaction preference is another variable that could influence participation. According to Wiertz and de Ruyter (2007) the preference for an online community is defined as the prevailing tendency of an individual to interact with relative strangers in an online environment. For such, an individual that enjoys interacting and exchanging with other members and likes to actively participate in discussions is more willing to participate in SNS. Individuals high in extroversion are expected to be more willing to provide online participation. Specifically, extravert individuals are more able to satisfy their desire for social interaction participating in SNS since they are sociable, energetic, talkative, outgoing, and enthusiastic (Thoms et al. 1996). All the above considerations lead to regard identification, satisfaction, and degree of influence as external variables of TAM, which together with extroversion and interaction preference are determinants of participation in SNS. Hence, the following hypotheses are suggested:

H2: Identification (H2a), satisfaction (H2b), degree of influence (H2c), ease of use (H2d), usefulness (H2e), extroversion (H2f), interaction preference (H2g) have a positive influence on participation.

Attitude towards using technology has been omitted in the final model (Igbaria et al. 1995) and perceived usefulness and perceived ease of use link directly to intention. Participation in SNS is regarded as a direct outcome of TAM. An individual that perceived usefulness and ease of use SNS is more willing to interact and establish social contact through SNS (Teo et al. 2003). Therefore, the following hypotheses are proposed

H3: Identification (H3a), satisfaction (H3b), and degree of influence (H3c) have a positive effect on ease of use.

H4: Identification (H4a), satisfaction (H4b), and degree of influence (H4c) have a positive effect on usefulness.

Finally, when a member is engaged and actively participates in a community, then she/he is more willing to be loyal, provide positive word-of-mouth (e.g., Algesheimer et al. 2005), and tends to use SNS frequently, spends more time on

SNS, and exerts herself/himself to SNS (actual use). Consequently, the following hypotheses are suggested:

H5: Participation has a positive effect on loyalty.
H6: Participation has a positive effect on actual use.
H7: Participation has a positive effect on word-of-mouth.

3 Method

Drawing from literature review, the research model of antecedents and outcomes of participation in SNS is shown in Fig. 1. The questionnaire was pre-tested by ten young adults (five full students and five working students) personally interviewed. Then, 336 young adults fulfilled the questionnaire during November and December 2010. The online survey questionnaire was distributed through Universities in Northern and Central of Portugal. The behaviours of students are a good proxy for behaviours of actual online use of social networking sites, since statistics show that young people are the dominant Internet users, and most (80 %) of the young Internet users (16–24 years) in the European Union post messages in chat sites, blogs or social networking sites (Seybert and Lööf 2010). The sample consisted of 48.2 % of male participants and 51.8 % of female participants. Most of them prefer Facebook. The 11 latent constructs in this study were measured by means of multi-item scales with 5 point Likert-type scale (1 = strongly disagree, 5 = strongly agree). Identification, satisfaction, degree of influence, and participation were adapted from Von Loewenfeld (2006). Ease of use, usefulness, and actual use are based on Thong et al. (2002). Interaction preference is based on Wiertz and de Ruyter (2007). Extroversion was adopted from Gosling et al. (2003). Finally, word-of-mouth and loyalty are assessed based on Algesheimer et al. (2005) and Loureiro et al. (2010).

4 Results

In this study a PLS approach is used, which employs a component-based approach for estimation purposes (Lohmoller 1989). The PLS model is analysed and interpreted in two stages. First, the adequacy of the measures is assessed by evaluating the reliability of the individual measures and the discriminant validity of the constructs (Hulland 1999). Then, the structural model is appraised. Thus, composite reliability is used to analyse the reliability of the constructs and to determinate convergent validity, we used the average variance of variables extracted by constructs (AVE) that should be at least 0.5. In this way, all factor loadings of reflective constructs approached or exceeded 0.707, which indicates that more than 50 % of the variance in the manifest variable is explained by the construct (Carmines and Zeller 1979). All constructs in the model satisfied the requirements for reliability

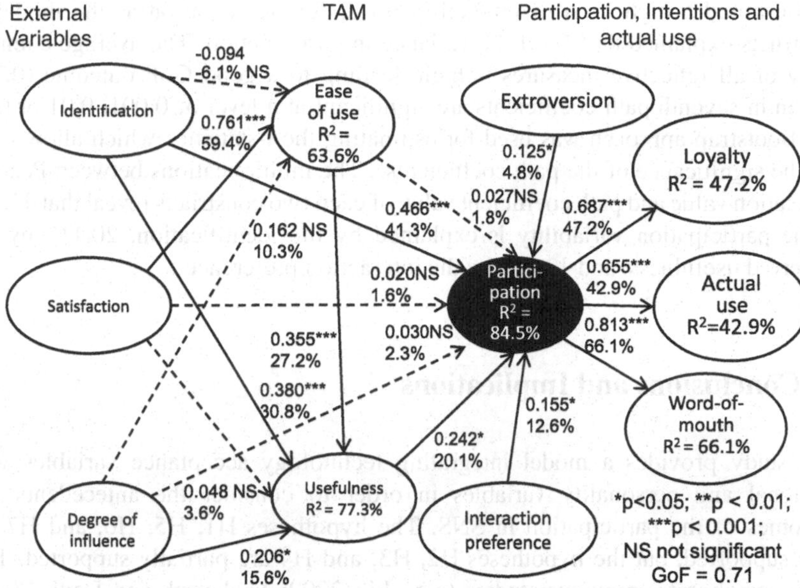

Fig. 1 Structural results of the antecedents and outcomes of participation in SNS

Table 1 Measurement results

Construct	Mean	Cronbach's alpha	Composite reliability	AVE*
Participation	2.9	0.897	0.924	0.708
Ease of use	3.3	0.724	0.876	0.780
Usefulness	3.2	0.878	0.916	0.732
Identification	2.8	0.907	0.926	0.643
Satisfaction	3.0	0.845	0.906	0.763
Degree of influence	2.8	0.782	0.873	0.697
Extroversion	3.4	1.000	1,000	1,000
Interaction preference	2.8	0.789	0.877	0.704
Loyalty	2.7	1.000	1.000	1.000
Actual use	3.2	0.822	0.916	0.846
Word-of-mouth	3.0	0.895	0.934	0.826

* AVE Average variance extracted

(composite reliability greater than 0.70) and convergent validity (Table 1). The measures also demonstrate discriminant validity, because the rule that the square root of AVE should be greater than the correlation between the construct and other constructs in the model (Fornell and Larcker 1981) was met in this study.

Central criteria for the evaluation of the inner model comprise R^2 and the Goodness-of-Fit (GoF) index by Tenenhaus et al. (2005). All values of the cross-validated redundancy measure Q^2 (Stone-Geisser-Test) were positive, indicating high predictive power of the exogenous constructs (Chin 1998). The model demonstrated a high level of predictive power for participation as the modelled constructs explained 84.5 % of the variance in participation. The average communality of all reflective measures is high, leading to a good GoF outcome (0.71). Eleven in seventh path coefficients are significant at a level of 0.001, 0.01 or 0.05 (the Bootstrap approach was used for estimating the t-statistics which allow us to test the significance of the path coefficients). The multiplications between Pearson correlation value and path coefficient value of each two constructs reveal that 41.3 % of the participation variability is explained by the identification, 20.1 % by the perceived usefulness, and 12.6 % by the interaction preference.

5 Conclusions and Implications

This study provides a model integrating technology acceptance variables with relational and personality variables in order to consider the antecedents and outcomes of the participation in SNS. The hypotheses H1, H5, H6, and H7 are fully supported, but the hypotheses H2, H3, and H4 are partially supported. This study corroborates previous studies (e.g., Lin 2009; Venkatesh and Davis 2000), according to which the ease of use positively influences the perceived usefulness. The proposed determinates of participation are good predictors of participation on social networks, like Facebook. Therefore, identification, perceived usefulness, interaction preference, and extroversion are the most important antecedents that influence the participation. The degree of influence and identification have an indirect effect on participation through perceived usefulness. Participation in social networking sites, in turn, is linked to higher levels of loyalty, actual use, and word-of-mouth. Satisfaction with the online social network does not have a direct and significant influence on participation, but has a direct and significant influence on the ease of using the network. Managers that decide put his/her brands in social networking sites should be aware of these insights, i.e., the more identified, (the feel of belong and the perception of similarity between an individual and the SNS group), perceived usefulness (in achieving what a member want), perceived influence (the influence exercised over other members), and extroverted (energetic and enthusiastic) a community members is, the more willing he/she will be to keep the community alive through interacting with other members. The more active and participants members are more willing to speak about the group where they are included to others and could be a good vehicle to promote brands and corporations. Active members tend to be natural leaders that lead the group around a theme, an interest, a cause, or a brand. In the future, different profile of members should be categorised in order to better understand the characteristics of leaders and followings of members in SNS. The findings suggest that the TAM model need to be extended with relational and personality variables in order to provide a more comprehensive explanation of determinants and consequents of the participation in SNS. The sample for this study was collected in several of the the largest universities in Portugal, however, in future the survey should be extended to all other universities in

Portugal and even in other countries. According to European Union, 80 % of young Internet users (aged 16–24) in the European Union are active on social media (in Portugal the rate is 90 %) (Seybert and Lööf 2010), so students could be regarded as a good target population for study the determinants of participation in SNS. However, in the future should be collected more data regarding this and other group ages in order to compare the model between age groups and between gender.

References

Algesheimer, R., Dholakia, U. M., & Herrmann, A. (2005). The social influence of brand community: evidence from European car clubs. *Journal of Marketing, 69*(7), 19–34.

Bagozzi, R. P., & Dholakia, U. D. (2006). Antecedents and purchase consequences of customer participation in small group brand communities. *International Journal of Research in Marketing, 23*, 45–61.

Bandura, A. (1986). *Social foundations of thought and action*. Englewood Cliffs: Prentice Hall.

Bhattacharya, C. B., Rao, H., & Glynn, M. A. (1995). Understanding the bond of identification: an investigation of its correlates among art museum members. *Journal of Marketing, 59*, 46–57.

Bhattacharya, C. B., & Sen, S. (2003). Consumer-Company identification: a framework for understanding consumers' relationships with companies. *Journal of Marketing, 67*, 76–88.

Boyd, D. M., & Ellison, N. B. (2007). Social network sites: definition, history, and scholarship. *Journal of Computer-Mediated Communication, 13*(1), 210–230.

Carmines, E. G., & Zeller, R. A. (1979). *Reliability and validity assessment*. London: Sage.

Chin W. W. (1998). The partial least squares approach to structural equation modeling. In G. A. Marcoulides (Ed.), *Modern methods for business research* (pp. 295–358). Mahwah: Lawrence Erlbaum Associates .

Chung, J., & Tan, F. B. (2004). Antecedent of perceived playfulness: an exploratory study on users acceptance of general information-searching web sites. *Information & Management, 4*, 869–881.

Chavis, D., Hogge, J., McMillan, D., & Wandersman, A. (1986). Sense of community through Brunswick's lens. *Journal of Community Psychology, 14*, 24–40.

Davis, F. D. (1989). Perceived usefulness, perceived ease of use, and user acceptance of information technology, *MIS Quarterly, 13*, 319–340.

Fornell C., & Larcker, D. F. (1981). Evaluating structural models with unobservables variables and measurement error. *Journal of Marketing Research, 28*, 39–50.

Gosling, S. D., Rentfrow, P. J., & Swann, W. B., Jr. (2003). A very brief measure of the big-five personality domains. *Journal of Research in Personality, 37*(6), 504–528.

Igbaria, M., Guimaraes, T., & Davis, G. B. (1995). Testing the determinants of microcomputer usage via a structural model. *Journal of Management Information Systems, 11*(4), 87–114.

Hulland, J. (1999). Use of partial least squares (PLS) in strategic management research: a review of four recent studies. *Strategic Management Journal, 20*(2), 195–204.

Lee, Y., Kozar, K. A., & Larsen, K. R. T. (2003). The technology acceptance model: past, present, and future. *Communications of the Association for Information Systems, 12*, 752–780.

Lin, H.-F. (2009). Examination of cognitive absorption influencing the intention to use a virtual community. *Behaviour & Information Technology, 28*(5), 421–431.

Lohmoller, J. B. (1989). The PLS program system: latent variables path analysis with partial least squares estimation. *Multivariate Behavioral Research, 23*(1), 125–127.

Loureiro, S., Lienbacher, E., & Walter, E. (2010). Effects of customer value on internet banking corporate reputation and satisfaction: a comparative study in Portugal and Austria. In E. Y. Kim (Ed.), *Proceeding of 2010 global marketing conference-marketing in a turbulent environment* (pp.1977–1990). Seoul: Chungbuk National University (Republic of Korea) (ISSN: 1976-8699).

Mano, H., & Oliver, R. L. (1993). Assessing the dimensionality and structure of the consumption experience: evaluation, feeling and satisfaction. *Journal of Consumer Research, 20*(3), 451–466.

McDougall, G., & Levesque, T. (2000). Customer satisfaction with services: putting perceived value into the equation. *Journal of Services Marketing, 14*(5), 392–410.

Moon, J. W., & Kim, Y. G. (2001). Extending the TAM for a world-wide-web context. *Information & Management, 38*, 217–230.

Muniz, A. M., Jr., & O'Guinn, T. C. (2001). Brand community. *Journal of Consumer Research, 27*, 412–432.

Obst, P., Zinkiewicz, L., & Smith, S. (2002). An exploration of sense of community, part 3: dimensions and predictors of psychological sense of community in geographical communities. *Journal of Community Psychology, 30*(1), 119–133.

Oliver, R. (1980). A cognitive model of antecedents and consequences of satisfaction decisions. *Journal of Marketing Research, 17*, 460–469.

Seybert, H., & Lööf, A. (2010). Internet usage in 2010 - households and individuals: Eurostat data. http://epp.eurostat.ec.europa.eu/cache/ITY_OFFPUB/KS-QA-10-050/EN/KS-QA-10-050-EN.PDF.

Tenenhaus, M., Vinzi, V. E., Chatelin, Y.-M., & Lauro, C. (2005). PLS path modeling. *Computational Statistics & Data Analysis, 48*, 159–205.

Teo, H. H., Chan, H. C., Wei, K. K., & Zhang, Z. (2003). Evaluating information accessibility and community adaptivity features for sustaining virtual learning communities. *International Journal of Human-Computer Studies, 59*, 671–697.

Thoms, P., Moore, K. S., & Scott, K. S. (1996). The relationship between self-efficacy for participating in self-managed work groups and the big five personality dimensions. *Journal of Organizational Behavior, 17*(4), 349–362.

Thong, J. Y. L., Hong, W., & Tam, K. Y. (2002). Understanding user acceptance of digital libraries: what are the roles of interface characteristics, organizational context, and individual differences? *International Journal of Human-Computer Studies, 57*(3), 215–242.

Venkatesh, V., & Davis, F. D. (2000). A Theoretical Extension of the Technology Acceptance Model: Four Longitudinal Field Studies. *Management Science, 46*(2), 186–204.

Von Loewenfeld, F. (2006). *Brand communities*. Wiesbaden: Gabler.

Wiertz, C., & de Ruyter, K. (2007). Beyond the call of duty: why customers contribute to firm-hosted commercial online communities. *Organization Studies, 28*(3), 347–376.

Woisetschlägera, D. M., Hartlebb, V., & Bluta, M. (2008). How to make brand communities work: antecedents and consequences of consumer participation. *Journal of Relationship Marketing, 7*(3), 237–256.

User-Generated Content for Image Clustering and Marketing Purposes

Diana Schindler

Abstract The analysis of images for different purposes – particularly image clustering – has been the subject of several research streams in the past. Since the 1990s query by image content and, somewhat later, content-based image retrieval have been topics of growing scientific interest. A literature review shows that research on image analysis, so far, is primarily related to computer science. However, since the advent of Flickr and other media-sharing platforms there is an ever growing data base of images which reflects individual preferences regarding activities or interests. Hence, these data is promising to observe implicit preferences and complement classical efforts for several marketing purposes (see, e.g., Van House, Int J Hum-Comput Stud 67:1073–1086, 2009 or Baier D, Daniel I (2011) Image clustering for marketing purposes. In: W. Gaul, A. Geyer-Schulz, L. Schmidt-Thieme (eds) Challenges concerning the data analysis – Computer science – Optimization, vol. 43). Against this background, the present paper investigates options for clustering images on the basis of personal image preferences, e.g. to use the results for marketing purposes.

1 Introduction

A typical task in image retrieval is to find similar images based on similar attributes (see, e.g., Corridoni et al. 1999; Rui et al. 1998; Shen et al. 2009; Wang et al. 2010). There are two main research streams that deal with algorithms to optimize the respective image query: text-based image retrieval and content-based image retrieval (CBIR). Search attributes of the former are mostly given tags, e.g. a textual description of the visual content. On the other hand search attributes of the latter are

D. Schindler (✉)
Department of Business Administration and Economics, Bielefeld University, 33501 Bielefeld, Germany
e-mail: dschindler@wiwi.uni-bielefeld.de

B. Lausen et al. (eds.), *Algorithms from and for Nature and Life*, Studies in Classification, Data Analysis, and Knowledge Organization, DOI 10.1007/978-3-319-00035-0_48,

colors, textures or shapes, i.e. the visual content of the object (Wang et al. 2008). Before the advent of sharing digital images and other meta data, such as filenames or tags on the web, the data base of sufficiently tagged images was restricted. Therefore CBIR became more popular than text-based image retrieval.

However, since the advent of social media websites like Flickr the above-mentioned research streams reinvest in optimizing text-based image retrieval. With regard to this researcher extract online uploaded images automatically in combination with corresponding tags and images are clustered on the basis of these tags. Finally, all images are complemented by tags of similar images (see, e.g. Kennedy et al. 2007; Matusiak 2006; Monagahn and O'Sullivan 2007; Wang et al. 2008; Wu et al. 2011; Yang et al. 2010). Concerning this development Yeh et al. (2004) were one of the first to create a hybrid image-and-key-word searching technique, which combines user-generated tags and uploaded images. Research interest in this field is still growing (Sigurbjoernsson and Van Zwol 2008). Nevertheless, research in image analysis is still more related to computer science than to business sciences (Baier and Daniel 2011).

Recently, Van House (2009), for example, stated that on photo sharing platforms people are sharing their stories and their lives. Consequently, uploaded images reflect basis for lifestyle segmentation. However this data source of implicit preferences is still widely unused for marketing purposes. In this regard, Baier and Daniel (2011) investigate the question if image clustering can be used for marketing purposes, especially market segmentation. They clustered several holiday pictures, based on low-level features (i.e. content-based or to be precise based on RGB color histograms), and compared the results with collected preference data. The results were correlated and therefore promising regarding their use for market segmentation.

Apart from using visual content to cluster similar images it is a promising approach to cluster similar images on the basis of the similar associations they trigger. Since the advent of photo sharing websites this information is also provided online by users, namely in the form of tags and comments, which are made by both owners and users. Current research effort concerning this is primarily related to computer science in the very recent field of Emotion-Based Image Retrieval and emotional image tagging. For example, Kim et al. (2009) designed an image retrieval system, which uses human emotion and search images to improve retrieval systems. Moreover Schmidt and Stock (2009) analyzed emotional image tagging and found prototypical images for given emotions (see also Hastings et al. 2007).

Nevertheless, the question of using user-generated content (i.e. online posted associations in form of comments) is also interesting for marketing purposes or the aim of market segmentation. For example, if one has to prepare a promotion campaign. There are no restrictions but the promotional photograph should trigger the feeling of "freedom". The target customers are young students between 20 and 25 years. In this situation a great benefit would be to know about those people among the target customers who would associate several images with the feeling of "freedom".

Hence, the main object of this paper is to investigate the issue whether similar user-generated content (UGC) leads to clusters of similar images. Later on the results are compared with those of content-based image clustering in order to illustrate some advantages of image clustering based on UGC.

The content-based results are taken from a research project together with Brandenburg University of Technology Cottbus (see acknowledgments). In the following clustering based on UGC will be labeled User-Generated Content Clustering (UGCC) to differentiate it from the results based on RGB color historgrams (see Baier and Daniel 2011). The latter will be labeled Color-Based Image Clustering (CBIC). This paper focusses on comments which have been written by viewers. Even so, in the long view, our research efforts address both image associations that have been commented on by image owners and by viewers. Further research regarding owners and their comments on their images is still in progress.

The paper proceeds as follows. Data collection and pre-processing of the data are briefly illustrated in the next section. The main part of the paper is about the presentation of the results (Sect. 3). The paper concludes with some final remarks regarding the use of UGCC for image clustering. This will also include a brief forecast regarding the applicability of UGCC for marketing purposes (Sect. 4).

2 Survey and Data Pre-processing

Apart from objective descriptions images may be associated with feelings, for example, *hope*, *freedom*, *love* or *pleasure*. The bases of the online survey were images that triggered those typical associations – caused by memories of one's own experiences or events reported in the media. Half of the images show places of interest of five cities (subset "city"), namely Berlin, Paris, London, New York and Rio de Janeiro. The other images show scenes of living or events in these cities (subset "living"). It is assumed that some images have different effects when they are presented in color or in black and white, e.g. black and white images may have a distant effect. Therefore, totally the database includes 100 images – 50 in color and the same 50 in black and white.

The respondents (image viewers) were recruited on the social media platforms *facebook.com*, *fotocommunity.de* and *flickr.de*. The survey period lasted three weeks. Totally, 788 respondents participated in the survey and 330 completed it.

In the main part of the survey the respondents were asked to comment on the shown images and to suggest filenames. Furthermore they were asked to rate the images by stars from one to five, i.e. they were asked how much the respective image meet their approval. Each respondent had to comment on and rate four color and four black and white images. They were chosen randomly. Half of the images were from the subset "city" and the other from the subset "living".

With respect to the above-mentioned aim, the paper will concentrate on the analysis of comments, i.e. UGCC. The raw output of the comments are textual phrases including, e.g., stop-words like "and" or "or" which are not important for

UGCC and therefore were removed. The remaining textual phrases are reduced to their stems by Porter stemming algorithm (PSA) (see Porter 1980). The main idea of PSA is that terms with a common stem usually have similar meanings and can be seen as the same term (e.g. connect, connected, connecting). Porter stemming is still a standard pre-processing step in text analysis (see, e.g., Lee and Bradlow 2011, who also recently used PSA to analyze texts for marketing purposes).

The stemmed terms are transferred in a $(d_{ik} \times t_d)$ document-term matrix D with d_{ik} = UGC of image i commented on by respondent k and t_d = stemmed terms that occur in all documents d_{ik}.

The final document-term matrix D contains the frequencies of all the stemmed terms t_d with respect to the text documents d_{ik}. For further analyses the term frequencies are weighted by *term frequency-inverse document frequency* scheme (*tf-idf*). *tf-idf*-weighting is often used in the field of image retrieval to interpret the significance of a term. For example the term "image" is less important for UGCC than the term "freedom". The term "image" is only a definition of the object whereas the term "freedom" reflects an individual association (see, e.g., Manning et al. 2008 or Weiss et al. 2005).

To find out which images are similar *cosine similarity* measure is applied. This implies that two images are similar if similar terms were used in the comments, i.e. two images are identical if identical terms are posted ($sim = 1$) and they are dissimilar if they do not have common terms ($sim = 0$), where $sim \in (0, 1)$. Both *tf-idf*-weighting and *cosine similarity* are standard measures in the field of image retrieval and text mining (see, e.g., Manning et al. 2008 or Weiss et al. 2005).

3 Results

For the process of image clustering several hierarchical cluster algorithms were applied. In this case, using *complete linkage clustering* produced the best results. Therefore they are presented in the following. Figure 1 is an illustration of a cluster of images of Berlin. They are approximately grouped as expected. The images in cluster (a) (left hand side of the dendrogram) are grouped according to scenes that trigger similar associations, e.g. the Berlin Wall today and the original event of the Fall of the (Berlin) Wall. These images were mostly associated with the term "freedom" or "pleasure". The images in cluster (b) (right hand side of the dendrogram) are grouped on the basis of places of interest that are associated with Berlin. Here respondents wrote comments with respect to the respective place, building or landmark and additionally the city, i.e. "Berlin".

A closer look at the dendrogram shows that all pairs of cluster solutions on the lowest level are "color-black and white" pairs, i.e. a color image and the respective black and white image.[1] This result is representative for approximately the whole

[1]The last two numbers of the image ID have to be read as 01 = color and 02 = black and white image.

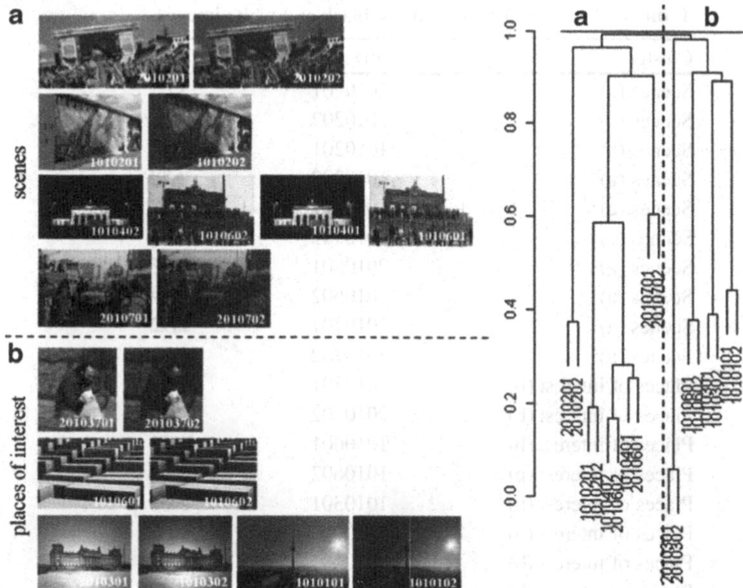

Fig. 1 Cluster "Berlin" – example result of using UGCC

data set. Obviously associations of color images are not significantly different from those of black and white images. Another noticeable result is that the levels of aggregation are mostly high. This points out that the reaction triggered by the images are very subjective.

As mentioned in Sect. 1 it was assumed that color images trigger different associations than black and white images. This assumption can neither be verified on the basis of comments nor on the basis of average star ratings. Table 1 shows the comparison of average star ratings for the respective "color-black and white" pairs.

In the following the results of UGCC are compared with those of CBIC.[2]

The most impressive result is that nearly all color and black and white images are classified separately. Here, CBIC results differ most from UGCC results. However, this effect occurs because of the applied method which uses RGB color histograms for feature extraction, Euclidean distance and Ward's algorithm for clustering. Separate analysis of color and black and white images would probably offer better results and is still planned.

Figure 2 shows an example of differences between UGCC and CBIC. A visual comparison of the images illustrated in Fig. 2 directly shows that these images are similar just because of their low-level features but they would not trigger similar associations. This is also confirmed by clustering the same subset based on associations. Applying UGCC, all four images were grouped in different clusters.

[2]Because analysis of the respective data set regarding the low-level features is still in progress in this paper only a few results are illustrated, namely those of the subset "city".

Table 1 Average rating – color vs. black and white images

Cluster	Image	Average rating
Scenes (a)	2010201	**3.04**
Scenes (a)	2010202	2.26
Scenes (a)	1010201	**2.94**
Scenes (a)	1010202	2.29
Scenes (a)	1010401	2.68
Scenes (a)	1010402	**3.33**
Scenes (a)	2010601	**2.95**
Scenes (a)	2010602	2.87
Scenes (a)	2010701	**2.92**
Scenes (a)	2010702	2.71
Places of interest (b)	2010301	2.26
Places of interest (b)	2010302	**2.79**
Places of interest (b)	1010601	3.33
Places of interest (b)	1010602	**3.69**
Places of interest (b)	1010301	**3.65**
Places of interest (b)	1010302	2.81
Places of interest (b)	1010101	**3.09**
Places of interest (b)	1010102	**2.90**

Fig. 2 Using CBIC for the subset "city" – example images

4 Implications and Outlook

In this paper some results of text-based image clustering were shown. Especially, the paper focused on association-based texts. With respect to UGCC complete linkage clustering offered acceptable results. In comparison to this CBIC results were not that plausible (see Fig. 2). Even so, for reasonable CBIC results color and black and white images have to be analyzed separately and more effective cluster algorithms have to be applied.

In particular and regarding UGCC, it could be demonstrated that images, which were assumed to trigger similar associations, are classified within the same cluster.

This implies that the results are promising in terms of their use for marketing efforts. With respect to the marketing campaign example it can be mentioned that, e.g., pictures that should trigger "freedom" can be searched for by the word "freedom". Nevertheless, this also requires that a user or viewer on, e.g., *flickr.com*, had commented on the image or had tagged it.

Nonetheless, the presented idea is also subject to some restrictions. In general, it should be mentioned, that a real-world database of commented or tagged images is not that sufficient until now. This is why an artificially data set was used. Moreover results cannot be generalized. The artificial data base was very heterogeneous which leads to heterogeneous and very subjective comments and, hence, high levels of aggregation concerning UGCC. Additionally, there may be more effective stemming algorithms than PSA for pre-processing, because PSA includes the threat of, e.g., losing semantically relevant content. Even so, subjectivity is still a key hindrance in text analysis (see, e.g. Liu 2010). Last but not least, until now only a part of UGC was examined. In order to indeed determine UGC comments of the image owners have to be analyzed as well.

Even though these challenging extensions have to be left for further research the current results make further research effort worthwhile. Recent investigations in hybrid image retrieval, i.e. automatically extracting images and UGC from photo sharing websites, are seemingly promising in order to observe implicit preferences and use UGCC to identify clusters of homogenous associations and, with respect to market segmentation, homogenous user or viewers of images. As mentioned before research efforts that were shown in this paper are supported by a cooperation project. In this project both research streams are followed to analyze the applicability of images for marketing purposes – CBIC as well as UGCC – and analyses are still in progress.

Acknowledgements This research is supported by the BMBF-ForMaT-Projekt "Multimediale Aehnlichkeitssuche zum Matchen, Typologisieren und Segmentieren".

Additionally I have to thank Sebastian Fruend, Bielefeld University, for auxiliary assistance in preparation and support of the online survey.

References

Baier, D., & Daniel, I. (2011). Image clustering for marketing purposes. In W. Gaul, A. Geyer-Schulz, & L. Schmidt-Thieme (Eds.), *Challenges concerning the data analysis – Computer science – Optimization* (Vol. 43), Proceedings of the 34th annual conference of the Gesellschaft für Klassifikation, Karlsruhe, July 21–23, 2010. Heidelberg/Berlin: Springer (Planned).

Corridoni, J. M., Del Bimbo, A., & Pala, P. (1999). Image retrieval by color semantics. *Multimedia Systems, 7*(3), 175–183.

Hasting, S. K., Iyer, H., Neal, D., Rorissa, A., & Yoon, J. W. (2007). Social computing, folksonomies, and image tagging – Reports from the research front. In *Proceedings of the 70th annual meeting of the american society for information and technology*, Milwaukee (Vol. 44, pp. 1026–1029). Wiley-Blackwell: Hoboken.

Kennedy, L., Naaman, M., Ahern, S., Nair, R., & Rattenbury, T. (2007). How Flickr helps us make sense of the world: Context and content in community-contributed media collections. In *Proceedings of ACM international conference on multimedia*, Augsburg (pp. 631–640). New York: Association for Computing Machinery.

Kim, Y., Shin, Y., Kim, S.-J., Kim, E. Y., & Shin, H. (2009). EBIR – Emotion-based image retrieval. In *Proceedings of international conference on consumer electronics*, Las Vegas (pp. 1–2). Piscataway, NJ: IEEE.

Lee, T. Y., & Bradlow E. T. (2011). Automated marketing research using online customer reviews. *Journal of Marketing Research, 48*(5), 881–894.

Liu, B. (2010). Sentiment analysis and subjectivity. In N. Indurkhya & F. J. Damerau (Eds.), *Handbook of natural language processing* (pp. 627–666). Boca Raton, FL: CRC.

Manning, C. D., Raghavan, P., & Schütze, H. (2008). *Introduction to information retrieval*. New York: Cambridge University Press.

Matusiak, K. (2006). Towards user-centered indexing in digital image collections. *OCLC Systems & Services, 22*(4), 283–298.

Monagahn, F., & O'Sullivan, D. (2007). Leveraging ontologies, context and social networks to automate photo annotation. In B. Falcidieno et al. (Eds.), *Semantic multimedia* (Lecture notes in computer science, Vol. 4816, pp. 252–255). Berlin/Heidelberg: Springer.

Porter, M. F. (1980). An algorithm for suffix stripping. *Program, 14*(3), 130–137.

Rui, Y., Huang, T. S., Ortega, M., & Mehrotra, S. (1998). Relevance feedback – A power tool for interactive content-based image retrieval. *IEEE Transactions on Circuits and Systems for Video Technology, 8*(5), 644–655.

Schmidt, S., & Stock, W. G. (2009). Collective indexing of emotions in images. A study in emotional information retrieval. *Journal of the American Society for Information Science and Technology, 60*, 863–876.

Shen, H. T., Jiang, S., Tan, K. L., Huang, Z., & Zhou X. (2009). Speed up interactive image retrieval. *The VLDB Journal, 18*(1), 329–343.

Sigurbjoernsson, B., & van Zwol R. (2008). Flickr tag recommendation based on collective knowledge. In: *Proceedings of international conference on world wide web*, Beijing (pp. 327–336). New York: Association for Computing Machinery.

Van House, N. A. (2009). Collocated photo sharing, story-telling, and the performance of self. *International Journal of Human-Computer Studies, 67*, 1073–1086.

Wang, C., Zhang, L., & Zhang, H.-J. (2008). Learning to reduce the semantic gap in web image retrieval and annotation. In *Proceedings of international ACM SIGIR conference on research and development in information retrieval*, Singapore (pp. 355–362). New York: Association for Computing Machinery.

Wang X.-Y., Yu, Y.-J. & Yang, H. Y. (2010). An effective image retrieval scheme using color, texture and shape features. *Computer Standards and Interfaces, 33*(1), 59–68.

Weiss, S. M., Indurkhya, N., Zhang, T., & Damerau, F. J. (2005). *Text minig – Predictive methods for analyzing unstructured information*. New York: Springer.

Wu P., Chu-Hong Hoi, S., Zhao, P., & He, Y. (2011). Mining social images with distance metric learning for automated image tagging. In *Proceedings of ACM international conference on web search and data mining*, Hong Kong (pp. 197–206). New York: Association for Computing Machinery.

Yang, Y., Huang, Z., Shen, H. T., & Zhou, X. (2010). Mining multi-tag association for image tagging. *World Wide Web, 14*(2), 133–156.

Yeh, T., Tollmar, K., & Darrel, T. (2004). Searching the web with mobile images for location recognition. In *Proceedings of IEEE computer society conference on computer vision and pattern recognition* (Part Vol. 2, pp. 76–81) Los Alamitos, CA: IEEE.

Logic Based Conjoint Analysis Using the Commuting Quantum Query Language

Ingo Schmitt and Daniel Baier

Abstract Recently, in computer science, the commuting quantum query language (CQQL) has been introduced for ranking search results in a database (Schmitt I, VLDB Very Large Data Base J 17(1):39–56, 2008). A user is asked to express search conditions and to rank presented samples, then, the CQQL system matches select-conditions with attribute values. Schmitt's (VLDB Very Large Data Base J 17(1):39–56, 2008) approach is based on von Neumann's (Grundlagen der Quantenmechanik. Springer, Berlin, 1932) quantum mechanics and quantum logic and is adapted in this paper to the usual conjoint analysis setting in marketing research where respondents are asked to evaluate selected attribute level combinations for choice prediction purposes. The results of an application show that the new approach competes well with the traditional approach.

1 Introduction

For modeling preferences and predicting choices of individuals w.r.t. objects of interest (e.g. hotels), conjoint analysis has proven to be a useful group of data collection and analysis methods. The main assumption is that preferential evaluations of presented objects (e.g. a sample of hotels) can be decomposed into partial evaluations w.r.t attribute levels of these objects in general (e.g. different price, quality levels). So, e.g., from the fact that a respondent prefers a 4-star-hotel for 100 Euro/night over a 3-star-hotel for 80 Euro/night one concludes that the

I. Schmitt (✉)
Institute of Computer Science, Information and Media Technology, BTU Cottbus, D-03013 Cottbus, Germany
e-mail: schmitt@tu-cottbus.de

D. Baier
Institute of Business Adminstration and Economics, BTU Cottbus, D-03013, Cottbus, Germany
e-mail: daniel.baier@tu-cottbus.de

B. Lausen et al. (eds.), *Algorithms from and for Nature and Life*, Studies in Classification, Data Analysis, and Knowledge Organization, DOI 10.1007/978-3-319-00035-0_49,

individual rates hotels of higher quality generally higher than hotels of lower quality even if the price is higher. If such a preference model is calibrated, it can be used to predict preferences and choices of this individual w.r.t. hotels in general, a tool that can be fruitfully used for designing and pricing product offerings in competitive markets.

However, recently, new preference learning methodologies have emerged in computer science. Schmitt's (2008) Commuting Quantum Query Language (CQQL) is one of these new methodologies. Compared to the usual conjoint analysis methodology it flexibly relates ratings to combinations of attribute levels. So, e.g., it allows to select subsets of observations and relevant attributes as well as to link partial ratings using various formulas. In this paper we develop the so-called Logic Based Conjoint Analysis (LBCA) from CQQL (Sects. 3, 4, and 5) and compare it with traditional Metric Conjoint Analysis (MCA) (Sect. 6). MCA is introduced in Sect. 2. The more advanced and popular Adaptive Conjoint Analysis (ACA) or Choice Based Conjoint Analysis (CBC) settings are neglected for simplicity reasons.

2 Metric Conjoint Analysis (MCA)

For modeling an individual's preferences, MCA is a widely spread method in marketing theory and practice (see, e.g., Green and Rao 1971 for first applications and Baier and Brusch 2009 for a recent review): Preferential evaluations w.r.t. attribute level combinations (stimuli) are collected and regressed according to

$$y_j = \underbrace{\mu + \sum_{k=1}^{K}\sum_{l=1}^{L_k} \beta_{kl} x_{jkl}}_{=:\,\hat{y}_j} + e_j \quad \text{with} \quad \begin{array}{ll} y_j: & \text{observed preference value for stimulus } j, \\ \mu: & \text{mean preference value across all stimuli,} \\ \beta_{kl}: & \text{partworth for level } l \text{ of attribute } k, \\ x_{jkl}: & \begin{cases} 1 & \text{if stimulus } j \text{ has level } l \text{ of attribute } k, \\ 0 & \text{else,} \end{cases} \\ e_j: & \text{measurement error for stimulus } j. \end{array}$$

The example in Table 1 is used to illustrate this data collection and analysis approach. It deals with a bus tour to Paris. Given are the descriptions of 16 training stimuli and the respective preference values of the first four – of a total of 100 – respondents.

From these observations model parameters μ and β_{kl} can be obtained using OLS, resulting in a preference model with parameters μ and β_{kl} for each respondent. In order to control the quality of calibration, Spearman's rank correlation coefficient can be used for relating observed and predicted preferences w.r.t. the training stimuli. In our example this results in values of 0.982, 0.944, 0.915, and 0.987 for the first four respondents and an average value of 0.976 across all 100 respondents. Additionally, for controlling the (predictive) validity, the respondents were asked to rank some additional (holdout) stimuli which are not used for parameter estimation. Table 2 shows them and the observed preference values from the first four respondents. Again, Spearman's rank correlation coefficient can be used to compare these observed with predicted preferences, resulting in Spearman's rank correlations

Table 1 The 16 stimuli of the bus tour example and corresponding preferences (1 = lowest preference,..., 16 = highest preference) of the first four – of 100 – respondents (R1,...,R4)

ID	Tr. time	Bus quality	Hotel quality	Price	Add. prog.	Disneyland	R1	R2	R3	R4
0	Day	Comfort	3 stars	249 Euro	Musical	Without	7	6	2	6
1	Day	Luxury	2 stars	209 Euro	Without	Without	2	14	16	8
2	Night	Komfort	2 stars	209 Euro	Sightseeing	With	12	7	3	11
3	Day	Luxus	1 star	249 Euro	Sightseeing	With	5	1	4	1
4	Night	Double-decker	2 stars	249 Euro	Without	With	3	5	15	5
5	Night	Double-decker	3 stars	209 Euro	Sightseeing	Without	10	16	10	12
6	Night	Luxury	2 stars	209 Euro	Musical	Without	15	8	5	9
7	Day	Double-decker	1 star	209 Euro	Sightseeing	Without	9	10	6	3
8	Day	Double-decker	2 stars	169 Euro	Musical	With	8	13	7	15
9	Day	Comfort	2 stars	209 Euro	Sightseeing	With	11	11	8	10
10	Night	Luxury	1 star	209 Euro	Musical	With	6	2	9	2
11	Night	Comfort	1 star	169 Euro	Without	Without	1	3	12	4
12	Night	Luxury	3 stars	169 Euro	Sightseeing	Without	13	15	11	16
13	Night	Luxury	2 stars	249 Euro	Sightseeing	Without	14	4	1	7
14	Day	Luxury	2 stars	169 Euro	Sightseeing	Without	16	12	13	14
15	Day	Luxury	3 stars	209 Euro	Without	With	4	9	14	13

Table 2 The four holdout stimuli and corresponding preferences (1 = lowest preference,..., 4 = highest preference) of the first four – of 100 – respondents (R1,...,R4)

ID	Tr. time	Bus quality	Hotel quality	Price	Add. prog.	Disneyland	R1	R2	R3	R4
0	Night	Comfort	2 stars	209 Euro	Without	With	3	3	1	3
1	Night	Luxury	1 star	169 Euro	Sightseeing	Without	2	1	2	1
2	Day	Luxury	2 stars	209 Euro	Musical	Without	4	4	4	4
3	Day	Double-decker	3 stars	249 Euro	Sightseeing	With	1	2	3	2

of 1.0, 0.0, 0.8, and 0.4 for the first four and an average value of 0.758 across all 100 respondents, indicating quite a good predictive validity on average. It should be noted that in MCA besides OLS also non-metric estimation methods (like Kruskal's MONANOVA) can be used for model calibration (taking the rank nature of the collected data better into account), but experiments have shown that this results in similar estimates of the model parameters.

3 Commuting Quantum Query Language (CQQL)

The database query language CQQL (Schmitt 2008) extends the relational tuple calculus to include the notion of proximity. Based on the mathematics of quantum logic, CQQL allows for logical operators as conjunction, disjunction and negation on score values out of the interval $[0, 1] \subseteq I\!R$. A query can be seen as a complex score function mapping objects to [0, 1]. Score values[1] are obtained from evaluating

[1] 1.0 corresponds to *true* and 0.0 to *false*.

monotonic proximity predicates on database attribute values. For example, the unary proximity predicate *low_price* returns a high score value for low prices and vice versa. The formal semantics may be defined by the function $e^{-\frac{|p-minp|}{c}}$ where *minp* is the lowest possible price and c an appropriate constant.[2]

If a query contains more than one proximity predicate then we can give the operands of a conjunction ($\wedge_{\theta_1,\theta_2}$) or a disjunction ($\vee_{\theta_1,\theta_2}$) different weights ($\theta_1, \theta_2$) referring to values out of [0, 1]. They define how strong the influence of that operand on the result should be. A weight value of 1.0 means strongest weight and a 0.0 means to ignore completely the corresponding operand.

CQQL obeys the rules of a Boolean algebra if for any object attribute at most one proximity predicate is used within a query, see Schmitt (2008). Following query example searches for an inexpensive bus tour (*bt*) with a culture program and a high-quality hotel. Weight variables control the impact of a good hotel *gh* and a low price *lp*: $\{b \in bt | b.program = culture \wedge \left(lp(b.price) \wedge_{\theta_{lp},\theta_{gh}} gh(b.hotel)\right)\}$. Let $\{o|\varphi(o)\}$ be a safe CQQL query on object $o \in O$ where φ is a CQQL formula. For evaluation, we eliminate weighted operands by applying following rules: $\varphi_1 \wedge_{\theta_1,\theta_2} \varphi_2 \rightsquigarrow (\varphi_1 \vee \neg\theta_1) \wedge (\varphi_2 \vee \neg\theta_2)$ and $\varphi_1 \vee_{\theta_1,\theta_2} \varphi_2 \rightsquigarrow (\varphi_1 \wedge \theta_1) \vee (\varphi_2 \wedge \theta_2)$. Thus, weight variables are transformed to proximity constants (0-ary predicates). Next, formula $\varphi(o)$ needs to be transformed[3] into prenex normal form and its quantifier-free part into CQQL normal form, see Schmitt (2008) and Saretz (2010). After evaluating the atomic proximity predicates, the resulting score values need to be combined in correspondence to conjunction ($eval(\varphi_1 \wedge \varphi_2, o) = eval(\varphi_1, o) * eval(\varphi_2, o)$), disjunction ($eval(\varphi_1 \vee \varphi_2, o) = eval(\varphi_1, o) + eval(\varphi_2, o) - eval(\varphi_1 \wedge \varphi_2, o)$) and negation ($eval(\neg\varphi, o) = 1 - eval(\varphi, o)$). The quantifier $\exists$ and $\forall$ are evaluated by finding the maximum and the minimum, respectively. As overall result we obtain an arithmetic formula which assigns to every object a score value. Applied to a database of objects and after descendingly sorting them by their scores, we obtain the objects fulfilling the query best at highest rank position.

4 Learning of CQQL Formulae

In our approach we learn a CQQL formula from preference values on stimuli specified by respondents. Furthermore, we assume n mutually independent proximity predicates $\{c_j\}$ on the attributes being given. As input for our learning algorithm we exploit preferences between pairs of stimuli. Such a preference pair, that is one stimulus is preferred to another, is derived from preference values. The preference values produce a rank and we take all neighboring stimulus pairs as preference pairs, see Fig. 1 (left). They form a partially ordered set. Our learning problem can

[2]For simplicity, we assume *metric* attribute values. The corresponding proximity predicates compare the value closeness to a highest or a lowest value. Due to space restriction, we do not discuss here the non-metric case.

[3]Recall, that the rules of Boolean algebra hold.

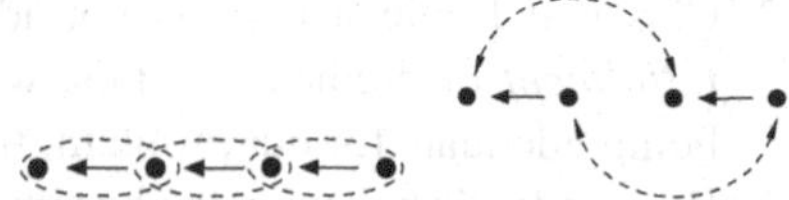

Fig. 1 *Left*: preference pairs extracted from a rank, *right*: protecting the transitivity chain

be interpreted as an optimization problem. Based on a set of atomic predicates, from all logical formulae fulfilling the preference pairs we want to find the best formula. Next, we explain what 'fulfilling formulas' and 'best formula' means. From Boolean algebra we know that every possible propositional formula can be transformed into the full disjunctive normal form $\bigvee_i(\bigwedge_j[\neg c_j \mid c_j])$[4] where the inner part $m_i = \bigwedge_j[\neg c_j \mid c_j]$ is called a minterm. Thus, every formula corresponds bijectively to a subset of 2^n minterms. For learning purposes, we assign to every minterm m_i a weight θ_i. For every preference pair $p_i : o_{i_1} \geq o_{i_2}$ we require $eval(w, o_{i_1}) - eval(w, o_{i_2}) \geq 0$. A feasible solution of the optimization problem is a function w which maps weight variables to values and fulfills all preference pair constraints. The arithmetic evaluation of the weighted full disjunctive formula provides[5] $eval^w(fdnf, o) = \sum_{i=0}^{2^n-1} w(\theta_i) \times eval(m_i, o)$ with $eval(m_i, o) = \Pi_{j=1}^{n}[1 - c_j(o) \mid c_j(o)]$. The best feasible solution w is the one which maximizes the sum of preference pair differences: $\sum_{i=1}^{m} (eval^w(fdnf, o_{i_1}) - eval^w(fdnf, o_{i_2})) \longrightarrow \max$. Due to compensation effects of the sum it is not a proper loss function. However, correctness is guaranteed by the demand for $eval(w, o_{i_1}) - eval(w, o_{i_2}) \geq 0$. It can be easily shown that the optimization problem is linear. Thus, the simplex algorithms can be applied. It solves the problem in average case in weak polynomial time. Please notice that due to the preference condition $o_{i_1} \geq o_{i_2}$ always a feasible solution exists when all weights θ_i have the same value. In this case the evaluation is independent from attribute values and returns the same score for all stimuli. In order to avoid such a meaningless solution we add a constraint which requires at least one preference inequality should be solved strictly: $\sum_{i=1}^{m} (eval^w(fdnf, o_{i_1}) - eval^w(fdnf, o_{i_2})) > 0$.

5 Logic Based Conjoint Analysis (LBCA)

Our LBCA data collection and analysis approach is based on following assumptions:

1. The mental model behind respondent's decisions on rating products can be expressed by means of a *logic* over a set of attribute predicates. In our bus tour example a respondent may think: 'If the price is high then I expect a good hotel otherwise a Disney tour must be included.'

[4]For simplicity, the expression $[a \mid b]$ means either a or b. This decision depends on the corresponding minterm.

[5]Please notice, that two different minterms differ always in the negation of at least one predicate. Therefore, disjunction of minterms is mapped to a simple sum.

2. Classical Boolean logic is not adequate since there is no support for *gradual fulfillment* of predicates. Thus, we assume a logic on truth values out of $[0, 1]$ being adequate for our problem. In our approach we use CQQL.
3. Respondent's behavior can be simulated by using a *compact* logical formula over a *small* set of predicates. We assume that most respondents consider only a small subset of all predicates for their decisions.
4. The product rank specified by a respondent can contain *flaws*. That is, we cannot guarantee that all preference pairs are consistent with the mental model. Thus, dropping some preference pairs may produce a better prediction and a more compact logical formula.

There is a tradeoff between compactness of a logical formula and the maximum number of allowed preference pair flaws. As described in the previous section, our starting point for learning is a weighted CQQL formula in full disjunctive form over a set of atomic predicates. All preference pairs correspond to inequality conditions on the evaluation of involved objects. Goal is to find a mapping w from weight variables to values out of the interval $[0, 1]$. The arithmetic evaluation together with the inequality conditions, the target function, the additional strictness constraint provide us a linear optimization problem. The simplex algorithm solves that problem. In the following three subsections, three aspects of the optimization problem are discussed.

5.1 Preference Pair Flaws

Not all preference pairs specified by a respondent are usually consistent with the respondent's mental model. Therefore, not all of them are of same priority, that is, some of them can be dropped. This topic is related to the problem of overfitting in machine learning: A too perfect learning of training data often produces low quality on test data and vice versa. Applied to our learning problem we can state, that less preference pairs may produce a CQQL formula which predicts the respondent's behavior better. Of course, too many drops result in a bad prediction.

When preference pairs need to be dropped the aspect of transitivity chain becomes relevant. Please recall that we generate preference pairs from neighbors of a total rank. For example, from the rank $\langle a, b, c, d \rangle$ we produce the preference pairs $a \leq b, b \leq c, c \leq d$. Thus, the relation $a \leq d$ can be derived from transitivity. Dropping $b \leq c$ breaks the transitivity chain and $a \leq d$ cannot be concluded anymore. In order to avoid such a break due to dropping one arbitrary preference pair, we produce not only direct neighbors but also 1-neighbors. In this way we obtain redundancy within the transitivity chain, see Fig. 1:right. We generate first preference pairs from 0-neighbors, i.e. $a \leq b, b \leq c, c \leq d$, and then from 1-neighbors, i.e. $a \leq c, b \leq d$. If two preference pairs are allowed to be dropped then we have to include also 2-neighbors and so on. In our approach, we use the software CPLEX which implements the simplex algorithm. If no weight values fulfilling the constraints can be found, i.e. an infeasible solution encounters, then CPLEX finds a preference pair constraint which cannot be fulfilled. In that case, we assume that this preference pair is a respondent's flaw and drop it. Then we start the simplex algorithm again. We repeat this cycle until a maximum number of predicate pairs is dropped or a feasible solution was found.

5.2 *Small Set of Predicates*

The full disjunctive normal form over n predicates contains 2^n minterms. Thus, the computational complexity explodes when the number of predicates increases. Due to our demand for compact logical formula we are looking for a compact formula based on few atomic predicates. However, a product may have many attributes and corresponding predicates. Our idea is to iterate over subsets of the set of all product predicates. Since we are looking for the most compact logical formula fulfilling the preference constraints we start with the smallest subsets and iterate to larger one. We stop if for a given subset a feasible solution is found. For choosing m predicates out of n predicates there exist $\binom{n}{m}$ subsets where every subset contains 2^m minterms.

5.3 *Result Interpretation and Binary Weight Values*

A feasible solution returns weight values for minterms which need to be interpreted. If the weight values were binary, i.e. 0 or 1, then we obtain a simple propositional formula which can be easily transformed into a form that a human understands. For example, a formula may be presented as implications. For example, the minterms $\overline{a} \wedge b, \overline{a} \wedge \overline{b}, a \wedge b$ are transformed to the implication $a \rightarrow b$ which can be well interpreted by humans.

There are different heuristics for mapping found weight values to discrete values. A straightforward approach is to use a threshold t: if $w(\theta_i) > t$ then $w(\theta_i) = 1$, otherwise if $w(\theta_i) < 1 - t$ then $w(\theta_i) = 0$ else $w(\theta_i) = 0.5$.

6 Application of LBCA and Comparison to MCA

Our experiments are based on the example introduced in Sect. 2. There are 100 respondents. Everyone ranked 16 training stimuli (sample bus tours) and 4 holdout stimuli. Every bus tour is characterized by six attribute values. The quality of learning (calibration) and prediction is measured by using Spearman's rank correlation coefficient which compares the given rank with the rank produced by a CQQL formula. All Spearman's rank coefficients shown below are averaged across the 100 respondents.

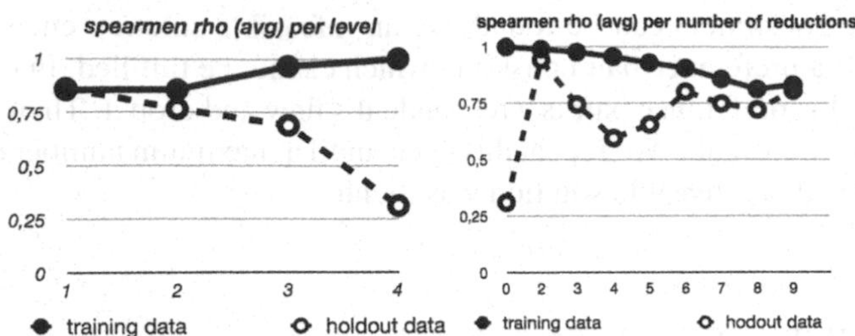

Fig. 2 Spearman's rank correlation coefficient in dependence of the number of predicates (diagram to the *left*) and the number of reductions (diagram to the *right*)

The diagram to the left of Fig. 2 shows the rank coefficients in dependence on the number of used atomic predicates. One can see, that the more predicates are involved the better the formula reproduces the training data. However, this does not hold for the holdout data. This observation confirms our assumption of a compact formula on few predicates. For two predicates we obtain a rank coefficient higher than 0.75 which is highly competitive in comparison to the existing MCA approach (see Sect. 2) where an average value of 0.758 was calculated.

The diagram to the right of Fig. 2 depicts the dependence on the number of dropped preference pairs. As expected, we obtain the best results for the training data if no preference pair is dropped. However, the results for the holdout data are bad. We see a sharp improvement w.r.t. the holdout data when a small number of preferences is dropped. Interestingly, if more preference pairs are dropped then first the prediction quality first goes down. Later, it increases again. So far, we cannot explain that phenomenon in detail. We need to perform more studies. One learning result of our example are the minterms $add.prog. \wedge disneyland$, $\overline{add.prog.} \wedge disneyland$, and $add.prog. \wedge \overline{disneyland}$, which are weighted by one. Those minterms can be interpreted as $\overline{add.prog.} \vee \overline{disneyland}$. That is, the corresponding user looks for a bus tour without disneyland option or without additional program.

7 Conclusion and Outlook

The experiments show that the new LBCA approach based on CQQL competes quite well with traditional conjoint analysis approaches like MCA. Further experiments in an ACA scenario are needed to clarify in which application fields and under which conditions the new approach has its practical advantages. The theoretical advantages w.r.t. logic modeling and simplified preference models are at least so far convincing. Further work is necessary to compare our approach with machine learning algorithms presented for example in Fürnkranz and Hüllermeier (2010). One approach from machine learning is the Choquet integral which is very similar to our presented approach but it lacks of a logical interpretation of a learn result.

References

Baier, D., & Brusch, M. (Eds.) (2009). *Conjointanalyse: Methoden – Anwendungen – Praxisbeispiele*. Berlin: Springer.

Fürnkranz, J., & Hüllermeier, E. (Eds.) (2010). *Preference learning*. Berlin: Springer.

Green, P. E., & RAO, V. (1971). Conjoint measurement for quantifying judgmental data. *Journal of Marketing Research, 8*(3), 355–363.

Saretz, S. (2010). *Evaluierung und Abgrenzung der Ähnlichkeitsanfragesprache CQQL gegenüber anderen DB-Retrieval-Verfahren*. Working paper series, Institute of Computer Science, Information and Media Technology, BTU Cottbus.

Schmitt, I. (2008). QQL: A DB&IR query language. *The VLDB (Very Large Data Base) Journal, 17*(1), 39–56.

Product Design Optimization Using Ant Colony And Bee Algorithms: A Comparison

Sascha Voekler, Daniel Krausche, and Daniel Baier

Abstract In recent years, heuristic algorithms, especially swarm intelligence algorithms, have become popular for product design, where problem formulations often are NP-hard (Socha and Dorigo, Eur J Oper Res 185:1155–1173, 2008). Swarm intelligence algorithms offer an alternative for large-scale problems to reach near-optimal solutions, without constraining the problem formulations immoderately (Albritton and McMullen, Eur J Oper Res 176:498–520 2007). In this paper, ant colony (Albritton and McMullen, Eur J Oper Res 176:498–520 2007) and bee colony algorithms (Karaboga and Basturk, J Glob Optim 39:459–471, 2007) are compared. Simulated conjoint data for different product design settings are used for this comparison, their generation uses a Monte Carlo design similar to the one applied in (Albritton and McMullen, Eur J Oper Res 176:498–520 2007). The purpose of the comparison is to provide an assistance, which algorithm should be applied in which product design setting.

1 Introduction

The purpose of product design optimization is to maximize profits, market share and sales or to minimize costs (Green et al. 1981; Gaul and Baier 2009; Gaul et al. 1995). Most products can be characterized using many attributes and their corresponding levels, so there is large number of possible product offerings (Socha and Dorigo 2008). That means, there is a very high complexity of complete enumeration (Kohli and Krishnamurti 1987). For this issue, swarm intelligence algorithms, where the behavior of foraging for food of social animals is copied, offer a good alternative

S. Voekler (✉) · D. Krausche · D. Baier
Institute of Business Administration and Economics, Brandenburg University of Technology Cottbus, D-03013 Cottbus, Germany
e-mail: sascha.voekler@TU-Cottbus.de; daniel.krausche@TU-Cottbus.de; daniel.baier@TU-Cottbus.de

B. Lausen et al. (eds.), *Algorithms from and for Nature and Life*, Studies in Classification, Data Analysis, and Knowledge Organization, DOI 10.1007/978-3-319-00035-0_50,

for the product design problem. In some cases and under certain conditions, they are the only way to find the real optimal solution to the underlying optimization problem (Albritton and McMullen 2007). In this paper, a new approach to the share-of-choice problem, the bee colony optimization (BCO), is introduced and compared to the already existing ant colony optimization (ACO) (Albritton and McMullen 2007; Camm et al. 2006; Teodorovic and Dell'Orco 2005; Karaboga and Basturk 2007; Teodorovic et al. 2011). Therefore, both algorithms were implemented with the statistical software *R*. Firstly, the problem formulation is introduced in Sect. 2. In Sects. 3 and 4, the methodologies for the ant colony and bee colony optimization are given. After that, in Sect. 5, the empirical comparison of both algorithms follows. At last, in Sect. 6, we give a short conclusion and an outlook.

2 Problem Formulation

To compare these two swarm intelligence algorithms, a common problem structure is needed. In this case, we are using the share-of-choice formulation for the underlying optimization problem (Camm et al. 2006). In the share-of-choice problem that product design is chosen, which satisfies most observed customers. That means, a binary linear integer program has to be solved (Camm et al. 2006). Considering a product with K attributes, L_k levels for each attribute k ($k = 1, 2, \ldots, K$) and a with S denoted number of respondents of the simulated conjoint study, the part-worth utility for a level j ($j = 1, 2, \ldots, L_k$) on attribute k for respondent s is given by u^s_{kj}. Then, a binary variable x_{kj} is introduced. This variable is set to $x_{kj} = 1$ if level j of attribute k is chosen and 0 otherwise. If the utility of a designed product exceeds the status quo utility for the s-th respondent (U_s), $y_s = 1$ and 0 otherwise. Also define a hurdle utility h_s for respondent s with $h_s = U_s + \epsilon$, where ϵ is a small real number. That leads to the following problem formulation:

$$\text{Max} \quad \sum_{s=1}^{S} y_s \tag{1}$$

$$\text{subject to} \quad \sum_{k=1}^{K} \sum_{j=1}^{L_k} u^s_{kj} x_{kj} \geq h_s y_s \qquad s = 1, 2, \ldots, S, \tag{2}$$

$$\sum_{j=1}^{L_k} x_{kj} = 1 \qquad k = 1, 2, \ldots, K. \tag{3}$$

Expression (1) represents the objective function which yields the number of respondents covered. The first constraint (2) ensures that only those respondents are counted in the objective function who met or exceeded the hurdle utility for a product. The purpose of the second constraint (3) is to make sure that exactly one level of each attribute is chosen (Camm et al. 2006).

3 Methodology for Ant Colony Optimization (ACO)

When a single ant is foraging for food, there are some local rules, that it follows. The first one is, that the ants follow that trail, where a high amount of pheromones lie. For the second rule, the ants leave pheromones on the recently taken path. Then, over time, the ants are taking more direct routes, because of the pheromone scent on the trail. In the initial state, the ants leave their nest randomly to forage for food. A single ant takes a certain path more likely, the higher the amount of pheromone is on this trail (Albritton and McMullen 2007).

The methodology for the ACO is given below and is based on Albritton and McMullen (2007).

Initialize probabilities: The value of $Prob^s_{kj}$ represents the relative desirability of selecting level k of attribute k. This probability stays constant throughout the search process and is calculated via roulette wheel selection:

$$Prob^s_{kj} = \frac{\sum_{s=1}^{S} u^s_{kj}}{\sum_{j*=1}^{L_k}\sum_{s=1}^{S} u^s_{kj*}} \quad \forall k = 1, 2, \ldots, K. \tag{4}$$

The objective function value of Z^* is set to zero:

$$Z^* = 0. \tag{5}$$

Initialize the history matrix: The history matrix **G** tracks all paths the ants have taken so far. All elements g_{kj} of matrix **G** are set to "1" at the beginning of the search.

Simulating the amount of pheromone: For simulating the amount of pheromone on a track, the global and local probabilities $GlobProb^s_{kj}$ and $LocProb^s_{kj}$ of selecting level j from attribute k are initialized to the relative desirability $Prob^s_{kj}$:

$$GlobProb^s_{kj} = Prob^s_{kj}, \tag{6}$$

$$LocProb^s_{kj} = Prob^s_{kj}. \tag{7}$$

Starting the search process: To find a solution to the optimization problem, an ant moves through the decision space from each level to the next level until the last attribute is reached. Therefore, a certain probability for each decision is constructed:

$$DecProb^s_{kj} = \frac{\gamma \cdot Prob^s_{kj} + \alpha \cdot LocProb^s_{kj} + \beta \cdot GlobProb^s_{kj}}{\sum_{j*=1}^{L_k} \gamma \cdot Prob^s_{kj*} + \alpha \cdot LocProb^s_{kj*} + \beta \cdot GlobProb^s_{kj*}} \quad \forall k = 1, 2, \ldots, K. \tag{8}$$

where

$$\alpha + \beta + \gamma = 1.$$

The three parameters α, β, γ are user-specified and represent the weights of the given probabilities.

Objective function: After a single ant completed its journey through the decision space, the objective function value Z is determined for the found solution:

$$Z = \sum_{s=1}^{S} u_{kj}^{s}, \quad \text{where } k, j \in \text{solution}. \tag{9}$$

Probabilistic updating: With a new solution to the problem, the objective function value of Z is compared to the former best solution of Z^*. If the new solution is better than the old one, a global update is performed, a local update otherwise:

$$\text{If } Z > Z^* \quad \text{global update,} \tag{10}$$

$$\text{If } Z \leq Z^* \quad \text{local update.} \tag{11}$$

The local update means, that the pheromone on this trail evaporates, so that new and better near optimal solutions can be found. Mathematically, this is expressed with a user-chosen value η:

$$LocProb_{kj}^{s} := LocProb_{kj}^{s} \cdot (1 - \eta). \tag{12}$$

On the other side, a global update is performed, to enhance the probability of selecting the path components of the recently found optimal solution. First, the history matrix **G**, is updated to represent the recently found solution:

$$g_{kj} := g_{kj} + 1, \quad \text{where } k, j \in \text{solution}. \tag{13}$$

Second, the global probability $GlobProb_{kj}^{s}$ is updated with the recently found solution:

$$GlobProb_{kj}^{s} = \frac{g_{kj}}{\sum_{j=1^*}^{L_k} g_{kj^*}}, \quad \text{where } k, j \in \text{solution}. \tag{14}$$

Iterations: These steps are repeated until the abortion criterion is met, e.g. the number of ants (=iterations) or the runtime of the algorithm.

4 Methodology for Bee Colony Optimization (BCO)

The following bee colony algorithm was not yet applied to the share-of-choice problem based on conjoint data. This is a **new approach** to solve the underlying optimization problem. Before developing the algorithm, the behavior of the artificial bees is described.

As for the ants, the food search of the bees is simulated. Collaboratively, the bees are searching for an optimal solution to difficult combinatorial optimization problems, where each bee creates one single solution to the problem. For foraging for food, there are two alternating phases for the bees' behavior: The forward and the backward pass. In the forward pass, the artificial bees visit a certain number of solution components (food sources). In the backward pass, the artificial bees save the information about the quality of their found solution (amount of nectar of the food source), fly back to the hive and communicate with the uncommitted bees. Back to the hive, each bee decides with a certain probability whether it stay loyal to its found solution or not. The better one solution, the higher are the chances, that this bee can advertise its found solution to the uncommitted bees (Teodorovic et al. 2011).

The methodology of BCO (Teodorovic and Dell'Orco 2005; Teodorovic et al. 2011) will be introduced below.

Forward pass: The bees do their first forward pass with selecting one level j of the first attribute k with roulette wheel selection according to Eq. (4).

Objective function: After the first forward pass, the found partial solutions are evaluated for each bee. Therefore, a new variable i is introduced, which represents the actual number of forward passes done. Note, the maximum value of i is the number of attributes ($i \leq K$). Hence, the objective function value of Z is determined by:

$$Z = \sum_{s=1}^{S} u_{ij}^{s}, \quad \text{where } i, j \in \text{(partial) solution.} \tag{15}$$

Backward pass: When the bees are back to the hive after the forward pass, each bee decides randomly whether to continue its own exploration and become a recruiter or to become a follower. Therefore, the bees have to make a loyalty decision. The notion B stands for the total number of bees. Mathematically, this formulation means, that the probability that the b-th bee ($b = 1, 2, \ldots, B$) is loyal to its previously generated (partial) solution is:

$$p_b^{i+1} = \exp\left(-\frac{O_{max} - O_b}{u}\right) \quad b = 1, 2, \ldots, B, \tag{16}$$

with

$$O_b = \frac{Z_{max} - Z_b}{Z_{max} - Z_{min}} \quad b = 1, 2, \ldots, B, \tag{17}$$

where O_b is the normalized value for the objective function of the partial solution created by the b-th bee and O_{max} is the maximum value of O_b.

If a bee is not loyal to its found (partial) solution, it become a follower. Hence, the total number of bees b is splitted into a certain number of recruiters R and followers F $(B = R + F)$. However, for each follower, a new solution from the recruiters is chosen by the roulette wheel (Teodorovic et al. 2011):

$$pf_b = \frac{O_b}{\sum_{r=1}^{R} O_r} \quad b = 1, 2, \ldots, B. \tag{18}$$

These steps are repeated until each bee has found a complete solution to the optimization problem.

Iterations: The loop described above, can be repeated for a certain number of iterations or is aborted by a former chosen runtime. For each iteration, the best found objective function value Z is saved.

5 Empirical Comparison

5.1 *Initialization of the Algorithms*

For initialization of both algorithms, simulated conjoint data is used. This is a common procedure in marketing to compare optimization algorithms for product design (Vriens et al. 1996). Therefore, a matrix **U** with the part-worths of S respondents is computed. One component u_{kj}^s of the matrix **U** represents the part-worth utility of the s-th respondent $(s = 1, 2, \ldots, S)$ with attribute k $(k = 1, 2, \ldots, K)$ and level j $(j = 1, 2, \ldots, L_k)$. To initialize these part-worth utilities, a lower (LB) and a upper bound (UB) have to be set from the researcher, as well as a uniformly distributed random variable (Φ_{kj}^s) in the range of [0, 1] (Albritton and McMullen 2007):

$$u_{kj}^s = LB + \Phi_{kj}^s \cdot (UB - LB). \tag{19}$$

In the next step, part-worth utility matrix **U** has to be transformed into a matrix **C** with the elements c_{kj}^s where the components are coded to "1" and "0". If a level of an attribute satisfies a customer, the part-worth utility is coded to "1" and if the part-worth utility has a negative value, it is coded to "0" (Steiner and Hruschka 2003):

$$c_{kj}^s = 0 \quad \text{if} \quad u_{kj}^s \leq 0, \tag{20}$$

$$c_{kj}^s = 1 \quad \text{if} \quad u_{kj}^s > 0. \tag{21}$$

Table 1 The means and maximum values of the marginal distribution of the objective function values for the Monte Carlo simulation of both algorithms

		ACO		BCO	
Factor	Level	Means	Max values	Means	Max values
Attributes	4	**212.88**	215.89	211.92	**216.44**
	8	**255.98**	**261.67**	250.86	255.89
	12	**150.64**	**156.22**	147.54	152.22
	16	**41.96**	**44.22**	41.00	43.33
	20	**27.77**	**29.56**	27.17	29.11
Levels	5	**139.11**	**141.80**	136.61	140.27
	8	**137.10**	**141.80**	135.49	139.07
	10	**137.31**	**140.93**	135.02	139.20
Customers	250	**66.63**	**69.47**	65.29	67.40
	500	**121.31**	**125.20**	118.92	123.27
	1000	**225.59**	**229.87**	222.88	227.53

5.2 *Test Configurations*

To compare the ACO and BCO, a Monte Carlo simulation is applied to both algorithms. In the first step, each algorithm solved one problem ten times in a predefined period, so that there is a big enough sample to make qualified statements about the algorithms' behavior. Altogether, there were 45 problems (5 attributes × 3 levels each × 3 customer set-ups), which means 45 different simulated conjoint matrices with uniformly distributed part-worths. For each problem, there are $|L_K|^{|K|}$ possible solutions. For capturing a solution, a certain number of corresponding attributes are needed to satisfy a customer, which depends on the number of attributes. After that, the means for each problem and algorithm were computed and based on that, the data shown in Table 1 were generated.

As one can see from Table 1, for comparison, the marginal distribution of the algorithms' settings is used. Even though, there are no significant differences between both algorithms (tested with Welsh's two sample t-test on the level of significance $\alpha := \{.1, .05, .01\}$), in almost each case (instead of the case for four attributes on the maximum value), the ACO performed better than the BCO. One possible explanation for these results is, that the mathematical structure of both algorithms is quite similar. A possible reason for the better performance of the ACO is, that this algorithm is more developed to the share-of-choice problem than the BCO. Another explanation could be, that the probability updates for the ACO are more enhanced than the ones from the BCO. However, both algorithms provided very good solutions in a reasonable amount of time, even to large problems and therefore the results are very satisfying. Unfortunately, the results could not be compared to the real optimal values, because complete enumeration would have taken too long.

6 Conclusion and Outlook

For the first application to the share-of-choice problem in conjoint analysis, the BCO algorithm worked very well. The solutions obtained were almost as good as the ones from the ACO. Nevertheless, there is a big potential for enhancing the BCO, e.g. through parameter modifications or other probability distributions for the bees' behavior.

For further research, we will try to apply genetic algorithms, particle swarm optimization, simulated annealing and other optimization heuristics to the share-of-choice problem.

Furthermore, another important goal is to apply these optimization methods to real world problems with real conjoint data and, of course, not just to product optimization in the single product case, but to whole product line problems.

References

Albritton, M. D., & McMullen, P. R. (2007). Optimal product design using a colony of virtual ants. *European Journal of Operational Research, 176*, 498–520.

Camm, J. D., Cochran, J. J., Curry, D. J., & Kannan, S. (2006). Conjoint optimization: An exact branch-and-bound algorithm for the share-of-choice problem. *Management Science, 52*, 435–447.

Gaul, W., & Baier, D. (2009). Simulations- und Optimierungsrechnungen auf Basis der Conjointanalyse. In D. Baier & M. Brusch (Hrsg.), Conjointanalyse – Methoden-Anwendungen-Praxisbeispiele. Berlin, Germany; Heidelberg, Germany: Springer.

Gaul, W., Aust, E., & Baier, D. (1995). Gewinnorientierte Produktliniengestaltung unter Beruecksichtigung des Kundennutzens. *Zeitschrift fuer Betriebswirtschaft, 65*(8).

Green, P. E., Carroll, J. D., & Goldberg, S. M. (1981). A general approach to product design optimization via conjoint analysis. *Journal of Marketing, 45*, 17–37.

Karaboga, D., & Basturk, B. (2007). A powerful and efficient algorithm for numerical function optimization: Artificial bee colony (ABC) algorithm. *Journal of Global Optimization, 39*, 459–471.

Kohli, R., & Krishnamurti, R. (1987). A heuristic approach to product design. *Management Science, 33*, 1523–1533.

Socha, K., & Dorigo, M. (2008). Ant colony optimization for continuous domains. *European Journal of Operational Research, 185*, 1155–1173.

Steiner, W., & Hruschka, H. (2003). Genetic algorithms for product design: how well do they really work? *International Journal of Market Research, 45*, 229–240.

Teodorovic, D., & Dell'Orco, M. (2005). Bee colony optimization – a cooperative learning approach to complex transportation problems. Advanced OR and AI methods in transportation. In *Proceedings of the 16th mini – EURO conference and 10th meeting of EWGT* (pp. 51–60).

Teodorovic, D., Davidovic, T., & Selmic, M. (2011). Bee colony optimization: The applications survey. *ACM Transactions on Computational Logic*, 1–20.

Vriens, M., Wedel, M., & Wilms, T. (1996). Metric conjoint segmentation methods: A Monte Carlo comparison. *Journal of Marketing Research, 33*, 73–85.

Part IX
Music Classification Workshop

Comparison of Classical and Sequential Design of Experiments in Note Onset Detection

Nadja Bauer, Julia Schiffner, and Claus Weihs

Abstract Design of experiments is an established approach to parameter optimization of industrial processes. In many computer applications however it is usual to optimize the parameters via genetic algorithms. The main idea of this work is to apply design of experiment's techniques to the optimization of computer processes. The major problem here is finding a compromise between model validity and costs, which increase with the number of experiments. The second relevant problem is choosing an appropriate model, which describes the relationship between parameters and target values. One of the recent approaches here is model combination.

In this paper a musical note onset detection algorithm will be optimized using design of experiments. The optimal algorithm parameter setting is sought in order to get the best onset detection accuracy. We try different design strategies including classical and sequential designs and compare several model combination strategies.

1 Introduction

Parameter optimization is an important issue in almost every industrial process or computer application. With an increasing number of parameters and an increasing function evaluation time it becomes infeasible to optimize the target variables in an acceptable period of time. Design of experiments allows to gain as much information as possible with minimal number of function evaluations and therefore helps to tackle this problem in the most effective way.

There are two types of experimental designs: classical and sequential. In classical designs all trial points are fixed in advance. In sequential designs the

N. Bauer (✉) · J. Schiffner · C. Weihs
Faculty of Statistics, Chair of Computational Statistics, TU Dortmund, Germany
e-mail: bauer@statistik.tu-dortmund.de; schiffner@statistik.tu-dortmund.de; weihs@statistik.tu-dortmund.de

B. Lausen et al. (eds.), *Algorithms from and for Nature and Life*, Studies in Classification, Data Analysis, and Knowledge Organization, DOI 10.1007/978-3-319-00035-0_51,

next trial point or the decision to stop the experiment depends on the results of previous experiments. The main challenge in design of experiments is the choice of the model, which describes the relationship between the target variable und the parameter vector. One promising approach to cope with this problem is model combination. The aim is to find the combination which is at least as good as the best single model. In this paper we introduce and test different model combination strategies and compare them to single models. Furthermore we will assess the influence of experimental design types on the evaluation results.

The next section provides a short overview of sequential parameter optimization and model combination strategies. Moreover, our experimental design types and model combination approaches are presented. The application problem, musical note onset detection, is discussed in Sect. 3. Finally in Sect. 4 we present the simulation results, summarize our work and provide points for future research.

2 Background and Research Proposal

2.1 Sequential Parameter Optimization

We assume that a non-linear, multimodal black-box function $f(\boldsymbol{P})$ of k numeric or integer parameters $\boldsymbol{P} = (P_1, P_2, \ldots, P_k)'$ is to optimize. The range of allowed values for parameter P_i is given by V_i. Let $V = V_1 \times V_2 \times \ldots \times V_k$ define the parameter space.

An experimental design is a scheme that prescribes in which order the trial points are to evaluate. In case of the classical approach this scheme depends on the assumed relationship between $f(\boldsymbol{P})$ and $\boldsymbol{P}$ (i.e. the model type) and a chosen optimization criterion (like A or D-optimality, Atkinson and Donev 1992) and a-priori specifies all trial points of the whole experiment. In case of a sequential approach only the initial design, whose dimension is usually much smaller than the total number of trials, has to be given in advance. The common procedure of sequential parameter optimization is as follows.

1. Let D denote the initial experimental design with $N_{initial}$ trial points and let $Y = f(D)$ be the vector of function values of points in D.
2. Do the following sequential step unless the termination criterion is fulfilled:
   2.1 Fit the model M with response Y and design matrix D;
   2.2 Find the next trial point d_{next} which optimizes the model prediction;
   2.3 Evaluate $y_{next} = f(d_{next})$ and update $D \longleftarrow (D \cup d_{next})$, $Y \longleftarrow (Y \cup y_{next})$.
3. Return the optimal value of the target variable y_{best} and the associated parameter setting d_{best}.

Usually the trial points for the initial design in step 1 are determined via Latin Hypercube Sampling (LHS, Stein 1987), which covers the interesting parameter space V uniformly. The major differences between the existing algorithms for sequential parameter optimization lie in steps 2.1 and 2.2. One popular approach here is a response surface methodology proposed by Jones et al. (1998): in step 2.1 a Kriging model (Krige 1951) is fitted and the next trial point in step 2.2 is chosen by maximizing the expected improvement criterion.

Another approach is given by Bartz-Beielstein et al. (2005): in step 2.1 a user-chosen model M is fitted and used to predict a new dataset: a sequential design D' (usually an LHS design) of size $N_{step} \gg N_{initial}$, which covers the parameter space uniformly. The next trial point in step 2.2 is the point of D' with the best model prediction value. Note that this optimization is not time-consuming because it requires merely prediction of the model M in D' but not evaluation of the function $f(D')$. There are several termination criteria to use in step 2: reaching the global optimum of f (if known), limitation of the function evaluations' number, time limitation or too small improvement.

For our research proposal we will use the above presented approach of Bartz-Beielstein et al. (2005) with D' of size $N_{step} = 20{,}000$. Two of the most important issues here are the initial design and the number of sequential steps. We will propose and test three algorithm settings, where the number of the influence parameters is assumed to be three.

The first setting (**Classic**) is a classical 3^3 factorial design with an additional inner "star" (Weihs and Jessenberger 1999, p. 250). The number of trial points here is 33 and just one further evaluation will be done according to the best model prediction (verification step). The total number of evaluations (34) should not be exceeded by all further designs in order to facilitate the comparability between them.

The second design (**SeqICC**) is given by an inscribed central composite initial design with 15 trial points (Weihs and Jessenberger 1999, p. 251) and 19 sequential steps. The third design (**SeqLHS**) is an LHS initial design with also 15 trial points and 19 sequential steps. The LHS initial design is commonly used in sequential parameter optimization of computer applications, while the central composite design is often applied to optimization of industrial processes. We employ both in order to assess which leads to better results.

2.2 Model Combination

The main idea behind model combination is to construct a combined model that yields better prediction accuracy than any single model.

Usually one of the following two approaches is applied: building an ensemble of one particular model type with different hyperparameter settings (Stanley and Miikkulainen 2002) or building an ensemble of heterogeneous models with fixed hyperparameter settings (Zhu 2010). Hyperparameters are parameters of a model or learner. A more challenging problem is developing an approach which handles different model types and optimizes their hyperparameter settings automatically. Such an algorithm is proposed for example by Gorissen et al. (2009) (island-model).

An important issue for this work is the combination of different model types. A good overview about some related approaches is given by Sharkey (1996). One of the most popular model combination methods is linear combination: a joint prediction for a particular trial point is obtained as simple or weighted average of the individual predictions. Model outputs could be weighted for example according to an accuracy criterion like the goodness of fit or prediction accuracy. Other approaches are the Dempster-Shafer belief-based method, supra Bayesian, stacked generalization etc. Gorissen et al. (2009) use e.g. merely the simple average approach for their algorithm in order to keep its complexity low.

The aim of this work is not only to compare different designs for sequential parameter optimization but also to test different model combination possibilities. Let us assume that m models (learners) $M_1, M_2, \ldots, M_m$ are given with Y as a response and a design matrix D, which includes the settings of the influential parameters. Let us further assume that we have a minimization problem: the minimum of Y is sought. For each model compute first a model prediction accuracy criterion $M_1^{acc}, M_2^{acc}, \ldots, M_m^{acc}$. We will use here the leave-one-out mean squared error estimator (LOO-MSE). Then calculate model predictions for each point $d_j, j = 1, \ldots, N_{step}$, of the sequential design D' and receive for each model a prediction vector of length N_{step}: $M_1^{pred}, M_2^{pred}, \ldots, M_m^{pred}$. The smaller the LOO-MSE is the better is the associated model.

As first model combination method we use the weighted average approach. In order to calculate the weights the model prediction accuracies are linearly rescaled into the interval from 1 to 2 where 1 corresponds to the worst model and 2 corresponds to the best model, leading to $scaled(M_i^{acc}), i = 1, \ldots, m$. The weighted average (**WeightAver**) then is defined as follows:

$$\boldsymbol{WeightedAverage}(d_j) = \sum_{i=1}^{m} \frac{M_i^{pred}(d_j)}{scaled(M_i^{acc})}, j = 1, \ldots, N_{step}.$$

In each sequential step the next evaluation is done in that point d_j, which minimizes the ***WeightedAverage*** function.

For the second combination approach (**BestModel**) we will just choose the best model according to the model prediction accuracy criterion. Then the function f is evaluated in that point d_j with the minimal model prediction value.

The third combination method (**Best2Models**) is similar to the second method but in each step we evaluate two points according to the predictions of the two best models. We take care that we do not more function evaluations than allowed (see the termination criterion in Sect. 2.1).

The last model combination approach is based on the ten best prediction points of each model (**Best10**). In the following this approach will be called best ten points approach. First, for each model M_i the best ten predicted values $best10(M_i^{pred})'$ are collected into a vector

$$\boldsymbol{Best} = (best10(M_1^{pred})', best10(M_2^{pred})', \ldots, best10(M_m^{pred})')'$$

of dimension $10 \cdot m$. The vector ***ScaledBest*** is obtained by rescaling ***Best*** into the interval from 1 to 2 (2 corresponds to the biggest value of ***Best*** and 1 to the smallest). The vector ***ModelWeight*** is defined as

$$\boldsymbol{ModelWeight} = (\underbrace{scaled(M_1^{acc})}_{\times 10}, \underbrace{scaled(M_2^{acc})}_{\times 10}, \ldots, \underbrace{scaled(M_m^{acc})}_{\times 10})'.$$

For each entry in ***Best*** its relative frequency in this vector is assessed and collected into the vector ***FrequencyWeight***. This is done because it could happen that many trial points belong to the best ten predictions of several model types and this makes them more influential. The final score for the best $10 \cdot m$ points is given by

$$\boldsymbol{Score}_l = \frac{\boldsymbol{ScaledBest}_l}{\boldsymbol{ModelWeight}_l} - \boldsymbol{FrequencyWeight}_l, l = 1, 2, \ldots, 10 \cdot m.$$

The next trial point for the function evaluation is that with the minimal score.

3 Application to a Musical Note Onset Detection Algorithm

A tone onset is the time point of the beginning of a musical note or another sound. Onset detection is an important step for music transcription and other applications like timbre or meter analysis (see an tutorial on onset detection by Bello et al. 2005).

Here the signal is analyzed merely on a low level: only the amplitude variations of the audio signal will be considered. The ongoing audio signal will be split up into windows of length $\boldsymbol{L}$ samples with overlap $\boldsymbol{U}$ samples. For each window the maximum of the absolute amplitude is calculated. An onset is detected in each window where the absolute amplitude maximum of this respective window is at least $\boldsymbol{S}$ times as large as the maximum of the previous window (see Bauer et al. 2010). Formally this model can be written down as follows:

$$O_T(\boldsymbol{L}, \boldsymbol{U}) = z\Big(\overbrace{\max(| x_t |)\Big|_{t=(T-1)\cdot(\boldsymbol{L}-\boldsymbol{U})+1}^{T\cdot \boldsymbol{L}-(T-1)\cdot \boldsymbol{U}} - \boldsymbol{S} \cdot \max(| x_t |)\Big|_{t=(T-2)\cdot(\boldsymbol{L}-\boldsymbol{U})+1}^{(T-1)\cdot \boldsymbol{L}-(T-2)\cdot \boldsymbol{U}}}^{\hat{O}_T(\boldsymbol{L},\boldsymbol{U},\boldsymbol{S})}\Big)$$

$$+e_T(\boldsymbol{L}, \boldsymbol{U}, \boldsymbol{S}), \text{ with}$$

- N – sample length of the ongoing signal,
- $t = 1, \ldots, N$ – sample index,
- x_t – amplitude of the ongoing signal in the tth sample,
- $T = 1, \ldots, \lfloor \frac{N-U}{L-U} \rfloor$ – window index,
- $O_T(\boldsymbol{L}, \boldsymbol{U})$ – vector of true onsets: $+1$, if onset in Tth window, 0, else,

- $\hat{O}_T(\boldsymbol{L},\boldsymbol{U},\boldsymbol{S})$ – vector of estimated onsets: $+1$, if onset in Tth window, 0, else,
- $O_1(\boldsymbol{L},\boldsymbol{U}) = 0$ (assumption),
- $z(x)$: $+1$, if $x > 0$, 0, else,
- $e_T(\boldsymbol{L},\boldsymbol{U},\boldsymbol{S})$ – model error,
- parameters to optimize: $\boldsymbol{L}$ – window length (in samples), $\boldsymbol{U}$ – overlap (in samples), $\boldsymbol{S}$ – threshold.

One of the most popular quality criteria in onset detection is the so called F-value: $F = \frac{2c}{2c+f^{+}+f^{-}}$, where c is the number of correctly detected onsets, f^{+} is the number of false detections and f^{-} represents the number of undetected onsets (Dixon 2006). Note that the F-value lies always between 0 and 1. The optimal F-value is 1.

Note that the optimal parameter setting could vary depending on e.g. music tempo, number of instruments or sound volume of an audio signal. Another important factor is whether there is a synthesized audio signal or a real piano recording. To take into account some of these points we decided to differentiate between six musical epochs where each epoch is represented by two famous European composers with one music piece respectively: Medieval (Perotin, A. de la Halle), Renaissance (O. di Lasso, H. L. Hassler), Baroque (C. Monteverdi, H. Schuetz), Classic (W. A. Mozart, F. J. Haydn), Romance (F. Chopin, R. Schumann), New Music (A. Schoenberg, I. Strawinski).

Music pieces were downloaded as MIDI-data from different internet archives and instruments of all music tracks were set to piano using the software Anvil-Studio.[1] After that the true onset times were extracted using the Matlab MIDI-Toolbox,[2] and then the MIDI-files were converted to WAV-files using the freely available software MIDI to WAVE Converter 6.1.[3] We converted (and then used) just the first 60 s of each music piece.

In design of experiments it is essential to define the region of interest for each parameter, i.e. its lower and upper boundaries. For the parameter $\boldsymbol{L}$ we will allow just the following powers of two: 256, 512, 1024, 2048, 4096 and 8192. The region of interest for $\boldsymbol{U}$ is given by an interval between 0 % (no overlap) and 50 %. Note that just 1 % steps are allowed. The lowest possible value for $\boldsymbol{S}$ is 1.1 and the largest 5.1 with step size 0.01.

In following we will model the relationship between the onset detection algorithm parameters ($\boldsymbol{L}$, $\boldsymbol{U}$ and $\boldsymbol{S}$) and the target variable (F-value). We actually have a maximization problem here, the sign of F will be reversed hence to get a minimization problem.

---

[1] http://www.anvilstudio.com (Version 2009.06.06), date: 01.07.2011.

[2] https://www.jyu.fi/hum/laitokset/musiikki/en/research/coe/materials/miditoolbox/,01.07.2011.

[3] http://www.heise.de/software/download/midi_to_wav_converter/53703,date:01.07.2011.

Table 1 Aggregated results

ID	Model type	Model combination type	Design	Place 1	Place 2	Place 3
1	**FSOM**	–	**Classic**	1	0	0
2	**KM**	–	**Classic**	1	0	0
3	**RF**	–	**Classic**	1	0	0
4	**SVM**	–	**Classic**	1	0	0
5	**NN**	–	**Classic**	1	0	0
6	**FSOM**	–	**SeqICC**	0	0	1
7	**KM**	–	**SeqICC**	4	2	0
8	**RF**	–	**SeqICC**	0	0	0
9	**SVM**	–	**SeqICC**	1	1	0
10	**NN**	–	**SeqICC**	5	0	0
11	**FSOM**	–	**SeqLHS**	0	0	0
12	**KM**	–	**SeqLHS**	0	0	0
13	**RF**	–	**SeqLHS**	0	0	0
14	**SVM**	–	**SeqLHS**	0	0	0
15	**NN**	–	**SeqLHS**	0	0	0
16	**COMB**	**WeightAver**	**Classic**	1	0	0
17	**COMB**	**BestModel**	**Classic**	1	0	0
18	**COMB**	**Best2Models**	**Classic**	1	0	0
19	**COMB**	**Best10**	**Classic**	1	0	0
20	**COMB**	**WeightAver**	**SeqICC**	0	0	1
21	**COMB**	**BestModel**	**SeqICC**	0	0	1
22	**COMB**	**Best2Models**	**SeqICC**	0	1	3
23	**COMB**	**Best10**	**SeqICC**	0	2	2
24	**COMB**	**WeightAver**	**SeqLHS**	0	0	0
25	**COMB**	**BestModel**	**SeqLHS**	0	0	0
26	**COMB**	**Best2Models**	**SeqLHS**	0	1	0
27	**COMB**	**Best10**	**SeqLHS**	0	3	1

4 Results and Conclusion

To compare different options for the sequential parameter optimization algorithm we generate an experimental scheme with three metaparameters: *model type*,[4] *model combination type* and *design*, which were discussed in previous sections.

Table 1 gives for each optimization strategy the amount of the best, the second best and the third best placements over all music pieces. As can be seen, the classical parameter optimization does not lead to appreciable results. The same is true for the sequential optimization via LHS initial design by using just single models. By using the model combination approaches nevertheless **SeqLHS** as good as **SeqICC**. However those can be found just for music pieces with rather worse onset error rates.

[4]FSOM: Full Second Order Model; KM: Kriging, RF: Random Forest; SVM: Support Vector Machines; NN: Neural Network.

An important issue is that neither of the model combination approaches is better than the best single model. One possible explanation is that all design points are equally taken into account when calculating the prediction accuracy. Instead, trial points near the optimum should be given higher weight. Regarding the model type the best results are achieved either with Kriging or Neural Network single models. The most appropriate design setting according to the simulation results is the sequential parameter optimization with classical initial design (**SeqICC**). Concerning the model combination methods the best two models and best ten trial points approaches seem to provide acceptable results.

As the relationship between characteristics of music pieces and optimal algorithm parameter settings for signal segmentation (onset detection) is important e.g. for improving algorithms for music processing in hearing aids, it is necessary to investigate this relationship in more detail. This is one of the aims for further work. Moreover, a parameter optimization of a more complex music signal analysis algorithm like an algorithm for music transcription is planned. Furthermore, other model combination strategies and experimental designs should be developed and tested.

Acknowledgements This work has been supported by the Collaborative Research Centre "Statistical Modelling of Nonlinear Dynamic Processes" (SFB 823) of the German Research Foundation (DFG), within the framework of Project C2, "Experimental Designs for Dynamic Processes".

References

Atkinson, A. C., & Donev, A. N. (1992). *Optimum experimental designs*. Oxford: Oxford University Press.

Bartz-Beielstein, T., Lasarczyk, C., & Preuß, M. (2005) Sequential parameter optimization. In B. McKay et al. (Eds.), *Proceedings 2005 congress on evolutionary computation (CEC'05)*, Edinburgh (Vol. 1, pp. 773–780). Piscataway: IEEE Press.

Bauer, N., Schiffner, J., & Weihs, C. (2010). Einsatzzeiterkennung bei polyphonen Musikzeitreihen. SFB 823 discussion paper 22/2010. http://www.statistik.tu-dortmund.de/sfb823-dp2010.html.

Bello, J. P., Daudet, L., Abdallah, S., Duxbury, C., Davies, M., & Sandler, M. B. (2005). A tutorial on onset detection in music signals. *IEEE Transactions on Speech and Audio Processing, 13*(5), 1035–1047.

Dixon, S. (2006). Onset detection revisited. In *Proceedings of the DAFx-06*, Montreal (pp. 133–137).

Gorissen, D., Dhaene, T., & De Turck, F. (2009). Evolutionary model type selection for global surrogate modeling. *Journal of Machine Learning Research, 10*, 2039–2078.

Jones, D., Schonlau, M., & Welch, W. (1998). Efficient global optimization of expensive black-box functions. *Journal of Global Optimization, 13*, 455–492.

Krige, D. G. (1951). A statistical approach to some basic mine valuation problems on the witwatersrand. *Journal of the Chemical, Metallurgical and Mining Society of South Africa, 52*(6), 119–139.

Sharkey, A. (1996). On combining artificial neural nets. *Connectionist Science, 8*(3), 299–314.

Stanley, K., & Miikkulainen, R. (2002). Evolving neural networks through augmenting topologies. *Evolutionary Computation, 10*(2), 99–127.

Stein, M. (1987). Large sample properties of simulations using Latin hypercube sampling. *Technometrics, 29*, 143–151.

Weihs, C., & Jessenberger, J. (1999). Statistische Methoden zur Qualitätssicherung und -optimierung in der Industrie. Weinheim: Wiley-VCH

Zhu, D. (2010). Hybrid approach for efficient ensembles. *Decision Support Systems, 48*, 480–487.

Recognising Cello Performers Using Timbre Models

Magdalena Chudy and Simon Dixon

Abstract In this paper, we compare timbre features of various cello performers playing the same instrument in solo cello recordings. Using an automatic feature extraction framework, we investigate the differences in sound quality of the players. The motivation for this study comes from the fact that the performer's influence on acoustical characteristics is rarely considered when analysing audio recordings of various instruments. We explore the phenomenon, known amongst musicians as the "sound" of a player, which enables listeners to differentiate one player from another when they perform the same piece of music on the same instrument. We analyse sets of spectral features extracted from cello recordings of five players and model timbre characteristics of each performer. The proposed features include harmonic and noise (residual) spectra, Mel-frequency spectra and Mel-frequency cepstral coefficients. Classifiers such as k-Nearest Neighbours and Linear Discrimination Analysis trained on these models are able to distinguish the five performers with high accuracy.

1 Introduction

Timbre, both as an auditory sensation and a physical property of a sound, although studied thoroughly for decades, still remains *terra incognita* in many aspects. Its complex nature is reflected in the fact that until now no precise definition of the phenomenon has been formulated, leaving space for numerous attempts at an exhaustive and comprehensive description.

The working definition provided by ANSI (1960) defines timbre in terms of a sound perceptual attribute which enables distinguishing between two sounds having

M. Chudy (✉) · S. Dixon
Centre for Digital Music, Queen Mary University of London, Mile End Road, London, E1 4NS, UK
e-mail: magdalena.chudy@eecs.qmul.ac.uk; simon.dixon@eecs.qmul.ac.uk

B. Lausen et al. (eds.), *Algorithms from and for Nature and Life*, Studies in Classification, Data Analysis, and Knowledge Organization, DOI 10.1007/978-3-319-00035-0_52,

the same loudness, pitch and duration. In other words, timbre is what helps us to differentiate whether a musical tone is played on a piano or violin.

But the notion of timbre is far more capacious than this simple distinction. Called in psychoacoustics *tone quality* or *tone color*, timbre not only categorises the source of sound (e.g. musical instruments, human voices) but also captures the unique sound identity of instruments/voices belonging to the same family (when comparing two violins or two dramatic sopranos for example).

The focus of this research is the timbre, or sound, of a player, a complex alloy of instrument acoustical characteristics and human individuality. What we perceive as a performer-specific sound quality is a combination of technical skills and perceptual abilities together with musical experience developed through years of practising and mastery in performance. Player timbre, seen as a specific skill, when applied to an instrument influences the physical process of sound production and therefore can be measured via acoustical properties of sound. It may act as an independent lower-level characteristic of a player. If timbre features are able to characterise a performer, then timbre dissimilarities can serve for performer discrimination.

2 Modelling Timbre

In order to find a general timbral profile of a performer, we considered a set of spectral features successfully used in music instrument recognition (Eronen 2001) and singer identification (Mesaros et al. 2007) applications. In the first instance, we turned our interest toward perceptually derived Mel filters as an important part of a feature extraction framework. The Mel scale was designed to mimic the entire sequence of pitches perceived by humans as equally spaced on the frequency axis. In reference to the original frequency range, it was found that we hear changes in pitch linearly up to 1 kHz and logarithmically above it (Stevens et al. 1937). Cepstrum transformation of the Mel scaled spectrum results in the Mel-frequency cepstrum whose coefficients (MFCCs) have become a popular feature for modelling various instrument timbres (e.g. Eronen 2003; Heittola et al. 2009) as well as for characterising singer voices (Tsai and Wang 2006).

We also investigated discriminant properties of harmonic and residual spectra derived from the additive model of sound (Amatriain et al. 2002). By decomposing an audio signal into a sum of sinusoids (harmonics) and a residual component (noise), this representation enables to track short time fluctuations of the amplitude of each harmonic and model the noise distribution.

Figure 1 illustrates consecutive stages of the feature extraction process. Each audio segment was analysed using the frame-based fast Fourier transform (FFT) with a Blackman-Harris window of 2,048-sample length and 87.5 % overlap which gave us 5.8 ms time resolution. The length of the FFT was set to 4,096 points resulting in a 10.76 Hz frequency resolution. The minimum amplitude value was set at a level of -100 dB.

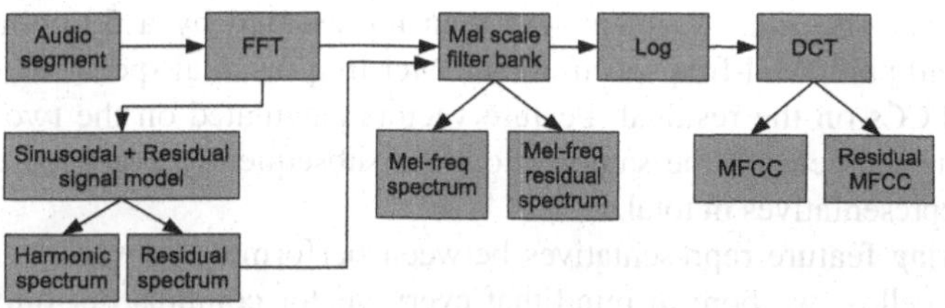

Fig. 1 Feature extraction framework

At the first stage, from each FFT frame, the harmonic and residual spectra were computed using the additive model. Then, all FFT frames, representing the full spectra at time points t, together with the residual counterparts, were sent to the Mel filter bank for calculating Mel-frequency spectra and residuals. Finally, MFCCs and residual MFCCs were obtained by logarithm and discrete cosine transform (DCT) operations on Mel-frequency spectra and Mel-frequency residual spectra respectively.

The spectral frames were subsequently averaged over time giving compact feature instances. Thus, the spectral content of each audio segment was captured by five variants of spectral characteristics: harmonic, Mel-frequency and Mel-frequency residual spectra, and MFCCs of the full and residual signals.

3 Experiment Description

For the purpose of this study we exploited a set of dedicated solo cello recordings made by five musicians who performed a chosen repertoire on two different cellos.[1] The recorded material consists of two fragments of Bach's *Cello Suite No.1*: *Prélude* (bars 1–22) and *Gigue* (bars 1–12). Each fragment was recorded twice by each player on each instrument, thus we collected 40 recordings in total. For further audio analysis the music signals were converted into mono channel .wav files with a sampling rate of 44.1 kHz and dynamic resolution of 16 bits per sample. To create a final dataset we divided each music fragment into six audio segments. The length of individual segments varied across performers giving approximately 11–12 s long excerpts from *Prélude* and 2–3 s long excerpts from *Gigue*. We intentionally differentiated the length of segments between the analysed music fragments. Our goal was to examine whether timbre characteristics extracted from shorter segments can be as representative for a performer as those extracted from the longer ones.

Having all 240 audio segments (24 segments per player performed on each cello) we used the feature extraction framework described in Sect. 2 to obtain sets

[1] The same audio database was used in the author's previous experiments (Chudy 2008; Chudy and Dixon 2010)

of feature vectors. Each segment was then represented by a 50-point harmonic spectrum, 40-point Mel-freq spectrum and Mel-freq residual spectrum, 40 MFCCs and 40 MFCCs on the residual. Feature vectors calculated on the two repetitions of the same segment on the same cello were subsequently averaged to give 120 segment representatives in total.

Comparing feature representatives between performers on various music segments and cellos, we bore in mind that every vector contains not only the mean spectral characteristics of the music segment (the notes played) but also spectral characteristics of the instrument, and then, on top of that, the spectral shaping due to the performer. In order to extract this "performer shape" we needed to suppress the influence of both the music content and the instrument. The simplest way to do this was to calculate the mean feature vector across all five players on each audio segment and subtract it from individual feature vectors of the players (*centering* operation).

When one looks at the spectral shape (whether of a harmonic or Mel-frequency spectrum) it exhibits a natural descending tendency towards higher frequencies as they are always weaker in amplitude. The so called *spectral slope* is related to the nature of the sound source and can be expressed by a single coefficient (slope) of the line-of-best-fit. Treating a spectrum as data of any other kind, if a trend is observed it ought to be removed accordingly for data decorrelation. Therefore subtracting the mean vector removes this descending trend of the spectrum.

Moreover, the spectral slope is related to the spectral centroid (perceptual *brightness* of a sound in audio analysis) which indicates the proportion of the higher frequencies in the spectrum. Generally, the steeper the spectral slope, the lower is the spectral centroid and less "bright" is the sound.

We noticed that performers' spectra have slightly different slopes, depending also on the cello and music segment. Expecting that it can improve differentiating capabilities of the features, we extended the centering procedure by removing individual trends first, and then subtracting the mean spectrum of a segment from the performers' spectra (*detrending* operation). Figures 2 and 3 illustrate individual trends and the centered spectra of the players after detrending operation. Our final performer-adjusted datasets consisted of two variants of features: centered and detrended-centered harmonic spectra, centered and detrended-centered Mel-frequency spectra and the residuals, centered MFCCs and the residuals.

The next step was to test the obtained performer profiles with a range of classifiers, which also would be capable to reveal additional patterns within the data if such exist. We chose the k-nearest neighbour algorithm (k-NN) for its simplicity and robustness to noise in training data.

k-Nearest Neighbours is a supervised learning algorithm which maps inputs to desired outputs (labels) based on *supervised* training data. The general idea of this method is to calculate the distance from the input vector to the training samples to determine the k nearest neighbours. Majority voting on the collected neighbours assigns the unlabelled vector to the class represented by most of its k nearest neighbours. The main parameters of the classifier are the number of neighbours k and distance measure *dist*.

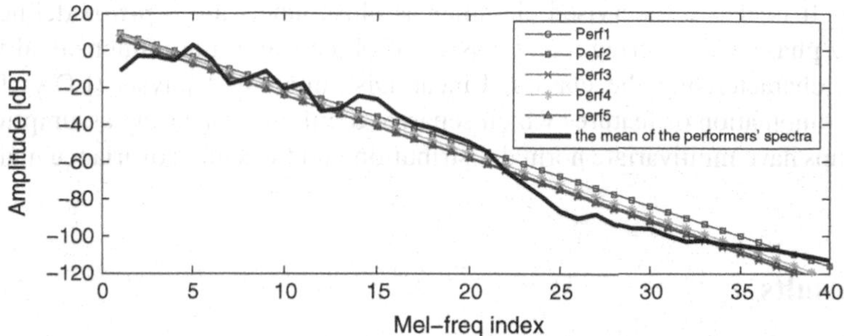

Fig. 2 Individual trends of five performers playing Segment1 of *Prélude* on Cello1 derived from Mel-frequency spectra

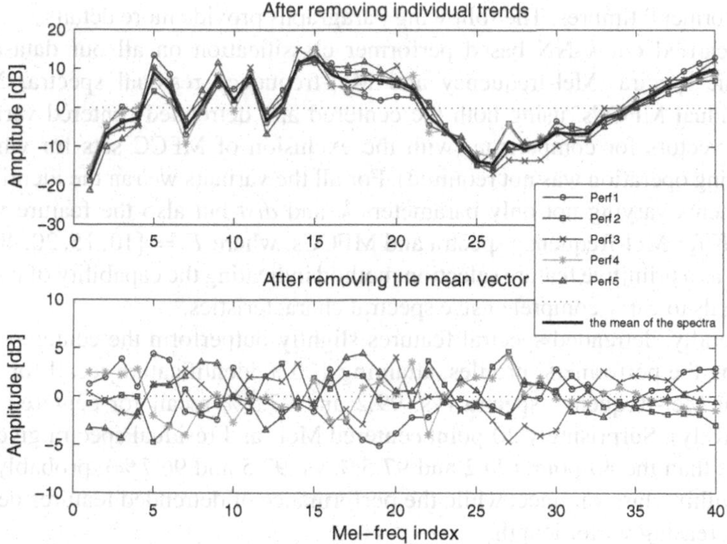

Fig. 3 Mel-frequency spectra of five performers playing Segment1 of *Prélude* on Cello1, after detrending and centering

We ran a classification procedure using exhaustive search for finding the neighbours, with k set from 1 to 10 and *dist* including the following measures: Chebychev, city block, correlation, cosine, Euclidean, Mahalanobis, Minkowski (with the exponent $p = 3, 4, 5$), standardised Euclidean, Spearman.

Classification performance can be biased if classes are not equally or proportionally represented in both training and testing sets. For each dataset, we ensured that each performer is represented by a set of 24 vectors calculated on 24 distinct audio segments (12 per each cello). To identify a performer p of a segment s, we used a leave-one-out procedure.

Amongst statistical classifiers Discriminant Analysis (DA) is one of the methods that build a parametric model to fit training data and interpolate to classify new objects. It is also a supervised classifier as class labels are *a priori* defined in a training phase. Considering many classes of objects and multidimensional feature vectors characterising the classes, Linear Discriminant Analysis (LDA) finds a linear combination of features which separate them under a strong assumption that all groups have multivariate normal distribution and the same covariance matrix.

4 Results

In general, all classification methods we examined produced highly positive results reaching even 100 % true positive rate (TP) in several settings, and showed a predominance of Mel-frequency based features in more accurate representation of the performers' timbres. The following paragraphs provide more details.[2]

We carried out k-NN based performer classification on all our datasets, i.e. harmonic spectra, Mel-frequency and Mel-frequency residual spectra, MFCCs and residual MFCCs, using both the centered and detrended-centered variants of feature vectors for comparison (with the exclusion of MFCC sets for which the detrending operation was not required). For all the variants we ran the identification experiments varying not only parameters k and *dist* but also the feature vectors' length F for Mel-frequency spectra and MFCCs, where $F = \{10, 15, 20, 40\}$. This worked as a primitive feature selection method indicating the capability of particular Mel-bands to carry comprehensive spectral characteristics.

Generally, detrended spectral features slightly outperform the centered ones in matching the performers' profiles, attaining 100 % identification recall for 20- and 40-point Mel-frequency spectra (vs. 99.2 and 97.5 % recall for centered spectra respectively). Surprisingly 20-point centered Mel- and residual spectra give higher TP rates than the 40-point (99.2 and 97.5 % vs. 97.5 and 96.7 %), probably due to lower within-class variance, while the performance of detrended features decreases with decreasing vector length.

What clearly emerges from the results is the choice of distance measures and their distribution between the two variants of features. Correlation and Spearman's rank correlation distances predominate within the centered spectra, while Euclidean, standardised Euclidean, cosine and correlation measures almost equally contribute to the best classification rates on detrended vectors. In regard to the role of parameter k, it seems that the optimal number of nearest neighbours varies with distance measure and the length of vectors but no specific tendency was observed.

It is worth noticing that the full spectrum features only slightly outperform the residuals (when comparing 100, 100, 97.5, 86.7 % recall of Mel-frequency detrended spectra with 99.2, 98.3, 92.5, 82.5 % recall of their residual counterparts for respective vector lengths = 40, 20, 15, 10). MFCCs and residual MFCCs in turn

[2]The complete overview of the results is available on http://www.eecs.qmul.ac.uk/~magdalenac/

perform better than the spectra especially in classifying shorter feature vectors giving 100, 100, 99.2, 95 % and 100, 99.2, 98.3, 90.8 % TP rates respectively.

For LDA-based experiments we used a standard stratified ten-fold cross validation procedure to obtain statistically significant estimation of the classifier performance. As previously, we exploited all five available datasets, also checking identification accuracy as a function of feature vector length.

For full length detrended-centered vectors of the harmonic, Mel-frequency and Mel-frequency residual spectra we were not able to obtain a positive definite covariance matrix. The negative eigenvalues related to the first two spectral variables (whether of the harmonic or Mel-frequency index) suggested that the detrending operation introduced a linear dependence into the data. In these cases, we carried out the classification discarding the two variables, bearing in mind that they might contain some important feature characteristics.

Similarly to the previous experiments, Mel-frequency spectra gave better TP rates then harmonic ones (100, 95.8, 89.2, 75 % for the vector length = 40, 20, 15, 10 vs 90, 85.8, 84.2, 75.8 % for the vector length = 50, 40, 30, 20 respectively) comparing centered features. Again, MFCCs slightly outperform the rest of features in classifying shorter feature vectors (99.2, 99.2, 99.2, 91.7 % recall for respective vector lengths = 40, 20, 15, 10). Detrended variants of spectra did not improve identification accuracy due to the classifier formulation and statistical dependencies occurring within the data. As previously, the residual Mel spectra (100, 93.3, 90, 79.2 %) and residual MFCCs (100, 98.3, 98.3, 90 %) produced worse TP rates than their counterparts, with the exclusion of the 100 % recall for 40 residual MFCCs.

5 Discussion

The most important observation that comes out from the results is that multidimensional spectral characteristics of the music signal are mostly overcomplete and therefore can be reduced in dimension without losing their discriminative properties. For example, taking into account only the first twenty bands of the Mel spectrum or Mel coefficients, the identification recall is still very high, reaching even 100 % depending on the feature variant and classifier.

This implied searching for more sophisticated methods of feature subspace selection and dimensionality reduction. We carried out additional LDA classification experiments on attributes selected by the greedy best-first search algorithm using centered and detrended Mel spectra. The results considerably outperformed the previous scores (e.g. 98.3 and 97.5 % recall for 13- and 10-point detrended vectors respectively), showing how sparse the spectral information is. What is interesting, from the Mel frequencies chosen by the selector, seven were identical for both feature variants indicating their importance and discriminative power.

As it was already mentioned, Mel spectra and MFCCs revealed their predominant capability to model the players' spectral profiles confirmed by high identification rates. Moreover, simple linear transformation of feature vectors by removing instrument characteristics and music context increased their discriminative properties. Surprisingly, the residual counterparts appeared as informative as full spectra, and this revelation is worth highlighting. It means that the residual (noise) part of signal contains specific transients, due to bow-string interaction (on string instruments), which seem to be key components of a player spectral characteristics.

Although we achieved very good classification accuracy on proposed features and classifiers (up to 100 %) we should also point out several drawbacks of the proposed approach: (i) working with dedicated recordings and experimenting on limited datasets (supervised data) makes the problem hard to generalise and non scalable; (ii) use of simplified parameter selection and data dimensionality reduction instead of other "smart" attribute selection methods such as PCA or factor analysis; (iii) the proposed timbre model of a player is not able to explain the nature of differences in sound quality between performers, but only confirms that they exist.

While obtaining quite satisfying representations ("timbral fingerprints") of each performer in the dataset, there is still a need for exploring temporal characteristics of sound production which can carry more information relating to physical actions of a player resulting in his/her unique tone quality.

References

Amatriain, X., Bonada, J., Loscos, A., & Serra, X. (2002). Spectral processing. In U. Zölzer (Ed.), *DAFX: Digital Audio Effects* (1st ed.). Chichester: Wiley.

American Standard Acoustical Terminology. (1960). *Definition 12.9*. New York: Timbre.

Chudy, M. (2008). Automatic identification of music performer using the linear prediction cepstral coefficients method. *Archives of Acoustics, 33*(1), 27–33.

Chudy, M., & Dixon, S. (2010). Towards music performer recognition using timbre features. In *Proceedings of the 3rd International Conference of Students of Systematic Musicology*, Cambridge (pp. 45–50).

Eronen, A. (2001). A comparison of features for musical instrument recognition. In *Proceedings of the IEEE Workshop on Applications of Signal Processing to Audio and Acoustics*, New Paltz (pp. 753–756).

Eronen, A. (2003). Musical instrument recognition using ICA-based transform of features and discriminatively trained HMMs. In *Proceedings of the 7th International Symposium on Signal Processing and its Applications*, Paris (Vol. 2, pp. 133–136).

Heittola, T., Klapuri, A., & Virtanen, T. (2009). Musical instrument recognition in polyphonic audio using source-filter model for sound separation. In *Proceedings of the 10th International Society for Music Information Retrieval Conference*, Kobe (pp. 327–332).

Mesaros, A., Virtanen, T., & Klapuri, A. (2007). Singer identification in polyphonic music using vocal separation and patterns recognition methods. In *Proceedings of the 8th International Conference on Music Information Retrieval*, Vienna (pp. 375–378).

Stevens, S. S., Volkman, J., Newman, E. B. (1937). A scale for the measurement of the psychological magnitude pitch. *Journal of the Acoustical Society of America, 8*(3), 185–190.

Tsai, W.-H., Wang, H.-M. (2006). Automatic singer recognition of popular recordings via estimation and modeling of solo vocal signals. *IEEE Transactions on Audio, Speech and Language Processing, 14*(1), 330–341.

A Case Study About the Effort to Classify Music Intervals by Chroma and Spectrum Analysis

Verena Mattern, Igor Vatolkin, and Günter Rudolph

Abstract Recognition of harmonic characteristics from polyphonic music, in particular intervals, can be very hard if the different instruments with their specific characteristics (overtones, formants, noisy components) are playing together at the same time. In our study we examined the impact of chroma features and spectrum on classification of single tone pitchs and music intervals played either by the same or different instruments. After the analysis of the audio recordings which produced the most errors we implemented two optimization approaches based on energy envelope and overtone distribution. The methods were compared during the experiment study. The results show that especially the integration of instrument-specific knowledge can significantly improve the overall performance.

1 Introduction

Harmony analysis of audio data provides a possibility to extract many high-level characteristics with aim to understand, describe and classify music. Many features with music theory background can be efficiently calculated from the score (Temperley 2007). However it is not always available. The music listener deals usually with mp3 or wave recordings. The automatic harmony analysis of the audio data, especially polyphonic content, is a very complex task. The recent works concentrate often on specific harmony features like chord recognition (Mauch 2010). The typical way is the analysis of the properties of many available low-level audio signal features (Lartillot and Toiviainen 2007; Theimer et al. 2008).

The original goal of our investigations was the identification of consonance and dissonance intervals in polyphonic audio recordings based on chroma features. During our experiments we have discovered that some of the intervals created by

V. Mattern (✉) · I. Vatolkin · G. Rudolph
Chair of Algorithm Engineering, TU Dortmund, Dortmund, Germany
e-mail: verena.mattern@udo.edu; igor.vatolkin@udo.edu; guenter.rudolph@udo.edu

B. Lausen et al. (eds.), *Algorithms from and for Nature and Life*, Studies in Classification, Data Analysis, and Knowledge Organization, DOI 10.1007/978-3-319-00035-0_53,

simple mixing of two instrument samples could not be identified correctly. Even the pitch of a single tone could not be derived from chroma in several cases because of the stronger overtones. Therefore we modified our first research target and tried to provide a deeper analysis of the problem cases and to apply some optimization techniques for the reduction of classification errors.

This remainder of this paper is organized as follows: At first we describe the used features. The following three sections correspond to the experimental studies: recognition of single tones, intervals played by the same instrument and finally intervals played by the different instruments. Two optimization approaches are introduced and applied: feature generation based on energy envelope and overtone analysis. In conclusion we give a summary of our work and discuss some ideas for ongoing research.

2 Features

Human perception of pitch shows a phenomenon that two pitches are perceived as similar in timbre if they differ by a whole number multiples of octave. Based on this observation the pitch can be divided into two components. Those components are the tone height and chroma (Bartsch and Wakefield 2005). In order to identify intervals we decided to use chroma features because of their strong relationship to harmonic characteristics. A commonly applied wrapped chroma vector consists of 12 elements which represent the octave semitones, which are created by the normalized sum of the energies of the corresponding tones across all octaves. Therefore we should reach our goal by the analysis of the strongest chroma values.

The HARMONIC PITCH CLASS PROFILE (HPCP) is based on the Pitch Class Profile (Fujishima 1999) with some more modifications introduced in Gómez (2006):

$$HPCP(n) = \sum_{i=1}^{nPeaks} w(n, f_i) \cdot a_i^2. \tag{1}$$

a_i represents the linear magnitude and f_i the frequency values of the peak number i. The energy is calculated from $i = 1$ to *nPeaks*, which is the number of the considered spectral peaks. $n \in \{1, \dots, size\}$ refers to the HPCP bin. $size$ corresponds to the number of HPCP vector dimensions and can be 12, 24 or another multiply of 12, representing semitones. $w(n, f_i)$ is the weight of the frequency f_i when considering the HPCP bin n.

Instead of one peak contributing to exactly one bin, the weighting function lets one peak i contribute to more than one bin. These bins have to be in a certain window around peak i. For each bin peak i is weighted by a cos^2 function in respect to the center frequency of the bin. The weight is influenced by the distance between the frequency f_i and the center frequency of the bin f_n. The distance is measered in semitones. With the help of this weighting procedure the estimation errors caused by tuning differences and inharmonicity should be minimized.

CHROMA ENERGY NORMALIZED STATISTICS (CENS) was introduced in Müller and Ewert (2010) and absorbs variations of acoustical properties such as dynamics, timbre and others, because it correlates to the short-time harmonic content of the audio signal.

At first, the audio signal is decomposed into 88 subbands for each semitone. After that the Short-Time Mean-Square Power (STMSP) or local energy is calculated in each subband:

$$STMSP(x) = \sum_{k \in [n - \lfloor \frac{t_w}{2} \rfloor : n + \lfloor \frac{t_w}{2} \rfloor]} |x(k)|^2. \tag{2}$$

x denotes a subband signal and $t_w \in \mathbb{N}$ a fixed window size. $STMSP(x)$ is calculated for each $n \in \mathbb{Z}$. A chroma representation is obtained by adding up the local energies of the subbands, which correspond to the same pitch class. All pitches that share the same chroma contribute to a pitch class. To enhance the robustness of the chroma representation, different modifications are performed. A first modification consists of normalizing a chroma vector v by calculating $\frac{v}{\|v\|_1}$, where $\| v \|_1$ denotes the l_1-norm:

$$\| v \|_1 = \sum_{i=1}^{12} |v(i)|. \tag{3}$$

The norm represents the relative distribution of the energy within the chroma bands. Afterwards each chroma energy distribution vector is quantized by applying $\tau : [0,1] \rightarrow \{0,1,2,3,4\}$ to each component. $\tau(a)$ assigns each component a value between 0 and 4 based on their energy (Müller and Ewert 2010). Next, the sequence of chroma vectors is convolved with a Hann window with a window size of t_w. In the last step the chroma vectors are downsampled by a factor d and the resulting vectors are normalized with respect to the euclidean norm.

3 Analysis of Single Tones

In the first experiment the pitch recognition of single tones was applied. The instrument samples have been taken from the McGill Master Sample collection (McGill University Master Samples): 35 acoustic guitar, 36 cello, 24 electric bass, 36 electric guitar, 36 flute, 36 piano, 24 sax and 24 trombone tones. HPCP was extracted within time frames of 4,410 samples without overlap, CENS within frames of 4,410 samples with 2,205 samples overlap. The extraction was done by AMUSE framework (Vatolkin et al. 2010) and Matlab. The estimated feature values were averaged across the audio samples and the pitch was identified as the strongest chroma vector dimension.

Even for this rather simple task it is hard to identify some pitches due to instrument-specific overtone distributions. Figure 1 illustrates HPCP and CENS progress for electric bass D♯2 tone and cello C♯2. The pitch of the bass tone corresponds to the highest chroma values at the very beginning of the recording; especially the second overtone A♯ seems to have the stronger amplitude after the

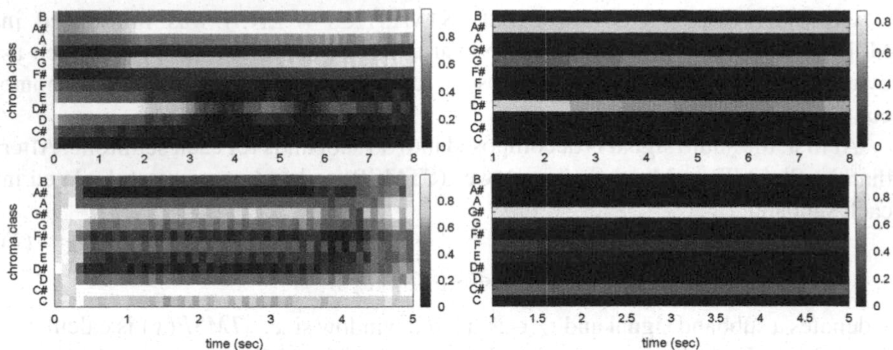

Fig. 1 *Top*: HPCP (*left*) and CENS (*right*) for e-bass D♯2; *Bottom*: HPCP (*left*) and CENS (*right*) for cello C♯2

time interval of appr. 2 s. The situation for the cello tone is slightly different: here the overtone G♯ corresponds to the highest CENS values over the complete sample duration but can be recognised correctly for the large interval within HPCP values.

Eronen proposed in Eronen (2009), that it is helpful to segment a note into attack and steady-state segments before extracting features. This proposal is based on the theory, that a note can be split into four stages (Park 2010): after the *attack* phase with the fast energy increase a short energy descent (*decay*) is observed following with the stable tone energy state (*sustain*). The *release* interval corresponds to the last evident sample sound with decreasing energy. However this theoretical background leads to very different observations depending on the instrument-specific influence of overtone distribution and formants on their timbre (Jourdain 1998). Also the none-harmonic noisy components such as violin bow strike have an impact on the frequency distribution.

Assuming that the energy envelope has some stable components for the different instruments, we implemented the ATTACK-ONSET-RELEASE-SPLITTER (AOR-Splitter) algorithm, which saves up to five feature series for each feature: from time frames which contain the beginning of the attack interval, the middle of the attack interval, the onset event, the middle of the release interval and the end of the release interval. The decay interval was omitted since the MIR Toolbox onset detection algorithm (Lartillot and Toiviainen 2007) could extract only the attack and release intervals. For the experiments with the short recordings such as single notes or intervals the analysis of time frames from the beginning of the attack phase and the end of the release obviously does not make sense, so we generated only three different feature series.

Table 1 summarizes the results. AOR-Splitter improves the performance of pitch recognition based on HPCP, if only the time frames with onset events are analyzed. It leads to error-free results. The different stages of the sound envelope have no beneficial influence on CENS. In general, HPCP analysis performs better than CENS without and with optimization, however it produces more errors if only the attack or release interval middles are used.

Table 1 Error numbers for single tone pitch recognition

	HPCP				CENS			
Instr./Oct.	NA	Att.	Ons.	Rel.	NA	Att.	Ons.	Rel.
Ac. guitar 3, 4, 5	0	0	0	8	0	0	0	3
Cello 2, 3, 4	0	7	0	0	2	3	3	4
E-bass 2, 3	3	0	0	5	3	3	3	5
E-guitar 3, 4, 5	0	0	0	1	1	0	0	1
Flute 4, 5, 6	0	4	0	8	0	0	0	0
Piano 3, 4, 5	0	0	0	1	0	2	2	1
Sax 3, 4	0	1	0	2	1	2	2	2
Trombone 3, 4	0	2	0	5	0	0	0	0
$\sum$ Errors	3	14	**0**	30	7	10	10	16
Relative error (%)	1.2	5.5	**0**	12	2.8	4	4	6.3

NA Naive Approach with averaged features, *Att.* only middles of attack intervals, *Ons.* only onset frames, *Rel.* only middles of release intervals

4 Recognition of Intervals Played by Equal Instruments

In this experiment a database of 273 intervals played by the equal instruments was used. To achieve this, we selected the single instrument samples from the McGill Collection and mixed them by using the own implementation based on Java Sound library. We chose three octaves for each instrument, except for electric bass, sax and trombone from which we took two octaves. Thirteen intervals ranging from the prime to the octave were generated for each instrument and octave. The interval was identified by the two strongest HPCP and CENS values.

The results without optimization are given by NA column in Table 2. As expected, the error number is higher for the increased complexity of the audio files. It can be observed that in some cases strong overtones are existing. Figure 2 gives an example of the CENS feature for C3-C♯3 interval: the tone G has a high magnitude even though it does not belong to the interval. Actually G is the second overtone of C – this observation supports our suggestion that overtones cause most of the errors.

The AOR-Splitter shows a similar behavior to experiment 1. As it can be seen in Table 2, the AOR-Splitter has a significant influence on the HPCP. We can also observe that CENS is more robust against this optimization approach. The smallest error rate is maintained by the analysis of the time frames from the onset stage with the HPCP feature. Using the averaged approach (NA column), CENS shows the better results of the both features. This time though, we do not have an error-free result. According to the suggestion that overtones cause the errors, we developed a method that recognises and eliminates overtones. This method will be explained in the following.

Knowing that the spectral peaks may exist for the overtone frequencies our approach is to determine whether a fundamental frequency has a set of overtones that have the significant energies. If that is the case we do not analyze the overtones. We consider only the magnitudes over an instrument-specific bound which will

Table 2 Error numbers for recognition of intervals played by the same instrument

	HPCP				CENS				
Instr./Oct.	NA	Att.	Ons.	Rel.	NA	Att.	Ons.	Rel.	Overtone Analysis
Ac. guitar 4, 5, 6	14	8	7	13	8	0	0	21	0
Cello 2, 3, 4	12	16	3	14	16	15	16	19	7
E-bass 2, 3	13	0	1	10	2	10	10	4	1
E-guitar 3, 4, 5	6	0	0	10	1	0	0	13	0
Flute 4, 5, 6	0	12	1	17	0	1	1	2	0
Piano 3, 4, 5	3	9	12	5	8	11	11	9	1
Sax 3, 4	1	3	0	5	2	6	3	4	0
Trombone 3, 4	3	6	0	17	8	13	12	15	0
$\sum$ Errors	52	54	24	91	45	56	53	87	**9**
Relative error (%)	19	20	8.8	33.5	16.5	20.5	19.5	32	**3.3**

NA Naive Approach with averaged features, *Att.* only middles of attack intervals, *Ons.* only onset frames, *Rel.* only middles of release intervals

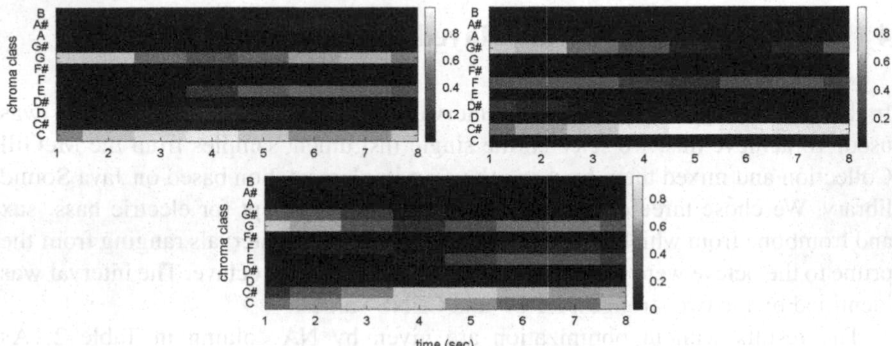

Fig. 2 CENS feature for: C3 piano tone (*top left*), C♯3 piano tone (*top right*) and the interval from the both tones (*bottom*)

be explained later on. For the remaining frequencies the integer multiplies are calculated. If one or more of the multiplies are part of the remaining magnitude peaks, they are eliminated. At the end, only the fundamental frequencies will remain in case no disharmonic content is part of the frequency spectrum.

The minimal and maximum bound magnitudes *minAmp* and *maxAmp* for each audio file are identified, so that the overtone analysis calculates error-free results. Two normalized bound values w_l and w_u are calculated by dividing the minimal magnitude by the absolute maximum magnitude of the sound and the maximum (correct) magnitude by the absolute maximum. After the calculation of w_l and w_u for each audio file, we take the minimal intersection for each octave of each instrument. Starting from the lower bound w_l, we search an optimal bound value w that generates the fewest errors over the octave increasing the bound by 0.01 in each step. The estimated w-values are listed in Table 3. The bound itself is calculated by $minAmp + (w * maxAmp)$. In all cases $minAmp = 0$, since there always exist

Table 3 *w* values for the intervals played by the equal instruments

Octave	A-guitar	Cello	E-bass	E-guitar	Flute	Piano	Sax	Trombone
2	x	0.021	0.02	x	x	x	x	x
3	0.04	0.24	0.02	0.06	x	0.08	0.04	0.05
4	0.14	0.24	x	0.06	0.08	0.29	0.04	0.05
5	0.14	x	x	0.06	0.08	0.29	x	x
6	x	x	x	x	0.08	x	x	x

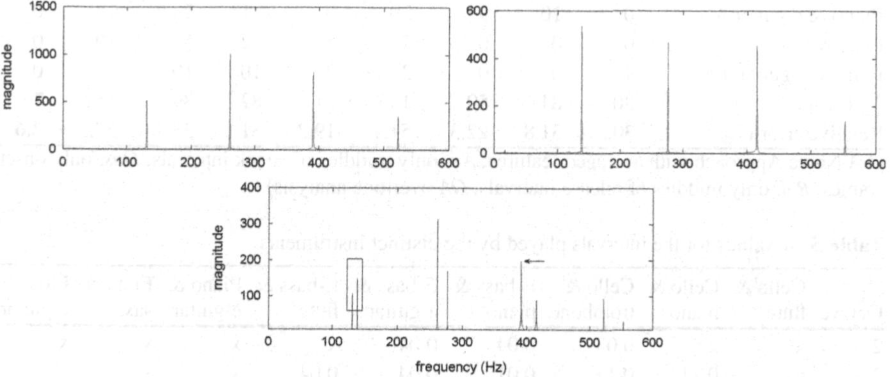

Fig. 3 Spectrum for: C3 piano tone (*top left*), C♯3 piano tone (*top right*) and the interval from the both tones (*bottom*)

some frequencies with zero magnitude. For example, the C4-D♯4 flute interval has the following values: $minAmp = 0$, $maxAmp = 3385$ and $w = 0.08$. The overtone analysis will only consider frequencies with magnitudes over the magnitude bound of 270.8.

The results of this approach are shown in the last column of Table 2. The overtone analysis reduces the relative error from 8.8 % (best result after AOR-Splitter optimization) to 3.3 %.

Figure 3 shows the spectrum for the interval piano C3-C♯3, which was not correctly identified before the overtone analysis. The peaks inside the square mark the magnitudes of the fundamental frequencies. The first overtones of the both single tones (two peaks right next to the square) have a higher magnitude. Since the first overtone is an octave above the fundamental frequency, it does not have an influence on the chroma vector. But the peak of the second overtone, marked by the arrow, guides to the erroneous identification of the interval without overtone analysis.

Summarizing the results it can be stated that the most problematic instrument is the cello. After the overtone analysis it still produces 7 of the 9 errors. All of them occur in octave 2. It may be explained by the fact that this octave is low and a large cello-specific overtone number with significant energies is distributed across several octaves (Jourdain 1998).

Table 4 Error numbers for recognition of intervals played by the different instruments

	HPCP				CENS				
Instr./Oct.	NA	Att.	Ons.	Rel.	NA	Att.	Ons.	Rel.	OA
Cello & flute 4,5	0	8	4	4	0	2	3	5	0
Cello & piano 3,4,5	20	25	23	33	12	18	19	32	1
Cello & trombone 2,3,4	16	16	7	23	15	15	15	22	3
E-bass & piano 2,3,4	11	5	4	10	12	8	8	13	2
E-bass & e guitar 2,3,4	13	8	0	22	5	0	0	22	0
E-bass & flute 3,4	16	12	12	13	0	26	27	23	1
Piano & e-guitar 5	0	10	9	20	0	1	2	18	0
Flute & sax 4	0	0	0	7	5	2	5	10	0
Flute & e-guitar 4,5	4	0	0	3	2	10	10	7	0
$\sum$ Errors	80	84	59	135	51	82	89	152	**7**
Relative error (%)	30.3	31.8	22.3	51.1	19.3	31	33.7	57.5	**2.6**

(*NA* Naive Approach with averaged features, *Att.* only middles of attack intervals, *Ons.* only onset frames, *Rel.* only middles of release intervals, *OA* overtone analysis)

Table 5 w values for the intervals played by the distinct instruments

Octave	Cello & flute	Cello & piano	Cello & trombone	E-bass & piano	E-bass & a-guitar	E-bass & flute	Piano & e-guitar	Flute & sax	Flute & e-guitar
2	x	x	0.03	0.04	0.04	x	x	x	x
3	x	0.11	0.12	0.01	0.04	0.04	x	x	x
4	0.08	0.11	0.02	0.01	0.02	0.04	x	0.04	0.04
5	0.08	0.09	x	x	x		0.04	x	0.04
6	x	x	x	x	x	x	x	x	x

5 Recognition of Intervals Played by Different Instruments

A set of 264 intervals played by the two distinct instruments was used. The instrument combinations include cello mixed with flute, piano and trombone. The electric bass was mixed with electric guitar, piano and flute. Further combinations are flute and sax, flute and electric guitar and also piano and electric guitar.

The results are shown in Table 4. The averaged approach produced again a lot of errors, the relative error is increased to 30 % for HPCP. CENS is an exception with a rather small relative error of 19.3 % in comparison to the HPCP. The application of the AOR-Splitter is also similar to the first and second experiments. Analysis of time frames of the onset stage with the HPCP shows the best results for this feature. It decreases the relative error from 30.3 to 22.3 %. However the best overall performance is achieved by the averaged CENS with 19.3 %. The error sources are also coming from the strong overtones. The AOR-Splitter has no positive impact on the CENS feature. It even increases the relative error from 19.3 % to over 30 %.

Before the overtone analysis, we had to find the lower and upper boundaries in search for w for each instrument (Table 5). After the overtone analysis we got the best overall performance across all approaches. We were able to decrease the relative error to 2.6 %, having only 7 errors. Again, the cello is the most problematic instrument, being part of intervals which produce 4 of the 7 errors.

6 Conclusion

In our work we developed and compared several methods for classification of single tone pitchs and intervals based on chroma and spectrum analysis. For all of the tasks the HPCP feature could be optimized very well by the AOR-Splitter. CENS filtered the different areas of the sound envelope well, but due to the existing strong overtones the performance was reduced. However the averaged CENS vector achieves the best classification results for the intervals of two different instruments, if no overtone analysis is applied. The octave and instrument-specific domain information incorporated into the overtone analysis led to the smallest errors. The drawback of this method is that the knowledge of instruments and octaves is required – it can be valueable to identify the instruments at first and run the harmony analysis afterwards. So one of the ongoing tasks is to run automatic instrument recognition for polyphonic recordings, which is not a trivial problem. Another promising direction is a more complex evaluation of the used features. The optimal parameters for feature extraction can be examined. The pitch and interval recognition can be also done by the application of data mining algorithms, which build classification models based on a number of features (Witten and Frank 2005).

Acknowledgements We thank the Klaus Tschira Foundation for the financial support.

References

Bartsch, M. A., & Wakefield, G. H. (2005). Audio thumbnailing of popular music using chroma-based representations. *IEEE Transactions on Multimedia, 7*(1), 96–104.

Eronen, A. (2009). *Signal processing methods for audio classification and music content analysis*. PhD thesis, Tampere University of Technology.

Fujishima, T. (1999). Realtime chord recognition of musical sound: A system using common lisp music. In *Proceedings of the international computer music conference (ICMC)*, Beijing (pp. 464–467).

Gómez, E. (2006). *Tonal description of music audio signals*. PhD thesis, Universitat Pompeu Fabra, Department of Technology.

Jourdain, R. (1998). *Music, the brain and ecstasy: How music captures our imagination*. New York: Harper Perennial.

Lartillot, O., & Toiviainen, P. (2007). MIR in Matlab (II): A toolbox for musical feature extraction from audio. In *Proceedings of the 8th international conference on music information retrieval (ISMIR)* (pp. 127–130).

Mauch, M. (2010). *Automatic chord transcription from audio using computational models of musical context*. PhD thesis, Queen Mary University of London.

McGill University Master Samples. http://www.music.mcgill.ca/resources/mums/html/.

Müller, M., & Ewert, S. (2010). Towards timbre-invariant audio features for harmony-based music. *IEEE Transactions on Audio, Speech, and Language Processing, 18*(3), 649–662.

Park, T. H. (2010). *Introduction to digital signal processing: Computer musically speaking* (1st Ed.). Singapore/Hackensack: World Scientific Publishing Co. Pte. Ltd.

Temperley, D. (2007). *Music and probability*. Cambridge, MA: MIT.

Theimer, W., Vatolkin, I., & Eronen, A. (2008). Definitions of audio features for music content description. Technical report TR08-2-001, University of Dortmund.

Vatolkin, I., Theimer, W., & Botteck, M. (2010). Amuse (advanced mUSic explorer) – a multitool framework for music data analysis. In *Proceedings of the 11th international society for music information retrieval conference (ISMIR)*, Utrecht (pp. 33–38).

Witten, I. H., & Frank, E. (2005). *Data mining: Practical machine learning tools and techniques*. Amsterdam/Boston: Morgan Kaufmann.

Computational Prediction of High-Level Descriptors of Music Personal Categories

Günther Rötter, Igor Vatolkin, and Claus Weihs

Abstract Digital music collections are often organized by genre relationships or personal preferences. The target of automatic classification systems is to provide a music management limiting the listener's effort for the labeling of a large number of songs. Many state-of-the art methods utilize low-level audio features like spectral and time domain characteristics, chroma etc. for categorization. However the impact of these features is very hard to understand; if the listener labels some music pieces as belonging to a certain category, this decision is indeed motivated by instrumentation, harmony, vocals, rhythm and further high-level descriptors from music theory. So it could be more reasonable to understand a classification model created from such intuitively interpretable features. For our study we annotated high-level characteristics (vocal alignment, tempo, key etc.) for a set of personal music categories. Then we created classification models which predict these characteristics from low-level audio features available in the AMUSE framework. The capability of this set of low level features to classify the expert descriptors is investigated in detail.

G. Rötter (✉)
Institute for Music and Music Science, TU Dortmund, Germany
e-mail: guenther.roetter@tu-dortmund.de

I. Vatolkin
Chair for Algorithm Engineering, TU Dortmund, Germany
e-mail: igor.vatolkin@tu-dortmund.de

C. Weihs
Chair for Computational Statistics, TU Dortmund, Germany
e-mail: claus.weihs@tu-dortmund.de

B. Lausen et al. (eds.), *Algorithms from and for Nature and Life*, Studies in Classification, Data Analysis, and Knowledge Organization, DOI 10.1007/978-3-319-00035-0_54,

1 Introduction

Organization of large music collections becomes more challenging with the growing amount of digitally available music and capacities of hard drives and mobile devices. The manual assignment of songs to some categories is very time consuming, so automatic classification is promising for such work (Weihs et al. 2007). The simplest possibility is to utilize some existing genre information available in mp3 tags or on the Internet. Unfortunately, there exists no commonly accepted genre taxonomy (Pachet and Cazaly 2000), and the community labels gathered e.g. from Last. FM are often noisy, not clear defined, not available for the given music or unacceptable for a certain user. Also, the labels can be learned in an unsupervised way, e.g. by clustering of the corresponding features or similarity analysis (Vembu and Baumann 2004; Pampalk et al. 2005). However such organization often does not correspond to the typical music genre taxonomies and can be even more far away from the personal preferences of the music listener. In contrast, a supervised approach builds models from a given song set labeled by human effort. A reasonable way to achieve high accuracy of models built from preferably small training sets is to incorporate the most predictive features which would capture the important music characteristics very well.

Therefore the first target of our study was to measure the impact of high-level descriptors inspired by music theory for the prediction of personal music categories. Previous work in this direction is often done for symbolic/MIDI data: The results of the study in Basili et al. (2004) underline the important role of instrumentation for genre classification and in de Leon and Inesta (2007) 28 melodic, harmonic and rhythmic high-level descriptors were used for music style detection. For general discussion of features based on music theory please refer to Temperley (2007) and Lomax (1968). Since a music score is not always available and also can not capture some characteristics like effects or dynamics, we concentrate on audio recordings in our work. The extraction of high-level descriptors from signal is not a straightforward task – e.g. for orchestral or other polyphonic recordings it becomes hard to separate the signals of the single instruments for further analysis. In the second part of our study we investigate the possibility to extract the high-level predictors from a large set of available audio features by supervised classification.

2 High-Level Music Descriptors

People often use music in a functional way, which means, that they do not hear music for its own sake, but to alter their mood, come to other thoughts, or use music to complete the atmosphere of a certain environment. In our experiment 18 subjects were asked to list 5 items they would hear in a particular situation of their choice. The subject group mentioned for example situations called "campfire", "music for rail travel" or "context switch". Five music experts carried out an intensive

instrumentation											singing														speech						melodics			
											solo								voice			choir							articulation		ambitus			
1. guitar	2. bass	3. drums	4. piano	5. strings	6. brass	7. saxophone	8. synthesizer	9. orchestra	10. chamber ens.	11. other	12. woman	13. man	14. polyphonic	15. unison	16. clear	17. rough	18. squawk	19. other	20. high	21. medium	22. low	23. male	24. female	25. both	26. german	27. english	28. scat	29. other	30. clearly	31. unclear	32. <= octave	33. > octave	34. volatile	35. linearly

harmony				rhythm								dynamics		effects					structure				level of activation		
36. major	37. minor	38. harmony simple	39. harmony complex	40. straight	41. shuffle	42. dotted	43. accentuation (1+3)	44. accentuation (2+4)	45. straight bar	46. odd bar	47. free	48. homogenous	49. heterogenous	50. hall	51. distortion	52. wahwah	53. chorus	54. flanger	55. strophic	56. free	57. intro	58. outro	59. high	60. medium	61. low

Fig. 1 High-level music descriptors

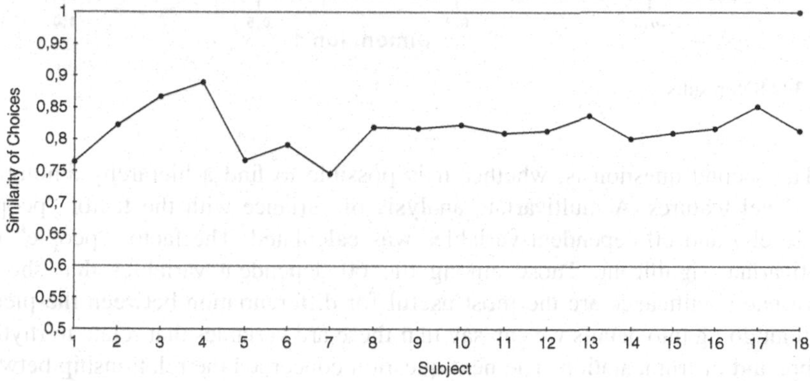

Fig. 2 Similarity of choices

discussion about the best features to be included for further investigations. As a result, Fig. 1 lists 61 acoustic and musical-theoretical criteria which have been found to describe functional music. The criterion "chamber ens." (chamber ensembles) was not taken into account in any of the following analyses since it did not show any variance.

In a next step, experts analyzed the 18 * 5 pieces with the aid of these criteria. This resulted in a data matrix with 5490 values of 0 or 1 (criterion applicable / or not). The first question was whether subjects select similar or different music for a particular situation. To answer this, we calculated the relative frequency of a high-level descriptor, e.g. 0.6 for the criterion string 11,100. This was done for all criteria. After that the harmonic mean was determined for every person. In theory, harmonic means can be distributed between 0.6 (no match) and 1 (perfect match). Figure 2 shows values between 0.75 and 0.9. This means, in a specific situation, listeners tend to choose more music of a similar style.

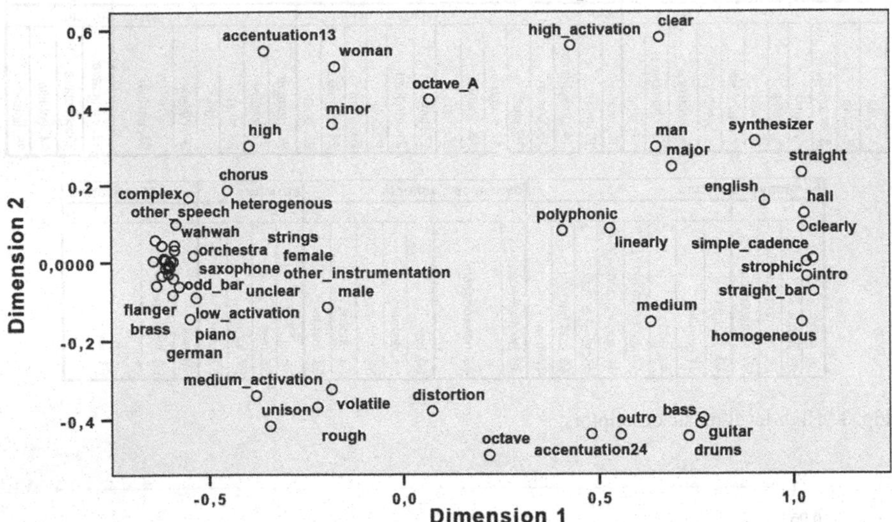

Fig. 3 MDS results

The second question is, whether it is possible to find a hierarchy among the high-level features. A multivariate analysis of variance with the factor "people" (18 levels) and 60 dependent variables was calculated. The factor "people" was multivariate significant. Those among the 60 dependent variables that showed univariate significance are the most useful for differentiation between the pieces. Without going into details we can say that these are variables that relate to rhythm, timbre and instrumentation. The next question concerned the relationship between the high-level features with each other. For this we used Multidimensional Scaling (MDS) because of the binary representation of high-level features. MDS orders data in a two-dimensional space. The analysis showed that contrasting criteria are spatially far away from each other such as "male" and "female". However no structure, such as "clouds", was found among the variables, so they are only sparsely interdependent (Fig. 3).

Another method to show correlations between the variables is cluster analysis of variables (Ward method). Two main clusters were found here. One of the main clusters refers more to simpler musical structures such as rock music. The second cluster, however, shows a more complex structure with a greater variety of instruments and more complex rhythms and forms. This cluster is more related to jazz music (classical music was hardly mentioned in this sample). The interpretation of smaller clusters at lower levels was unclear. This is again a sign of independence of the high-level features from each other. As a summary it may be stated that high-level features do not show any recognizable structure. Nevertheless, similarities of chosen pieces of music can be expressed by high-level features. There is a hierarchy among the features that indicates which criteria are best suitable for the distinction.

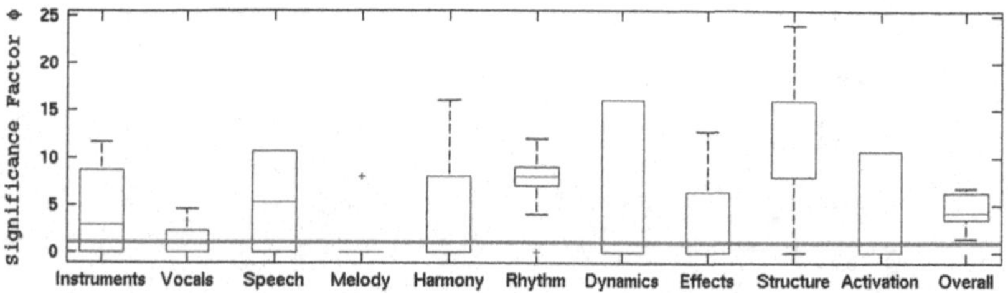

Fig. 4 Significance for high-level feature groups

Finally, we considered the significance of high-level descriptors compared to the artificial situation, where their values of 0 and 1 would be distributed completely randomly. For $N = 5$ songs of each category and $F = 61$ high-level features the expected number of high-level features F^1 which have value 1 for all N songs of a category i is:

$$\mathbf{E}\left[F^1(i)\right] = F \cdot \left(\frac{1}{2^N}\right) \tag{1}$$

Features with high expectation can be stated as significant for a category since they describe musical characteristics which are common for all example songs. Now we can estimate the significance factor ϕ which measures the relative occurence of such features:

$$\phi = \frac{F^1(i)}{\mathbf{E}\left[F^1(i)\right]} \tag{2}$$

$\phi = 1$ means that the feature is rather randomly distributed (and has no relevant impact for a category); larger ϕ values correspond to a higher significance of this feature. Figure 4 illustrates the ϕ values as boxplots for high-level feature groups. The overall statistics is given by the last boxplot. The proposed features seem to be explicitly significant compared to the randomly distributed values. Some of them seem to be more important as descriptors of a category (structure, dynamics and harmony) and some of them less important (melody and vocals).

3 Prediction of High-Level Descriptors from Audio Features

In the second part of our work we studied possibilities for the automatic extraction of high-level descriptors. For our song set we extracted 326 audio features available in AMUSE Vatolkin et al. (2010b) and combined them with the ground truth provided by music experts. The features were related to low-level spectral and time domain characteristics, tempo, chroma distribution etc. Only the time frames between the onset events were used for feature calculation since these frames correspond to the stable sound and this method has been found quite evident in a previous study Vatolkin et al. (2010a). Three different feature aggregation intervals were used

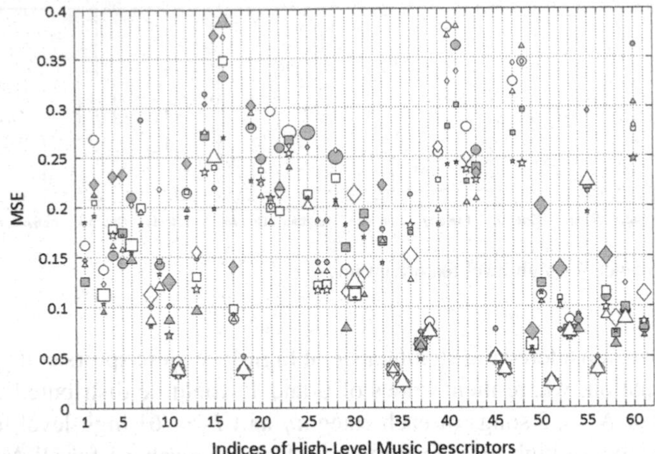

Fig. 5 Classification results with the smallest MSE for each classifier. Circles: C4.5; squares: RF; stars: *k*NN; diamonds: NB; triangles: SVM. Small markers: 4s aggregation intervals; middle markers: 12s intervals; large markers: complete song. Hollow markers: GMM1; filled markers: quartiles

(4s with 2s overlap, 12s/6s and complete song), and two aggregation methods (mean and standard deviation (GMM1) and minimum, maximum and boundaries between quartiles). Five classifiers were applied to learn the models: decision tree C4.5, Random Forest (RF), Naive Bayes (NB), *k*-Nearest Neighbors (*k*NN) and Support Vector Machine (SVM). Classification evaluation was done by the estimation of mean squared error during the ten-fold cross validation on the labeled feature vectors (652 features for GMM1 and 1,630 for quartiles aggregation).

Figure 5 illustrates the classifiation performance by Mean Square Error (MSE) for 58 of 61 high-level descriptors (for three of them the models could not be trained due to the very strong inbalance between the number of positive and negative instances). For each classifier only the best created model is shown. The easiest descriptors to predict (error averaged for five classifiers) are saxophon (lowest error of the best classifier: 0.0228), rhythm uneven (0.025) and orchester (0.0363). The hardest to predict are vocals position average (0.2695), melodic ambitus ≤ octave (0.2437) and > octave (0.2418).

It should be noted, that our aggregation method was rather simplified and labeled all song partitions; however it could more sophisticated e.g. for the vocal characteristics to select only the intervals which actually include a strong vocal rate. The drawback of such data preprocessing is that it requires a lot of manual labeling.

Table 1 summarizes the average share of the different algorithms in the best created models. The shares were averaged for all high-level descriptors (upper part of the table), the 12 hardest descriptors (middle part of the table) and the 12 easiest descriptors (bottom). *k*NN seems to perform at best: more than 15 % of the best models for all categories were created by *k*NN trained with GMM1 of 4s intervals.

Table 1 Average shares of the algorithm choice and interval aggregation parameters for the classification models with the lowest errors for each category. GMM1: mean and standard deviation; Compl.: feature aggregation over the complete song

Interval aggregation ↦	GMM1			Quartiles		
Interval size ↦	4s	12s	Compl.	4s	12s	Compl.
All high-level descriptors						
C4.5	0	1.72	0	1.72	3.45	0
RF	1.72	5.17	0	0	1.72	0
*k*NN	**15.52**	8.62	0	13.79	8.62	0
NB	0	0	0	0	0	1.72
SVM	8.62	5.17	1.72	12.07	8.62	0
12 hardest high-level descriptors						
C4.5	0	0	0	0	0	0
RF	0	0	0	0	0	0
*k*NN	16.67	16.67	0	**33.33**	8.33	0
NB	0	0	0	0	0	0
SVM	16.67	0	0	8.33	0	0
12 easiest high-level descriptors						
C4.5	0	0	0	0	0	0
RF	8.33	16.67	0	0	0	0
*k*NN	16.67	8.33	0	**25.0**	0	0
NB	0	0	0	0	0	8.33
SVM	0	0	0	8.33	8.33	0

The second place is occupied by SVM and the third by RF. C4.5 and NB provide the lowest performance shares. For the descriptors most difficult to predict, only *k*NN- and SVM-based models were chosen. For the rather simple high-level features RF performs also well; however the best method is again *k*NN.

For feature aggregation the definite conclusion is that aggregation of the complete song makes no sense – it relates to the smaller training set and also aggregated features from very different song segments (intro, verse etc.). The choice of the smaller classification instances seems to produce smaller errors in general. The differentiation between GMM1 model and quartiles is not so clear; while storing of mean and standard deviation performs better at average, sometimes it can be reasonable to save quartile statistics, e.g. for SVM and all high-level descriptors.

A final statement with regard to performance analysis is that the investigated study could not consider further optimization techniques due to limited time and computing resources. A very promising method would be to run feature selection before the training of the classification models since many irrelevant or noisy features may significantly reduce the performance of the classifier Vatolkin et al. (2011). Also an analysis of the optimal classifier parameters can be adequate for deeper algorithm comparison.

4 Summary and Future Work

In our study we explored the role of high-level music descriptors for music categories assigned to user-specific listening situations. It could be stated, that these features are significant and can be considered as a similarity criterion for classification and recommendation of appropriate songs. The proposed high-level characteristics have very sparse interdependency and may provide different subsets which are well suited for recognition of user-specific preferences. As it could be expected, some of the high-level descriptors can be identified automatically from a large set of actual audio features very well, whereas some of them are hard to recognize. So future research may concentrate on more intelligent extraction methods for such characteristics as well as on the further design of such features. Also, feature selection and hyperparameter tuning of classifiers may provide significant performance improvements. Especially the high-level descriptors which can be very well recognized can be used as input features for improved classification of large music collections into genres, music styles or user preferences.

Acknowledgements We thank the Klaus Tschira Foundation for the financial support. Thanks to Uwe Ligges for statistical support.

References

Basili, R., Serafini, A., & Stellato, A. (2004). Classification of musical genre: A machine learning approach. In Proceedings of the 5th international conference on music information retrieval Barcelona, Spain (pp. 505–508).

de Leon, P. P., & Inesta, J. (2007). Pattern recognition approach for music style identification using shallow statistical descriptors. *IEEE Transactions on Systems, Man, and Cybernetics, 37*(2), 248–257.

Lomax A. (1968). Folk song style and culture. Washington: American Association for the Advancement of Science.

Pachet, F., & Cazaly, D. (2000). A taxonomy of musical genres. In Proceedings content-based multimedia information access, Paris.

Pampalk, E., Flexer, A., & Widmer, G. (2005). Improvements of audio-based music similarity and genre classification. In Proceedings of 6th international conference on music information retrieval London, UK (pp. 628–633).

Temperley, D. (2007). Music and probability. Cambridge: MIT.

Vatolkin, I., Theimer, W., & Botteck, M. (2010a). Partition based feature processing for improved music classification. In Proceedings of the 34th annual conference of the German classification society, Karlsruhe.

Vatolkin, I., Theimer, W., & Botteck, M. (2010b). Amuse (Advanced MUSic Explorer) – A multitool framework for music data analysis. In Proceedings of the 11th international society for music information retrieval conference, Utrecht (pp. 33–38).

Vatolkin, I., Preuß, M., & Rudolph, G. (2011). Multi-objective feature selection in music genre and style recognition tasks. In Proceedings of the the 2011 genetic and evolutionary computation conference, Dublin (pp. 411–418).

Vembu, S., & Baumann, S. (2004). A self-organizing map based knowledge discovery for music recommendation systems. In Proceedings of the 2nd international symposium on computer music modeling and retrieval, Esbjerg.

Weihs, C., Ligges, U., Mörchen, F., & Müllensiefen, D. (2007). Classification in music research. *Advances in Data Analysis and Classification, 1*(3), 255–291. Springer.

High Performance Hardware Architectures for Automated Music Classification

Ingo Schmädecke and Holger Blume

Abstract Today, stationary systems like personal computers and even portable music playback devices provide storage capacities for huge music collections of several thousand files. Therefore, the automated music classification is a very attractive feature for managing such multimedia databases. This type of application enhances the user comfort by classifying songs into predefined categories like music genres or user-defined categories. However, the automated music classification, based on audio feature extraction, is, firstly, extremely computation intensive, and secondly, has to be applied to enormous amounts of data. This is the reason why energy-efficient high-performance implementations for feature extraction are required. This contribution presents a dedicated hardware architecture for music classification applying typical audio features for discrimination (e.g., spectral centroid, zero crossing rate). For evaluation purposes, the architecture is mapped on an Field Programmable Gate Array (FPGA). In addition, the same application is also implemented on a commercial Graphics Processing Unit (GPU). Both implementations are evaluated in terms of processing time and energy efficiency.

1 Introduction

Today, modern electronic devices are extremely popular for their capability of handling any multimedia processing. Typically, they provide huge storage capacities and therefore enable the creation of extensive databases, especially personal music collections that comprise several thousand music files. Retaining an overview over the various music files of such a database is at least complex and time-consuming.

I. Schmädecke (✉) · H. Blume
Institute of Microelectronic Systems, Appelstr. 4, 30167 Hannover, Germany
e-mail: schmaedecke@ims.uni-hannover.de; blume@ims.uni-hannover.de

B. Lausen et al. (eds.), *Algorithms from and for Nature and Life*, Studies in Classification, Data Analysis, and Knowledge Organization, DOI 10.1007/978-3-319-00035-0_55,

Thus, it exists an increasing interest in new applications for managing music databases. A suitable solution is the content-based automated music classification that, in contrast to social network based approaches like last.fm, do not require an internet access. This type of application allows to structure huge music collections into user defined groups like mood, genre, etc. By this way, it also enables the automated generation of music playlists. Depending on the amount of data to be analysed and the feasible implementation, the computation effort of the content-based music classification can be extremely time-consuming on stationary systems. Especially on current mobile devices, the time for analysing a database can take several hours, which is critical at least because of the limited battery life.

For this reason, two different approaches are introduced in this paper, which accelerate the content-based music classification and, in addition, provide a high energy efficiency compared to modern CPUs. The first approach is based on a GPU, that is dedicated for stationary systems. The second one is a dedicated hardware accelerator, which is suitable for mobile devices in particular. Both approaches are optimized to accelerate the most time consuming processing step of the automated classification and are evaluated in terms of their computation performance as well as their energy efficiency.

Section 2 gives an overview about recent works related to this topic. In Sect. 3, the basics of a content-based automated music classification are explained and its most time consuming step is highlighted. Fundamentals about the examined GPU architecture and GPU-based optimizations of the application are presented in Sect. 4. In Sect. 5, the proposed hardware based accelerator is introduced. Extensive benchmarks visualize the advantage of both realizations in Sect. 6 and conclusions are given in Sect. 7.

2 Related Work

In Schmidt et al. (2009), spectral-based features were implemented on an FPGA. For this purpose, a plugin for generating the VHDL code from a MATLAB reference was used. The resulting time for extracting the corresponding features from a 30 s long music clip with 50 % window overlap amounts to 33 ms. Another method to reduce the computation effort regarding the required computation time is presented in Friedrich et al. (2008), which is suitable for extracting features from compressed audio data. The presented approach is based on a direct conversion from the compressed signal presentation to the spectral domain. By this way, the computational complexity could be reduced from *O(NlogN)* to *O(N)*. A feature extraction implementation for mobile devices is given in Rodellar-Biarge et al. (2007). In this paper, the power consumption and the required hardware resources have been evaluated. However, the design is conform with the standard aurora for distributed speech recognition systems and hence is not dedicated to high-performance audio data processing.

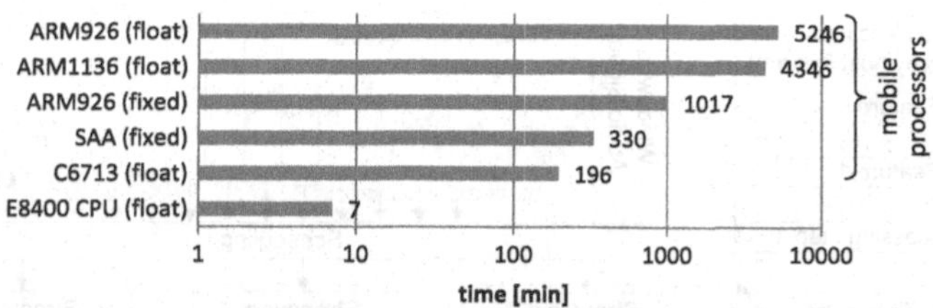

Fig. 1 Computation times for extracting nine different features from thousand music files with 3 min per song on mobile devices and a general purpose CPU

3 Content-Based Music Classification

The first step of the content-based music classification is the extraction of characteristical audio information, also called audio features, from the original time signal Fu et al. (2011). Therefore, the signal is divided into windows of equal size, which may overlap. A typical window size is about 512 audio samples. From each window a predefined set of audio features is computed. Since audio features are based on different signal representations, precomputations for transforming the time-domain based signal into other domains are required. Next, the evolution of each feature over time is abstracted to a reduced amount of data, i.e. statistical parameters like mean, variation. By this way, a music file specific feature vector is generated, whose dimension is independent from the audio length regarding the varying number of windows. Finally, an arbitrary classifier (e.g. SVM) assigns a music label to the computed feature vector on the basis of a predefined classification model. This model can be stationary or dynamically generated by a user defined categorization.

Normally, the audio feature extraction is the most computation intensive step within the overall music classification process and can be extremely time consuming Blume et al. (2011). This is especially the case for mobile devices as shown in Fig. 1 Blume et al. (2008). The computation time for extracting nine different features from a small music database is shown for typical mobile processors and a general purpose CPU. Here, a distinction is made between floating and fixed point implementations. While the feature extraction task is done within an acceptable time on the observed CPU, all other architectures require several hours for extracting all features from the music collection. Such computation times are insufficient for mobile devices because they exceed the available battery life in the case of completely utilized processor capacities. Moreover, the computation effort increases with further audio features and music files, which results in a significantly increased computation effort. Actually, todays' personal music collections can comprise considerably more than thousand songs and further audio features can be required for an accurate classification. Thus, even stationary systems with modern general purpose processors can be strongly utilized for a significant amount of time. That is the reason why an acceleration of the extraction process is demanded for stationary systems and mobile devices.

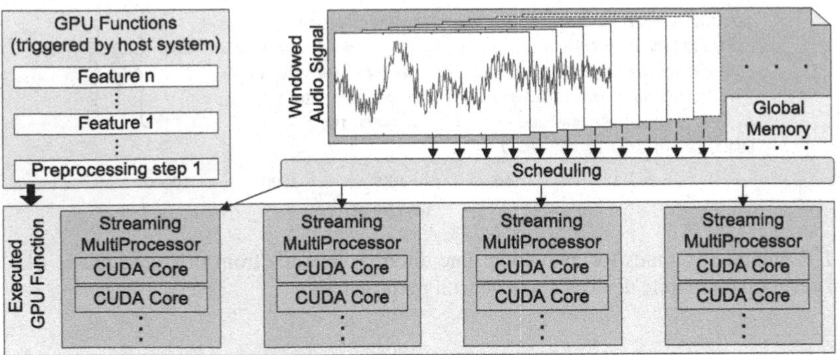

Fig. 2 CUDA based feature extraction based on a Single-Instruction-Multiple-Data approach

4 Feature Extraction on GPU

NVIDIA provides the Compute Unified Device Architecture (CUDA), which comprises a scalable GPU architecture and a suitable programming framework for data parallel processing tasks. On software level, extensions for standard programming languages like C enable the definition of program functions that are executed on a GPU in a Single-Instruction-Multiple-Data (SIMD) manner. Therefore, such GPU functions are executed several times in parallel by a corresponding number of threads, that are grouped into thread blocks. On hardware level, CUDA-enabled GPUs are based on a scalable number of streaming multiprocessors (SMs), which manage a set of processing units, also called CUDA cores. Here, each thread block is executed by one SM. Thereby, the thread blocks of one GPU function are required to be data independent, because only the CUDA cores of one SM are able to work cooperatively by data sharing and synchronizing each other. Moreover, the data to be processed must be preloaded from the underlying host systems' memory to the graphic board's global memory. The host system also controls and triggers the execution sequence of GPU functions.

The GPU-based audio feature extraction concept is shown in Fig. 2. Here, the first step is to load a music file into global memory so that the parallel processing capacities of CUDA can be utilized. Then, GPU functions are applied to the complete audio data to preprocess each window and afterwards to extract the features. Here, the main approach is to employ each thread block of a GPU function with the feature extraction from a window respectively from its transformed signal. This is reasonable because the audio windows can be processed independently, which is required for the thread blocks. Furthermore, the number of thread blocks corresponds to the number of audio windows on global memory. By this way, the windows can be processed concurrently and a scalable parallelization is achieved, which depends on the number of available SMs. In addition, the window specific processing is further accelerated by utilizing all CUDA cores of an SM at the same time Schmaedecke et al. (2011).

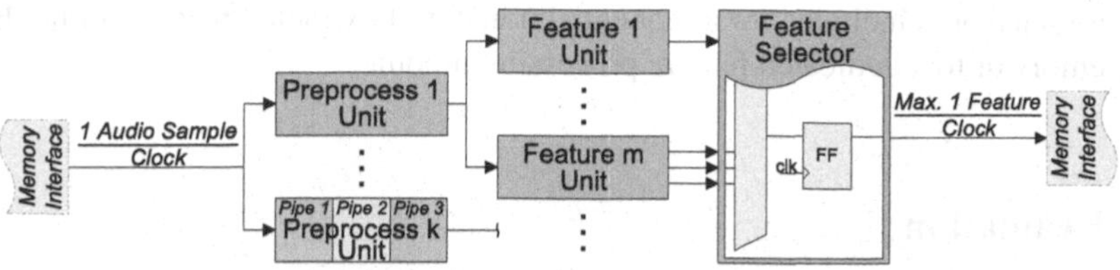

Fig. 3 A dedicated hardware architecture design for the audio feature extraction based on a Multiple-Instruction-Single-Data approach

5 Hardware Dedicated Feature Extraction

An application specific hardware accelerator has been identified as suitable for mobile devices, since the available computation power of modern mobile processors has to be kept free for the interaction with the user. Otherwise, the processor's capacities would significantly inhibit usability of the mobile device during the feature extraction. For the design of such a hardware accelerator, various aspects have to be respected like the distributable memory bandwidth respectively data rate.

Current mobile devices typically offer less memory bandwidths than stationary systems. This is especially the case if huge databases have to be processed because the required storage capacities are often realized by memory cards with data rates of only a few megabytes per second. Thus, a data parallel processing is not suitable for accelerating the extraction of audio features. Instead, a parallel extraction of features from the same data is more applicable, which reduces the processing time according to the number of extracted features. In this work, such an approach has been developed as a dedicated hardware design as illustrated in Fig. 3.

The hardware architecture corresponds to a Multiple-Instruction-Single-Data (MISD) concept. For a fast feature extraction the data input rate has been constituted to one audio sample per clock. On the one hand, this implies that all implemented processing modules must work continuously and parallel to the continuous data stream and therefore are developed for a pipeline-based computation. On the other hand, each processing module corresponds to a preprocessing step or feature function respectively algorithm. Thus, the computation time for extracting features is independent from the number of features to be extracted per window. Instead, the processing time only depends on the number of audio samples plus a fix latency, which is caused by the pipelining concept of the processing units. Nevertheless, this latency is only a negligible fraction of the overall time.

Furthermore, mathematical computations are realized with fix-point arithmetic. This results in a less exhaustion of hardware ressources compared to an equivalent floating-point implementation Underwood and Hemmert (2008). Thereby, the bit accuracy can be individually set for each calculation step so that a sufficiently precise computation for the subsequent classification can be achieved. Finally, a feature selector is included for a sequential transfer of extracted features either back to memory or to a dedicated feature processing module.

6 Evaluation

Before the presented approaches are evaluated, the underlying platforms and adopted benchmark settings are introduced. The evalutation of the GPU-based feature extraction has been performed on a C2050 graphic card, which provides a Fermi GPU with 14 SMs and and 32 cores per SM. In contrast, a development board (MCPA) with a Virtex-5 XC5VLX220T FPGA is applied to verify and emulate the hardware dedicated feature extraction. The presented approaches are rated against a personal computer as a reference system, which deploys a Core 2 Duo Processor. The specifications of the constituted platforms are listed in Table 1.

The Fermi GPU that is employed on the C2050 operates at a significantly higher clock rate than FPGA. In addition, the memory bandwidth available on the C2050 is 96.6 times higher compared to the MCPA board, which is required for a data parallel processing. Hence, the MCPA board consumpts only a sixteenth part of power compared to the GPU platform. Moreover, the reference system's CPU is clocked with 3 GHz and can access to 3 GB DDR2 RAM.

All platforms inspected here are able to extract five different feature types, which are specified in Table 2. In general, these features are used to get information about the timbre and energy characteristics of a signal.

The mel frequency cepstral coefficients and the audio spectrum envelope algorithm extract more than one feature. Thus, 24 features per window are extracted. From each feature the minimum, maximum and mean value are computed, which results in a 72 dimensional feature vector. Furthermore, the classification is done with a SVM classifier with a linear kernel. With this setting, 67 percent songs of the popular music database GTZAN Tzanetakis and Cook (2002) can be classified correctly.

6.1 Computation Performance

The computation performances are determined by extracting the implemented features from 1,000 non-overlapping audio windows with a size of 512 samples per window. The results are shown in Fig. 4a. As it can be seen, the required time for extracting features on the reference platform amounts to 31 ms. In contrast, with the completely optimized GPU code the C2050 outperforms the reference platform with a speed up of about 19. In addition, the hardware solution, which is mapped on the FPGA is six times faster compared to the reference. This demonstrates that the audio feature extraction benefits from both presented approaches. Furthermore,

Table 1 Platform specifications (TDP: Thermal design power)

Platform	System clock rate	Bandwidth	Memory	Power consumption
C2050	1,150 MHz	144 GB/s	3,072 MB	237 W (overall, TDP)
MCPA	100 MHz	1.49 GB/s	256 MB	15 W (overall)
Reference	3,000 MHz	6.4 GB/s	3,072 MB	65 W (only CPU, TDP)

Table 2 Implemented features

Feature name	Domain	Number of features
Zero crossing rate	Time	1
Root mean square	Time	1
Spectral centroid	Frequency	1
Audio spectrum envelope	Frequency	9
Mel frequency cepstral coefficients	Frequency	12

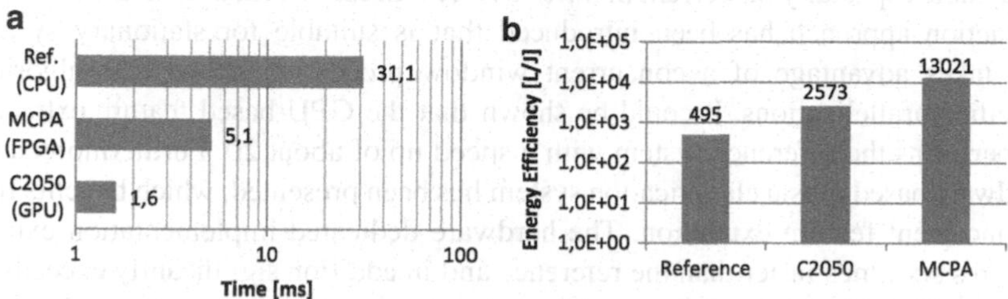

Fig. 4 Benchmark results: **(a)** Computation performances, **(b)** Energy efficiencies

it has to be considered that with an increasing number of features to be extracted the FPGA approach can even outperform the GPU-based implementation.

6.2 Energy Efficiency

In general, the energy efficiency is defined as the relation of processing performance to power consumption. The processing performance is frequently specified in million operations per second while the power consumption is measured in watt, which corresponds to joule (J) per second. Since, the computation effort for performing the feature extraction depends on the number of features, which are extracted from a window, it is more reasonable to determine the processing performance as the number of windows (L) per second. Thus, the energy efficiency can be defined as

$$Energy\ Efficiency = \frac{Processing\ Performance}{Power\ Consumption} = \frac{{}^{L}/_{s}}{{}^{J}/_{s}} = \frac{L}{J}. \quad (1)$$

Based on this definition, the results of the examined hardware architectures are illustrated in Fig. 4b. Here, the GPU is 5.2 times more energy efficient than the reference regarding the corresponding application, while the best result is achieved with the FPGA based approach with a 26 times better efficiency compared to the reference. There by, the power consumption of the dedicated hardware approach can be further reduced by realizing it as an application specific instruction circuit Kuon and Rose (2007). By this way, the presented solution becomes suitable for the usage in mobile devices.

7 Conclusions

The content based automated music classification becomes more and more interesting for managing increasing music collections. However, the required feature extraction can be extremely time consuming and hence this approach is hardly applicable especially on current mobiles devices. In this work, a GPU-based feature extraction approach has been introduced that is suitable for stationary systems and takes advantage of a concurrent window processing as well as algorithm specific parallelizations. It could be shown that the GPU-based feature extraction outperforms the reference system with a speed up of about 20. Furthermore, a full hardware based music classification system has been presented, which benefits from a concurrent feature extraction. The hardware dedicated implementation extracts features six times faster than the reference and in addition significantly exceeds the energy efficiency of the reference. Thus, the presented hardware approach is very attractive for mobile devices. In future work, a detailed examination of the dedicated hardware approach will be performed regarding the required computation accuracy for extracting features, which affects the hardware costs and classification results.

References

Rodellar-Biarge, V., Gonzalez-Concejero, C., Martinez De Icaya, E., Alvarez-Marquina, A., & Gómez-Vilda, P. (2007). Hardware reusable design of feature extraction for distributed speech recognition, proceedings of the 6th WSEAS international conference on applications of electrical engineering, Istanbul, Turkey (pp. 47–52).

Blume, H., Haller, M., Botteck, M., & Theimer, W. (2008). Perceptual feature based music classification a DSP perspective for a new type of application, proceedings of the SAMOS VIII conference (IC-SAMOS), (pp. 92–99).

Blume, H., Bischl, B., Botteck, M., Igel, C., Martin, R., Roetter, G., Rudolph, G., Theimer, W., Vatolkin, I., & Weihs, C. (2011). Huge music archives on mobile devices – Toward an automated dynamic organization. *IEEE Signal Processing Magazine, Special Issue on Mobile Media Search, 28*(4), 24–39.

Friedrich, T., Gruhne, M., & Schuller, G. (2008). A fast feature extraction system on compressed audio data, audio engineering society, 124th convention, Netherlands.

Fu, Z., Lu, G., Ting, K. M., & Zhang, D. (2011). A survey of audio-based music classification and annotation. *IEEE Transactions on Multimedia, 13*, 303–319.

Kuon, I., & Rose, J. (2007). Measuring the gap between FPGAs and ASICs, IEEE transactions on computer-aided design of integrated circuits and systems, Dortmund, Germany, Vol. 26 (pp. 203–215).

Schmaedecke, I., Moerschbach, & J., Blume, H. (2011). GPU-based acoustic feature extraction for electronic media processing, proceedings of the 14th ITG conference, Dortmund, Germany.

Schmidt, E., West, K., & Kim, Y. (2009). Efficient acoustic feature extraction for music information retrieval using programmable gate arrays, Kobe, Japan, ISMIR 2009.

Tzanetakis, G., & Cook, P. (2002). Musical genre classification of audio signals. *IEEE Transaction on Speech and Audio Processing, 10*(5), 293–302.

Underwood, K. D., & Hemmert, K. S. (2008). The implications of floating point for FPGAs. In S. Hauck and A. Dehon, (Eds.), Reconfigurable computing (pp. 671–695). Boston: Elsevier.